RF C

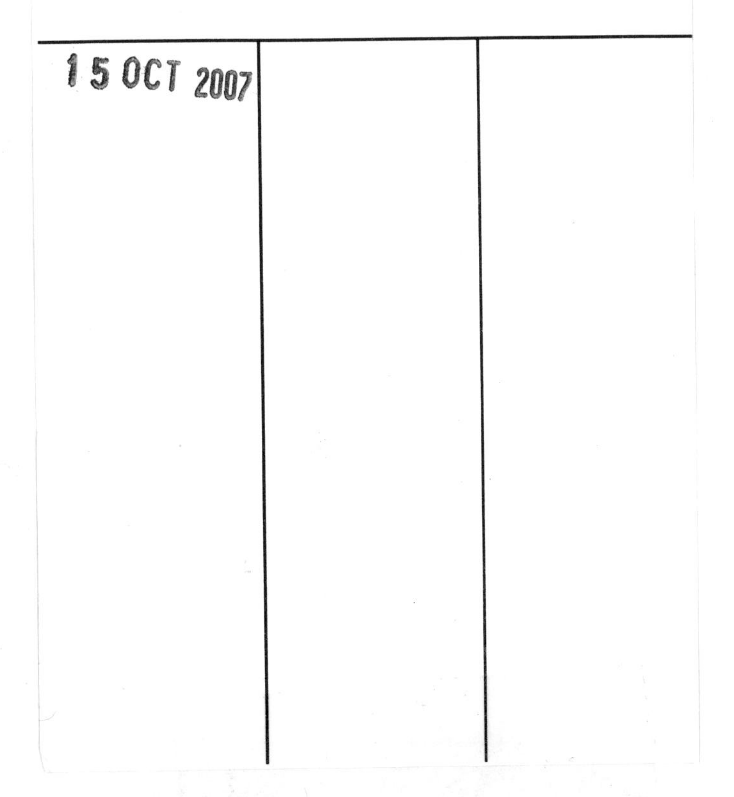

RF Components and Circuits

Joseph J. Carr

AMSTERDAM • BOSTON • HEIDELBERG • LONDON • NEW YORK • OXFORD
PARIS • SAN DIEGO • SAN FRANCISCO • SINGAPORE • SYDNEY • TOKYO
Newnes is an imprint of Elsevier

Newnes
An imprint of Elsevier
Linacre House, Jordan Hill, Oxford OX2 8DP
30 Corporate Drive, Burlington, MA 01803

First edition 2002
Reprinted 2005

British Library Cataloguing in Publication Data
A catalogue record for this book is available from the British Library

ISBN 0 7506 4844 9

Published in conjunction with Radio Society of Great Britain,
Lambda House, Cranborne Road, Potters Bar, Herts, EN6 3JE. UK
www.rsgb.org.uk

Cover illustration supplied by Coilcraft Europe Ltd

Composition by Genesis Typesetting, Rochester, Kent
Printed and bound in The Netherlands

Contents

Foreword

Remembering Joe Carr, K4IPV

This book represents the last published work of an extraordinary man – a teacher, mentor, engineer, husband, father, and friend.

To say that Joe's life, work, and loving spirit enriched many lives is an understatement indeed. A relentless communicator, both through the written word and through Amateur Radio, Joe redefined the word 'prolific' with 1,000 or more published articles and papers and nearly 100 books on topics ranging from science and technology to matters of history and faith.

Like his colleagues, family, and innumerable friends, his editors revered him. When Joe undertook a project, we could count on its being completed on time, in good shape, and exactly as promised. He met his deadlines, accepted criticism graciously, and lived each day as ready to learn from others as he was to teach.

He spoke ill of no one, cherished his professional and personal relationships, honored the work of others, and never failed to laugh at himself.

Writing these words means admitting that Joe is really gone. After 20 years of working with him, that is a difficult thing to do. But like so many other 'Silent Keys' who have preceded him, Joe will continue to educate and inform new generations of radio engineers and Amateurs worldwide through his rich legacy of written work. Rather than regret that these new enthusiasts could not have known him firsthand, we can hope they will come to know him through this book.

Dorothy Rosa
KA1LBO
Former Managing Editor, *ham radio magazine*

In memory of Joe Carr (1943–2000) the staff of Newnes would like to dedicate this book to Joe's family, friends and colleagues.

We would also like to acknowledge with thanks the work undertaken by Dave Kimber to prepare the manuscript for publication.

Matthew Deans
Newnes Publisher

Preface

This is a book on radio frequency (RF) circuits. RF circuits are different from other circuits because the values of stray or distributed capacitances and inductances become significant at RF. When circuit values are calculated, for example, these distributed values must be cranked into the equations or the answer will be wrong . . . perhaps by a significant amount. It is for this reason that RF is different from low frequency circuits.

Joseph J. Carr

Part 1 Introduction

1 Introduction to radio frequencies

This book grew out of a series in a British magazine called *Electronics World/Wireless World*. In fact, most of the material presented in this book first appeared, at least in general form, in that publication. Other material was written new for this book.

What are the 'radio frequencies'?

The radio frequencies (RF) are, roughly speaking, those which are above human hearing, and extend into the microwave spectrum to the edge of the infrared region. That means the RF frequencies are roughly 20 000 Hz to many, many gigahertz. In this book, we will assume the radio frequencies are up to about 30 GHz for practical purposes.

There are radio frequencies below 20 kHz, however. In fact, there are radio navigation transmitters operating in the 10 to 14 kHz region. The difference is that the waves generated by those stations are electromagnetic waves, not acoustical waves, so humans cannot hear them.

Why are radio frequencies different?

Why are radio frequencies different from lower frequencies? The difference is largely due to the fact that capacitive and inductive stray reactances tend to be more significant at those frequencies than they are at lower frequencies. At the lower frequencies, those stray or distributed reactances exist, but they can usually be ignored. Their values do not approach the amount required to establish resonance, or frequency responses such as high pass, low pass or bandpass. At RF frequencies, the stray or distributed reactances tend to be important. As the frequency drops into the audio range (1–20 kHz), and the ultrasonic range (20–100 kHz) the importance of stray reactances tends to diminish slightly.

What this book covers

We will look at a number of different things regarding RF circuits. But first, we will take a look at signals and noise. This sets the scene for a general look at radio receivers. Most radio frequency systems have one or more receivers, so they are an important type of circuit. They also include examples of many of the individual RF circuits we will be looking at. So Part 1 is an introduction.

It is hard to say very much about RF circuits without talking about the components, but the special RF components don't make much sense unless you already know something about the circuits. So which should come first? (Think of chickens and eggs!) The way round this is to put circuits in Part 2 and components in Part 3, but then think of them as running in parallel rather than one after the other. So you can swap between them, but because this is a paper-based book we have to print Part 3 after Part 2.

Part 2 looks at the various types of RF circuits in roughly the order a radio signal sees them as it goes through a normal superhet receiver. Many of these circuits are also used in transmitters, test equipment and other RF stuff but there isn't enough space to go into all that in this book.

Part 3 mainly deals with the sort of components you won't see in lower frequency or digital circuits. Radio frequency is a bit unusual because some components, mainly various types of inductors, can't always be bought 'off the shelf' from catalogues but instead have to be made from parts such as bits of ferrite with holes in them and lengths of wire. So I will give you design information for this.

We finish up by looking at some RF measurements and techniques in Part 4.

One big important RF topic has been left out of this book – antennas. This is because doing this properly would make the book much too long, but a quick look would not tell you enough information to be useful. You can learn about antennas from my book *Antenna Toolkit*, second edition (published by Newnes/RSGB, Oxford 2001). This includes a CD-ROM to help you with antenna calculations.

Now, let's get started . . .

2 Signals and noise

Types of signals

The nature of signals, and their relationship to noise and interfering signals, determines appropriate design all the way from the system level down to the component selection level. In this chapter we will take a look at signals and noise, and how each affects the design of amplification and other RF circuits.

Signals can be categorized several ways, but one of the most fundamental is according to *time domain* behaviour (the other major category is *frequency domain*). We will therefore consider signals of the form $v = f(t)$ or $i = f(t)$. The time domain classes of signals include: *static, quasistatic, periodic, repetitive, transient, random,* and *chaotic*. Each of these categories has certain properties that can profoundly influence appropriate design decisions.

Static and quasistatic signals

A *static* signal (Fig. 2.1A) is, by definition, unchanging over a very long period of time (T_{long} in Fig. 2.1A). Such a signal is essentially a DC level, so must be processed in low drift DC amplifier circuits. This type of signal does not occur at radio frequencies because it is DC, but some RF circuits may produce a DC level, e.g. a continuous wave, constant amplitude RF signal applied to an envelope detector.

The term *quasistatic* means 'nearly unchanging', so a quasistatic signal (Fig. 2.1B) refers to a signal that changes so slowly over long times that it possesses characteristics more like static signals than dynamic (i.e. rapidly changing) signals.

Periodic signals

A *periodic* signal (Fig. 2.1C) is one that exactly repeats itself on a regular basis. Examples of periodic signals include sine waves, square waves, sawtooth waves, triangle waves, and so forth. The nature of the periodic waveform is such that each waveform is identical at like points along the time line. In other words, if you advance along the time line by exactly one period (T), then the voltage, polarity and direction of change of the waveform will be repeated. That is, for a voltage waveform, $V(t) = V(t + T)$.

Repetitive signals

A *repetitive* signal (Fig. 2.1D) is quasiperiodic in nature, so bears some similarity to the periodic waveform. The principal difference between repetitive and periodic signals is

Figure 2.1 Various types of signal: (A) static.

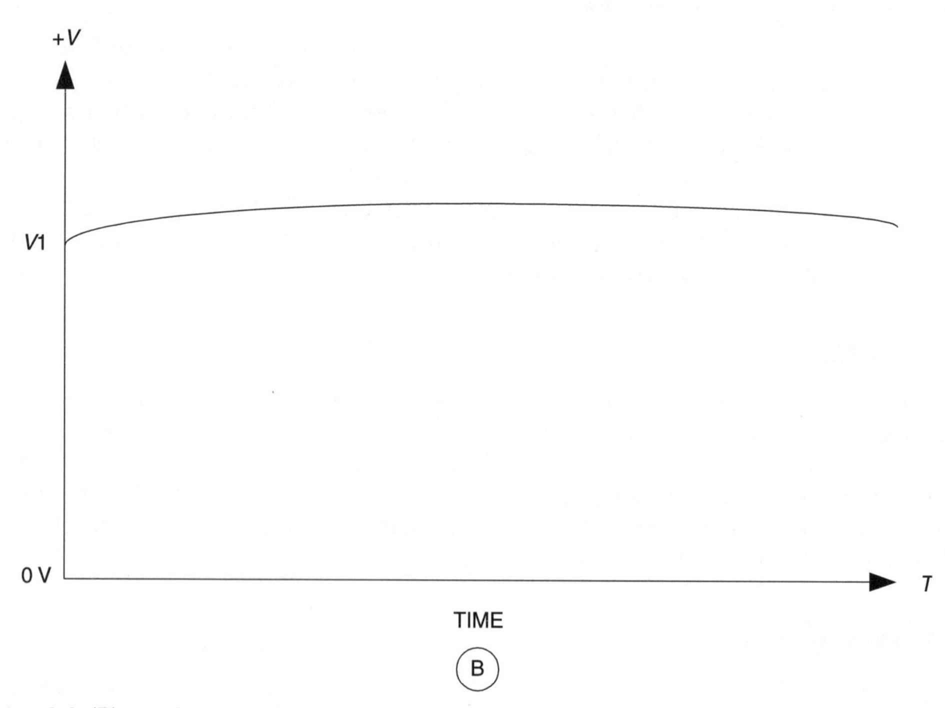

Figure 2.1 (B) quasistatic.

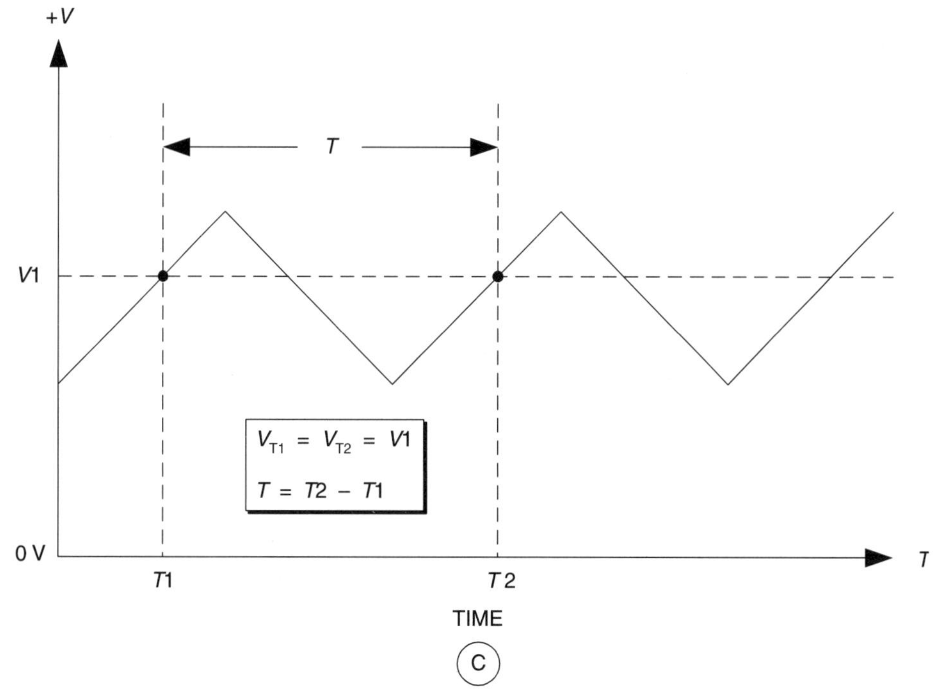

Figure 2.1 (C) periodic.

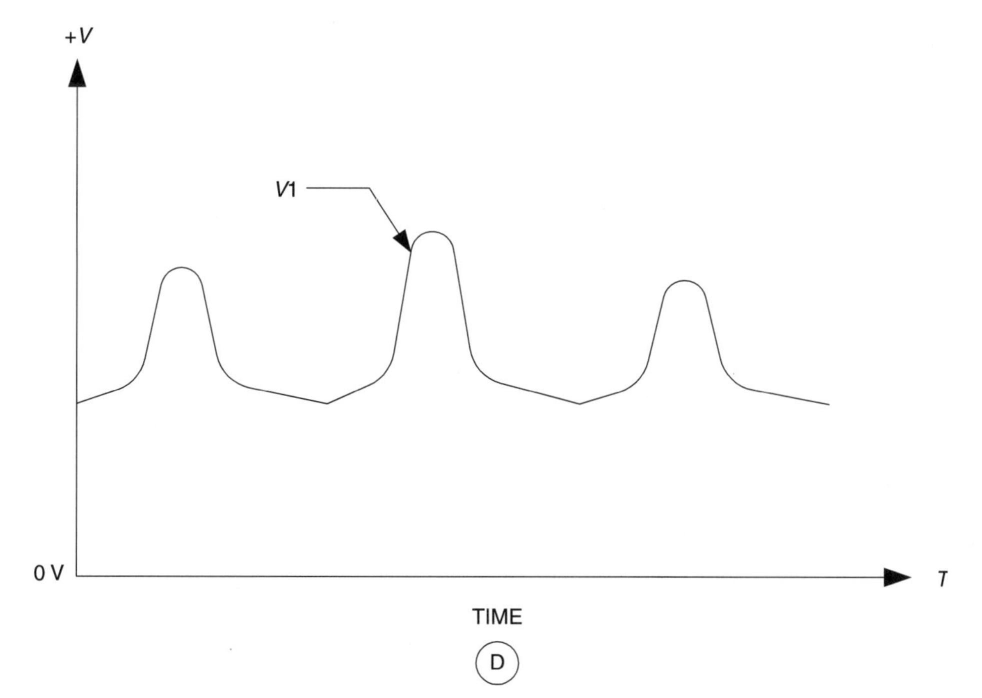

Figure 2.1 (D) quasiperiodic.

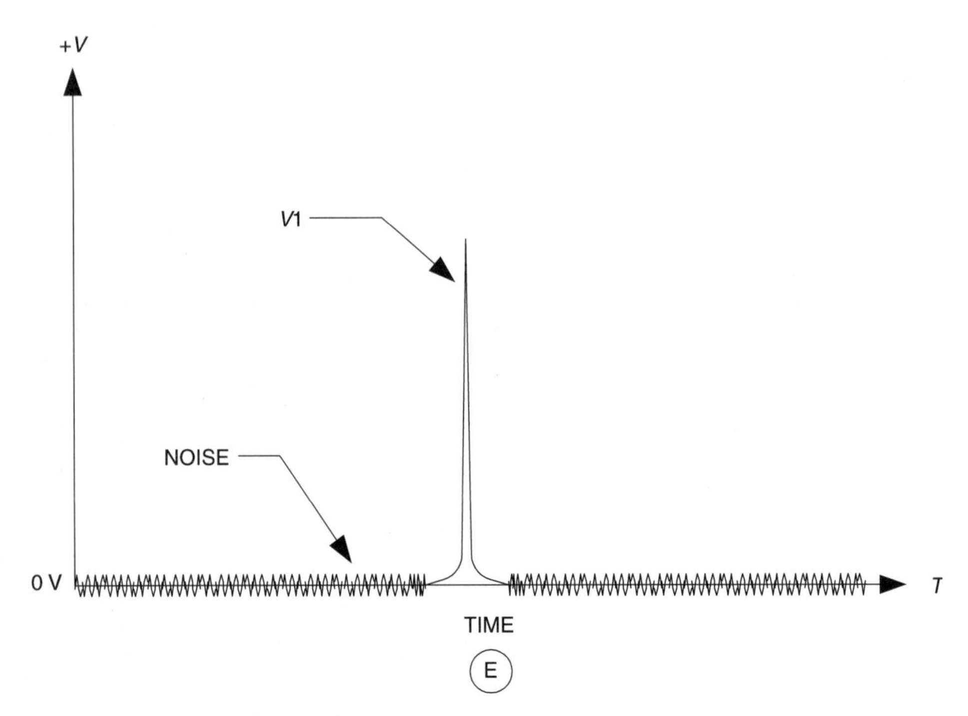

Figure 2.1 (E) spectrum of single frequency.

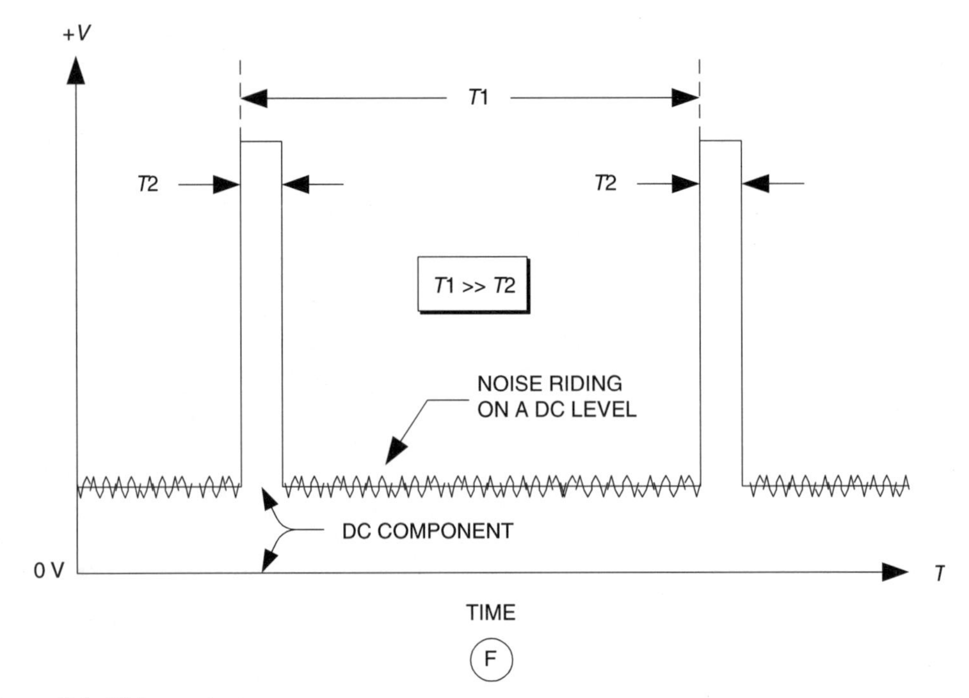

Figure 2.1 (F) two pulses.

seen by comparing the signal at $f(t)$ and $f(t + T)$, where T is the period of the signal. Unlike periodic signals, in repetitive signals these points might not be identical although they will usually be similar. The general waveshape is nearly the same. The repetitive signal might contain either transient or stable features that vary from period to period.

Transient signals and pulse signals

A *transient* signal (Fig. 2.1E) is either a one-time event, or a periodic event in which the event duration is very short compared with the period of the waveform (Fig. 2.1F). In terms of Fig. 2.1F, the latter definition means that $t_1 <<< t_2$. These signals can be treated as if they are transients. In RF circuits these signals might be intentionally generated as pulses (radar pulses resemble Fig. 2.1F), or a noise transient (Fig. 2.1E).

Fourier series

All continuous periodic signals can be represented by a fundamental frequency sine wave, and a collection of sine or cosine harmonics of that fundamental sine wave, that are

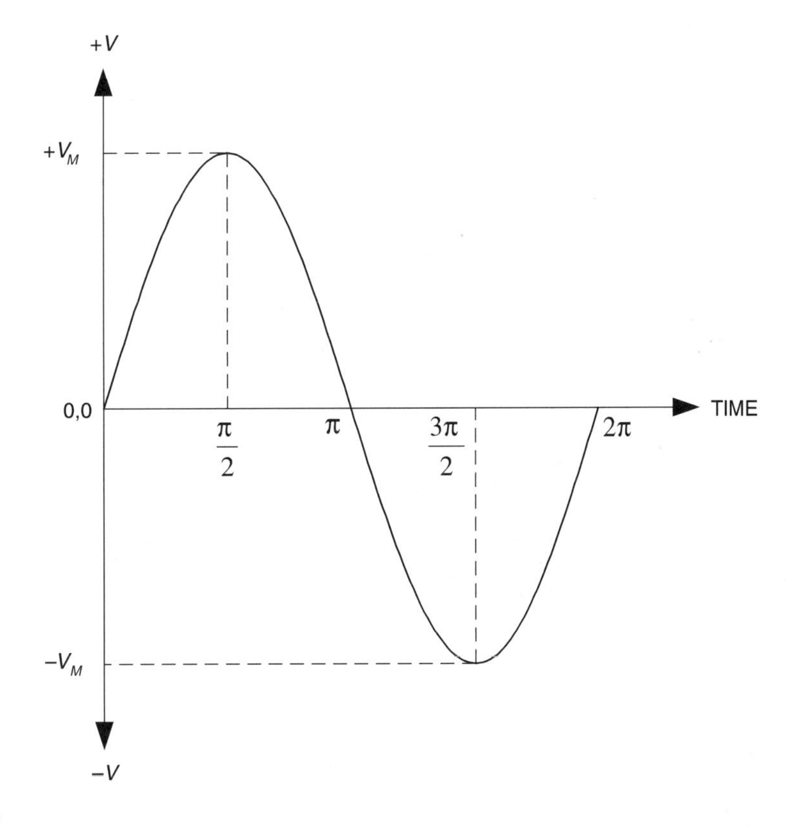

Figure 2.2 Sine wave.

summed together linearly. These frequencies comprise the *Fourier series* of the waveform. The elementary sine wave (Fig. 2.2) is described by:

$$v = V_m \sin(\omega t) \tag{2.1}$$

Where:
v is the instantaneous amplitude of the sine wave
V_m is the peak amplitude of the sine wave
ω is the angular frequency ($2\pi F$) of the sine wave
t is the time in seconds

The *period* of the sine wave is the time between repetition of identical events, or $T = 2\pi/\omega = 1/F$ (where F is the frequency in cycles per second).

The *Fourier series* that makes up a waveform can be found if a given waveform is decomposed into its constituent frequencies either by a bank of frequency selective filters, or a digital signal processing algorithm called the *fast Fourier transform* (FFT). The Fourier series can also be used to construct a waveform from the ground up. Figure 2.3 shows a triangular wave signal constructed from a fundamental sine wave and harmonics.

The Fourier series for any waveform can be expressed in the form:

$$f(t) = \frac{a_o}{2} + \sum_{n=1}^{\infty} [a_n \cos(n\omega t) + b_n \sin(n\omega t)] \tag{2.2}$$

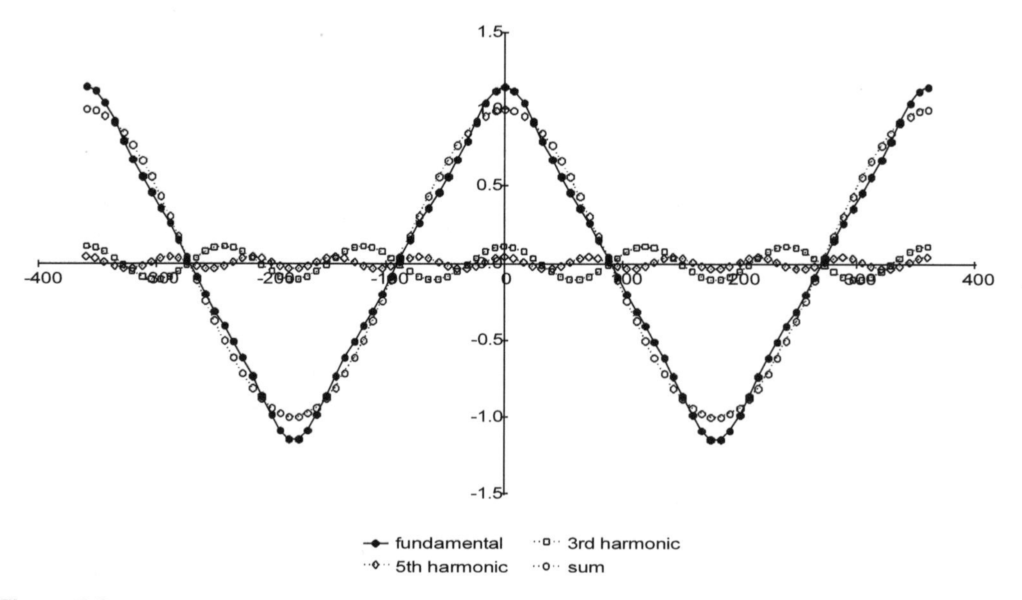

Figure 2.3

Where:
a_n and b_n are the amplitudes of the components (see below)
n is an integer ($n = 1$ is the fundamental)
Other terms are as previously defined

The amplitude coefficients (a_n and b_n) are expressed by:

$$a_n = \frac{2}{T}\int_0^T f(t)\cos(n\omega t)\,dt \tag{2.3}$$

and,

$$b_n = \frac{2}{T}\int_0^T f(t)\sin(n\omega t)\,dt \tag{2.4}$$

Because only certain frequencies, fundamental plus harmonics determined by integer n, are present the spectrum of the periodic signal is said to be *discrete*.

The term $a_o/2$ in the Fourier series expression (Eq. (2.2)) is the average value of $f(t)$ over one complete cycle (one period) of the waveform. In practical terms, it is also the *DC component* of the waveform. When the waveform possesses *half-wave symmetry* (i.e. the peak amplitude above zero is equal to the peak amplitude below zero at every point in t, or $+V_m = |-V_m|$), there is no DC component, so $a_o = 0$.

An alternative Fourier series expression replaces the $a_n \cos(n\omega t) + b_n \sin(n\omega t)$ with an equivalent expression of another form:

$$f(t) = \frac{2}{T}\sum_{n=1}^{\infty} C_n \sin(n\omega t - \phi_n) \tag{2.5}$$

Where:

$$C_n = \sqrt{a_n^2 + b_n^2}$$

$$\phi_n = \arctan\left(\frac{a_n}{b_n}\right)$$

All other terms are as previously defined

Waveform symmetry

One can infer certain things about the Fourier spectrum of a waveform by examination of its *symmetries*. One would conclude from the above equations that the harmonics extend to infinity on all waveforms. Clearly, in practical systems a much less than infinite

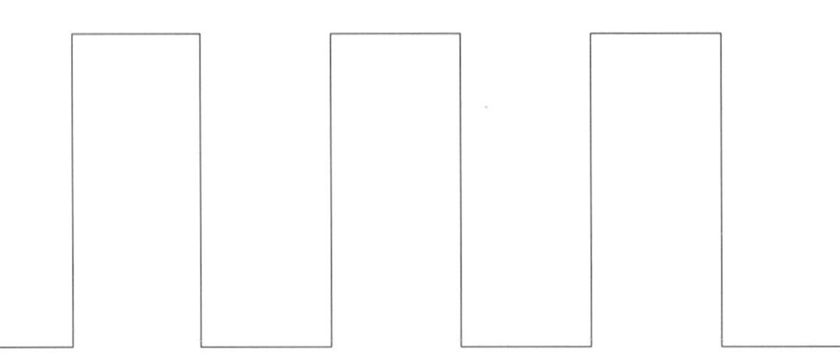

Figure 2.4 Square wave.

bandwidth is found, so some of those harmonics will be removed by the normal action of the electronic circuits. Also, it is sometimes found that higher harmonics might not be truly significant, so can be ignored. As n becomes larger, the amplitude coefficients a_n and b_n tend to become smaller. At some point, the amplitude coefficients are reduced sufficiently that their contribution to the shape of the wave is either negligible for the practical purpose at hand, or are totally unobservable in practical terms. The value of n at which this occurs depends partially on the *rise time* of the waveform. Rise time is usually defined as the time required for the RF pulse waveform to rise from 10 per cent to 90 per cent of its final amplitude.

Figure 2.4 shows an RF pulse waveform based on a square impulse. The square wave represents a special case because it has an extremely fast rise time. Theoretically, the square wave contains an infinite number of harmonics, but not all of the possible harmonics are present. For example, in the case of the square wave only the odd harmonics are found (e.g. 3, 5, 7). According to some standards, accurately reproducing the square wave requires 100 harmonics, while others claim that 1000 harmonics are needed. Which standard to use may depend on the specifics of the application.

Another factor that determines the profile of the Fourier series of a specific waveform is whether the function is *odd* or *even*. Figure 2.5A shows an even-function square wave, and Fig. 2.5B shows an odd-function square wave. The even function is one in which $f(t) = f(-t)$, while for the odd function $-f(t) = f(-t)$. In the even function only cosine harmonics are present, so the sine amplitude coefficients b_n are zero. Similarly, in the odd function only sine harmonics are present, so the cosine amplitude coefficients a_n are zero.

Both *symmetry* and *asymmetry* can occur in several ways in a waveform (Fig. 2.6), and those factors can affect the nature of the Fourier series of the waveform. In Fig. 2.6A we see the case of a waveform with a DC component. Or, in terms of the Fourier series equation, the term a_o is non-zero. The DC component represents a case of asymmetry in a signal. This offset can seriously affect instrumentation electronic circuits that are DC-coupled.

Two different forms of symmetry are shown in Fig. 2.6B. *Zero-axis symmetry* occurs when, on a point-for-point basis, the waveshape and amplitude above the zero baseline is equal to the amplitude below the baseline (or $|+V_{\rm m}| = |-V_{\rm m}|$). When a waveform possesses zero-axis symmetry it will usually not contain even harmonics, only odd harmonics are present; this situation is found in square waves, for example (Fig. 2.7A). Zero-axis symmetry is not found only in sine and square waves, however, as the sawtooth waveform in Fig. 2.6C demonstrates.

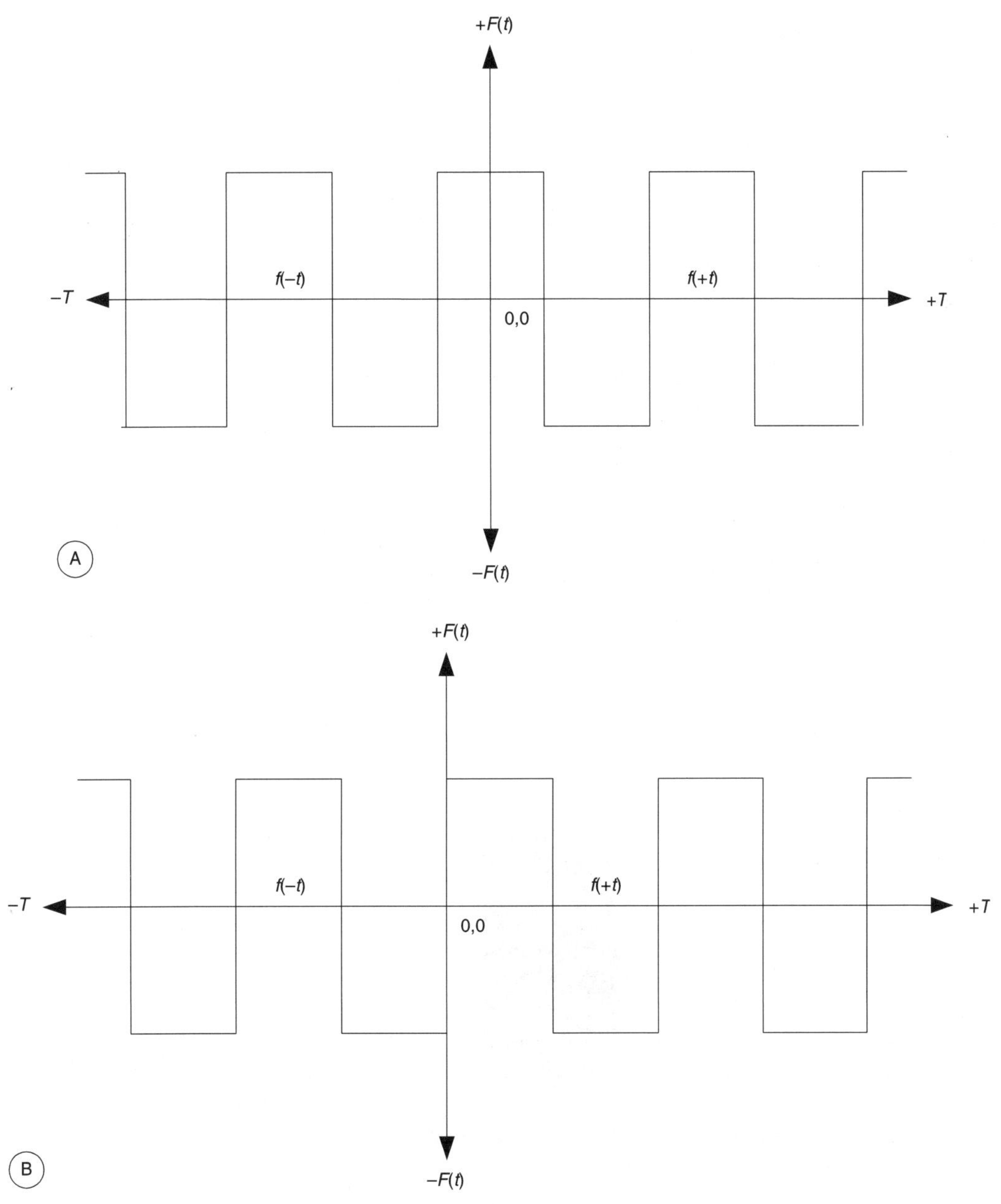

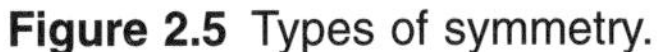
Figure 2.5 Types of symmetry.

An exception to the 'no even harmonics' general rule is that there will be even harmonics present in the zero-axis symmetrical waveform (Fig. 2.7B) if the *even harmonics are in-phase with the fundamental sine wave*. This condition will neither produce a DC component, nor disturb the zero-axis symmetry.

Also shown in Fig. 2.6B is *half-wave symmetry*. In this type of symmetry the *shape* of the wave above the zero baseline is a mirror image of the shape of the waveform below the

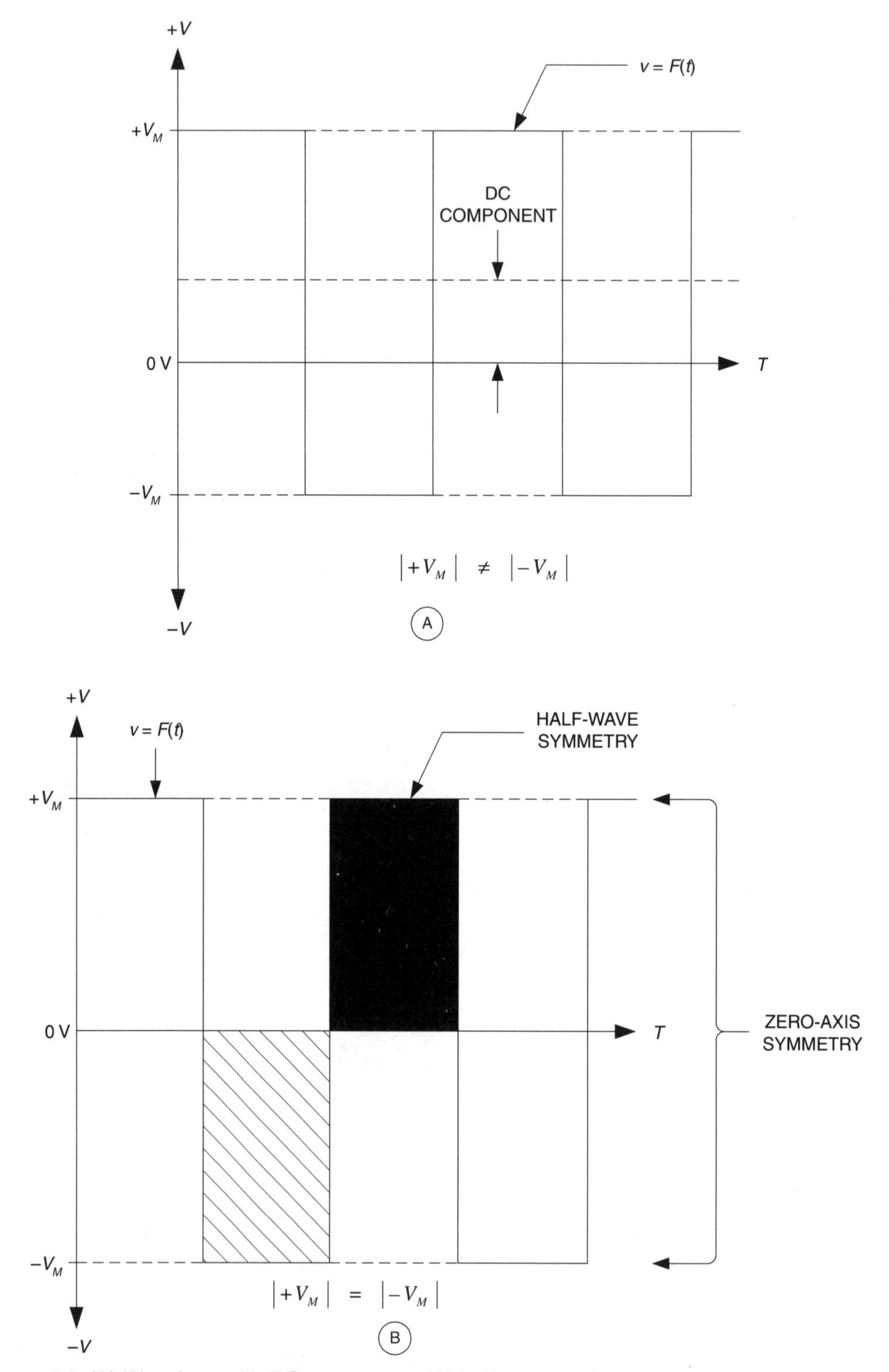

Figure 2.6 (A) Waveform with DC component; (B) half-wave and zero-axis symmetry;

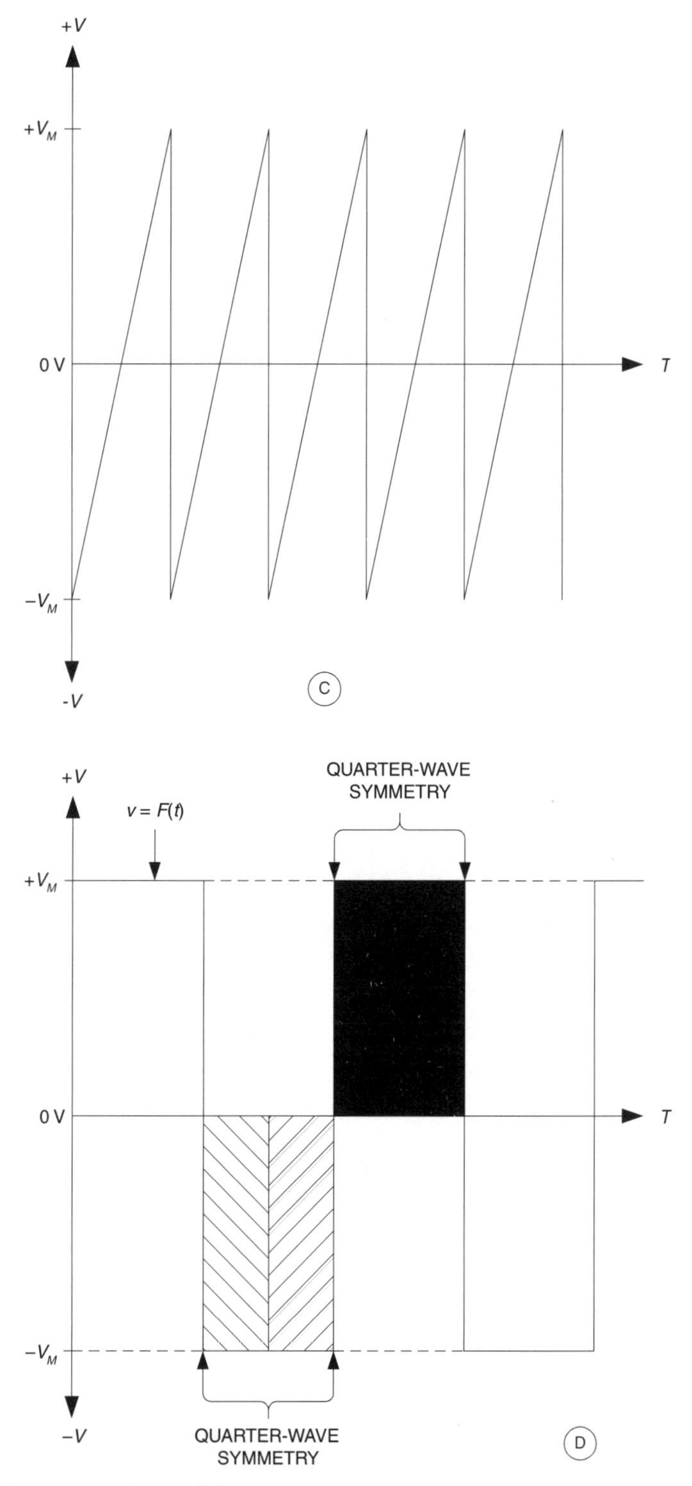

Figure 2.6 (C) triangle waveform; (D) quarter-wave symmetry.

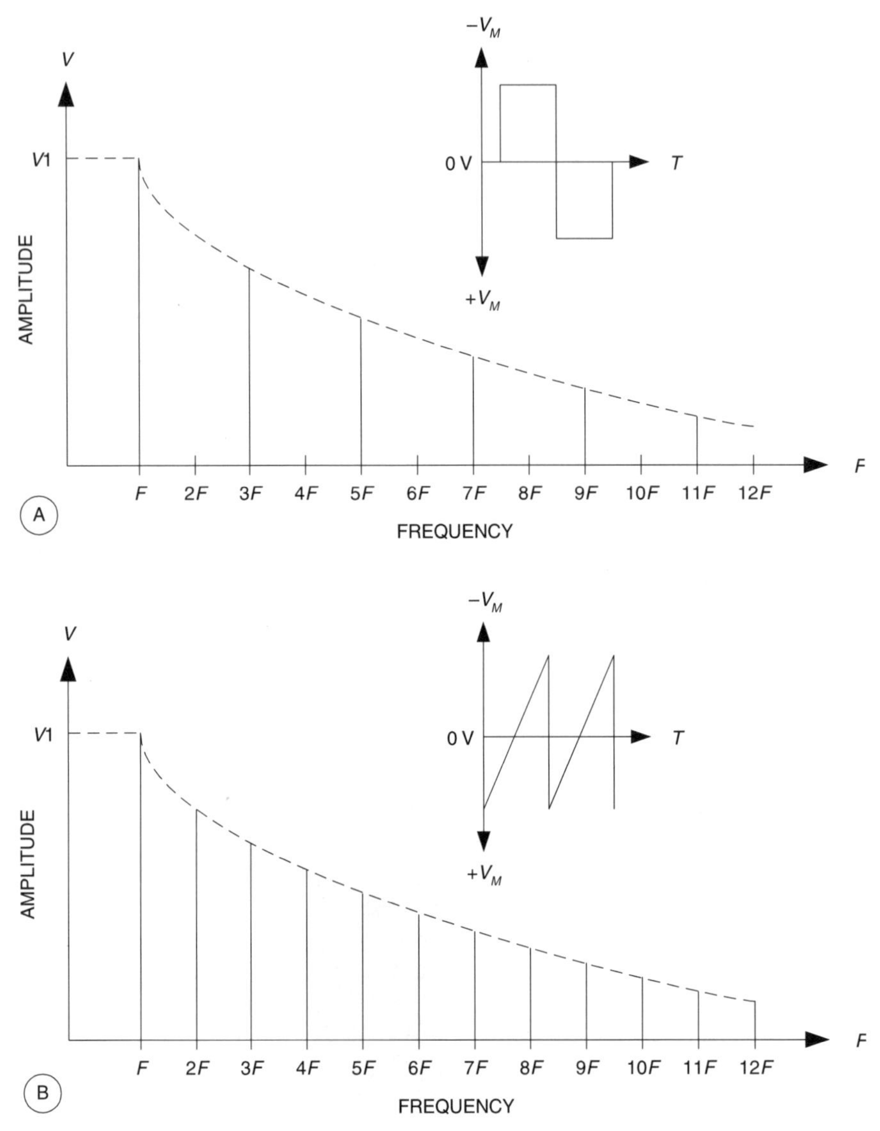

Figure 2.7 Spectrum of two waveforms.

baseline (shaded region in Fig. 2.6B). Half-wave symmetry also implies a lack of even harmonics.

Quarter-wave symmetry (Fig. 2.6D) exists when the left-half and right-half sides of the waveforms are mirror images of each other on the same side of the zero-axis. Note in Fig. 2.6D, that above the zero-axis the waveform is like a square wave, and indeed the left- and right-hand sides are mirror images of each other. Similarly, below the zero-axis the rounded waveform has a mirror image relationship between left and right sides. In this case, there is a full set of even harmonics, and any odd harmonics that are present are in-phase with the fundamental sine wave.

Transient signals

A transient signal is an event that occurs either once only, or occurs randomly over a long period of time, or is periodic but has a very short duration compared with its period (i.e. it is a very short duty cycle event). Many pulse signals fit the latter criterion even though mathematically they are actually periodic.

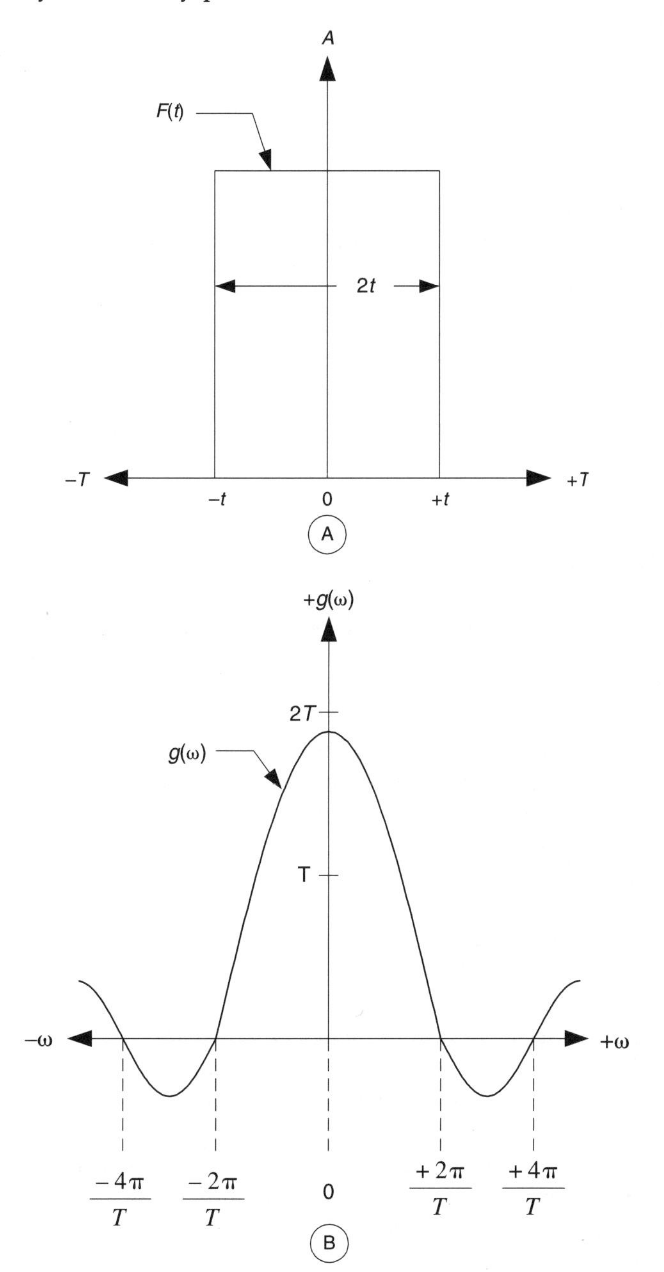

Figure 2.8

Transient signals are not represented properly by the Fourier series, but can nonetheless be represented by sine waves in a spectrum. The difference is that the spectrum of the transient signal is *continuous* rather than discrete. Consider a transient signal of period 2T, such as Fig. 2.8A. The *spectral density*, $g(\omega)$, is:

$$g(\omega) = \int_{-\infty}^{+\infty} f(t) e^{-j\omega wt}\, dt \tag{2.6}$$

Given a spectral density the original waveform can be reconstructed from:

$$f(t) = \frac{1}{2\pi} \int_{-\infty}^{+\infty} g(\omega) e^{j\omega wt}\, d\omega \tag{2.7}$$

The shape of the spectral density is shown in Fig. 2.8B. Note that the negative frequencies are a product of the mathematics, and do not have physical reality. The shape of Fig. 2.8B is expressed by:

$$g(\omega) = \frac{\sin \omega t}{\omega t} \tag{2.8}$$

The general form sin x/x is used also for repetitive pulse signals as well as the transient form shown in Fig. 2.8B.

Sampled signals

The digital computer is incapable of accepting analogue input signals, but rather requires a digitized representation of that signal. The *analogue-to-digital* (A/D) *converter* will convert an input voltage (or current) to a representative binary word. If the A/D converter is either clocked or allowed to run asynchronously according to its own clock, then it will take a continuous string of samples of the signal as a function of time. When combined, these signals represent the original analogue signal in binary form.

But the sampled signal is not exactly the same as the original signal, and some effort must be expended to ensure that the representation is as good as possible. Consider Fig. 2.9. The waveform in Fig. 2.9A is a continuous voltage function of time, $V(t)$; in this case a triangle waveform is seen. If the signal is sampled by another signal, $p(t)$, with frequency F_s and sampling period $T = 1/F_s$, as shown in Fig. 2.9B, and then later reconstructed, the waveform may look something like Fig. 2.9C. While this may be sufficiently representative of the waveform for many purposes, it would be reconstructed with greater fidelity if the sampling frequency (F_s) is increased.

Figure 2.10 shows another case in which a sine wave, $V(t)$ in Fig. 2.10A, is sampled by a pulse signal, $p(t)$ in Fig. 2.10B. The sampling signal, $p(t)$, consists of a train of equally spaced narrow pulses spaced in time by T. The sampling frequency F_s equals $1/T$. The resultant is shown in Fig. 2.10C, and is another pulsed signal in which the amplitudes of the pulses represent a sampled version of the original sine wave signal.

The sampling rate, F_s, must by *Nyquist's theorem* be twice the maximum frequency (F_m) in the Fourier spectrum of the applied analogue signal, $V(t)$. In order to reconstruct the original signal after sampling, it is necessary to pass the sampled waveform through a

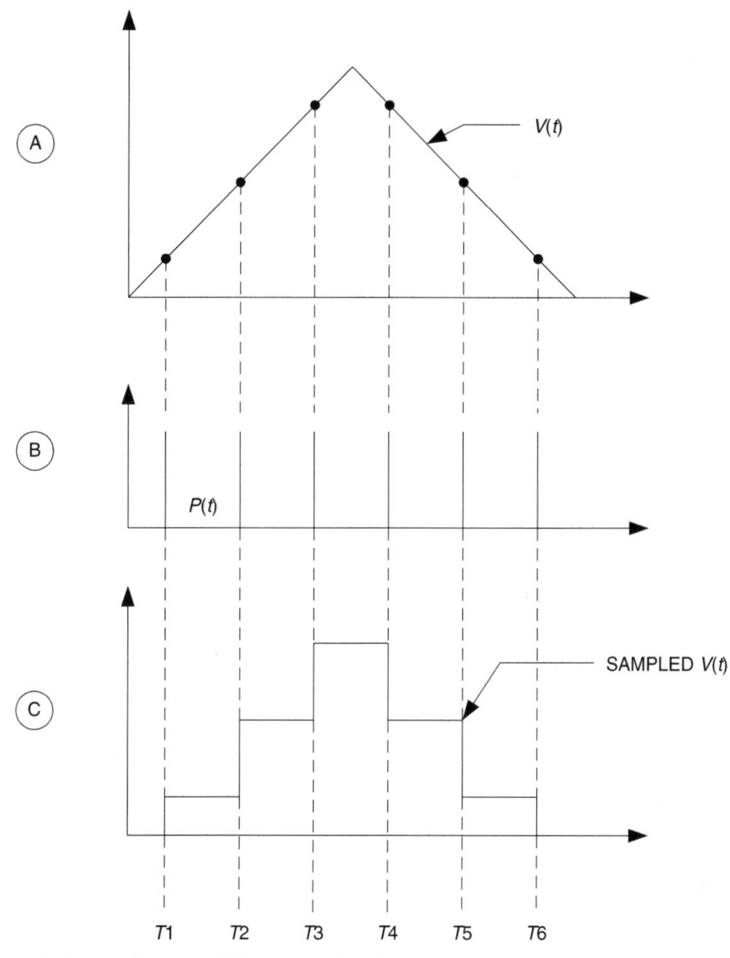

Figure 2.9 Sampled waveform and its reconstruction.

low-pass filter that limits the pass band to F_s. In practical RF systems, you will find that many engineers determine that the minimum Nyquist rate is insufficient for good fidelity reproductions of the sampled waveform, so will specify a faster rate. Also, some oversampling methods are used to dramatically reduce noise.

The sampling process is analogous to a form of *amplitude modulation* (AM), in which $V(t)$ is the modulating signal, with spectrum from DC to F_m, and $p(t)$ is the carrier frequency. The resultant spectrum is shown partially in Fig. 2.11, and resembles the double sideband with carrier AM spectrum. The spectrum of the modulating signal appears as 'sidebands' around the 'carrier' frequency, shown here as F_o. The actual spectrum is a bit more complex, as shown in Fig. 2.12. Like an unfiltered AM radio transmitter, the same spectral information appears not only around the fundamental frequency (F_s) of the carrier (shown at zero in Fig. 2.12), but also at the harmonics spaced at intervals of F_s up and down the spectrum.

Providing that the sampling frequency $F_s \geq 2F_m$, the original signal is recoverable from the sampled version by passing it through a low-pass filter with a cut-off frequency F_c, set

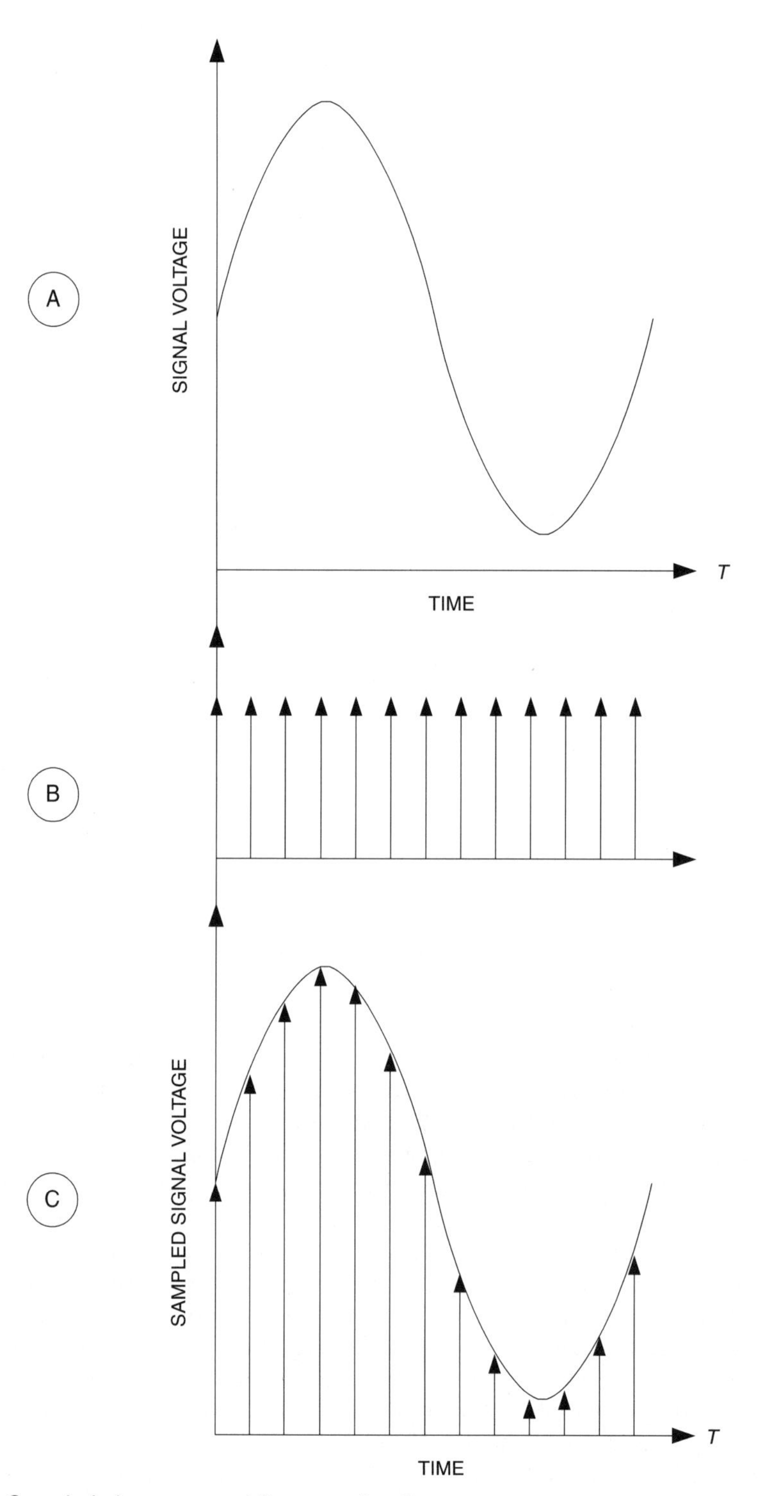

Figure 2.10 Sampled sine wave and its reconstruction.

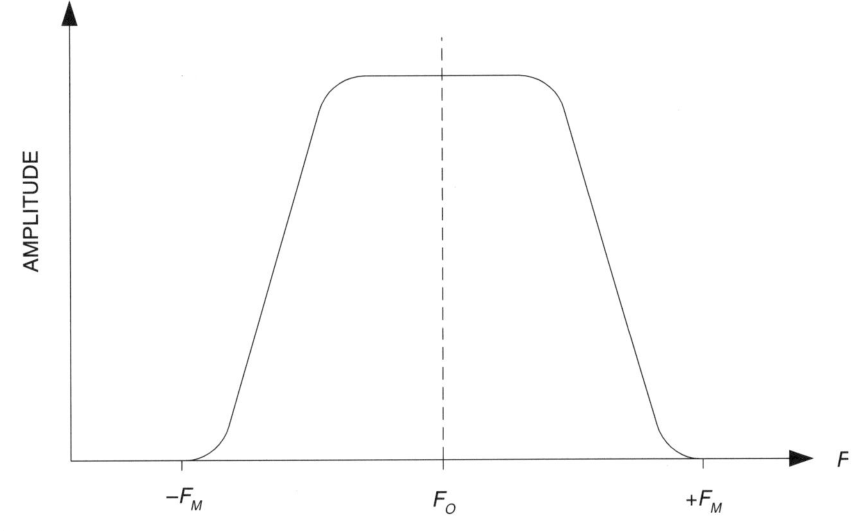

Figure 2.11 Spectrum.

to pass only the spectrum of the analog signal – but not the sampling frequency. This phenomenon is shown with the dashed line in Fig. 2.12.

When the sampling frequency $F_s < 2F_m$, then a problem occurs (see Fig. 2.13). The spectrum of the sampled signal looks similar to before, but the regions around each harmonic overlap such that the value of $-F_m$ for one spectral region is less than $+F_m$ for the next lower frequency region. This overlap results in a phenomenon called *aliasing*. That is, when the sampled signal is recovered by low-pass filtering it will produce not the original sine wave frequency F_o but a lower frequency equal to $(F_s - F_o)$ . . . and the information carried in the waveform is thus lost or distorted.

The solution, for accurate sampling of the analogue waveform for input to a computer, is to:

1 Bandwidth limit the signal at the input of the sampler or A/D converter with a low-pass filter with a cut-off frequency F_c selected to pass only the maximum frequency in the waveform (F_m) and not the sampling frequency (F_s).
2 Set the sampling frequency F_s *at least* twice the maximum frequency in the applied waveform's Fourier spectrum, i.e. $F_s \geq 2F_m$.

Noise

An ideal electronic circuit produces no noise of its own, so the output signal from the ideal circuit contains only the noise that was in the original signal. But real electronic circuits and components do produce a certain level of inherent noise. Even a simple fixed value resistor is noisy. Figure 2.14A shows the equivalent circuit for an ideal, noise-free resistor. The inherent noise is represented in Fig. 2.14B by a noise voltage source, V_n, in series with the ideal, noise-free resistance, R_i. At any temperature above *absolute zero* (0 K or about

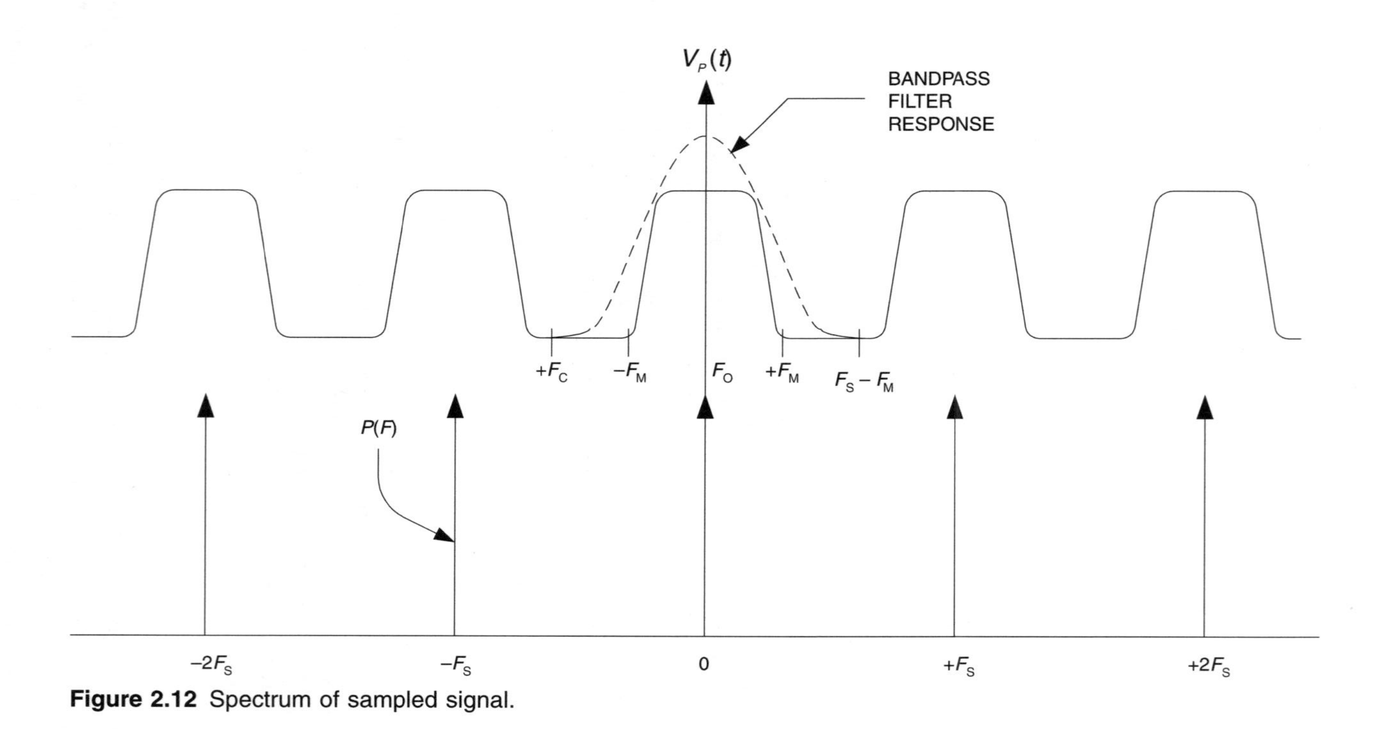

Figure 2.12 Spectrum of sampled signal.

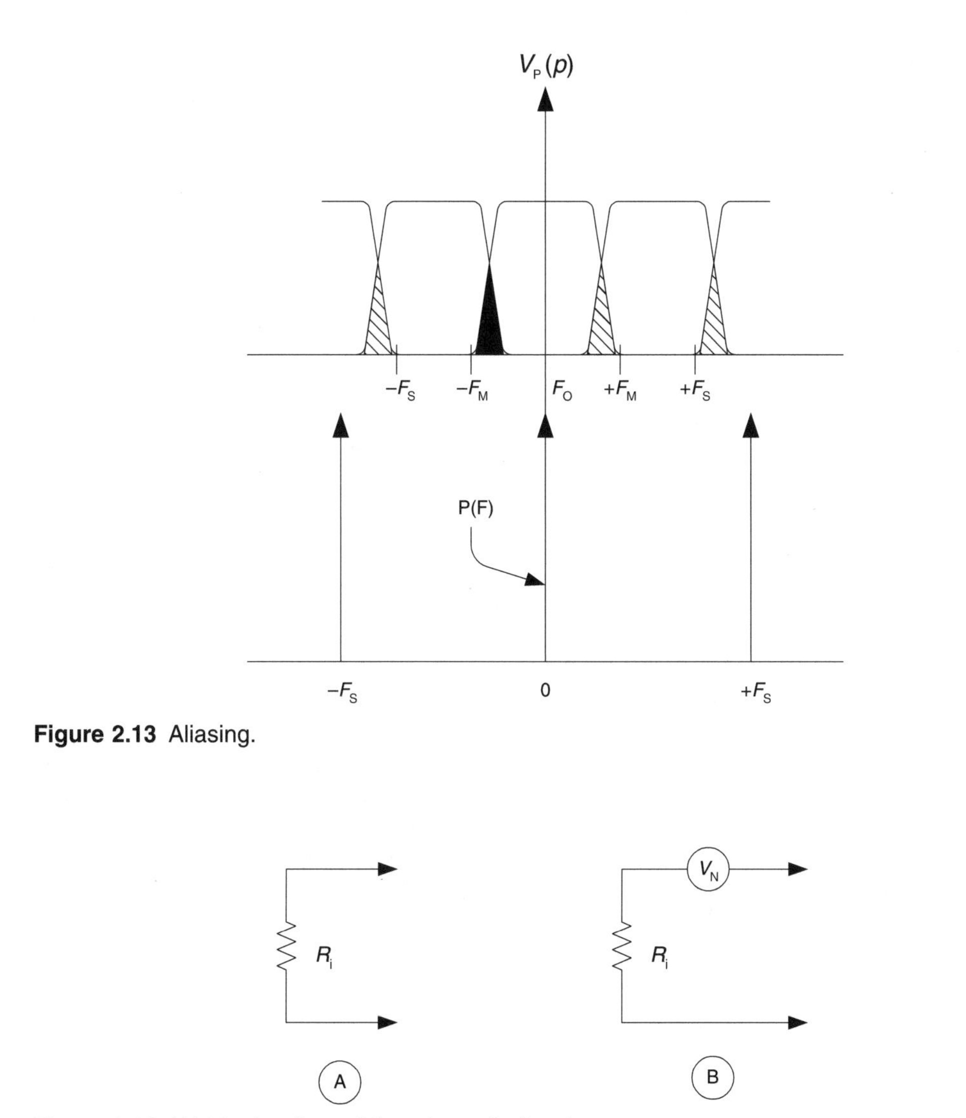

Figure 2.13 Aliasing.

Figure 2.14 (A) Ideal resistor; (B) real practical resistor.

–273°C) electrons in any material are in constant random motion. Because of the inherent randomness of that motion, however, there is no detectable current in any one direction. In other words, electron drift in any single direction is cancelled over short time periods by equal drift in the opposite direction. Electron motions are therefore statistically decorrelated. There is, however, a continuous series of random current *pulses* generated in the material, and those pulses are seen by the outside world as a noise signal. This signal is called by several names: *thermal agitation noise, thermal noise,* or *Johnson noise.*

Johnson noise is a so-called 'white noise' because it has a very broadband (nearly gaussian) spectral density. The thermal noise spectrum is essentially flat. The term 'white noise' is a metaphor developed from white light, which is composed of all visible colour frequencies. The expression for Johnson noise is:

$$V_{\mathrm{n}} = \sqrt{4KTRB}\ \mathrm{V} \tag{2.9}$$

Where:
V_{n} is the noise voltage (V)
K is Boltzmann's constant (1.38×10^{-23} J/K)
T is the temperature in kelvin (K)
R is the resistance in ohms (Ω)
B is the bandwidth in hertz (Hz)

At normal room temperature, with the constants collected, and the resistance normalized to 1 kΩ, Eq. (2.9) reduces to:

$$V_{\mathrm{n}} = 4\sqrt{\frac{R}{1\ \mathrm{k}\Omega}}\ \frac{\mathrm{nV}}{\sqrt{\mathrm{Hz}}} \tag{2.10}$$

The evaluated solution of Eq. (2.10) is normally read *nanovolts (nV) per square root hertz.* In this equation, a 1 megohm resistor will have a thermal noise of $126\,\mathrm{nV}/\sqrt{\mathrm{Hz}}$.

Several other forms of noise are present in linear ICs and other semiconductor amplifiers to one extent or another. For example, because current flow at the quantum level is not smooth and predictable, an intermittent burst phenomenon is sometimes seen. This noise is called *popcorn noise,* and consists of pulses of many milliseconds duration. Another form of noise is *shot noise* (also called *Schottky noise*). The name 'shot' is derived from the fact that the noise sounds like a handful of B-B shot thrown against a metal surface. Shot noise is a consequence of DC current flowing in any conductor, and is found from:

$$I_{\mathrm{n}} = \sqrt{2qIB}\ \frac{\mathrm{A}}{\sqrt{\mathrm{Hz}}} \tag{2.11}$$

Where:
I_{n} is the noise current in amperes (A)
q is the elementary electric charge (1.6×10^{-19} coulombs)
I is the current in amperes (A)
B is the bandwidth in hertz (Hz)

Finally, there is *flicker noise,* also called *pink noise* or *1/f noise.* The latter name applies because flicker noise is predominantly a low frequency (<1000 Hz) phenomenon. This type of noise is found in all conductors, and becomes important in IC devices because of manufacturing defects.

The noise spectrum in any given instrumentation system will contain elements of several kinds of noise, although in some systems one form or another may dominate the others. It is common to characterize noise from a single source using the *root mean square* (rms) value of the voltage amplitudes:

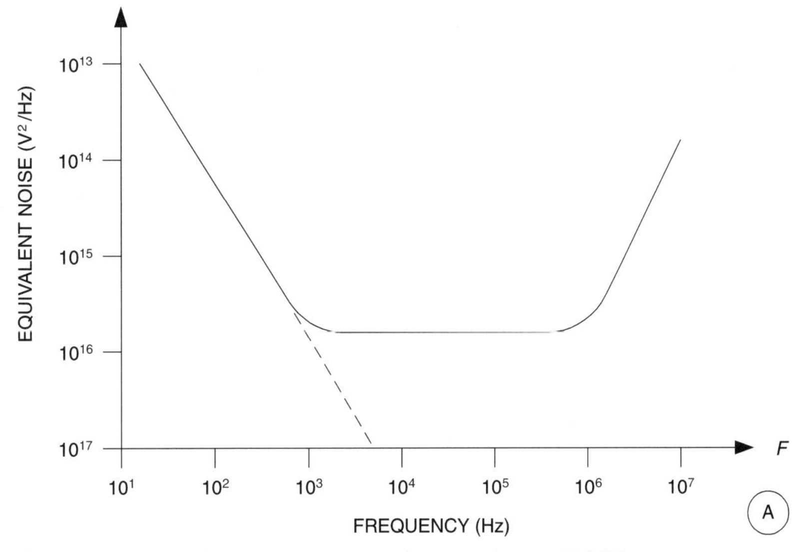

Figure 2.15 (A) Equivalent noise temperature at frequencies to 10 MHz.

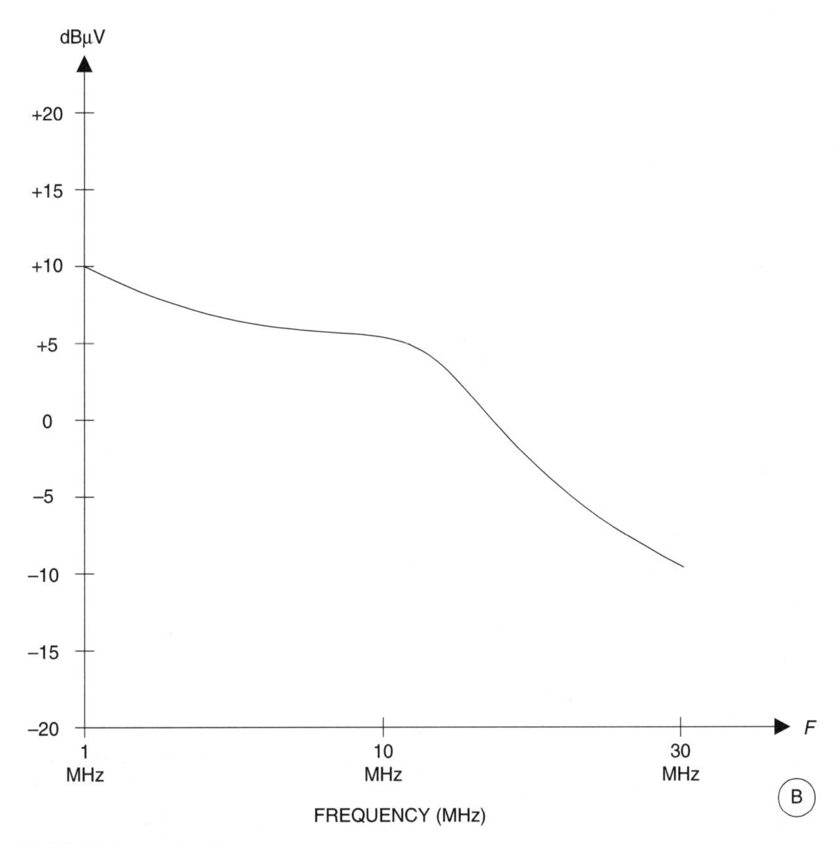

Figure 2.15 (B) Universal noise sources.

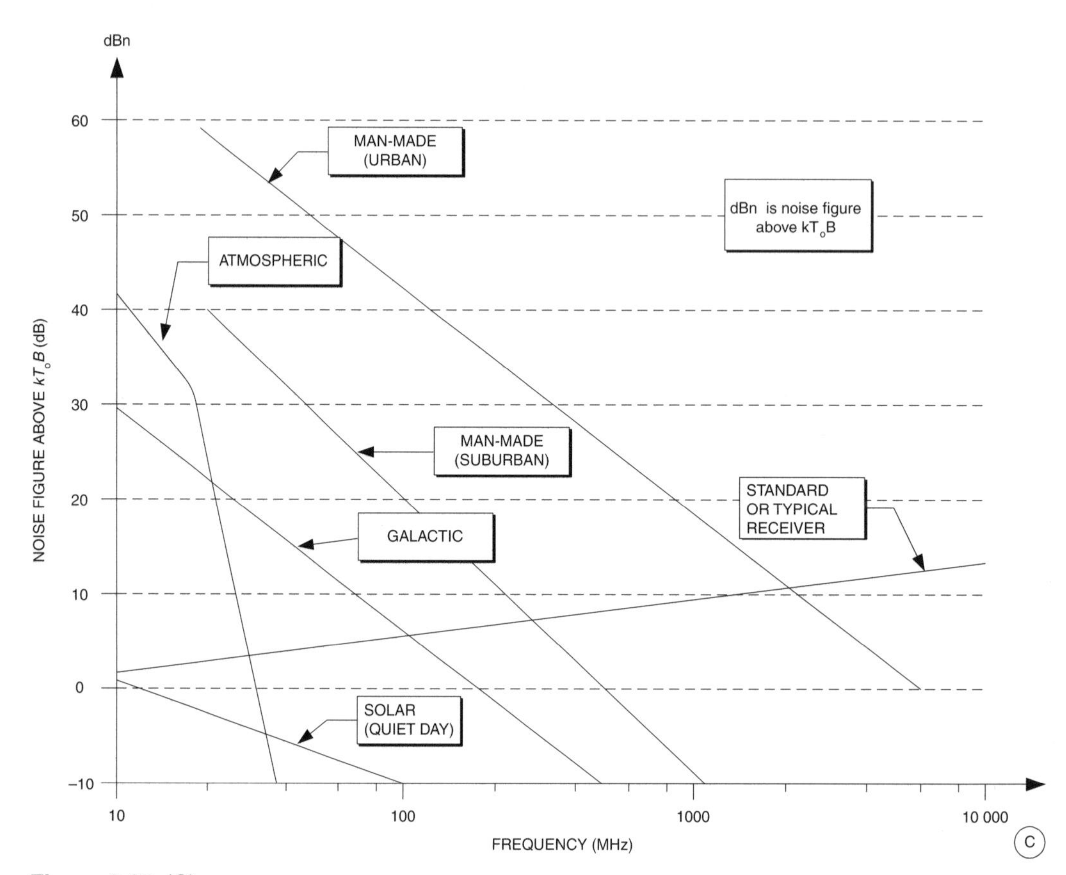

Figure 2.15 (C).

$$V_{n\,(rms)} = \sqrt{\frac{1}{T}\int_0^T [F(t)^2]\,dt} \tag{2.12}$$

Figure 2.15A shows the noise spectrum profile for a typical system that contains $1/F$ noise, thermal or 'white' noise, and some high frequency noise. Noise comes in a number of different guises, but in the case of radio reception we can divide noise into two classes: *sources external to the receiver* and *sources internal to the receiver.* There is little one can do about the external noise sources, for they consist of natural and man-made electromagnetic signals that fall within the passband of the receiver. Figure 2.15B shows an approximation of the external noise situation from the middle of the AM broadcast band to the low end of the VHF region. A somewhat different view, which captures the severe noise situation seen by receivers, is shown in Fig. 2.15C. One must select a receiver that can cope with external noise sources, especially if the noise sources are strong.

Some natural external noise sources are extraterrestrial. It is these signals that form the basis of radio astronomy. For example, if you aim a beam antenna at the eastern horizon prior to sunrise, a distinct rise of noise level occurs as the Sun slips above the horizon, especially in the VHF region (the 150–152 MHz band is used to measure solar flux). The reverse occurs in the west at sunset, but is less dramatic, probably because atmospheric ionization decays much slower than it is generated. During World War II, it was reported that British radar operators noted an increase in received noise level any time the Milky Way was above the horizon, decreasing the range at which they could detect in-bound German bombers. There is also some well-known, easily observed noise from the planet Jupiter in the 18 to 30 MHz range.

Signal-to-noise ratio (SNR or S_n)

Amplifiers can be evaluated on the basis of *signal-to-noise ratio* (S/N or 'SNR'), denoted S_n. The goal of the circuit or instrument designer is to enhance the SNR as much as possible. Ultimately, the minimum signal level detectable at the output of an amplifier is that level which appears above the noise floor level. Therefore, the lower the system noise floor, the smaller the *minimum allowable signal*. Although often thought of as a radio receiver parameter, SNR is applicable in other amplifiers where signal levels are low.

Noise resulting from thermal agitation of electrons is measured in terms of *noise power* (P_n), and carries the units of power (watts or its sub-units). Noise power is found from:

$$P_n = KTB \tag{2.13}$$

Where:
P_n is the noise power in watts (W)
K is Boltzmann's constant (1.38×10^{-23} J/K)
T is the temperature in kelvin (K)
B is the bandwidth in hertz (Hz)

Notice in Eq. (2.13) that there is no centre frequency term, only the bandwidth (B). Thus, in bandwidth limited systems, such as a practical amplifier or network, the total noise power is related to temperature and bandwidth. We can conclude that a 3000 Hz bandwidth centred on 1 MHz produces the same thermal noise level as a 3000 Hz bandwidth centred on 60 MHz or any other frequency.

Noise sources can be categorized as either *internal* or *external*. The internal noise sources are due to thermal currents in the semiconductor material resistances. It is the noise component contributed by the amplifier under consideration. If noise, or S/N ratio, is

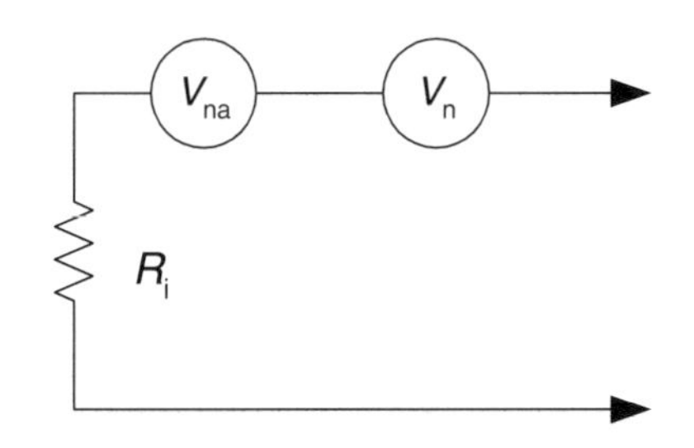

Figure 2.16 Noise and signal voltages.

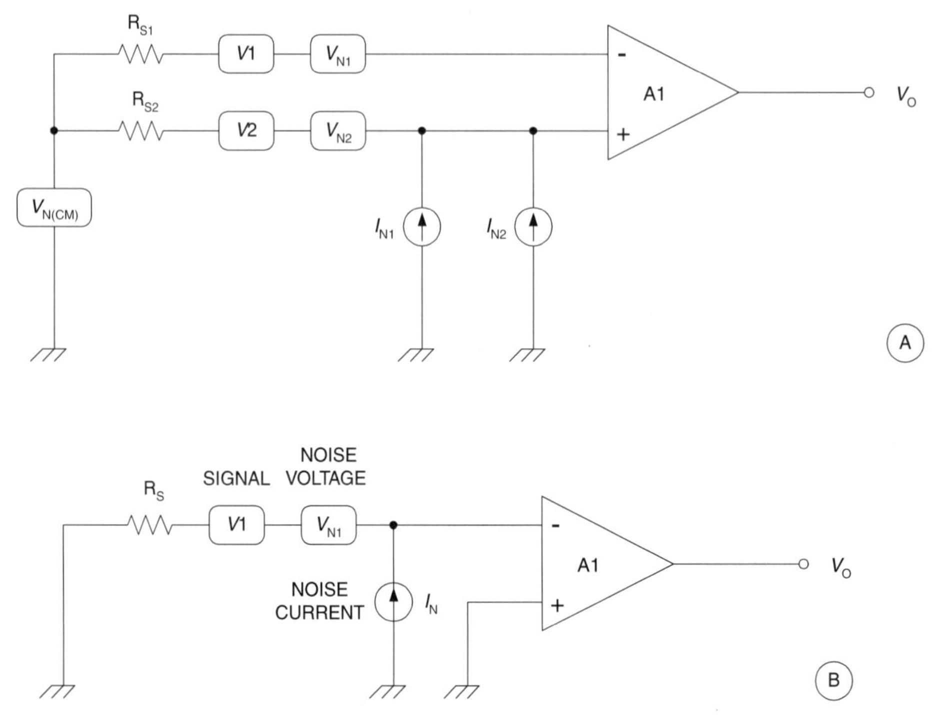

Figure 2.17 Various noise sources in amplifier circuit.

measured at both input and output of an amplifier, the output noise is greater. The internal noise of the device is the difference between output noise level and input noise level.

External noise is the noise produced by the signal source, so is often called *source noise*. Part of this noise signal is due to thermal agitation currents in the signal source. In fact, the simple terminated noise level might be higher than V_n because of component construction. For example, the noise signal produced by a carbon composition resistor has an additional noise source modelled as V_{na} in Fig. 2.16. This noise generator is a function of resistor construction and manufacturing defects.

Figure 2.17A shows a circuit model showing that several voltage and current noise sources exist in an op-amp. The relative strengths of these noise sources, hence their overall contribution, varies with op-amp type. In an FET-input op-amp, for example, the current noise sources are tiny, but voltage noise sources are significant. For bipolar op-amps the opposite situation applies.

All of the noise sources in Fig. 2.17A are uncorrelated with respect to each other, so one cannot simply add noise voltages; only noise power can be added. To characterize noise voltages and currents they must be added in the *root sum squares* (RSS) manner.

Models such as Fig. 2.17A are too complex for most situations, so it is standard practice to lump all of the voltage noise sources into one source, and all of the current noise sources into another source. The composite sources have a value equal to the RSS voltage (or current) of the individual sources. Figure 2.17B is such a model in which only a single current source and a single voltage source are used. The *equivalent AC noise* in Fig. 2.17B is the overall noise, given a specified value of source resistance, R_s, and is found from the RSS value of V_n and I_n:

$$V_{nt} = \sqrt{V_n^2 + (I_n R_s)^2} \tag{2.14}$$

Noise factor, noise figure and noise temperature

The noise of a system or network can be defined in three different but related ways: *noise factor* (F_n) – a simple ratio; *noise figure* (NF) – a decibel ratio; and *equivalent noise temperature* (T_e).

Noise factor (F_n)

For components such as resistors, the noise factor is the ratio of the noise produced by a real resistor to the simple thermal noise of a perfect resistor. The noise factor of a system is the ratio of output noise power (P_{no}) to equivalent input noise power (P_{ni}):

$$F_n = \left. \frac{P_{no}}{P_{ni}} \right|_{T = 290\,\text{K}} \tag{2.15}$$

In order to make comparisons easier the noise factor is always measured at the standard temperature (T_o) 290 K (standardized room temperature).

The equivalent input noise power P_{ni} is defined as the product of the source thermal noise at standard temperature (T_o) and the amplifier gain (G):

$$P_{ni} = GKBT_o \tag{2.16}$$

It is also possible to define noise factor F_n in terms of output and input S/N ratio:

$$F_n = \frac{S_{ni}}{S_{no}} \tag{2.17}$$

which is also:

$$F_n = \frac{P_{no}}{KT_oBG} \tag{2.18}$$

Where:
S_{ni} is the input signal-to-noise ratio
S_{no} is the output signal-to-noise ratio
P_{no} is the output noise power
K is Boltzmann's constant (1.38×10^{-23} J/K)
T_o is 290 kelvin
B is the network bandwidth in hertz
G is the amplifier gain

The noise factor can be evaluated in a model that considers the amplifier ideal, and therefore only amplifies through gain G the noise produced by the 'input' noise source:

$$F_n = \frac{KT_oBG + \Delta N}{KT_oBG} \tag{2.19}$$

or,

$$F_n = 1 + \frac{\Delta N}{KT_oBG} \tag{2.20}$$

Where:
ΔN is the noise added by the network or amplifier
All other terms are as defined above

Noise figure (NF)

The noise figure is a frequently used measure of an amplifier's 'goodness', or its departure from *goodness*. Thus, it is a *figure of merit*. The noise figure is the noise factor converted to decibel notation:

$$NF = 10 \log_{10} F_n \tag{2.21}$$

Where:
NF is the noise figure in decibels (dB)
F_n is the noise factor

Noise temperature (T_e)

The noise temperature is a means for specifying noise in terms of an equivalent temperature. The idea is to pretend that the noise added by the amplifier comes instead from thermal noise from the source. The noise temperature is how hot a perfect input resistor would have to be to give rise to the output noise. It is a particularly useful concept in situations where the source noise does not come from thermal noise at room temperature, e.g. a UHF or microwave antenna.

Note that the equivalent noise temperature T_e is *not* the physical temperature of the amplifier. The noise temperature is related to the noise factor by:

$$T_e = (F_n - 1)T_o \tag{2.22}$$

and to the noise figure by:

$$T_e = [10^{NF/10} - 1] \times T_o \tag{2.23}$$

Now that we have noise temperature T_e, we can also define noise factor and noise figure in terms of noise temperature:

$$F_n = \frac{T_e}{T_o} + 1 \tag{2.24}$$

and,

$$NF = 10 \log \left[\frac{T_e}{T_o} + 1 \right] \tag{2.25}$$

The total noise in any amplifier or network is the sum of internally generated and externally generated noise. In terms of noise temperature:

$$P_{n(total)} = GKB(T_o + T_e) \tag{2.26}$$

Where:
$P_{n(total)}$ is the total noise power
All other terms are as previously defined

Noise in cascade amplifiers

A noise signal is treated by any following amplifier as a valid input signal. Thus, in a cascade amplifier the final stage sees an input signal that consists of the original signal and noise amplified by each successive stage. Each stage in the cascade chain amplifies signals and noise from previous stages, and also contributes some noise of its own. The overall noise factor for a cascade amplifier can be calculated from *Friis' noise equation*:

$$NF = F_1 + \frac{F_2 - 1}{G_1} + \frac{F_3 - 1}{G_1 G_2} + \ldots + \frac{F_N - 1}{G_1 G_2 \ldots G_{N-1}} \tag{2.27}$$

Where:
NF is the overall noise factor of N stages in cascade
F_N is the noise factor of the Nth stage
G_N is the gain of the Nth stage

As you can see from Eq. (2.27), the noise factor of the entire cascade chain is dominated by the noise contribution of the first stage or two. Later stages are less important where noise is concerned, provided that the input stages have sufficient gain.

Noise reduction strategies

Although noise is a serious problem for the designer, especially where low signal levels are experienced, there are a number of common sense approaches to minimize the effects of noise on a system. In this section we will examine several of these methods. For example:

1 Keep the source resistance and the amplifier input resistance as low as possible. Using high value resistances will increase thermal noise voltage.

2 Total thermal noise is also a function of the bandwidth of the circuit. Therefore, reducing the bandwidth of the circuit to a minimum will also minimize noise. But this job must be done mindfully because signals have a Fourier spectrum that must be preserved for faithful reproduction or accurate measurement. The solution is to match the bandwidth to the frequency response required for the input signal.
3 Prevent external noise from affecting the performance of the system by appropriate use of grounding, shielding and filtering.
4 Use a low noise amplifier (LNA) in the input stage of the system.
5 For some semiconductor circuits, use the lowest DC power supply potentials that will do the job.

Noise reduction by signal averaging

If a signal is either periodic or repetitive, or can be made so, then it is possible to enhance signal-to-noise ratio (S_n) by signal averaging. The basis for this simple signal processing technique is the assumption that noise meets the definition of either random or chaotic processes. If so, then noise tends to integrate to zero or near-zero over time. If time-averaging integration is performed in a coherent manner, then a repetitive signal tends to build in value, while noise levels (being decorrelated) decrease. If we assume that the signal-to-noise ratio is:

$$S_n = 20 \log \left(\frac{V_{in}}{V_n} \right) \tag{2.28}$$

Then, for systems where $V_i < V_n$, the noise reduction by time averaging is:

$$\bar{S}_n = 20 \log \left(\frac{V_{in}}{V_n/\sqrt{N}} \right) \tag{2.29}$$

Where:
S_n is the time-averaged SNR
V_n is the unprocessed SNR
N is the number of repetitions of the signal

Example

An RF amplifier processes a 5 μV signal in the presence of a 500 μV random noise level. Calculate the unprocessed SNR, the processed SNR for 1000 repetitions of the signal, and the processing gain.

Solution:

A. *Unprocessed SNR*:

$$S_{no} = 20 \log(V_i/V_n)$$
$$= 20 \log(5\,\mu V/500\,\mu V) = -40\,dB$$

B. *Processed SNR*:

$$\bar{S}_n = S_n/[N]^{1/2}$$
$$= (-40\,\text{dB})/[1000]^{1/2} = -1.3\,\text{dB}$$

C. *Processing gain*:

$$G_p = S_n - \bar{S}_{no}$$
$$= (-1.3\,\text{dB}) - (-40\,\text{dB}) = +38\,\text{dB}$$

The effect of time averaging is to increase the time required to collect data, so time averaging is effectively a means of decreasing the bandwidth of the system (by $F = 1/T$).

Coherency is maintained in a system by ensuring that repetitive data points are processed in a consistent time relationship with respect to each other. The averaging will be triggered by a repetitive event, and that action starts the process. Data points are always matched to other data points taken at the same elapsed time after the trigger for previous iterations. For example, the ith datum point following a current sweep is paired with all other ith points from previous sweeps, and none other.

An example of signal averaging used to extract weak signals from larger noise signals is found in radar systems where multiple reflected pulses are integrated (i.e. time averaged) to improve SNR.

3 Radio receivers

Most RF systems include one or more radio receivers. A receiver typically includes a number of RF circuits, so getting an understanding of radio receivers is a good way to begin. In this chapter you will find the basic theories of radio receiver design and their specifications.

Signals, noise and reception

No matter how simple or fancy the system may be, the basic function of a radio receiver is the same: *to distinguish signals from noise*. The concept 'noise' covers both man-made and natural radio frequency signals. The man-made signals include all signals in the passband other than the one being sought.

In communications systems the signal is some form of modulated (AM, FM, PM, etc.) periodic sine wave propagating as an electromagnetic (i.e. radio) wave. The 'noise', on the other hand, tends to be a random signal that sounds like the 'hiss' heard between stations on a radio. The spectrum of such noise signals appears to be gaussian ('white noise') or pseudogaussian ('pink noise' or 'bandwidth limited' noise).

In radio astronomy and satellite communications systems the issue is complicated because the signals are also noise. The radio emissions of Jupiter and the Sun are very much like the gaussian or pseudogaussian signals that are, in other contexts, nothing but useless noise. In fact, in the early days of radar the galactic noise tended to mask returns from incoming enemy aircraft, so to the radar operators these signals were noise of the worst kind. Yet to a radio astronomer, those signals are the goal! In satellite communications systems the 'signals' of the radio astronomer are limitations and annoyances at best and devastating at worse. The trick is to separate out the noise you want from the noise you don't.

Figure 3.1A shows an amplitude-vs-time plot of a typical noise signal, while Fig. 3.1B shows a type of regular radio signal that could be generated by a transmitter. Notice the difference between the two. The signal is regular and predictable. Once you know the frequency and period you can predict the amplitude at other points along the time line. The noise signal, on the other hand, is unpredictable. Knowing the cycle-to-cycle amplitude and duration (there is no true 'period') does not confer the ability to predict anything at all about the following cycles.

In some receivers, especially those designed for pulse reception, the differences highlighted between Figs 3.1A and 3.1B are used to increase the performance of the receiver by averaging over time as described in Chapter 2.

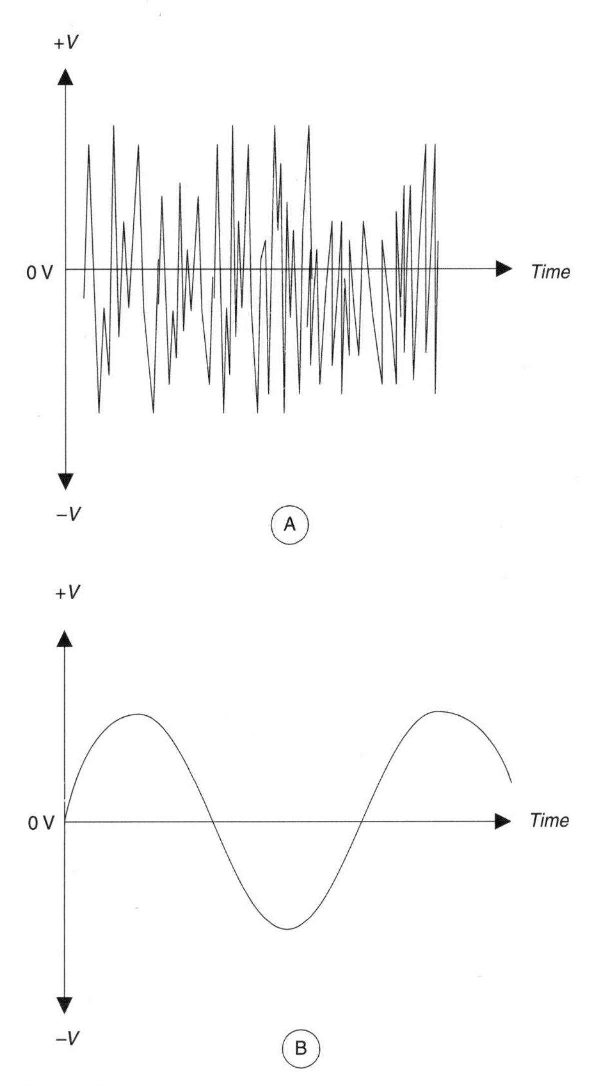

Figure 3.1 (A) Noise signal; (B) single frequency sine wave.

The reception problem

Figure 3.2 shows the basic problem of radio reception, especially in cases where the signal is very weak. The signal in Fig. 3.2A is embedded in noise that is relatively high amplitude. This signal is lower than the noise level, so is very difficult (perhaps impossible) to detect. The signal in Fig. 3.2B is easily detectable because the signal amplitude is higher than the noise amplitude. It becomes difficult when the signal is only slightly stronger than the average noise power level.

The *signal-to-noise ratio* (SNR) of a receiver system tells us something about the chance of success in detecting the signal. The SNR is normally quoted in *decibels* (dB), which is defined as:

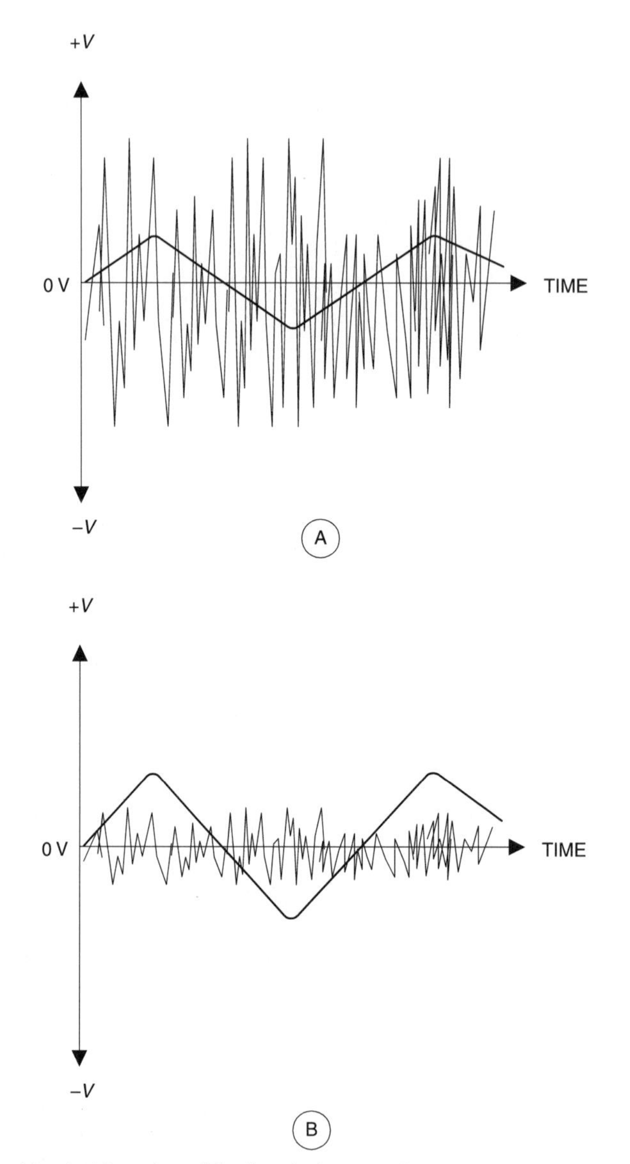

Figure 3.2 (A) Signal buried in noise; (B) signal above noise.

$$SNR = 10 \log \left[\frac{P_s}{P_n} \right] \text{dB} \tag{3.1}$$

Where:
SNR is the signal-to-noise ratio in decibels (dB)
P_s is the signal power level
P_n is the noise power level

How high an SNR is required? That depends on a lot of subjective factors when a human listener is present. Skilled radio operators can detect signals with an SNR of less than one decibel . . . but the rest of us cannot even hear that signal. Most radio operators can detect 3 dB SNR signals, but for 'comfortable' listening 10 dB SNR is usually specified. For digital systems the noise performance is usually defined by the acceptable *bit error rate* (BER).

Strategies

A number of strategies can be used to improve the SNR of a system. First, of course, is to buy or build a receiver that has a low internal 'noise floor'. High quality receivers may have very low noise, but there is sometimes some creative spec writing in the advertisements. For example, different bandwidths are used for the measurement, and only the most favourable value (which may not be the bandwidth that matches your needs) is reported.

By common sense we see that there are two approaches to SNR improvement: either increase the signal or decrease the noise. Most successful systems do both, but it must be done carefully.

One approach to SNR improvement is to use a preamplifier ahead of the receiver antenna terminals. This approach may or may not work, and under some situations may make the situation worse. The problem is that the preamplifier adds noise of its own, and will amplify noise from outside (received through the antenna) and the desired signal equally. If you have an amplifier with a gain of, say, 20 dB, then the external noise is increased by 20 dB and the signal is increased by 20 dB. The result is that the absolute numbers are bigger but the SNR is the same. If the amplifier produces any significant noise of its own, then the SNR will degrade. The key is to use a very *low noise amplifier* (LNA) for the preamplifier.

Another trick is to use a *preselector* ahead of the receiver. A preselector is either a tuned circuit or bandpass filter placed in the antenna transmission line ahead of the receiver antenna terminals. A *passive preselector* has no amplification (uses L–C elements only), while an *active preselector* has a built-in amplifier. The amplifier should be an LNA type. The reason why the preselector can improve the system is that it amplifies the signal by a fixed amount, but only the noise within the passband is amplified the same amount as the signal. Improvement comes from bandwidth limiting the noise but not the signal.

Another practical approach is to use a directional antenna. This method works especially well when the unwanted noise is other man-made signal sources. An omnidirectional antenna receives equally well in all directions. As a result, both natural and man-made external noise sources operating within the receiver's passband will be picked up. But if the antenna is made highly directional, then all noise sources that are not in the direction of interest are suppressed.

Directional antennas have *gain*, so the signal levels in the direction of interest are increased. Although the noise also increases in that direction, the rest of the noise sources (in other directions) are suppressed. The result is that SNR is increased by both methods.

When designing a communications system the greatest attention should usually be paid to the antenna, then to an LNA or low noise preselector, and then to the receiver. Generally speaking, money spent on the antenna gives more SNR improvement for a given investment than the same money spent on amplifiers and other attachments.

Radio receiver specifications

Radio receivers are at the heart of nearly all communications activities. In this chapter we will discuss the different types of radio receivers that are on the market. We will also learn how to interpret receiver specifications. Later, we will look at specific designs for specific applications.

Origins

The very earliest radio receivers were not receivers at all, in the sense that we know the term today. Early experiments by Hertz, Marconi and others used spark gaps and regular telegraph instruments of the day. Range was severely limited because those devices have a terribly low sensitivity to radio waves. Later, around the turn of the twentieth century, a device called a Branly coherer was used for radio signal detection. This device consisted of a glass tube filled with iron filings placed in series between the antenna and ground. Although considerably better than earlier apparatus, the coherer was something of a dud for weak signal reception. In the first decade of the twentieth century, however, Fleming invented the diode vacuum tube, and Lee DeForest invented the triode vacuum tube. These devices made amplification possible and detection a lot more efficient.

A receiver must perform two basic functions:

1 It must respond to, detect and demodulate desired signals.
2 It must not respond to, detect, or be adversely affected by undesired signals.

If it fails in either of these two functions, then it is a poorly performing design.

Both functions are necessary. Weakness in either function makes a receiver a poor bargain unless there is some mitigating circumstance. The receiver's performance

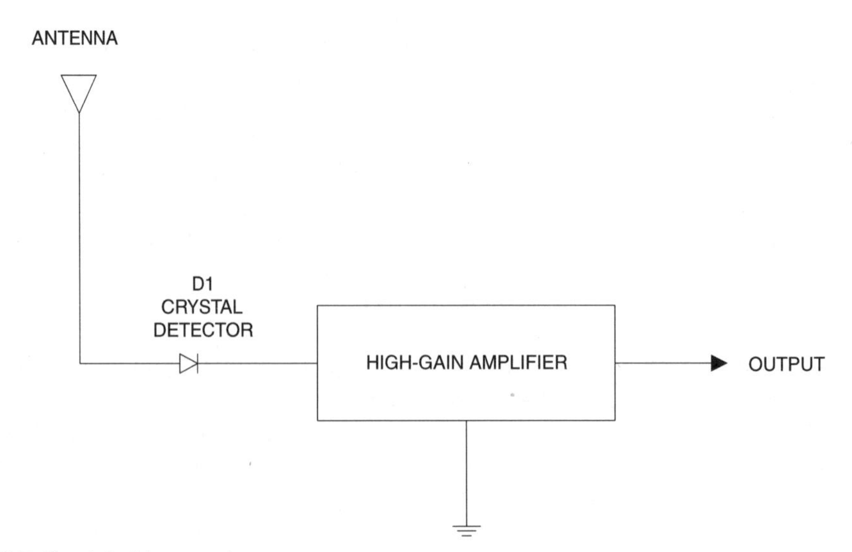

Figure 3.3 Crystal video receiver.

specifications tell us how well the manufacturer claims that their product does these two functions.

Crystal video receivers

Crystal video receivers (Fig. 3.3) grew out of primordial crystal sets, but are sometimes used in microwave bands even today. The original crystal sets (pre-1920) used a naturally occurring PN junction 'diode' made from a natural lead compound called *galena crystal* with an inductor–capacitor (L–C) tuned circuit. Later, crystal sets were made using germanium or silicon diodes. When vacuum tubes became generally available, it was common to place an audio amplifier at the output of the crystal set. Modern crystal video receivers use silicon or gallium arsenide microwave diodes and a wideband video amplifier (rather than the audio amplifier). Applications include some speed radar receivers, aircraft warning receivers, and some communications receivers (especially short range).

Tuned radio frequency (TRF) receivers

The tuned radio frequency (TRF) radio receiver uses an L–C resonant circuit in the front end, followed by one or more radio frequency amplifiers ahead of a detector stage. Two varieties are shown in Figs 3.4 and 3.5. The version in Fig. 3.4 is called a *tuned gain-block receiver*. It is commonly used in monitoring very low frequency (VLF) signals to detect solar flares and sudden ionospheric disturbances (SIDs). Later versions of the TRF concept use multiple tuned RF circuits between the amplifier stages. These designs are also used in VLF solar flare/SID monitoring. Early models used independently tuned L–C circuits, but those proved to be very difficult to tune without oscillating. Later versions mechanically linked ('ganged') the tuned circuits to operate from a single tuning knob.

Superheterodyne receivers

Figure 3.6 shows the block diagram of a superheterodyne receiver. We will use this hypothetical receiver as the basic generic framework for evaluating receiver performance.

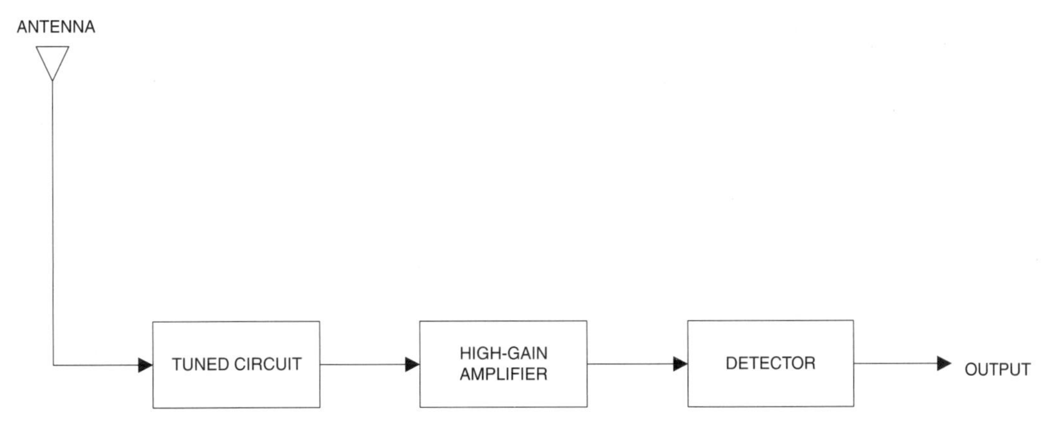

Figure 3.4 TRF circuit.

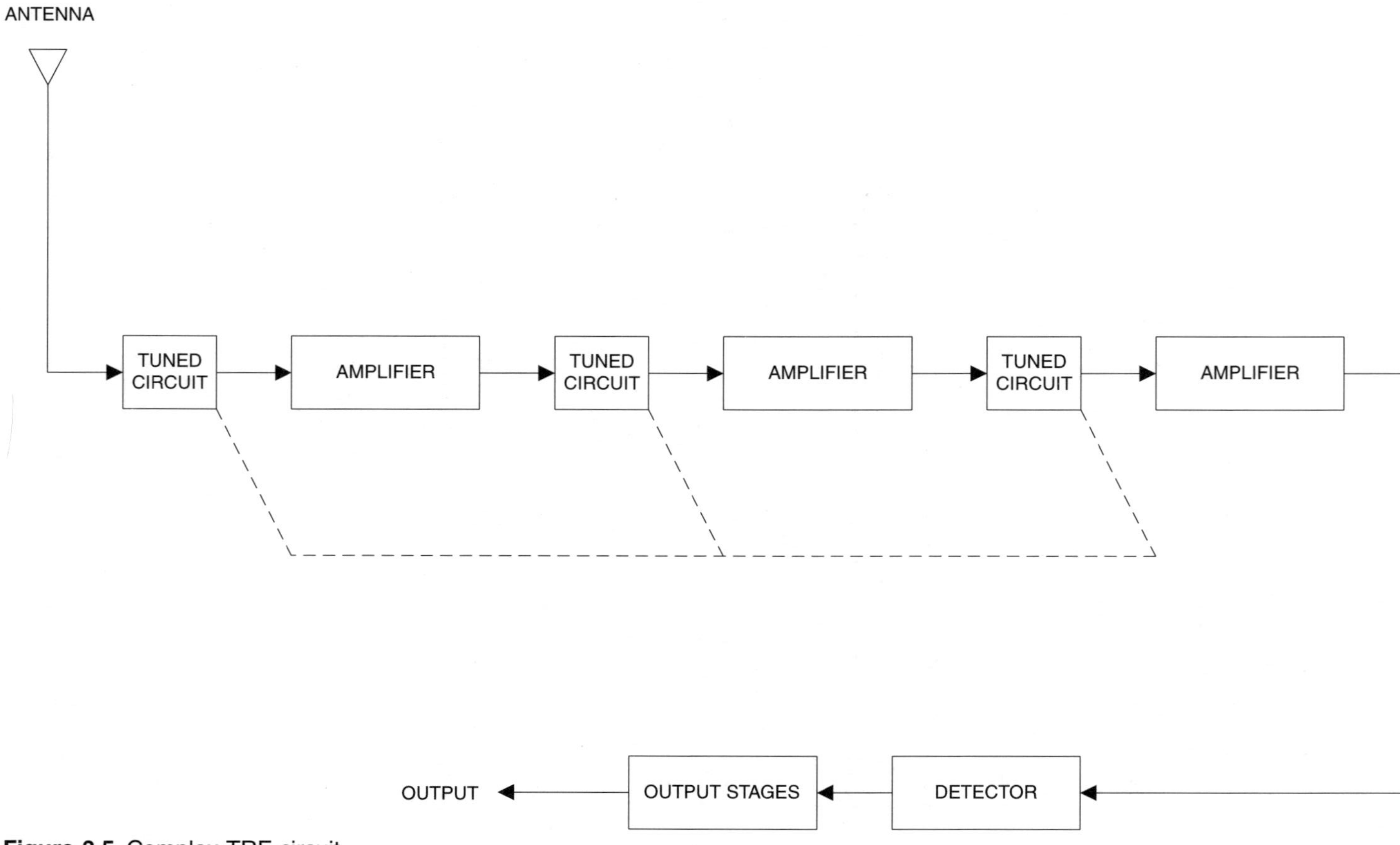

Figure 3.5 Complex TRF circuit.

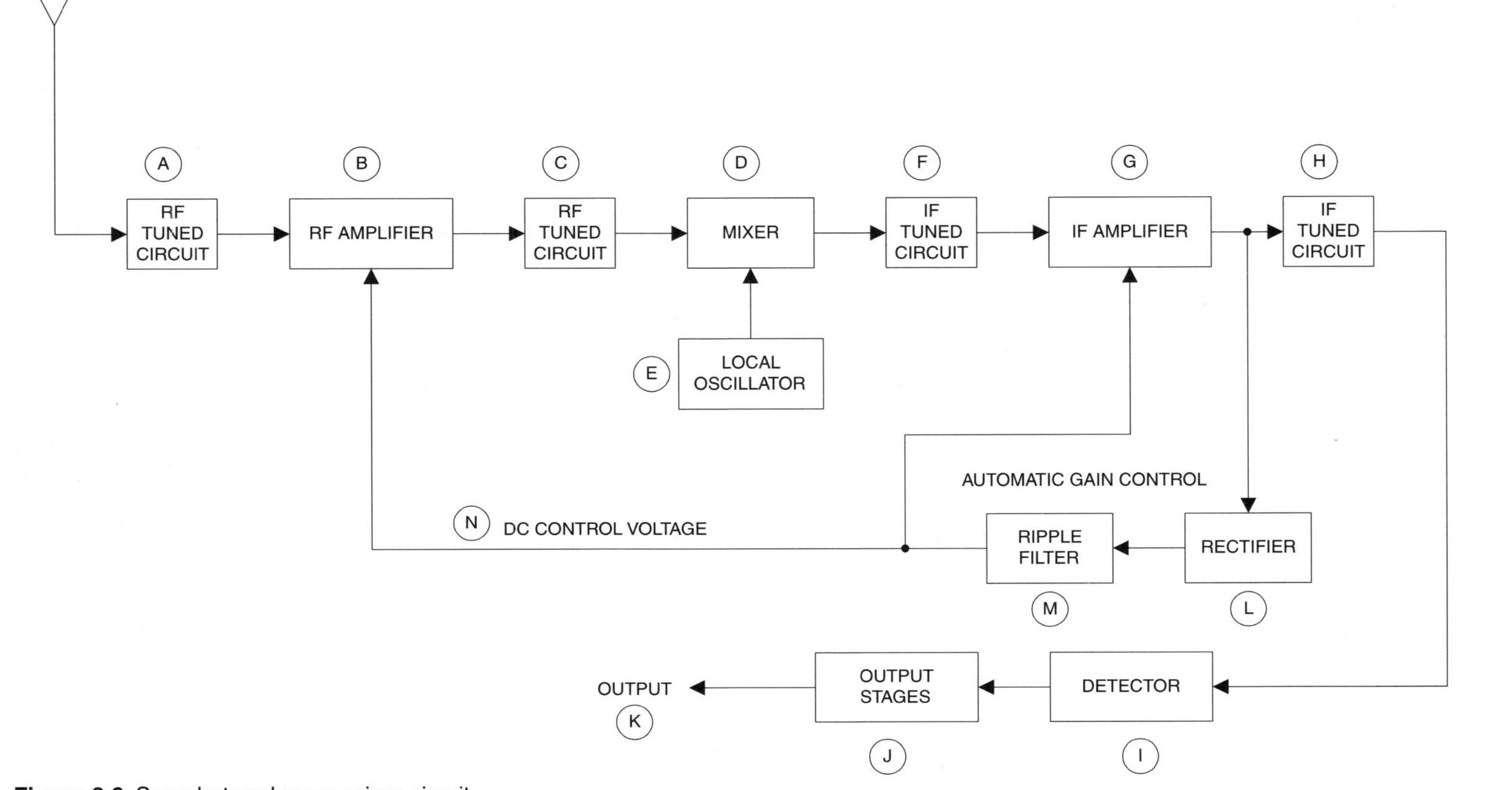

Figure 3.6 Superheterodyne receiver circuit.

The superhet design represents the largest class of radio receivers; it covers the vast majority of receivers on the market.

The block diagram of Fig. 3.6 is typical of many receivers. The purpose of a superheterodyne is to convert the incoming RF frequency to a single frequency where most of the signal processing takes place. The front-end section of the receiver consists of the *radio frequency* (RF) amplifier and any RF tuning circuits that may be used (A–B–C in Fig. 3.6). In some cases, the RF tuning is very narrow, and basically tunes one frequency. In other cases, the RF front-end tuning is broadband. In that case, bandpass filters are used.

The frequency translator section (D and E) is also considered part of the front-end in most textbooks, but here we will label it as a separate entity. The translator consists of a frequency mixer and a local oscillator. This section does the heterodyning, which is discussed in more detail below. The output of the frequency translator is called the *intermediate frequency* (IF).

The translator stage is followed by the intermediate frequency amplifier. The IF amplifier (F–G–H) is basically a radio frequency amplifier tuned to a single frequency. The IF can be higher or lower than the RF frequency, but it will always be a single frequency.

A sample of the IF amplifier output signal is applied to an *automatic gain control* (AGC) section (L–M). The purpose of this section is to keep the signal level in the output more or less constant. The AGC circuit consists of a rectifier and ripple filter that produces a DC control voltage. The DC control voltage is proportional to the input RF signal level (N). It is applied to the IF and RF amplifiers to raise or lower the gain according to signal level. If the signal is weak, then the gain is forced higher, and if the signal is strong the gain is lowered. The end result is to smooth out variations of the output signal level.

The detector stage (I) is used to recover any modulation that is on the input RF signal. The type of detector depends on the type of modulation used for the incoming signal. Amplitude modulation (AM) signals are generally handled in an *envelope detector.* In some cases a special variant of the envelope detector called a *square law detector* is used. Single sideband (SSB), double sideband suppressed carrier (DSBSC), and keyed CW signals will use a *product detector,* while FM and PM need a frequency or phase sensitive detector.

The output stages (J–K) are used to amplify and deliver the recovered modulation to the user. If the receiver is for broadcast use, then the output stages are audio amplifiers and loudspeakers. In some radio astronomy and instrumentation telemetry receivers the output stages consist of integrator circuits and DC amplifiers.

Heterodyning

The main attribute of the superheterodyne receiver is that it converts the radio signal's RF frequency to a standard frequency for further processing. Although today the new frequency, called the *intermediate frequency* or IF, may be either higher or lower than the RF frequencies, early superheterodyne receivers always down-converted the RF signal to a lower IF frequency. The reason was purely practical, for in those days higher frequencies were more difficult to process than lower frequencies. Even today, because variable tuned circuits still tend to offer different performance over the band being tuned, converting to a single IF frequency, and obtaining most of the gain and selectivity functions at the IF, allows more uniform overall performance over the entire range being tuned.

A superheterodyne receiver works by frequency converting ('heterodyning' – the 'super' part is 1920s vintage advertising hype) the RF signal. This occurs by non-linearly mixing the incoming RF signal with a *local oscillator* (LO) signal. When this process is done, disregarding noise, the output spectrum will contain a large variety of signals according to:

$$F_O = mF_{RF} \pm nF_{LO} \tag{3.2}$$

Where:
F_{RF} is the frequency of the RF signal
F_{LO} is the frequency of the local oscillator
m and n are either zero or integers $(0, 1, 2, 3, \ldots, n)$

Equation (3.2) means that there will be a large number of signals at the output of the mixer, although for the most part the only ones that are of immediate concern to understanding superheterodyne operation are those for which m and n are either 0 or 1. Thus, for our present purpose, the output of the mixer will be the fundamentals (F_{RF} and F_{LO}), and the second-order products ($F_{LO} - F_{RF}$ and $F_{LO} + F_{RF}$), as seen in Fig. 3.7. Some mixers, notably those described as *double-balanced mixers* (DBM), suppress F_{RF} and F_{LO} in the mixer output, so only the second-order sum and difference frequencies exist with any appreciable amplitude. This case is simplistic, and is used only for this present discussion. Later on, we will look at what happens when third-order ($2F1 \pm F2$ and $2F2 \pm F1$) and fifth-order ($3F1 \pm 2F2$ and $3F2 \pm 2F1$) products become large.

Note that the local oscillator frequency can be either higher than the RF frequency (*high-side injection*) or lower than the RF frequency (*low-side injection*). In some cases there may be a practical reason to prefer one over the other.

The candidates for IF are the sum (LO + RF) and difference (LO – RF) second-order products found at the output of the mixer. A high-Q tuned circuit following the mixer will select which of the two are used. Consider an example. Suppose an AM broadcast band

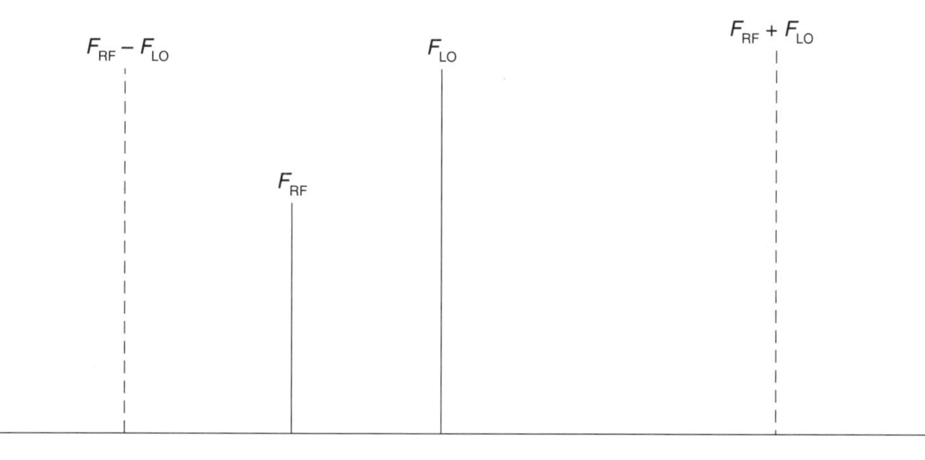

Figure 3.7 Relationship of LO, RF and IF frequencies.

superhet radio has an IF frequency of 455 kHz, and the tuning range is 540 to 1700 kHz. Because the IF is lower than any frequency within the tuning range, it will be the difference frequency that is selected for the IF. The local oscillator is set to give high-side injection, so will tune from (540 + 455) = 995 kHz, to (1700 + 455) = 2155 kHz.

Front-end circuits

The principal task of the front-end and frequency translator sections of the receiver in Fig. 3.6 is to select the signal and convert it to the IF frequency. But in many radio receivers there may be additional functions. In some cases (but not all), an RF amplifier will be used ahead of the mixer. Typically, these amplifiers have a gain of 3 to 10 dB, with 5 to 6 dB being very common. The tuning for the RF amplifier is sometimes a broad bandpass fixed frequency filter that admits an entire band. In other cases, it is a narrow band, but variable frequency, tuned circuit.

Intermediate frequency (IF) amplifier

The IF amplifier is responsible for providing most of the gain in the receiver, as well as the narrowest bandpass filtering. It is a high gain, usually multi-staged, single frequency tuned radio frequency amplifier. For example, one HF shortwave receiver block diagram lists 120 dB of gain from antenna terminals to audio output, of which 85 dB are provided in the 8.83 MHz IF amplifier chain. In the example of Fig. 3.6, the receiver is a *single conversion* design, so there is only one IF amplifier section.

Detector

The detector demodulates the RF signal, and recovers whatever audio (or other information) is to be heard by the listener. In a straight AM receiver, the detector will be an ordinary half-wave rectifier and ripple filter, and is called an *envelope detector*. In a *product detector* a second local oscillator operating near the IF frequency, usually called a *beat frequency oscillator* (BFO), is heterodyned with the IF signal. The resultant difference signal is the recovered audio. That type of detector is used for double sideband suppressed carrier (DSBSC), single sideband suppressed carrier (SSBSC or SSB), or continuous wave (CW or Morse telegraphy) signals. When used for suppressed carrier signals, the BFO is sometimes called a *carrier insertion oscillator* (CIO).

Audio amplifiers

The audio amplifiers are used to finish the signal processing. They also boost the output of the detector to a usable level to drive a loudspeaker or set of earphones. The audio amplifiers are sometimes used to provide additional filtering. It is quite common to find narrow band filters to restrict audio bandwidth, or notch filters to eliminate interfering signals that make it through the IF amplifiers intact.

Receiver performance factors

There are three basic areas of receiver performance that must be considered. Although interrelated, they are sufficiently different to merit individual consideration: *noise*, *static attributes* and *dynamic attributes*. We will look at all of these areas, but first let's look at the units of measure that we will use in this section.

Units of measure

Input signal voltage

Input signal level, when specified as a voltage, is typically stated in either *microvolts* (μV) or *nanovolts* (nV). The volt is simply too large a unit for practical use on radio receivers. Signal input voltage (or sometimes power level) is often used as part of the *sensitivity* specification, or as a test condition for measuring certain other performance parameters.

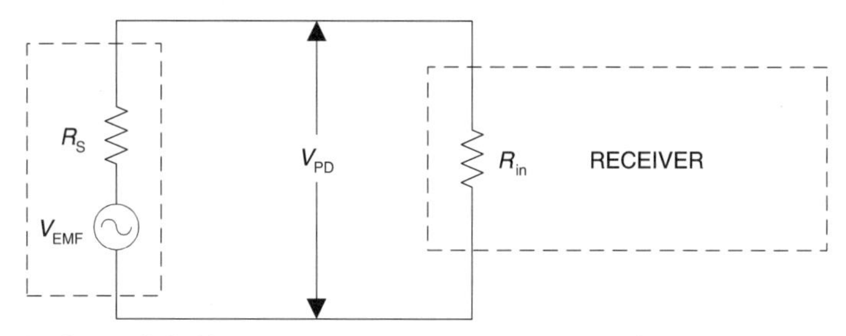

Figure 3.8 Input voltage definitions.

There are two forms of signal voltage that are used for input voltage specification: *source voltage* (V_{EMF}) and *potential difference* (V_{PD}), as illustrated in Fig. 3.8. The source voltage (V_{EMF}) is the open terminal (no load) voltage of the signal generator or source, while the potential difference (V_{PD}) is the voltage that appears across the receiver antenna terminals with the load connected (the load is the receiver antenna input impedance, R_{in}). When $R_S = R_{in}$, the preferred 'matched impedances' case in radio receiver systems, the value of V_{PD} is one-half V_{EMF}. This can be seen in Fig. 3.8 by noting that R_S and R_{in} form a voltage divider network driven by V_{EMF}, and with V_{PD} as the output.

dBm

These units refer to *decibels relative to one milliwatt (1 mW) dissipated in a 50 ohm resistive impedance* (defined as the 0 dBm reference level), and is calculated from:

$$dBm = 10 \log \left[\frac{P_{\text{watts}}}{0.001} \right] \tag{3.3}$$

or,

$$dBm = 10 \log(P_{\text{mW}}) \tag{3.4}$$

For example, 0.05 μV in 50 ohms, the power is $V^2/50$, or 5×10^{-17} watts, which is 5×10^{-14} mW. In dBm notation, this value is $10 \log(5 \times 10^{-14})$, or -133.0 dBm.

dBmV

This unit is used in television receiver systems in which the system impedance is 75 ohms, rather than the 50 ohms normally used in other RF systems. It refers to the signal voltage, measured in decibels, with respect to a signal level of one millivolt (1 mV) across a 75 ohm resistance (0 dBmV). In many TV specs, 1 mV is the full quieting signal that produces no 'snow' (i.e. noise) in the displayed picture. Note: 1 mV = 1000 μV.

dBμV

This unit refers to a signal voltage, measured in decibels, relative to one microvolt (1 μV) developed across a 50 ohm resistive impedance (0 dBμV). For our example signal voltage, the level is 0.05 μV which is the same as –26.0 dBμV. But the voltage used for this measurement is usually the V_{EMF} rather than V_{PD} so a correction factor of two, or 6 dB, may need to be applied.

It requires only a little algebra to convert signal levels from one unit of measure to another. This job is sometimes necessary when a receiver manufacturer mixes methods in the same specifications sheet. In the case of dBm and dBμV, 0 dBμV is 1 μV V_{EMF}, or a V_{PD} of 0.5 μV, applied across 50 ohms, so the power dissipated is 5×10^{-15} watts, or –113 dBm.

Rule of thumb: To convert dBμV to dBm, subtract 113 dB; i.e. 100 dBμV = (100 dBμV – 113 dB) = –13 dBm.

Noise

A radio receiver must detect signals in the presence of noise. The *signal-to-noise ratio* (SNR) is the key here because a signal must be above the noise level before it can be successfully detected and used.

Noise comes in a number of different guises, but for the sake of this discussion we can divide them into two classes: *sources external to the receiver* and *sources internal to the receiver.* There is little one can do about the external noise sources, for they consist of natural and man-made electromagnetic signals that fall within the passband of the receiver. Figures 2.15B and 2.15C (pages 25–26) show an approximation of the external noise situation which a receiver may have to face.

Signal-to-noise ratio (SNR or S_n)

Receivers are evaluated for quality on the basis of *signal-to-noise ratio* (S/N or 'SNR'), sometimes denoted S_n. The goal of the designer is to enhance the SNR as much as possible. Ultimately, the minimum signal level detectable at the output of an amplifier or radio receiver is that level which appears just above the noise floor level. Therefore, the lower the system noise, the smaller the *minimum allowable signal.*

The matter of signal-to-noise ratio (S/N) is sometimes treated in different ways that each attempt to crank some reality into the process. The signal-plus-noise-to-noise ratio (S + N/N) is found quite often. As the ratios get higher, the S/N and S + N/N converge (only about 0.5 dB difference at ratios as little as 10 dB). Still another variant is the SINAD (signal-plus-noise-plus-distortion-to-noise) ratio. The SINAD measurement takes into account most of the factors that can deteriorate reception.

Receiver noise floor

The *noise floor* of the receiver is a statement of the amount of noise produced by the receiver's internal circuitry, and directly affects the *sensitivity* of the receiver. The noise floor is typically expressed in dBm. The noise floor specification is evaluated as follows: *the more negative the better.* The best receivers have noise floor numbers of less than –130 dBm, while some very good receivers offer numbers of –115 dBm to –130 dBm.

The noise floor is directly dependent on the bandwidth used to make the measurement. Receiver advertisements usually specify the bandwidth, but be careful to note whether or not the bandwidth that produced the very good performance numbers is also the bandwidth that you'll need for the mode of transmission you want to receive. If, for example, you are interested only in weak 6 kHz wide AM signals, and the noise floor is specified for a 250 Hz CW filter, then the noise floor might be too high for your use.

Static measures of receiver performance

The two principal static levels of performance for radio receivers are *sensitivity* and *selectivity*. The sensitivity refers to the level of input signal required to produce a usable output signal (variously defined). The selectivity refers to the ability of the receiver to reject adjacent channel signals (again, variously defined). Let's take a look at both of these factors. Keep in mind, however, that in modern high performance radio receivers the static measures of performance, although frequently cited, may also be the least relevant compared with the dynamic measures (especially in environments with high interference levels).

Sensitivity

Sensitivity is a measure of the receiver's ability to pick up ('detect') signals, and is often specified in microvolts (μV). A typical specification might be '0.5 μV sensitivity'. The question to ask is: 'relative to what?' *The sensitivity number in microvolts is meaningless unless the test conditions are specified.* For most commercial receivers, the usual test condition is the sensitivity required to produce a 10 dB signal-plus-noise-to-noise (S + N/N) ratio in the mode of interest. For example, if only one sensitivity figure is given, one must find out what bandwidth is being used. Typical bandwidths are 5 to 6 kHz for AM, 2.6 to 3 kHz for single sideband, 1.8 kHz for radioteletype or 200 to 500 Hz for CW.

Indeed, one of the places where 'creative spec writing' takes place for commercial receivers is that the advertisements will enthusiastically cite the sensitivity for a narrow bandwidth mode (e.g. CW), while the other specifications are cited for a more commonly used wider bandwidth mode (e.g. SSB). In one particularly misleading example, an advertisement claimed a sensitivity number that was applicable to the 270 Hz CW mode only, yet the 270 Hz CW filter was an expensive accessory not normally included!

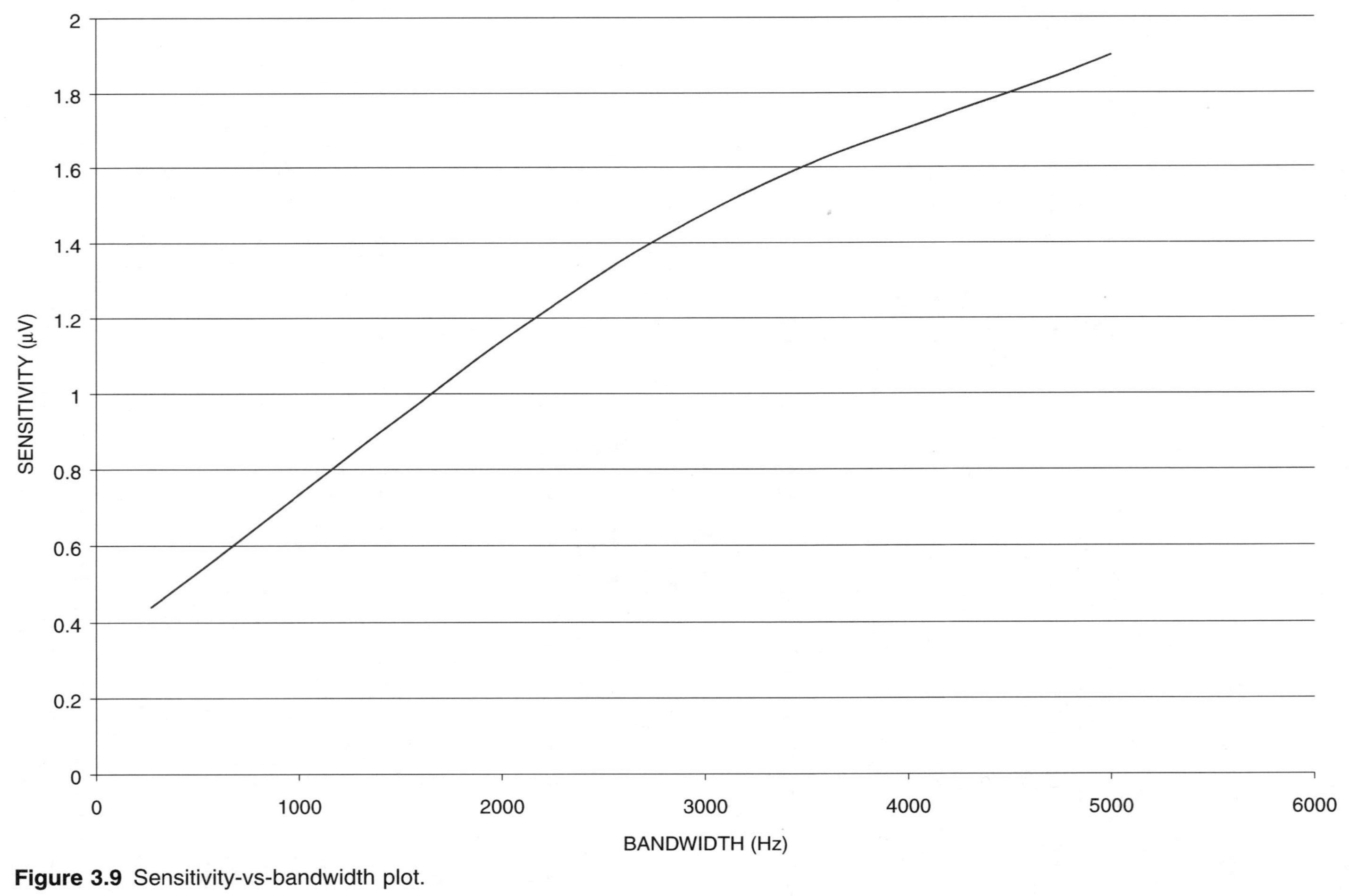

Figure 3.9 Sensitivity-vs-bandwidth plot.

Consider the case where the main mode for a high frequency (HF) shortwave receiver is AM (for international broadcasting), and the sensitivity is 1.9 μV for 10 dB SNR, and the bandwidth is 5 kHz. If the bandwidth were reduced to 2.8 kHz for SSB or 270 Hz for CW, then the sensitivity improves by the square root of the bandwidth ratio. The 1.9 μV AM sensitivity therefore translates to 1.42 μV for SSB and 0.44 μV for CW. If only the CW version is given, then the receiver might be made to look a whole lot better than it is, even though the typical user may never use the CW mode (note differences in Fig. 3.9).

The sensitivity differences also explain why weak SSB signals can be heard under conditions when AM signals of similar strength have disappeared into the noise, or why the CW mode has as much as 20 dB advantage over SSB, other things being equal.

In some receivers, the difference in mode (AM, SSB, RTTY, CW, etc.) can result in sensitivity changes that are more than the differences in the associated bandwidths. The reason is that there is sometimes a 'processing gain' associated with the type of detector circuit used to demodulate the signal at the output of the IF amplifier. A simple AM envelope detector is lossy because it consists of a simple diode and an R–C filter (a passive circuit without amplification). Other detectors (product detector for SSB, synchronous AM detectors) have their own signal gain, so may produce better sensitivity numbers than the bandwidth suggests.

Another indication of sensitivity is *minimum detectable signal* (MDS), which is usually specified in dBm. This signal level is the signal power at the antenna input terminal of the receiver required to produce some standard S+N/N ratio, such as 3 dB or 10 dB (Fig. 3.10). In radar receivers, the MDS is usually described in terms of a single pulse return and a

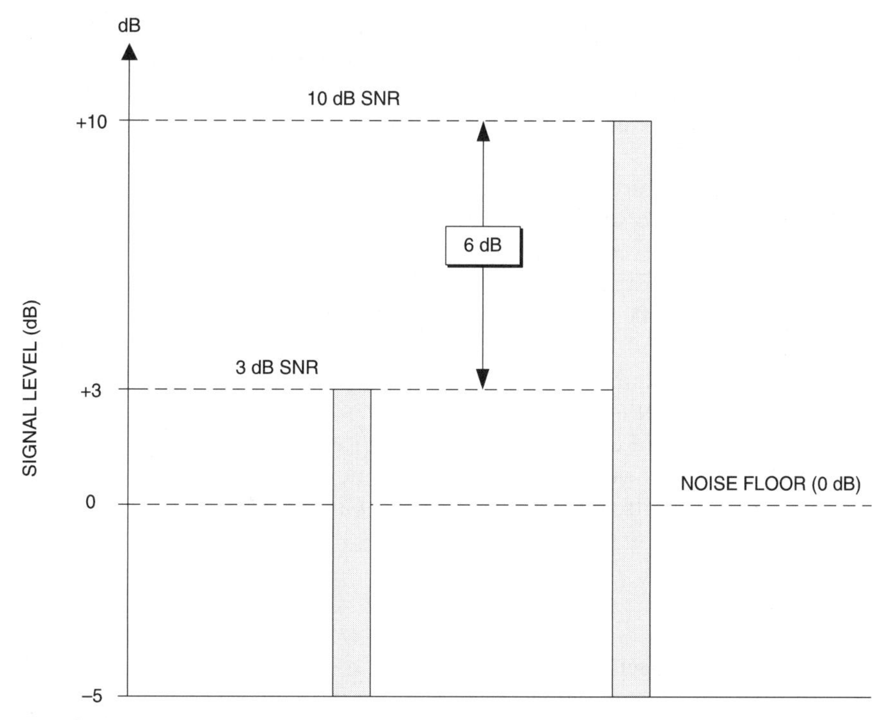

Figure 3.10 SNR situation for minimum and good reception.

specified S + N/N ratio. Also, in radar receivers, the sensitivity can be improved by integrating multiple pulses. If N return pulses are integrated, then the sensitivity is improved by a factor of N if coherent detection is used, and $\sqrt{N}$ if non-coherent detection is used.

For modulated signals, it is common to specify the conditions under which the measurement is made. For example, in AM receivers the sensitivity to achieve 10 dB SNR is measured with the input signal modulated 30 per cent by a 400 or 1000 Hz sinusoidal tone.

An alternate method is sometimes used for AM sensitivity measurements, especially in servicing consumer radio receivers (where SNR may be a little hard to measure with the equipment normally available to technicians who work on those radios). This is the 'standard output conditions' method. Some manuals specify the audio signal power or audio signal voltage at some critical point, when a 30 per cent modulated signal is present. In one automobile radio receiver, the sensitivity was specified as 'X μV to produce 400 mW across 8 ohm resistive load substituted for the loudspeaker when the signal generator is modulated 30 per cent with a 400 Hz audio tone'. The cryptic note on the schematic showed an output sine wave across the loudspeaker with the label '400 mW in 8 Ω (1.79 volts), @30% mod. 400 Hz, 1 μV RF.' What is missing is mention of the level of total harmonic distortion (THD) that is permitted.

The sensitivity is sometimes measured essentially the same way, but the specification will state the voltage level that will appear at the top of the volume control, or output of the detector/filter, when the standard signal is applied. Thus, there are two ways seen for specifying AM sensitivity: *10 dB SNR* and *standard output conditions*.

There are also two ways to specify FM receiver sensitivity. The first is the 10 dB SNR method discussed above, i.e. the number of microvolts of signal at the input terminals required to produce a 10 dB SNR when the carrier is modulated by a standard amount. The measure of FM modulation is *deviation* expressed in kilohertz. Sometimes, the full deviation for that class of receiver is used, while for others a value that is 25 to 35 per cent of full deviation is specified.

The second way to measure FM sensitivity is the level of signal required to reduce the quiescent (i.e. no signal) noise level by 20 dB. This is the *20 dB quieting sensitivity* of the receiver. If you tune between signals on an FM receiver you will hear a loud 'hiss', especially in the VHF/UHF bands. Some of that noise is externally generated, while some is internally generated. When an FM signal appears in the pass band, the hiss is suppressed, even if the FM carrier is unmodulated. The *quieting sensitivity* of an FM receiver is a statement of the number of microvolts required to produce some standard quieting level, usually 20 dB.

Pulse receivers, such as radar and pulse communications units, often use the *tangential sensitivity* as the measure of performance, which is the amplitude of pulse signal required to raise the noise level by its own rms amplitude (Fig. 3.11).

Selectivity

Although no receiver specification is unimportant, if one had to choose between sensitivity and selectivity, the proper choice most of the time would be to take selectivity.

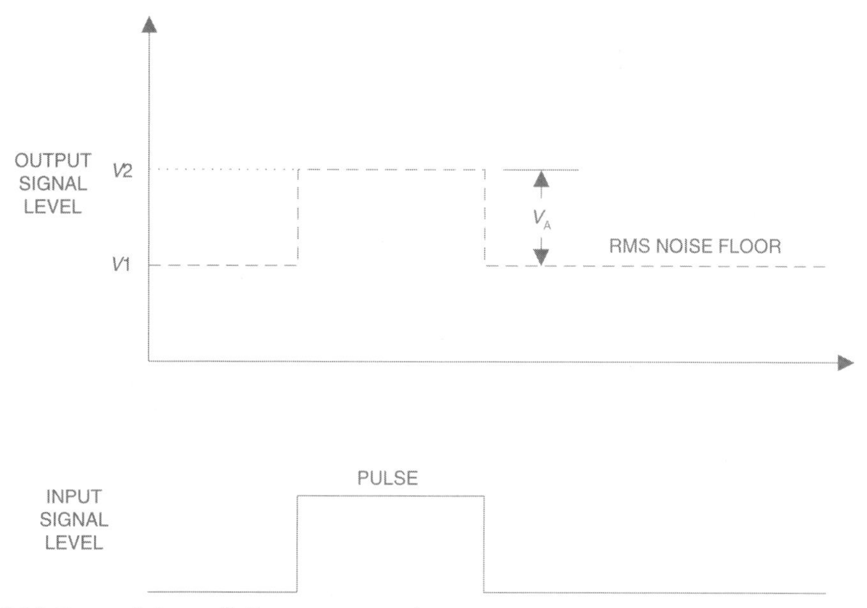

Figure 3.11 Tangential sensitivity measurement.

Selectivity is the measure of a receiver's ability to reject adjacent channel interference. Or put another way, it's the ability to reject interference from signals on frequencies close to the desired signal frequency.

In order to understand selectivity requirements, one must first understand a little bit of the nature of radio signals. An unmodulated radio carrier theoretically has an infinitesimal (near-zero) bandwidth (although all real unmodulated carriers have a very narrow, but non-zero, bandwidth because they are modulated by noise and other artifacts). As soon as the radio signal is modulated to carry information, however, the bandwidth spreads. Even an on/off telegraphy (CW) or pulse signal spreads out either side of the carrier frequency an amount that is dependent on the sending speed and the shape of the keying waveform.

An AM signal spreads out an amount equal to twice the highest audio modulating frequencies. For example, a communications AM transmitter will have audio components from 300 to 3000 Hz, so the AM waveform will occupy a spectrum that is equal to the carrier frequency (F) plus/minus the audio bandwidth ($F \pm 3000$ Hz in the case cited). An FM signal spreads out according to both the audio bandwidth and the *deviation;* a good approximation is to add them together. For example, a narrow-band FM mobile transmitter with 3 kHz audio bandwidth and 5 kHz deviation spreads out ±8 kHz, while FM stereo broadcast transmitters with 75 kHz deviation spread out ±120 kHz.

An implication of the fact that radio signals have bandwidth is that the receiver must have sufficient bandwidth to recover virtually all of the signal. Otherwise, information may be lost and the output is distorted. On the other hand, allowing too much bandwidth increases the noise picked up by the receiver and thereby deteriorates the SNR. The goal of the selectivity system of the receiver is to match the bandwidth of the receiver to that of the signal. That is why receivers will use 270 or 500 Hz filters for CW, 2 to 3 kHz for SSB and 4 to 6 kHz for AM signals. This allows matching the receiver bandwidth to the transmission type.

The selectivity of a receiver has a number of aspects that must be considered: *front-end bandwidth, IF bandwidth, IF shape factor,* and the *ultimate* (distant frequency) *rejection*.

Front-end bandwidth

The 'front-end' of a modern superheterodyne radio receiver is the circuitry between the antenna input terminal and the output of the first mixer stage. One reason why front-end selectivity is important is to keep out-of-band signals from affecting the receiver. Transmitters located nearby can easily overload a poorly designed receiver. Even if these signals are not heard by the operator, they can desensitize a receiver, or create harmonics and intermodulation products that show up as 'birdies' or other types of interference. Strong local signals can take up a lot of the receiver's dynamic range, and thereby make it harder to hear weak signals.

Front-end selectivity also helps improve a receiver's *image rejection* and *1st IF rejection* capabilities.

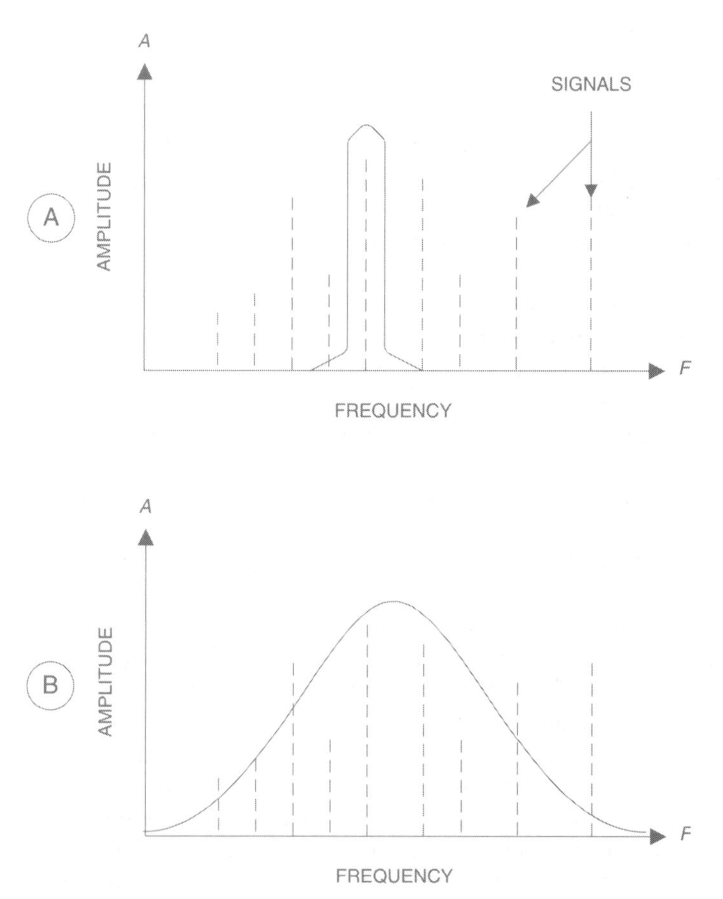

Figure 3.12 Narrow vs wide bandwidth.

In some crystal video microwave receivers the front-end might be wide open without any selectivity at all, but in nearly all other receivers there will be some form of frequency selection present.

Two forms of frequency selection are typically found. A designer may choose to use either or both (together, or as alternatives under operator selection). These forms can be called the *resonant frequency filter* (Fig. 3.12A) and *bandpass filter* (Fig. 3.12B) approaches.

The resonant frequency approach uses L–C elements tuned to the desired frequency to select which RF signals reach the mixer. In some receivers, these L–C elements are designed to track with the local oscillator that sets the operating frequency. That's why you see two-section variable capacitors for AM broadcast receivers. One section tunes the local oscillator and the other section tunes the tracking RF input. In other designs, a separate tuning knob ('preselector' or 'antenna') is used.

The other approach uses a suboctave bandpass filter to admit only a portion of the RF spectrum into the front-end. For example, a shortwave receiver that is designed to take the HF spectrum in 1 MHz pieces may have an array of RF input bandpass filters that are each 1 MHz wide (e.g. 9 to 10 MHz).

Image rejection

An *image* in a superheterodyne receiver is a signal that appears at *twice the IF distance from the desired RF signal,* and *located on the opposite side of the LO frequency from the desired RF signal.* In Fig. 3.13, a superheterodyne operates with a 455 kHz (i.e. 0.455 MHz) IF, and is turned to 24.0 MHz (F_{RF}). Because this receiver uses low-side LO injection, the LO frequency F_{LO} is 24.0 – 0.455, or 23.545 MHz. If a signal appears at twice the IF below the RF (i.e. 910 kHz below F_{RF}), and reaches the mixer, then it too has a difference frequency of 455 kHz, so will pass right through the IF filtering as a valid signal. The image rejection specification tells how well this image frequency is suppressed. Normally, anything over about 70 dB is considered good.

Tactics to reduce image response vary with the design of the receiver. The best approach, at design time, is to select an IF frequency that is high enough that the image frequency will fall well outside the passband of the receiver front-end. Some HF receivers

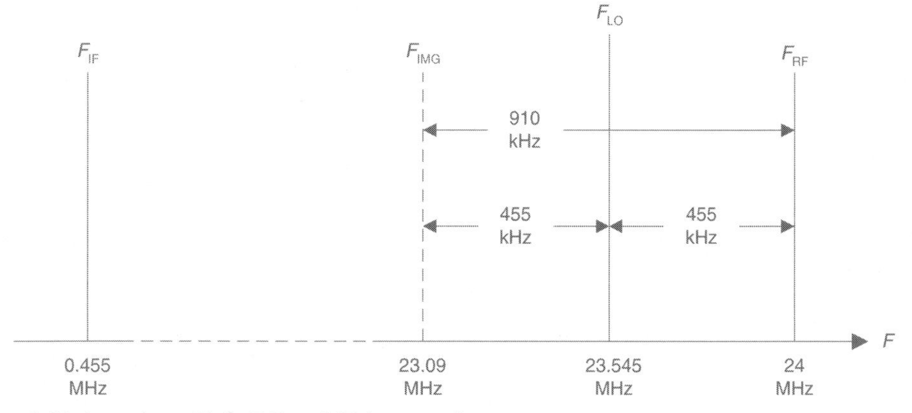

Figure 3.13 Location of LO, RF and IF frequencies.

use an IF of 8.83 MHz, 9 MHz, 10.7 MHz or something similar, and for image rejection these frequencies are considerably better than 455 kHz receivers in the higher HF bands. However, a common design trend is to do *double conversion*. In most such designs, the first IF frequency is considerably higher than the RF, being in the range 35 to 60 MHz (50 MHz is common in HF receivers, 70 MHz in microwave receivers).

The high IF makes it possible to suppress the VHF images with a simple low-pass filter. If the 24.0 MHz signal (above) were first up-converted to 50 MHz (74 MHz LO), for example, the image would be at 124 MHz. The second conversion brings the IF down to one of the frequencies mentioned above, or even 455 kHz. The lower frequencies are preferable to 50 MHz for selectivity reasons because good quality crystal, ceramic or mechanical filters in the lower frequency ranges are easily available.

1st IF rejection

The 1st IF rejection specification refers to how well a receiver rejects radio signals operating on the receiver's first IF frequency. For example, if your receiver has a first IF of 50 MHz, it must be able to reject radio signals operating on that frequency when the receiver is tuned to a different frequency. The front-end selectivity affects how well the receiver performs rejects 1st IF signals, although the shielding of the receiver is also an issue.

If there is no front-end selectivity to discriminate against signals at the IF frequency, then they arrive at the input of the mixer unimpeded. Unless rejected by the mixer (e.g. by balancing), they then may pass directly through to the high gain IF amplifiers and be heard in the receiver output.

IF bandwidth

Most of the selectivity of the receiver is provided by the filtering in the IF amplifier section. The filtering might be L–C filters (especially if the principal IF is a low frequency like 50 kHz), a ceramic resonator, a crystal filter or a mechanical filter. Of these, the

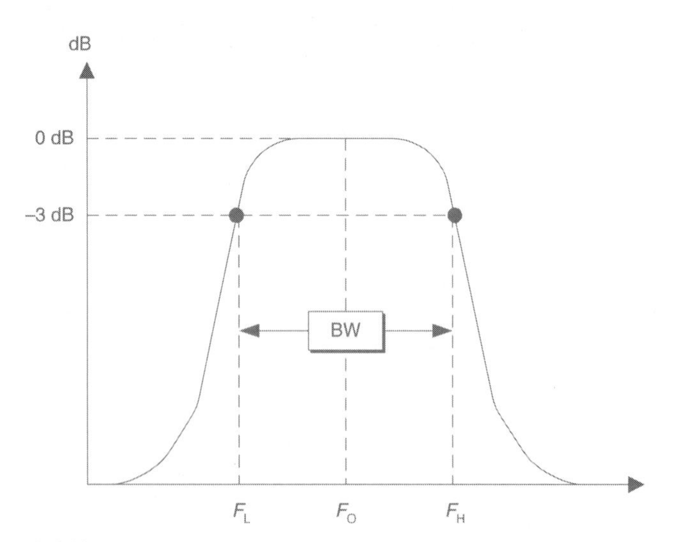

Figure 3.14 IF bandwidth response.

mechanical filter is usually regarded as best for narrow bandwidths, with the crystal filter and ceramic filters coming in next.

The IF bandwidth is measured from the points on the IF frequency response curve where gain drops off –3 dB from the mid-band value (Fig. 3.14). This is why you will sometimes see selectivity specified in terms such as '6 kHz between –3 dB points'.

The IF bandwidth must be matched to the bandwidth of the received signal for best performance. If a wider bandwidth is selected, then SNR deteriorates and the received signal may be noise. If too narrow, then you might experience difficulties recovering all of the information that was transmitted. For example, an AM broadcast band radio signal has audio components out to 5 kHz, so the signal occupies up to 10 kHz of spectrum space ($F \pm 5$ kHz). If a 2.8 kHz SSB IF filter is selected, then it will sound 'mushy' and distorted.

IF passband shape factor

The shape factor is a measure of the steepness of the skirts of the receiver's IF selectivity, and is the ratio of the bandwidth at –6 dB to the bandwidth at –60 dB (Fig. 3.15A). The general rule is that the closer this ratio is to 1:1, the better the receiver. Anything in the 1:1.5 to 1:1.9 region can be considered high quality, while anything worse than 1:3 is not worth looking at for 'serious' receiver uses. If the numbers are between 1:1.9 and 1:3, then the receiver could be regarded as being middling, but useful.

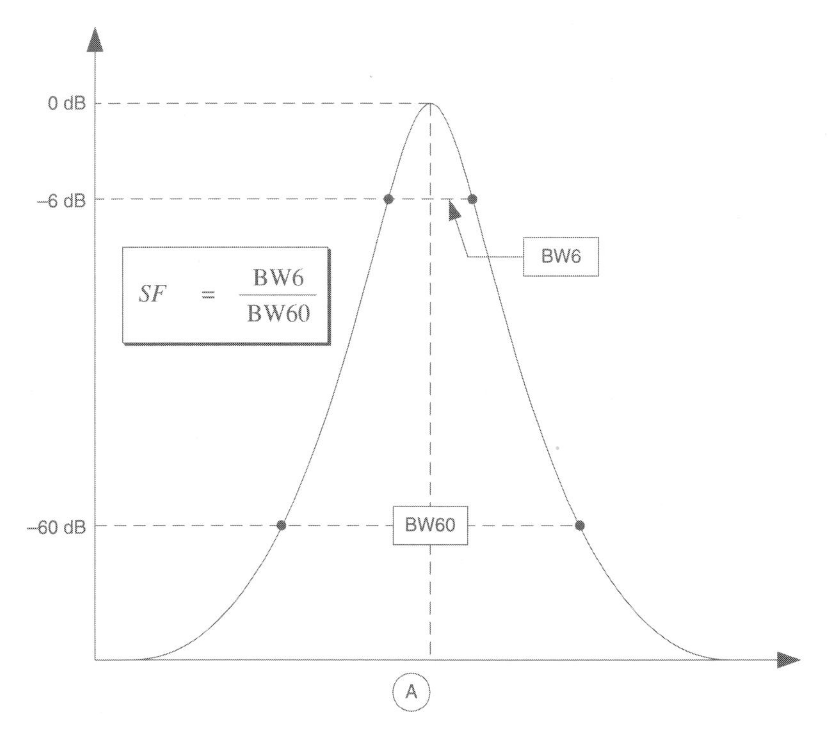

Figure 3.15 (A) Shape factor defined.

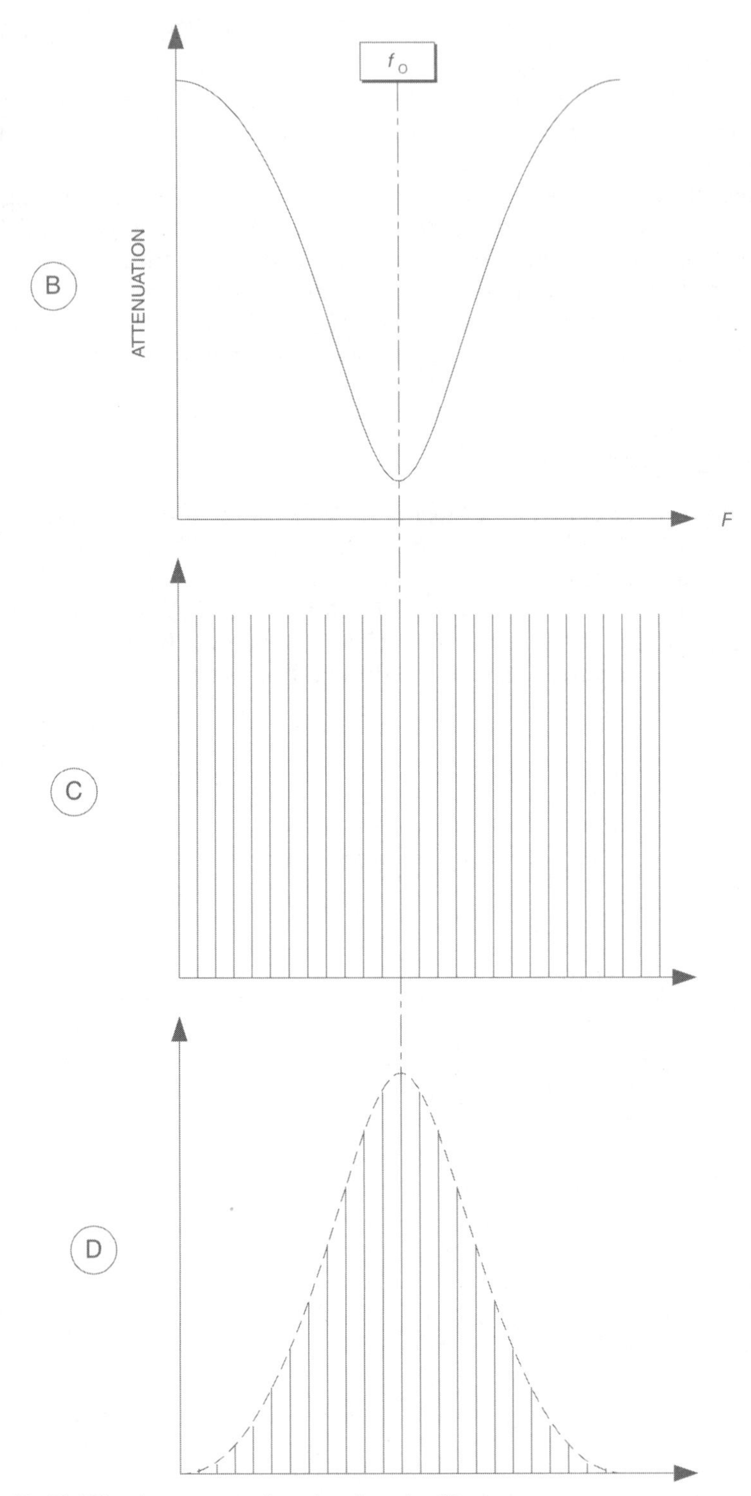

Figure 3.15 (B–D) What happens to the signal under filtering.

The importance of shape factor is that it modifies the notion of bandwidth. The cited bandwidth (e.g. 2.8 kHz for SSB) does not take into account the effects of strong signals that are just beyond those limits. Such signals can easily 'punch through' the IF selectivity if the IF passband skirts are not steep. After all, the steeper they are, the closer a strong signal can be without messing up the receiver's operation. The situation is illustrated in Fig. 3.15B. This curve inverts Fig. 3.15A by plotting attenuation-vs-frequency. Assume that equal amplitude signals close to f_o are received (Fig. 3.15C), the relative post-filtering amplitudes will match Fig. 3.15D. Thus, selecting a receiver with a shape factor as close to the 1:1 ideal as possible will result in a more usable radio.

Distant frequency ('ultimate') rejection

This specification tells something about the receiver's ability to reject very strong signals that are located well outside the receiver's IF passband. This number is stated in negative decibels, and the bigger (i.e. more negative) the number the better. An excellent receiver will have values in the −60 to −90 dB range, a middling receiver will see numbers in the −45 to −60 dB range, and a terrible receiver will be −44 or worse.

Stability

The *stability* specification measures how much the receiver frequency drifts as time elapses or temperature changes. The LO drift sets the overall stability of the receiver. This specification is usually given in terms of *short-term drift* and *long-term drift* (e.g. from LO crystal aging). The short-term drift is important in daily operation, while the long-term drift ultimately affects general dial calibration.

If the receiver is VFO controlled, or uses partial frequency synthesis (which combines VFO with crystal oscillators), then the stability is dominated by the VFO stability. In fully synthesized receivers the stability is governed by the master reference crystal oscillator. If either an *oven-controlled crystal oscillator* (OCXO), or a *temperature compensated crystal oscillator* (TCXO), is used for the master reference, then stability on the order of 1 part in 10^8/°C is achievable.

For most users the short-term stability is what is most important, especially when tuning SSB, ECSS or RTTY signals. A common specification value for a good receiver will be 50 Hz/hour after a three hour warm-up, or 100 Hz/hour after a 15 minute warm-up. The smaller the drift, the better the receiver.

The foundation of good stability is at design time. The local oscillator, or VFO portion of a synthesizer, must be operated in a cool or constant temperature location within the equipment, and must have the correct type of components. Capacitor temperature coefficients are selected in order to cancel out temperature related drift in inductance values.

Later modifications may also help, but these are less likely to be possible today than in the past. The chief cause of drift problems is heat. In the days of vacuum tube (valve) oscillators, the internal cathode heater produced lots of heat that in turn created drift.

Another stability issue comes from mechanical frequency shifts. Although not seen on most modern receivers (even some very cheap designs), it was once a serious problem on less costly models. This problem is usually seen on VFO controlled receivers in which vibration to the receiver cabinet imparts movement to either the inductor (L) or capacitor (C) element in an L–C VFO. Mechanically stabilizing these components will work wonders.

AGC range and threshold

Modern communications receivers must be able to handle signal strengths over a dynamic range of about 1 000 000:1. Tuning across a band occupied by signals of widely varying strengths is hard on the ears and hard on the receiver's performance. As a result, most receivers have an *automatic gain control* (AGC) circuit to reduces these changes. The AGC will reduce gain for strong signals, and increase it for weak signals. Most HF communications receivers allow the AGC to be turned off.

The AGC range is the change of input signal (in dBμV) from some reference level (e.g. $1\,\mu V_{EMF}$) to the input level that produces a 2 dB change in output level. Ranges of 90 to 110 dB are commonly seen.

The AGC threshold is the signal level at which the AGC begins to operate. If set too low, then the receiver gain will respond to noise, and irritate the user. If set too high, then the user will experience irritating shifts of output level as the band is tuned. AGC thresholds of 0.7 to 2.5 μV are common, with the better receivers being in the 0.7 to 1 μV range.

Another AGC specification sometimes seen deals with the speed of the AGC. Although sometimes specified in milliseconds, it is also frequently specified in subjective terms like 'fast' and 'slow'. This specification refers to how fast the AGC responds to changes in signal strength. If set too fast, then rapidly keyed signals (e.g. on/off CW) or noise transients will cause unnerving large shifts in receiver gain. If set too slow, then the receiver might as well not have an AGC. Many receivers provide two or more selections in order to accommodate different types of signals.

In order to avoid overloading, it is usually necessary for the AGC to respond quickly to strong signals by reducing the gain. The subsequent recovery of gain when the signal disappears or is moved outside the pass band may be much slower.

Dynamic performance

The dynamic performance specifications of a radio receiver are those which deal with how the receiver performs in the presence of very strong signals on adjacent channels. Until about the 1960s, dynamic performance was somewhat less important than static performance for most users. However, today the role of dynamic performance is probably more critical than static performance because of crowded band conditions.

There are at least two reasons for this change in outlook. First, in the 1960s receiver designs evolved from tubes to solid state. The new solid-state amplifiers were somewhat easier to drive into non-linearity than tube designs. Second, there has been a tremendous increase in radio frequency signals on the air. There are far more transmitting stations than ever before, and there are far more sources of electromagnetic interference (EMI – pollution of the air waves) than in prior decades. With the advent of new and expanded wireless services available to an ever widening market, the situation can only worsen. For this reason, it is now necessary to pay more attention to the dynamic performance of receivers than in the past.

Intermodulation products

Understanding the dynamic performance of the receiver requires knowledge of *intermodulation products* (IP) and how they affect receiver operation. Whenever two signals at frequencies $F1$ and $F2$ are mixed together in a non-linear circuit, a number of products are created according to the $mF1 \pm nF2$ rule, where m and n are either integers or zero (0, 1, 2, 3, 4, 5, . . .). Mixing can occur in either the mixer stage of a receiver front end, or in the RF amplifier (or any outboard preamplifiers used ahead of the receiver) if the RF amplifier is overdriven by a strong signal.

It is also possible for corrosion on antenna connections, or even rusted antenna screw terminals to create IPs under certain circumstances. One even hears of alleged cases where a rusty downspout on a house rain gutter caused re-radiated intermodulation signals.

The *order* of an IP product is given by the sum $(m + n)$. Given input signal frequencies of $F1$ and $F2$, the main IPs are:

Second order:	$F1 \pm F2$	$2F1$	$2F2$	
Third order:	$2F1 \pm F2$	$2F2 \pm F1$	$3F1$	$3F2$
Fifth order:	$3F1 \pm 2F2$	$3F2 \pm 2F1$	$5F1$	$5F2$

When an amplifier or receiver is overdriven, the second-order content of the output signal increases as the square of the input signal level, while the third-order responses increase as the cube of the input signal level.

Consider the case where two HF signals, $F1 = 10\,\text{MHz}$ and $F2 = 15\,\text{MHz}$ are mixed together. The second-order IPs are 5 and 25 MHz; the third-order IPs are 5, 20, 35 and 40 MHz; and the fifth-order IPs are 0, 25, 60 and 65 MHz. If any of these are inside the passband of the receiver, then they can cause problems. One such problem is the emergence of 'phantom' signals at the IP frequencies. This effect is seen often when two strong signals ($F1$ and $F2$) exist and can affect the front-end of the receiver, and one of the IPs falls close to a desired signal frequency, F_d. If the receiver were tuned to 5 MHz, for example, a spurious signal would be found from the $F1$–$F2$ pair given above.

Another example is seen from strong in-band, adjacent channel signals. Consider a case where the receiver is tuned to a station at 9610 kHz, and there are also very strong signals at 9600 kHz and 9605 kHz. The near (in-band) IP products are:

Third-order:	9595 kHz ($\Delta F = 15\,\text{kHz}$)	
	9610 kHz ($\Delta F = 0\,\text{kHz}$)	(ON CHANNEL!)
Fifth-order:	9590 kHz ($\Delta F = 20\,\text{kHz}$)	
	9615 kHz ($\Delta F = 5\,\text{kHz}$)	

Note that one third-order product is on the same frequency as the desired signal, and could easily cause interference if the amplitude is sufficiently high. Other third- and fifth-order products may be within the range where interference could occur, especially on receivers with wide bandwidths.

The IP orders are theoretically infinite because there are no bounds on either m or n. However, because the higher order IPs have smaller amplitudes only the second-order, third-order and fifth-order products usually assume any importance. Indeed, only the third-order is normally used in receiver specification sheets because they fall close to the RF signal frequency.

There are a large number of IMD products from just two signals applied to a non-linear medium. But consider the fact that the two-tone case used for textbook discussions is rarely encountered in actuality. A typical two-way radio installation is in a signal rich environment, so when dozens of signals are present the number of possible combinations climbs to an unmanageable extent.

–1 dB compression point

An amplifier produces an output signal that has a higher amplitude than the input signal. The transfer function of the amplifier (indeed, any circuit with output and input) is the ratio *OUT/IN*, so for the power amplification of a receiver RF amplifier it is P_o/P_{in} (or, in terms of voltage, V_o/V_{in}). Any real amplifier will saturate given a strong enough input signal (see Fig. 3.16). The dotted line represents the theoretical output level for all values of input signal (the slope of the line represents the gain of the amplifier). As the amplifier saturates (solid line), however, the actual gain begins to depart from the theoretical at some level of input signal. The –1 dB compression point is that output level at which the actual gain departs from the theoretical gain by –1 dB.

The –1 dB compression point is important when considering either the RF amplifier ahead of the mixer (if any), or any outboard preamplifiers that are used. The –1 dB compression point is the point at which signal distortion becomes a serious problem. Harmonics and intermodulation are generated at high levels when an amplifier goes into compression.

Third-order intercept point

It can be claimed that the *third-order intercept point* (TOIP) is the single most important specification of a receiver's dynamic performance because it predicts the performance as regards intermodulation, cross-modulation and blocking desensitization.

Third-order (and higher) intermodulation products (IP) are normally very weak, and don't exceed the receiver noise floor when the receiver is operating in the linear region. As input signal levels increase, forcing the front-end of the receiver toward the saturated non-linear region, the IP emerge from the noise and begin to cause problems. When this happens, new spurious signals appear on the band and self-generated interference begins to arise.

Look again at Fig. 3.16. The dotted gain line continuing above the saturation region shows the theoretical output that would be produced if the gain did not clip. It is the nature of third-order products in the output signal to emerge from the noise at a certain input level, and increase as the cube of the input level. Thus, the third-order line increases 3 dB for every 1 dB increase in the response to the fundamental signal. Although the output response of the third-order line saturates similarly to that of the fundamental signal, the gain line can be continued to a point where it intersects the gain line of the fundamental signal. This point is the *third-order intercept point* (TOIP).

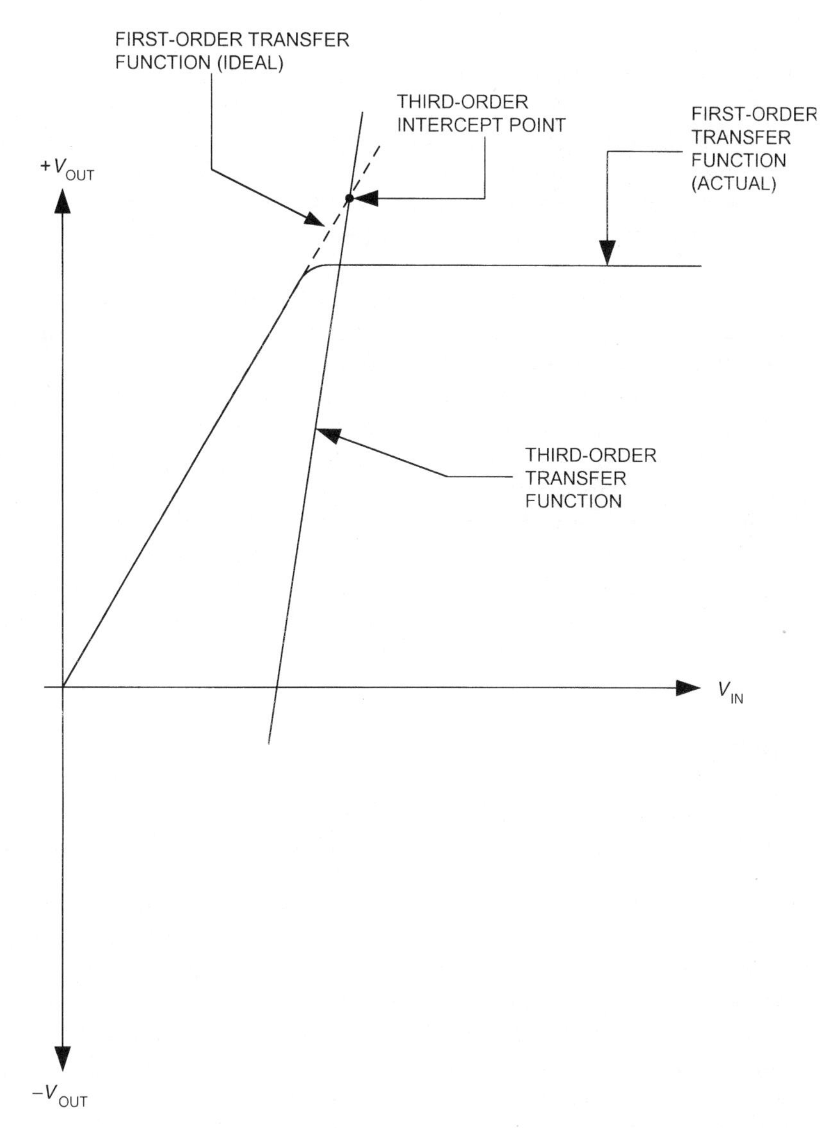

Figure 3.16 Third-order intercept point.

Interestingly, one receiver feature that can help reduce IP levels is the use of a *front-end attenuator* (or *input attenuator*). In the presence of strong signals even a few dB of input attenuation is often enough to drop the IPs back into the noise, while afflicting the desired signals only a small amount.

Other effects that reduce the overload caused by a strong signal also help. Situations arise where the apparent third-order performance of a receiver improves dramatically when a lower gain antenna is used. This effect can be easily demonstrated using a spectrum analyser for the receiver. This instrument is a swept frequency receiver that displays an output on an oscilloscope screen that is *amplitude-vs-frequency,* so a single signal shows as a spike. In one test, a local VHF band repeater came on the air every few seconds, and one could observe the second- and third-order IPs along with the fundamental repeater signal. There were also other strong signals on the air, but just outside the band. Inserting a 6 dB barrel attenuator in the input line eliminated the IP products, showing just the actual signals. Rotating a directional antenna away from the direction of the interfering signal will also accomplish this effect in many cases.

Preamplifiers are popular receiver accessories, but can often reduce rather than enhance performance. Two problems commonly occur (assuming the preamp is a low noise device). The best known problem is that the preamp amplifies noise as much as signals, and while it makes the signal louder it also makes the noise louder by the same amount. Since it's the signal-to-noise ratio that is important, this does not improve the situation. Indeed, if the preamp is itself noisy, it will deteriorate the SNR. The other problem is less well known, but potentially more devastating. If the increased signal levels applied to the receiver push the receiver into non-linearity, then IPs will emerge.

When evaluating receivers, a TOIP of +5 to +20 dBm is excellent performance, while up to +27 dBm is relatively easily achievable, and +35 dBm has been achieved with good design; anything greater than +50 dBm is close to miraculous (but attainable). Receivers are still regarded as good performers in the 0 to +5 dBm range, and middling performers in the –10 to 0 dBm range. Anything below –10 dBm is not usually acceptable. A general rule is to buy the best third-order intercept performance that you can afford, especially if there are strong signal sources in your vicinity.

Dynamic range

The *dynamic range* of a radio receiver is the range (measured in decibels) from the *minimum discernible signal* to the *maximum allowable signal*. Several definitions of dynamic range are used.

One definition of dynamic range is that it is the input signal difference between the sensitivity figure (e.g. 0.5 μV for 10 dB S + N/N) and the level that drives the receiver far enough into saturation to create a certain amount of distortion in the output. This definition was common on consumer broadcast band receivers at one time (especially automobile radios, where dynamic range was somewhat more important due to mobility). A related definition takes the range as the distance in dB from the sensitivity level and the –1 dB compression point. Another definition, the *blocking dynamic range*, is the range of signals from the sensitivity level to the blocking level (see below).

A problem with the above definitions is that they represent single signal cases, so do not address the receiver's dynamic characteristics. The *spurious free dynamic range* (SFDR) is the range of signals over which dynamic effects (e.g. intermodulation) do not exceed the noise floor of the receiver. This is often taken to be two-thirds of the difference between the noise floor in a 3 kHz bandwidth and the third-order intercept point.

Many US amateur radio magazine product reviews use a measurement procedure that produces a similar result. Two equal strength signals are input to the receiver at the same time. The frequency difference has traditionally been 20 kHz for HF and 30 to 50 kHz for VHF receivers, but modern band crowding may indicate a need for a reduction to 5 kHz on HF. The amplitudes of these signals are raised until the third-order distortion products are raised to the noise floor level. For 20 kHz spacing, using the two-signal approach, anything over 90 dB is an excellent receiver, while anything over 80 dB is at least decent.

The difference between the single-signal and two-signal (dynamic) performance is not merely an academic exercise. Besides the fact that the same receiver can show as much as 40 dB difference between the two measures (favouring the single-signal measurement), the worst effects of poor design show up most in the dynamic performance.

Blocking

The blocking specification refers to the ability of the receiver to withstand very strong off-tune signals that are at least 20 kHz away from the desired signal, although some use 100 kHz separation. When very strong signals appear at the input terminals of a receiver, they may desensitize the receiver, i.e. reduce the apparent strength of desired signals below what they would be if the interfering signal were not present.

When a strong signal is present, it takes up more of the receiver's resources than normal, so there is not enough of the output power budget to accommodate the weaker desired signals. But if the strong undesired signal is turned off, then the weaker signals receive a full measure of the unit's power budget.

The usual way to measure blocking behaviour is to input two signals, a desired signal at 60 dBμV and another signal 20 (or 100) kHz away at a much stronger level. The strong signal is increased to the point where blocking causes a 3 dB drop in the output level of the desired signal. A good receiver will show at least 90 dBμV, with many being considerably better. An interesting note about modern receivers is that the blocking performance is so good, that it's often necessary to specify the input level difference (dB) that causes a 1 dB drop, rather than 3 dB drop, of the desired signal's amplitude.

The phenomenon of blocking leads us to an effect which may seem puzzling at first. Many receivers are equipped with front-end attenuators that permit fixed attenuation values of 1 dB, 3 dB, 6 dB, 12 dB or 20 dB (or some subset) to be inserted into the signal path ahead of the active stages. When a strong signal that is capable of causing desensitization is present, *adding attenuation can increase the level of the desired signals in the output*, even though overall gain is apparently reduced. This occurs because the overall signal that the receiver front end is asked to handle is below the threshold where desensitization occurs.

Cross-modulation

Cross-modulation is an effect in which amplitude modulation (AM) from a strong undesired signal is transferred to a weaker desired signal. Testing is usually done (in HF receivers) with 20 kHz spacing between the desired and undesired signals, a 3 kHz IF bandwidth on the receiver, and the desired signal set to 1000 μV_{EMF} (–53 dBm). The undesired signal has 30 per cent amplitude modulation. This undesired AM signal is increased in strength until an unwanted AM output 20 dB below the desired signal is produced.

A cross-modulation specification of greater than 100 dB would be considered decent performance. This figure is often not given for modern HF receivers, but if the receiver has a good third-order intercept point, then it is likely also to have good cross-modulation performance.

Cross-modulation may also occur naturally, especially in transpolar and North Atlantic radio paths where the effects of the aurora are strong. Something called the 'Radio Luxembourg effect' was reported in the 1930s. Modulation from a very strong broadcaster (BBC) appeared on the Radio Luxembourg signal received in North America. The cause was said to be ionospheric cross-modulation, which apparently occurs when the strong station is within 175 miles of the great circle path between the desired station and the receiver site.

Reciprocal mixing

Reciprocal mixing occurs when noise sidebands from the local oscillator (LO) signal in a superheterodyne receiver mix with a strong undesired signal that is close to the desired signal. Every oscillator produces noise, and that noise will modulate the oscillator's output signal. It will thus form sidebands either side of the LO signal. The production of phase noise in all LOs is well known, but in more recent designs the digitally produced synthesized LOs are prone to additional noise elements as well. The noise is usually measured in –dBc (decibels below carrier, or, in this case, dB below the LO output level).

If a strong unwanted signal is present, then it might mix with the noise sidebands of the LO, to reproduce the noise spectrum at the IF frequency. In the usual test scenario, the reciprocal mixing is defined as the level of the unwanted signal (dB) at 20 kHz required to produce a noise sidebands 20 down from the desired IF signal in a specified bandwidth (usually 3 kHz on HF receivers). Figures of –90 dBc or better are considered good.

The importance of the reciprocal mixing specification is that it can degrade the practical selectivity of the receiver, yet is not detected in the normal static measurements made of selectivity because it is a 'dynamic selectivity' problem. When the LO noise sidebands appear in the IF, the distant frequency attenuation (>20 kHz off-centre of a 3 kHz bandwidth filter) can deteriorate by 20 to 40 dB.

The reciprocal mixing performance of receivers can be improved by reducing the noise in the oscillator signal. Although this sounds simple, in practice it is often quite difficult. A tactic that works well is to add high-*Q* filtering between the LO output and the mixer input. The narrow bandwidth of the high-*Q* filter prevents wideband noise from getting to the mixer, but will not help the close-in performance.

IF notch rejection

If two signals fall within the passband of a receiver they will both compete to be heard. They will also heterodyne together in the detector stage, producing an audio tone equal to their carrier frequency difference. For example, suppose we have an AM receiver with a 5 kHz bandwidth and a 455 kHz IF. If two signals appear on the band such that one appears at an IF of 456 kHz and the other is at 454 kHz, then both are within the receiver passband and both will be heard in the output. However, the 2 kHz difference in their carrier frequency will produce a 2 kHz heterodyne audio tone difference signal in the output of the AM detector.

In some receivers, a tunable high-*Q* (narrow and deep) notch filter is in the IF amplifier circuit. This tunable filter can be turned on and then adjusted to attenuate the unwanted interfering signal, reducing the irritating heterodyne. Attenuation figures for good receivers vary from −35 to −65 dB or so (the more negative the better).

There are some trade-offs in notch filter design. First, the notch filter *Q* is more easily achieved at low IF frequencies (such as 50 kHz to 500 kHz) than at high IF frequencies (e.g. 9 MHz and up). Also, the higher the *Q* the better the attenuation of the undesired squeal, but the touchier it is to tune. Some happy middle ground between the irritating squeal and the touchy tune is mandated here.

Some receivers use audio filters rather than IF filters to help reduce the heterodyne squeal. In the AM broadcast band, channel spacing is 9 or 10 kHz (depending on the part of the world), and the transmitted audio bandwidth is 5 kHz. Designers of AM broadcast receivers may insert an R–C low-pass filter with a −3 dB point just above 4 or 5 kHz right after the detector in order to suppress the audio heterodyne. This R–C filter is called a 'tweet filter' in the slang of the electronic service/repair trade.

Another audio approach is to sharply limit the bandpass of the audio amplifiers. Although the shortwave bands typically only need 3 kHz bandwidth for communications, and 5 kHz for broadcast, the tweet filter and audio roll-off might not be sufficient. In receivers that lack an effective IF notch filter, an audio notch filter can be provided.

Internal spurii

All receivers produce a number of internal spurious signals that sometimes interfere with the operation. Both old and modern receivers have spurious signals from assorted high-order mixer products, from power supply harmonics, parasitic oscillations, and a host of other sources. Newer receivers with synthesized local oscillators and digital frequency readouts can produce noise and spurious signals in abundance. With appropriate filtering and shielding, it is possible to hold the 'spurs' down to −100 dB relative to the main maximum signal output, or within about 3 dB of the noise floor, whichever is lower.

Part 2 Circuits

4 RF amplifiers

In this chapter we will take a look at small signal radio frequency amplifiers and preamplifiers. These circuits are used to amplify radio signals from antennas prior to input to the mixer, in order to improve signal-to-noise ratio and front-end selectivity. A *preamplifier* is simply an RF amplifier which is external to the receiver, rather than being built in.

The performance of some radio receivers can be improved by the use of either a *preselector* or preamplifier between the antenna and the receiver. Most low priced receivers (and some high priced ones as well) suffer from performance problems that are a direct result of the trade-offs the manufacturers have to make in order to produce a low cost model. In addition, older receivers often suffer the same problems, as do many homebrew radio receiver designs. Chief among these are sensitivity, selectivity and image response, which we looked at in Chapter 3.

A cure for all of these problems is a little circuit called the *active preselector*. A preselector can be either active or passive. In both cases, however, the preselector includes a resonant circuit that is tuned to the frequency that the receiver is tuned to. The preselector is connected between the antenna and the receiver antenna input connector (Fig. 4.1). Therefore, it adds a little more selectivity to the front-end of the radio to help discriminate against unwanted signals.

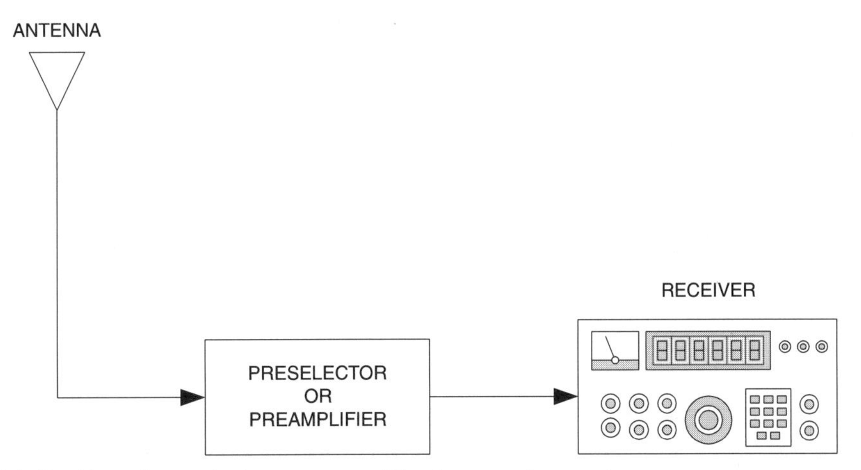

Figure 4.1 Position of preselector or preamplifier in a receiver system.

The difference between the active and passive designs is that the active design contains an RF amplifier stage, while the passive design does not. Thus, the active preselector also deals with the sensitivity problem of the receiver.

The difference between a preamplifier and the amplifying variety of preselector is that the preselector is tuned to a specific frequency or narrow band of frequencies. The wideband preamplifier amplifies all signals coming into the front-end, with no discrimination, and therein lies an occasional problem.

A possible problem with any amplifier ahead of the receiver is that this might deteriorate performance, rather than make it better. The preamplifier gain will use up part of the receiver's dynamic range, so it must be able to improve other parameters by a sufficient amount to make this loss worthwhile.

Always use a preamplifier or preselector that has a noise figure that is better than the receiver being served. The Friis equation for noise (Eq. 2.27, page 31) demonstrates that the noise figure of the system is dominated by the noise figure of the first amplifier. So make sure that the amplifier is a low noise amplifier (LNA), and has a noise figure a few dB less than the receiver's noise figure.

Noise and preselectors/preamplifiers

The weakest radio signal that you can detect on a receiver is determined mainly by the *noise level* in the receiver. Some noise arrives from outside sources, while other noise is generated inside the receiver. At the VHF/UHF/microwave range, the internal noise is predominant, so it is common to use a *low noise preamplifier* ahead of the receiver. The preamplifier will reduce the noise figure for the entire receiver.

The low noise amplifier (LNA) should be mounted on the antenna if it is wideband, and at the receiver if it is tunable but cannot be tuned remotely. (Note: the term *preselector* only applied to tuned versions, while *preamplifier* could denote either tuned or wideband models.) Of course, if your receiver is used only for one frequency, then it may also be mounted at the antenna. The reason for mounting the preamplifier right at the antenna is to build up the signal and improve the signal-to-noise ratio (SNR) *prior* to feeding the signal into the transmission line where losses cause it to weaken somewhat.

Amplifier configurations

Most RF amplifiers use bipolar junction transistors (BJT) or field effect transistors (FET). These may be discrete, or part of an integrated circuit.

Transistor gain

There are actually several popular ways to denote bipolar transistor current gain, but only two are of interest to us here: *alpha* (α) and *beta* (β). Alpha gain (α) can be defined as the ratio of collector current to emitter current:

$$\alpha = \frac{I_c}{I_e} \tag{4.1}$$

Where:
α is the alpha gain
I_c is the collector current
I_e is the emitter current

Alpha has a value less than unity (1), with values between 0.7 and 0.99 being the typical range.
The other representation of transistor gain, and the one that seems more often favoured over the others, is the beta (β) which is defined as the ratio of collector current to base current:

$$\beta = \frac{I_c}{I_b} \tag{4.2}$$

Where:
β is the beta gain
I_c is the collector current
I_b is the base current

Alpha (α) and beta (β) are related to each other, and one can use the equations below to compute one when the other is known.

$$\alpha = \frac{\beta}{1 + \beta} \tag{4.3}$$

and,

$$\beta = \frac{\alpha}{1 - \alpha} \tag{4.4}$$

The values given above are for static DC situations. In AC terms you will see *AC alpha* gain (H_{fb}) defined as:

$$H_{fb} = \frac{\Delta I_c}{\Delta I_e} \tag{4.5}$$

and *AC beta* gain (H_{fe}) is defined as:

$$H_{fe} = \frac{\Delta I_c}{\Delta I_b} \tag{4.6}$$

In both equations above, the Greek letter *delta* (Δ) indicates a *small change in* the parameter it is associated with. Thus, the term ΔI_c denotes a small change in collector current I_c.

Classification by common element

This method of classifying amplifier circuits revolves around noting which element (collector, base or emitter) is common to both input and output circuits. Although technically incorrect, this is sometimes referred to as the *grounded* element, i.e. 'grounded emitted amplifier'. We tend to use *common* and *grounded* interchangeably, so bear with us if you are a purist. Figure 4.2 shows the different entries into this class.

Common emitter circuits

The circuit shown in Fig. 4.2A is the *common emitter* circuit. The input signal is applied to the transistor between the base and emitter terminals, while the output signal is taken across the collector and emitter terminals, i.e. the emitter is common to both input and output circuits.

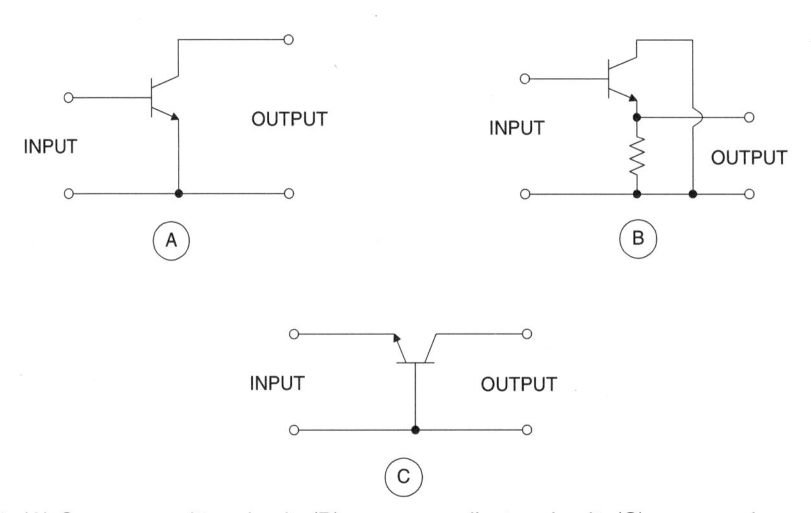

Figure 4.2 (A) Common emitter circuit; (B) common collector circuit; (C) common base circuit.

The common emitter circuit offers high current amplification – the beta rating of the transistor. This circuit can also offer a substantial amount of voltage gain if a series impedance is placed between the collector terminal and the collector DC power supply. The current gain is H_{fe}, but the voltage gain depends on other factors as well. Later, you will see that voltage gain depends on the R_L/R_E ratio in some circuits, and the product of that ratio and the beta in other cases.

The input impedance of the common emitter amplifier is medium ranged, or in the 1000 ohms range. The output impedance, though, is typically high (up to 50 kohms). Values will be determined by the specific type of circuit, but there are some approximations that can be made. For most common emitter amplifiers, Z_{in} is equal to the product of the emitter resistor R_E and the H_{fe} of the transistor. The output impedance is essentially the value of the collector load resistor and will range from 5 kohms to about 50 kohms.

The output signal in the common emitter circuit is 180 degrees out of phase with the input signal. This means that the common emitter amplifier is an *inverter* circuit. The output signal will be negative going for a positive-going input signal, and vice versa. The common emitter transistor amplifier is probably the most often used circuit configuration.

Common collector circuits

This configuration is shown in Fig. 4.2B. In the common collector circuit the collector terminal of the transistor is common to both input and output circuits. This circuit is also sometimes called the *emitter follower* circuit. The common collector circuit offers little or no voltage gain. Most of the time the voltage gain is actually less than unity (1), but the current gain is considerably higher ($\approx H_{fe} + 1$).

There is no phase inversion between input and output in the emitter follower circuit. The output voltage is in phase with the input signal voltage.

The input impedance of this circuit tends to be high, sometimes greater than 100 kohms at frequencies less than 100 kHz. The output impedance is very low, perhaps as low as 5–50 ohms. This situation leads us to one of the primary applications of the emitter follower: *impedance transformation*. The circuit is often used to connect a high impedance source to an amplifier with low input impedance.

The emitter follower is also frequently used as a *buffer amplifier*, which is an intermediate stage used to isolate two circuits from each other. One example of this is in the output circuit of oscillator circuits. Many oscillators will 'pull', or change frequency, if the load impedance changes. Yet some of the very circuits used with oscillators naturally provide a changing impedance situation. The oscillator proves a lot more stable under these conditions if an emitter follower buffer amplifier is used between its output and its load.

Common base circuits

Common base amplifiers use the base terminal of the transistor as the common element between input and output circuits (Fig. 4.2C); the output is taken between the collector and base.

The voltage gain of the common base circuit is high, on the order of 100 or more; however, the current gain is low, usually less than unity. The input impedance is also low, usually less than 100 ohms. On the other hand, the output impedance is quite high. Again, there is no phase inversion between input and output circuits.

The principal use of the common base circuit is in VHF and UHF RF amplifiers in receivers. The base acts as a shield between the emitter and collector elements, which reduces the effect of internal capacitances. These would otherwise provide a feedback signal, which can reduce gain or lead to instability. This makes it superior to common emitter circuits at high frequencies.

Transistor biasing

Biasing sets the operating characteristics of any particular transistor circuit, and is usually set by the current conditions at the base terminal of the device. There are two different bias networks commonly seen in simple transistor circuits, and these are summarized below.

Collector-to-base bias

In this type of bias network the resistor supplying bias current to the base (R_B) is connected to the collector of the transistor (see Fig. 4.3A). A feature of this circuit is that the quiescent (no signal) conditions are stabilized somewhat by DC negative feedback. Thus, when I_c tries to increase, the voltage drop across R_L increases, and because $V_{ce} = V_{cc} - V_{RL}$, the value of V_{ce} decreases. This action, in turn, reduces I_b so, by $I_c = H_{fe}I_b$, the collector current decreases. A similar action takes place when I_c tries to decrease. The end result in both cases is that I_c tends to stabilize around the quiescent value.

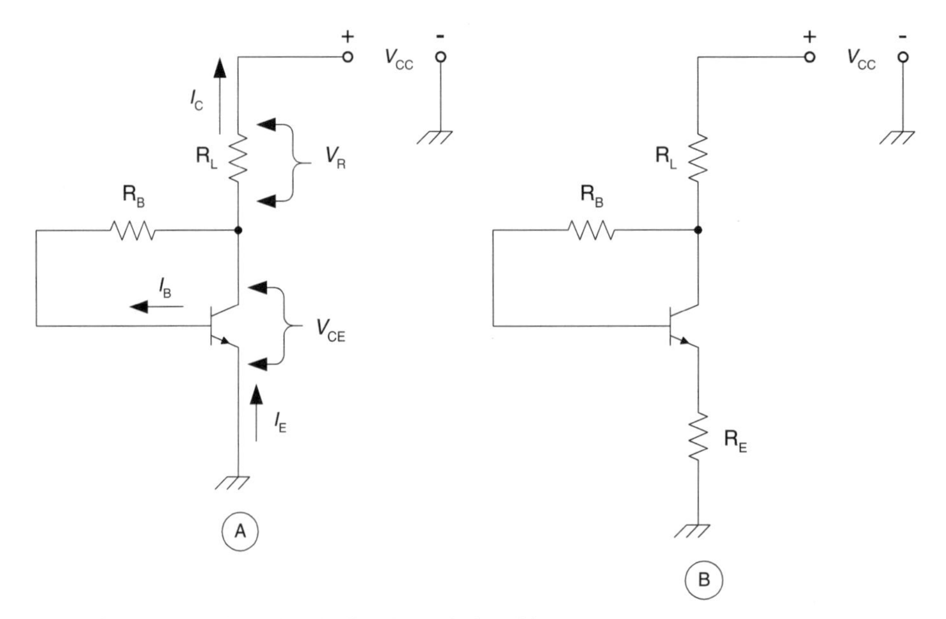

Figure 4.3 Common emitter circuit showing relationships.

It is sometimes prudent to use an emitter resistor to gain further stability, as in Fig. 4.3B. For the circuit of Fig. 4.3B:

$$Z_o = R_L$$

$$Z_{in} = R_E H_{fe}$$

$$A_I = H_{fe}$$

$$A_v = R_L H_{fe}/R_E$$

Emitter bias or 'self-bias'

Figure 4.4 is recognized as the most stable configuration for transistor amplifier stages. This circuit uses a resistor voltage divider ($R1/R2$) to set a fixed bias voltage (V_B) on the transistor. As a general rule, the best stability usually occurs when $R1||R2 \approx R_E$. Because

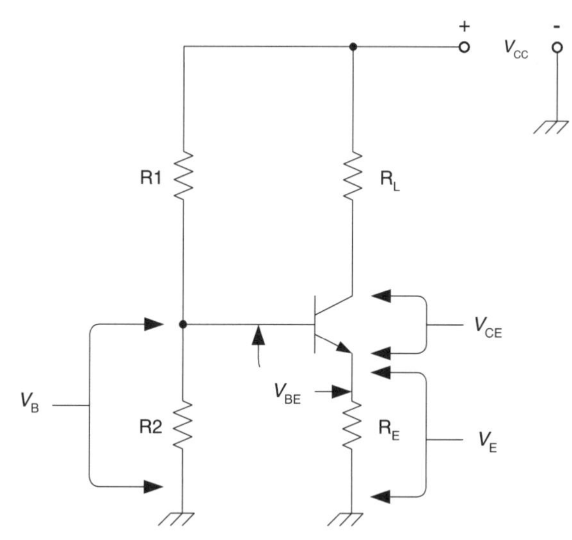

Figure 4.4 Common emitter circuit with emitter resistor.

there is a substantial voltage drop across R_E, the V_{cc} voltage required for Fig. 4.4 is a bit higher than for the previous circuit.

Frequency characteristics

Transistors, like most other electron devices, operate only over a limited frequency range. There are three frequencies that may interest us: f_α, f_β, f_T.

f_α is the frequency at which the common base AC current gain h_{fb} drops to a level 3 dB below its low frequency (usually 1000 Hz) gain.

f_β is similarly defined as the frequency where the common emitter AC beta h_{fe} drops 3 dB relative to its 1000 Hz value. In general, this frequency is lower than the alpha cut-off, but is considered somewhat more representative of a transistor's performance.

The frequency specification that seems to be quoted most often is the beta cut-off frequency, which is given the symbol f_T. This is the frequency at which h_{fe} drops to unity, and is relevant for transistors operated in the common emitter configuration

If f_β is known, then f_T may be approximated from

$$f_T = f_\beta \times h_{feo} \tag{4.7}$$

JFET and MOSFET connections

Figure 4.5 shows the JFET and MOSFET configurations that are similar to the Fig. 4.2 connections for bipolar transistors. Figure 4.5A shows a common source circuit, which is similar to the common emitter circuit. Figure 4.5B shows the common drain circuit, which is similar to the common collector circuit. Finally, Fig. 4.5C shows the common gate circuit, which is similar to the common base circuit in bipolar technology.

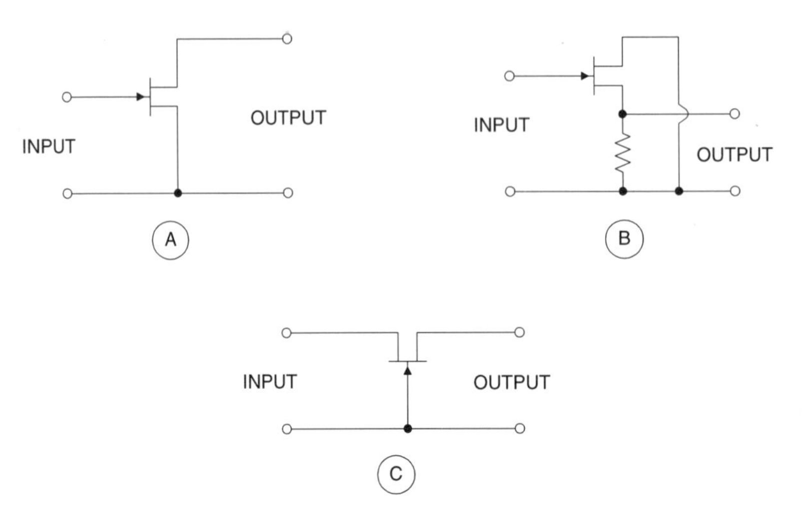

Figure 4.5 (A) Common source circuit; (B) common drain circuit; (C) common gate circuit.

JFET preselector

Figure 4.6 shows the basic form of JFET preselector. This circuit will work into the low VHF region. This circuit is in the common source configuration, so the input signal is applied to the gate and the output signal is taken from the drain. Source bias is supplied by the voltage drop across resistor R2, and drain load by a series combination of a resistor (R3) and a radio frequency choke (RFC1). RFC1 should be 1 mH at the AM broadcast band and HF (shortwave), and 100 μH in the low VHF region (>30 MHz). At VLF frequencies below the broadcast band use 2.5 mH for RFC1, and increase all 0.01 μF capacitors to 0.1 μF. All capacitors are either disk ceramic, or one of the newer dielectric capacitors (*if* rated for VHF service . . . be careful not all are!).

The input circuit is tuned to the RF frequency, but the output circuit is untuned. The reason for the lack of output tuning is that tuning both input and output permits the JFET to oscillate at the RF frequency . . . and that we don't want. Other possible causes of oscillation include poor layout, and a *self-resonance* frequency of RFC1 that is too near the RF frequency (select another choke).

The input circuit consists of an RF transformer that has a tuned secondary (L2/C1). The variable capacitor (C1) is the tuning control. Although the value shown is the standard 365 pF 'AM broadcast variable', any form of variable can be used if the inductor is tailored to it. These components are related by:

$$f = \frac{1}{2\pi\sqrt{L2 \times C1}} \tag{4.8}$$

Where:
f is the frequency in hertz
L is the inductance in henrys
C is the capacitance in farads

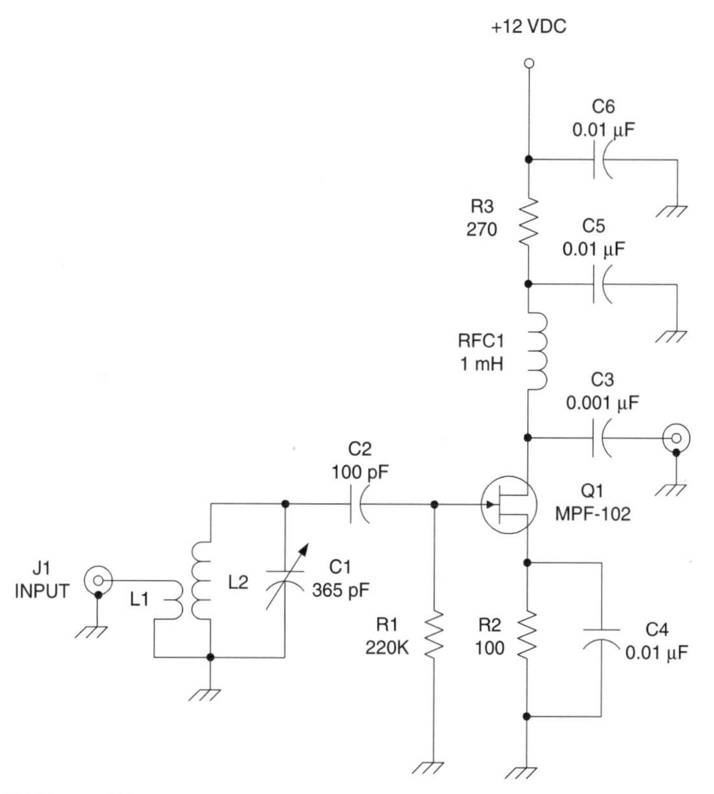

Figure 4.6 JFET RF amplifier.

Be sure to convert inductances from microhenrys to henrys, and picofarads to farads. Allow approximately 10 pF to account for stray capacitances, although keep in mind that this number is a guess that may have to be adjusted (it is a function of your layout, among other things). We can also solve Eq. (4.8) for either L2 or C1:

$$L2 = \frac{1}{39.5f^2C1} \tag{4.9}$$

Space does not warrant making a sample calculation, but we can report results for you to check for yourself. I wanted to know how much inductance is required to resonate 100 pF (90 pF capacitor plus 10 pF stray) to 10 MHz WWV. The solution, when all numbers are converted to hertz and farads, results in 0.00000253 H, or 2.53 μH. Keep in mind that the calculated numbers are close, but are nonetheless approximate . . . and the circuit may need tweaking on the bench.

Be careful when making JFET or MOSFET RF amplifiers in which both input and output are tuned. If the circuit is a common source circuit, there is the possibility of accidentally turning the circuit into a dandy little oscillator. Sometimes, this problem is alleviated by tuning the input and output L–C tank circuits to slightly different frequencies. In other cases, it is necessary to neutralize the stage. It is a common practice to make at least one end of the amplifier, usually the output, untuned in order to overcome this problem (although at the cost of some gain).

Figure 4.7 shows two methods for tuning both the input and output circuits of the JFET transistor. In both cases the JFET is wired in the common gate configuration, so signal is applied to the source and output is taken from the drain. The dotted line indicates that the output and input tuning capacitors are ganged to the same shaft.

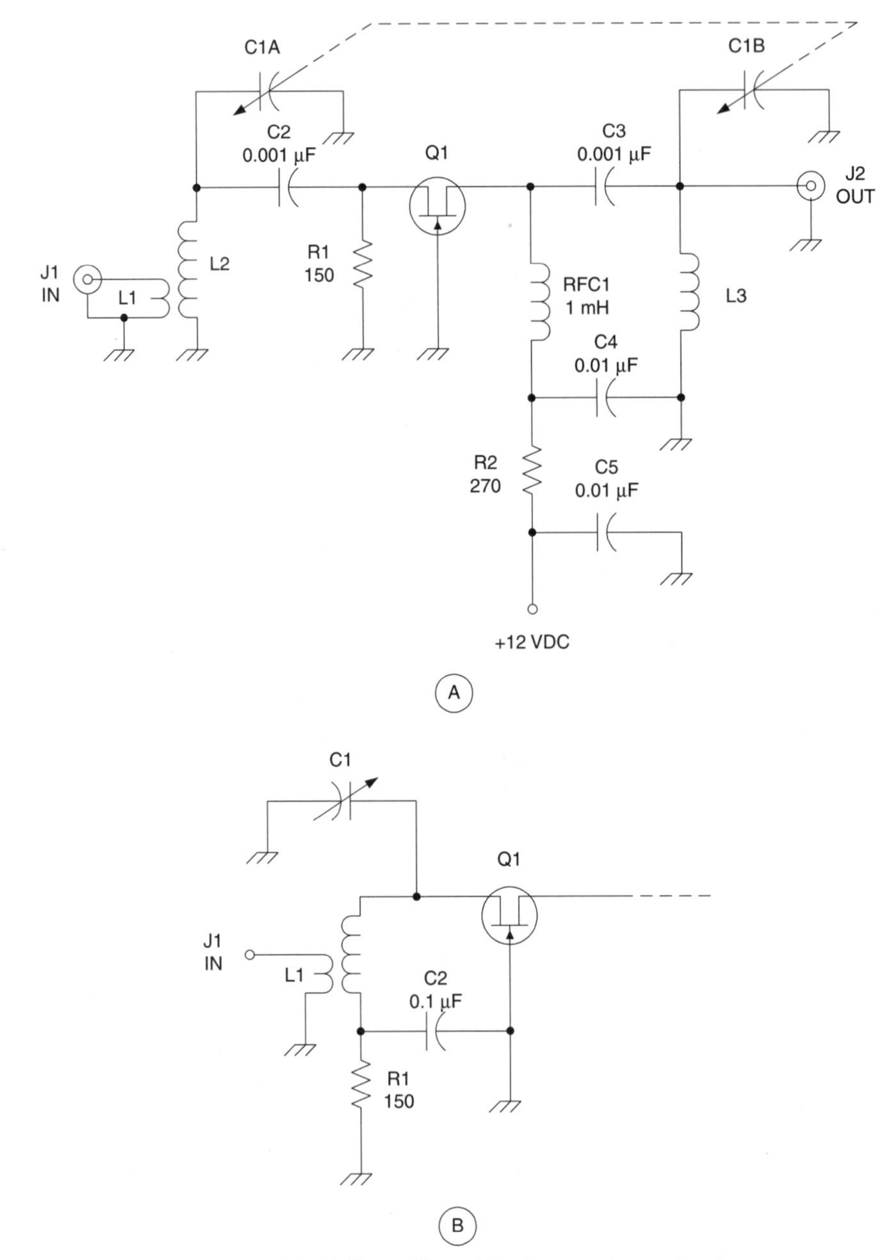

Figure 4.7 (A) Common base JFET RF amplifier; (B) alternate input circuit.

The source circuit of the JFET is low impedance, so some means must be provided to match the circuit to the tuned circuit. In Fig. 4.7A a link inductor is used for L1 for the lower impedance (50 ohms typically) of the source. In Fig. 4.7B a similar but slightly different configuration is used. In this example there is a bias resistor in the circuit, and it is bypassed by C2. This keeps the potential for DC, but sets the AC impedance to ground.

VHF receiver preselector

The circuit in Fig. 4.8 is a VHF preamplifier that uses two JFET devices connected in *cascode*, i.e. the input device (Q1) is in common source and is direct coupled to the common gate output device (Q2). In order to prevent self-oscillation of the circuit a *neutralization capacitor* (C3) is provided. This capacitor is adjusted to keep the circuit from oscillating at any frequency within the band of operation. In general, this circuit is tuned to a single channel by the action of L2/C1 and L3/C2.

MOSFET preselector

A *dual-gate MOSFET* is used in the preselector circuit of Fig. 4.9. One gate can be used for amplification and the other for DC-based gain control. Signal is applied to gate G1, while gate G2 is either biased to a fixed positive voltage or connected to a variable DC voltage

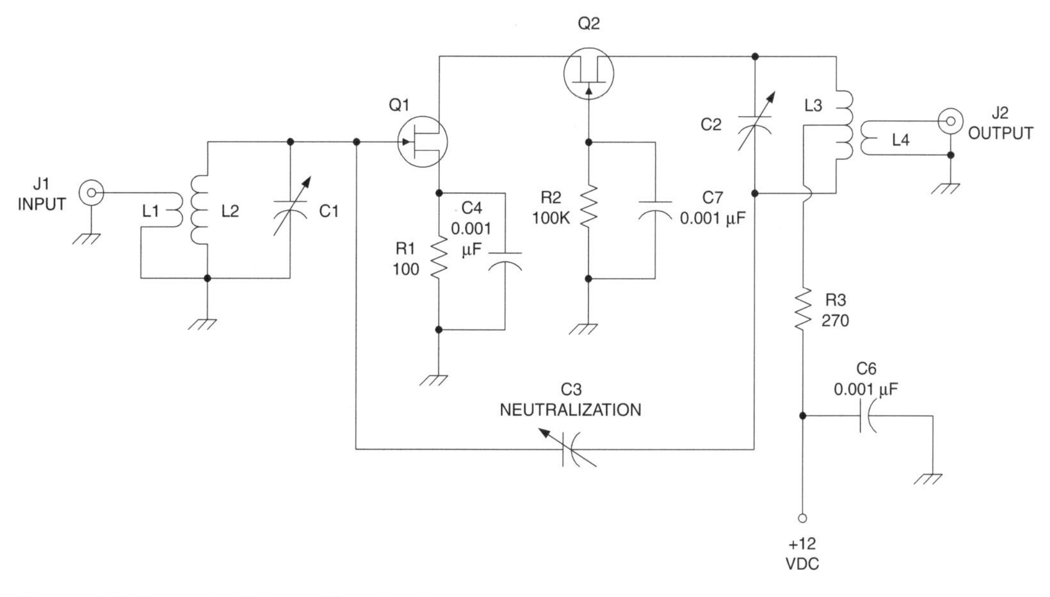

Figure 4.8 Cascode RF amplifier circuit.

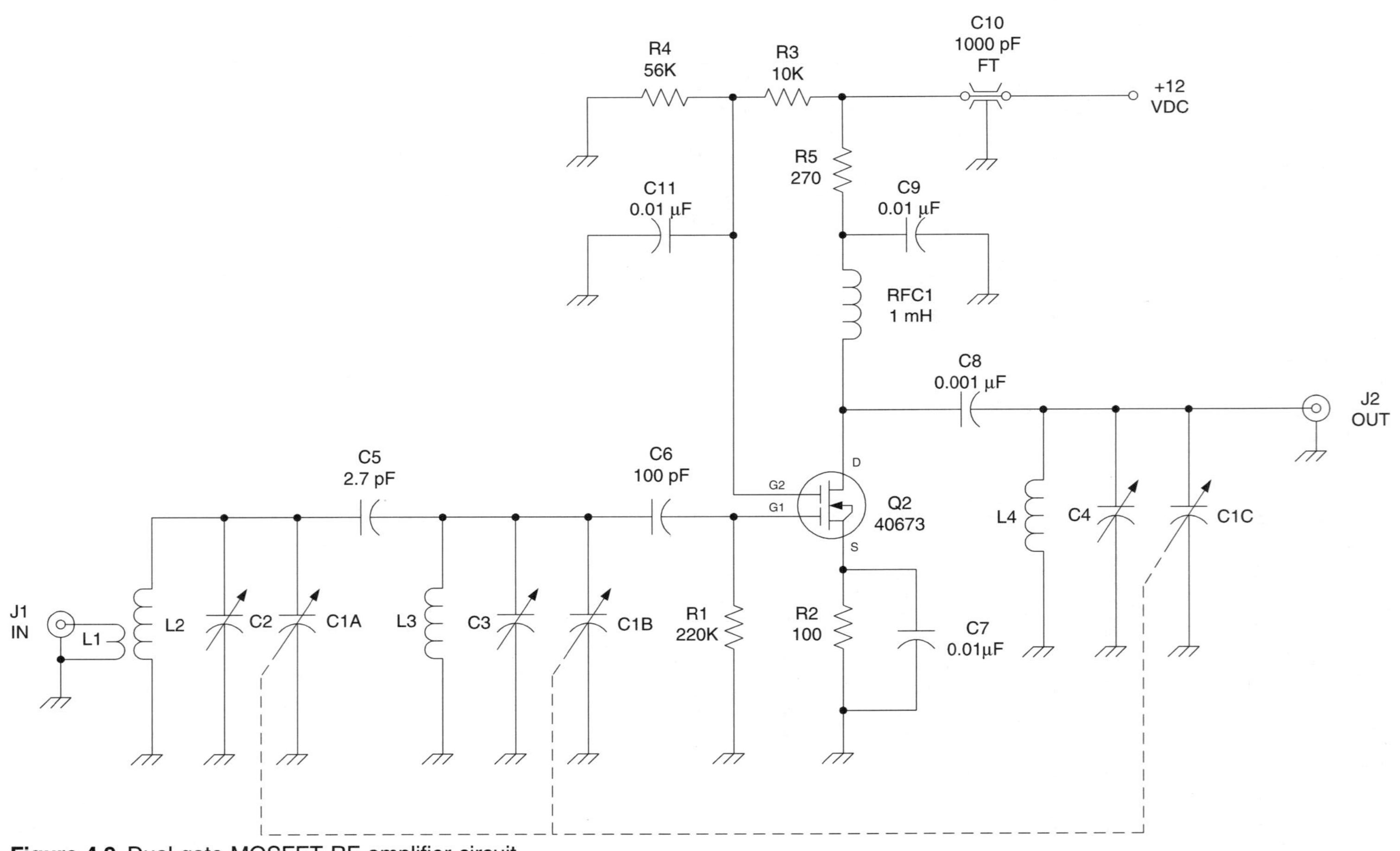

Figure 4.9 Dual-gate MOSFET RF amplifier circuit.

that serves as a gain control signal. The DC network is similar to that of the previous (JFET) circuits, with the exception that a resistor voltage divider (R3/R4) is needed to bias gate G2.

There are three tuned circuits for this preselector project, so it will produce a large amount of selectivity improvement and image rejection. The gain of the device will also provide additional sensitivity. All three tuning capacitors (C1A, C1B and C1C) are ganged to the same shaft for 'single-knob tuning'. The trimmer capacitors (C2, C3 and C4) are used to adjust the tracking of the three tuned circuits (i.e. ensure that they are all tuned to the same frequency at any given setting of C1A–C).

The inductors are of the same sort as described above. It is permissible to put L1/L2 and L3 in close proximity to each other, but these should be separated from L4 in order to prevent unwanted oscillation due to feedback arising from coil coupling.

Voltage-tuned receiver preselector

The circuit in Fig. 4.10 is a little different. In addition to using only input tuning (which lessens the potential for oscillation), it also uses *voltage tuning*. The hard-to-find variable capacitors are replaced with *varactor diodes*, also called *voltage variable capacitance diodes* (D1). These PN junction diodes exhibit a capacitance that is a function of the applied reverse bias potential, V_T. Although the original circuit was built and tested for the AM broadcast band (540 kHz to 1700 kHz), it can be changed to any band by correct selection of the inductor values. The varactor offers a capacitance range of 440 pF down to 15 pF over the voltage range 0 to +18 VDC.

The inductors may be either 'store-bought' types or wound over toroidal cores. I used a toroid for L1/L2 (forming a fixed inductance for L2) and 'store-bought' adjustable inductors for L3 and L4. There is no reason, however, why these same inductors cannot be used for all three uses. Unfortunately, not all values are available in the form that has a low impedance primary winding to permit antenna coupling.

In both of the MOSFET circuits the fixed bias network used to place gate G2 at a positive DC potential can be replaced with a variable voltage circuit. The potentiometer in Fig. 4.11 can be used as an RF gain control to reduce gain on strong signals, and increase it on weak signals. This feature allows the active preselector to be custom set to prevent overload from strong signals.

Broadband RF preamplifier for VLF, LF and AM BCB

There are many situations where a broadband RF amplifier is needed. Typical applications include boosting the output of RF signal generators (which tend to be normally quite low level), antenna preamplification, loop antenna amplifiers, and in the front-ends of receivers. There are a number of different circuits published, including some by me, but one failing that I've noted on most of them is that they often lack response at the low end of the frequency range. Many designs offer –3 dB frequency response limits of 3 to 30 MHz, or 1 to 30 MHz, but rarely are the VLF, LF or even the entire AM broadcast band (540 kHz to 1700 kHz) covered.

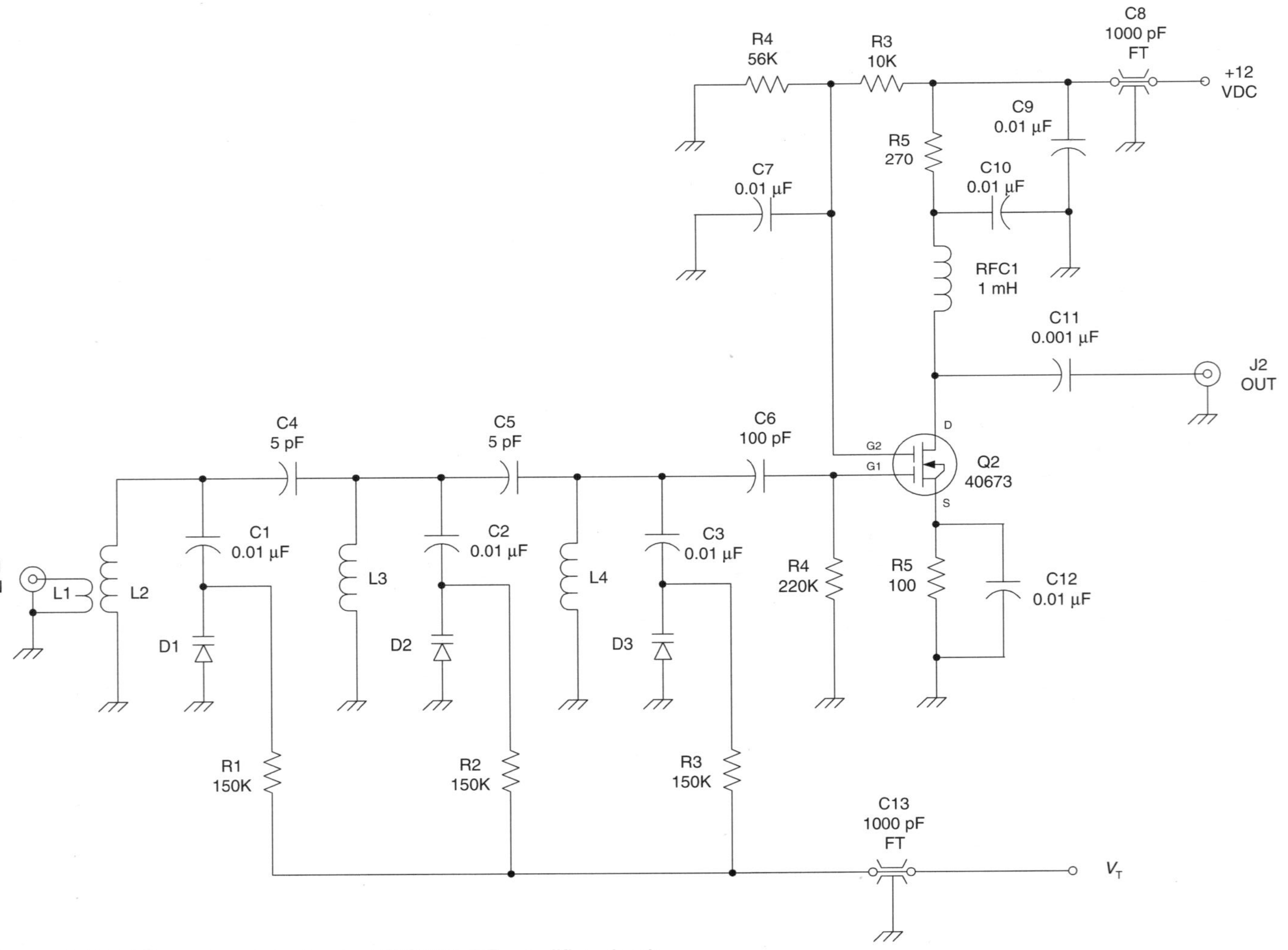

Figure 4.10 Voltage-tuned dual-gate MOSFET RF amplifier circuit.

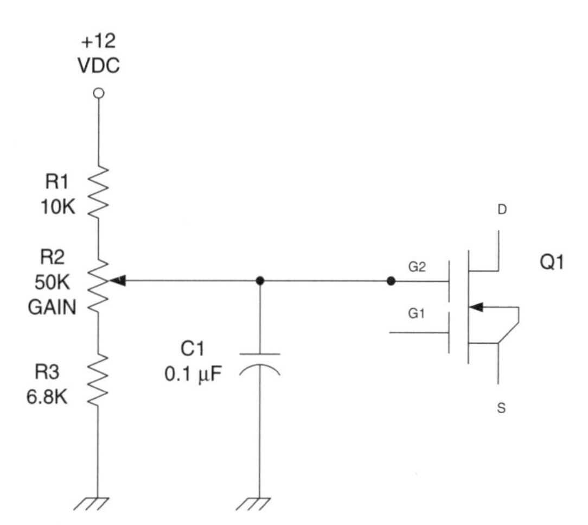

Figure 4.11 RF gain control arrangement.

The original need for this amplifier was that I needed an amplifier to boost AM BCB signals. Many otherwise fine communications or entertainment grade 'general coverage' receivers operate from 100 kHz to 30 MHz, or so, and that range initially sounds really good to the VLF through AM BCB owner. But when examined closer it turns out that the receiver lacks sensitivity on the bands below either 2 or 3 MHz, so it fails somewhat in the lower end of the spectrum. While most listening on the AM BCB is to powerful local stations (where receivers with no RF amplifier and a loopstick antenna will work nicely), those who are interested in DXing are not well served. In addition to the receiver, I wanted to boost my signal generator 50 ohm output to make it easier to develop some AM and VLF projects that I am working on, and to provide a preamplifier for a square loop antenna that tunes the AM BCB.

Several requirements were developed for the RF amplifier. First, it had to retain the 50 ohm input and output impedances that are standard in RF systems. Second, it had to have a high dynamic range and third-order intercept point in order to cope with the bone crunching signal levels on the AM BCB. One of the problems of the AM BCB is that those sought-after distant stations tend to be buried under multi-kilowatt local stations on adjacent channels. That's why high dynamic range, high intercept point and loop antennas tend to be required in these applications. I also wanted the amplifier to cover at least two octaves (4:1 frequency ratio), and in fact achieved a decade (10:1) response (250 kHz to 2500 kHz).

Furthermore, the amplifier circuit had to be easily modifiable to cover other frequency ranges up to 30 MHz. This last requirement would make the amplifier more useful to others, as well as extending its usefulness to me.

There are a number of issues to consider when designing an RF amplifier for the front-end of a receiver. The dynamic range and intercept point requirements were mentioned above. Another issue is the amount of distortion products (related to third-order intercept point) that are generated in the amplifier. It does no good to have a high capability on the preamplifier, only to overload the receiver with a lot of extraneous RF energy it can't handle ... energy that was generated by the preamplifier, not from the stations being received. These considerations point to the use of a *push-pull RF amplifier* design.

Push-pull RF amplifiers

The basic concept of a push-pull amplifier is demonstrated in Fig. 4.12. This type of circuit consists of two identical amplifiers that each processes half the input signal power, but in antiphase. In the circuit shown this job is accomplished by using a centre tapped transformer at the input to split the signal, and another at the output to recombine the signals from the two transistors. Because of normal transformer action, the signal polarity at end 'A' will be opposite that at end 'B' when the centre tap ('CT') is grounded. Thus, the two amplifiers are driven 180 degrees out of phase with each other. This is similar to the output stage of an audio amplifier, except that an RF preamplifier must operate strictly

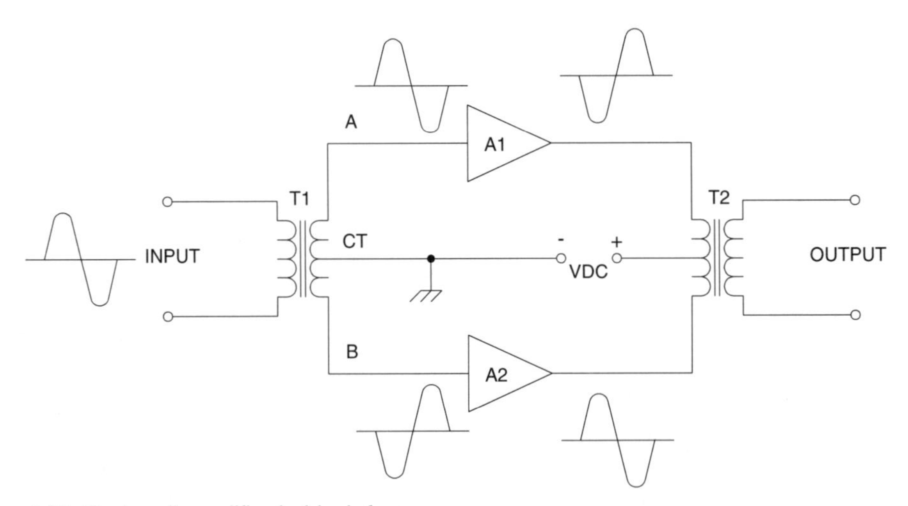

Figure 4.12 Push-pull amplifier in block form.

in Class A.

The push-pull amplifier circuit is balanced, and as a result it has a very interesting property: even-order harmonics are cancelled in the output, so the amplifier output signal will be cleaner than for a single-ended amplifier using the same active amplifier devices.

Types of push-pull RF amplifiers

There are two general categories of push-pull RF amplifiers: tuned amplifiers and wideband amplifiers. The tuned amplifier will have the inductance of the input and output transformers resonated to some specific frequency. In some circuits the non-tapped winding may be tuned, but in others a configuration such as Fig. 4.13 might be used. In this circuit both halves of the tapped side of the transformer are individually tuned to the desired resonant frequency. Where variable tuning is desired, a split-stator capacitor might be used to supply both capacitances.

The broadband category of circuit is shown in Fig. 4.14A. In this type of circuit a special transformer is usually needed. The transformer must be a broadband RF transformer, which means that it must be wound on a suitable core such that the windings are bifilar

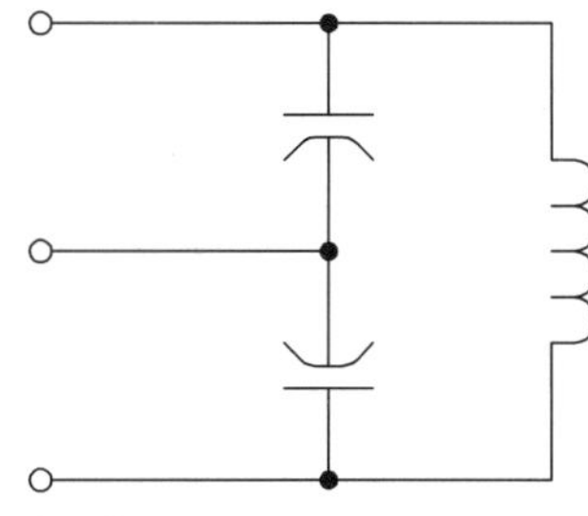

Figure 4.13 Push-pull output tuned network.

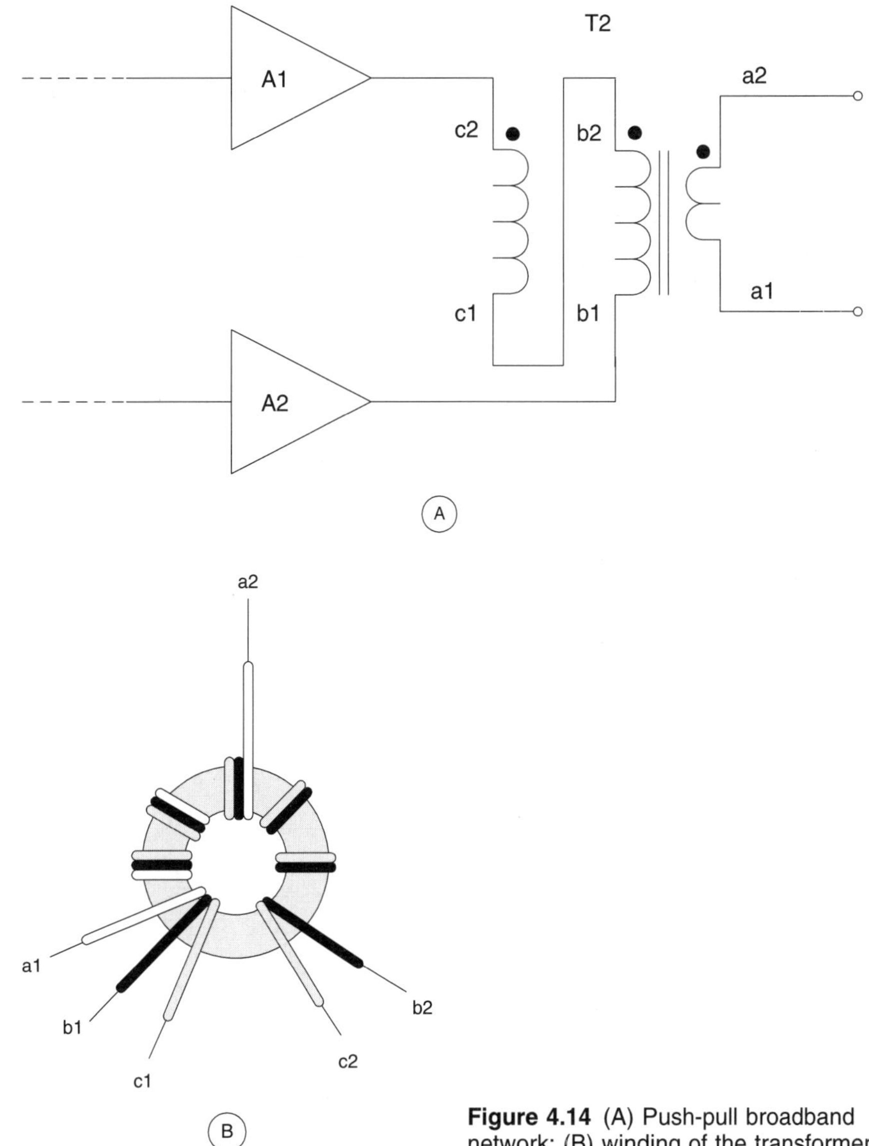

Figure 4.14 (A) Push-pull broadband network; (B) winding of the transformer.

or trifilar. The particular transformer in Fig. 4.14A has three windings, of which one is much smaller than the others. These must be trifilar wound for part of the way, and bifilar the rest of the way. This means that all three windings are kept parallel until no more turns are required of the coupling link, and then the remaining two windings are kept parallel until they are completed. Figure 4.14B shows an example for the case where the core of the transformer is a ferrite or powdered iron *toroid*.

Actual circuit details

The actual RF circuit is shown in Fig. 4.15. The active amplifier devices are junction field effect transistors (JFET) intended for service from DC to VHF. The device selected can be the MPF-102, or some similar device. Also useful is the 2N4416 device. The particular device that I used was the NTE-451 JFET transistor. This device offers a transconductance of 4000 microsiemens (1 μsiemen = 1 μMho), a drain current of 4 to 10 mA, and a power dissipation of 310 mW, with a noise figure of 4 dB maximum.

The JFET devices are connected to a pair of similar transformers, T1 and T2. The source bias resistor (R1) for the JFETs, and its associated bypass capacitor (C1), are connected to the centre tap on the secondary winding of transformer T1. Similarly, the +9 volt DC power supply voltage is applied through a limiting resistor (R2) to the centre tap on the primary of transformer T2.

Take special note of those two transformers. These transformers are known generally as wideband transmission line transformers, and can be wound on either toroid or binocular ferrite or powdered iron cores. For the project at hand, because of the low frequencies involved, I selected a type BN-43-202 binocular core. The type 43 material used in this core

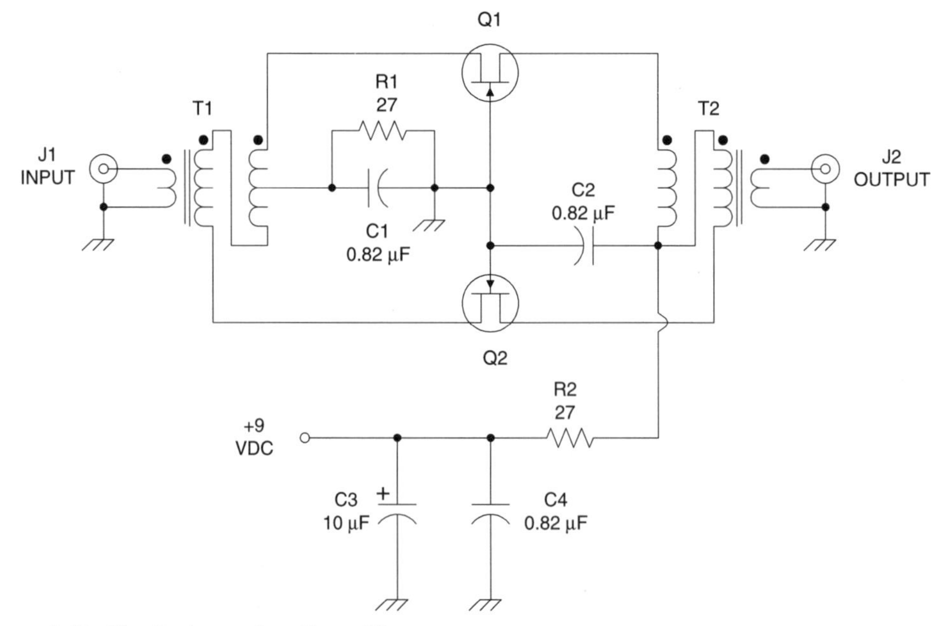

Figure 4.15 Circuit of a push-pull amplifier.

is a good selection for the frequency range involved. There are three windings on each transformer. In each case, the 'B' and 'C' windings are 12 turns of #30 AWG enamelled wire wound in a bifilar manner. The coupling link in each is winding 'A'. The 'A' winding on transformer T1 consists of four turns of #36 AWG enamelled wire, while on T2 it consists of two turns of the same wire. The reason for the difference is that the number of turns in each is determined by the impedance matching job it must do (T1 has a 1:9 primary/secondary ratio, while T2 has a 36:1 primary/secondary ratio). Neither the source nor drain impedances of this circuit are 50 ohms (the system impedance), so there must be an impedance transformation function.

The detail for transformers T1 and T2 is shown in Fig. 4.16. I elected to build a header of printed circuit perforated board for this part; the board holes are on 0.1 inch centres. The PC type of perf board has a square or circular printed circuit soldering pad at each hole. A section of perf board was cut with a matrix of five holes by nine holes. *Vector Electronics* push terminals are inserted from the unprinted side, and then soldered into place. These terminals serve as anchors for the wires that will form the windings of the transformer. Two terminals are placed at one end of the header, and three at the opposite end.

The coupling winding is connected to pins 1 and 2 of the header, and is wound first on each transformer. Strip the insulation from a length of #36 AWG enamelled wire for about $\frac{1}{4}$ inch from one end. This can be done by scraping with a scalpel of *X-acto* knife, or by burning with the tip of a soldering pencil. Ensure that the exposed end is tinned with solder, and then wrap it around terminal no. 1 of the header. Pass the wire through the first hole of the binocular core, across the barrier between the two holes, and then through

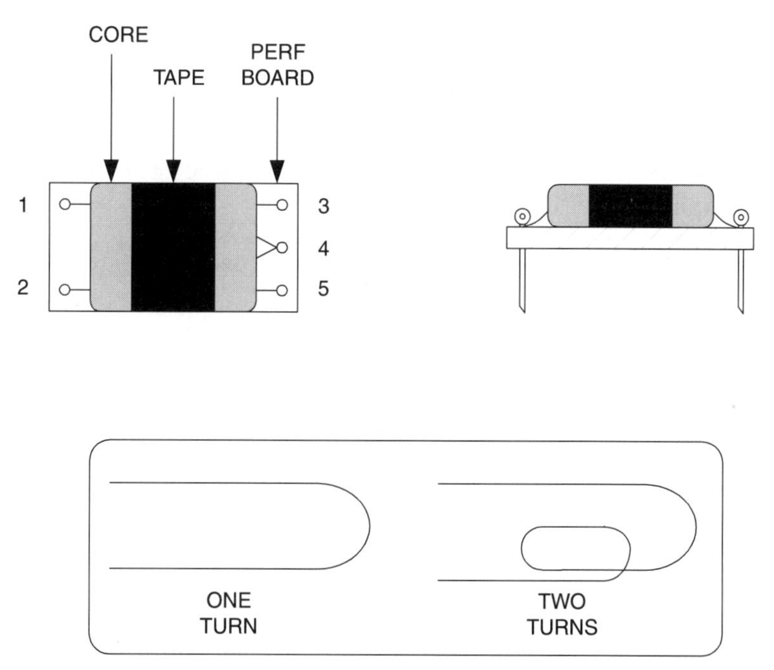

Figure 4.16 Binocular BALUN physical implementation.

the second hole. This 'U'-shaped turn counts as one turn. To make transformer T1 pass the wire through both sets of holes three more times (to make four turns). The wire should be back at the same end of the header as it started. Cut the wire to allow a short length to connect to pin no. 2. Clean the insulation off this free end, tin the exposed portion and then wrap it around pin no. 2 and solder. The primary of T1 is now completed.

The two secondary windings are wound together in the bifilar manner, and consist of 12 turns each of #30 AWG enamelled wire. The best approach seems to be twisting the two wires together. I use an electric drill to accomplish this job. Two pieces of wire, each 30 inches long, are joined together and chucked up in an electric drill. The other ends of the wire are joined together and anchored in a bench vice, or some other holding mechanism. I then back off, holding the drill in one hand, until the wire is nearly taut. Turning on the drill causes the two wires to twist together. Keep twisting them until you obtain a pitch of about eight to 12 twists per inch.

It is *very important* to use a drill that has a variable speed control so that the drill chuck can be made to turn very slowly. It is also *very important* that you follow certain safety rules, especially as regards your eyesight, when making twisted pairs of wire. *Be absolutely sure to wear either safety glasses or goggles while doing this operation.* If the wire breaks, and that is a common problem, then it will whip around as the drill chuck turns. While #36 wire doesn't seem to be very substantial, at high speed it can severely injure an eye.

To start the secondary windings, scrap all of the insulation off both wires at one end of the twisted pair, and tin the exposed ends with solder. Solder one of these wires to pin no. 3 of the header, and the other to pin no. 4. Pass the wire through the hole of the core closest to pin no. 3, around the barrier, and then through the second hole, returning to the same end of the header as where you started. That constitutes one turn. Now do it 11 more times until all 12 turns are wound. When the 12 turns are completed, cut the twisted pair wires off to leave about $\frac{1}{2}$ inch free. Scrap and tin the ends of these wires.

Connecting the free ends of the twisted wire is easy, but you will need an ohmmeter or continuity tester to see which wire goes where. Identify the end that is connected at its other end to pin no. 3 of the header, and connect this wire to pin no. 4. The remaining wire should be the one that was connected at its other end to pin no. 4 earlier; this wire should be connected to pin no. 5 of the header.

Transformer T2 is made in the identical manner as transformer T1, but with only two turns on the coupling winding rather than four. In this case, the coupling winding is the secondary, while the other two form two halves of the primary. Wind the two-turn secondary first, as was done with the four-turn primary on T1.

The amplifier can be built on the same sort of perforated board as was used to make the headers for the transformers. Indeed, the headers and the board can be cut from the same stock. The size of the board will depend somewhat on the exact box you select to mount it in.

Broadband RF amplifier (50 ohm input and output)

This project (Fig. 4.17) is a highly useful RF amplifier that can be used in a variety of ways. It can be used as a preamplifier for receivers operating in the 3 to 30 MHz shortwave band. It can also be used as a postamplifier following filters, mixers and other devices that have an attenuation factor. It is common, for example, to find that mixers and crystal filters have a signal loss of 5 to 8 dB (this is called 'insertion loss'). An amplifier following these devices will overcome that loss. The amplifier can also be used to boost the output level of signal generator and oscillator circuits. In this service it can be used either alone, in its own shielded container, or as part of another circuit containing an oscillator circuit.

The transistor (Q1) is a 2N5179 broadband RF transistor. It can be replaced by the NTE-316 or ECG-316 devices, if the original is not available to you. The NTE and ECG devices are intended for service and maintenance replacement applications, so tend to be found in local electronic parts distributors.

There are two main features to this amplifier: the degenerative feedback in the emitter circuit, and the negative feedback from collector to base. Degenerative, or negative, feedback is used in amplifiers to reduce distortion (i.e. make it more linear) and to stabilize the amplifier. In this case, the combination of two types of feedback sets the gain and the input and output impedances of the amplifier.

The emitter resistance consists of two resistors, R5 is 10 ohms and R6 is 100 ohms. In most amplifier circuits the emitter resistor is bypassed by a capacitor to set the emitter of

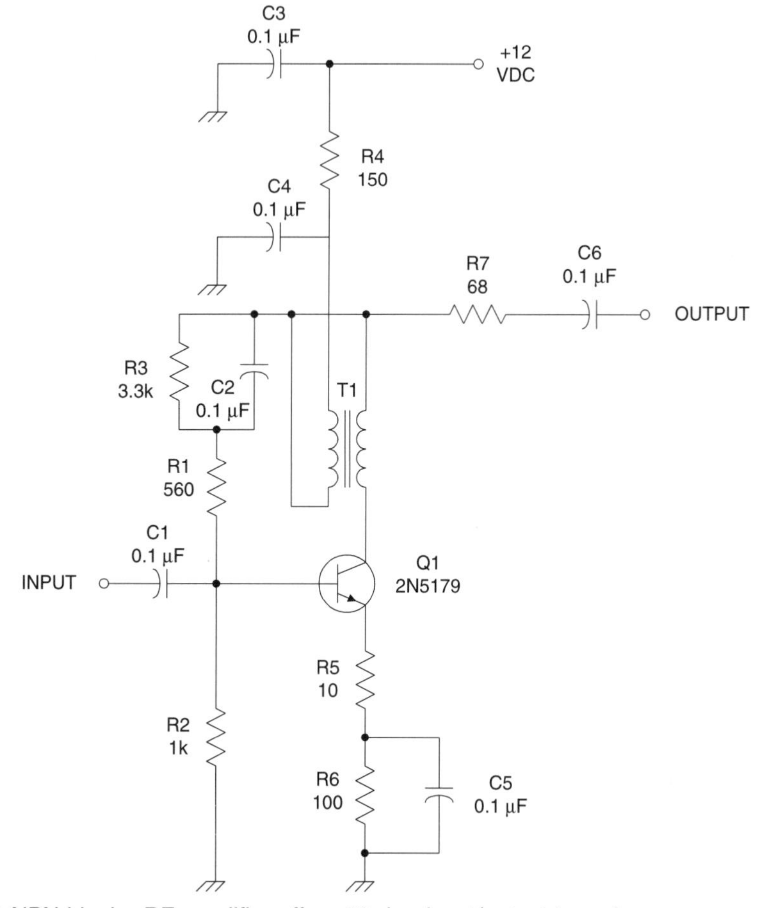

Figure 4.17 NPN bipolar RF amplifier offers 50 ohm input/output impedance.

the transistor at ground potential for RF signals, while keeping it at the DC level set by the resistance. In normal situations, the reactance of the capacitor should be not more than one-tenth the resistance of the emitter resistor. The 10 ohm portion of the total resistance is left unbypassed to provide negative feedback.

The collector-to-base feedback is accomplished by two means. First, a resistor/capacitor network (R1/R3/C2) is used; second, a 1:1 broadband RF transformer (T1) is used. This transformer can be home-made. Wind 15 bifilar turns of #26 enamelled wire on a toroidal core such as the T-50-2 (RED) or T-50-6 (YEL); smaller cores can also be used.

The circuit can be built on perforated wire-board that has a grid of holes on 0.100 inch centres. You can use a homebrew RF transformer made on a small toroidal core. Use the size 37 core, with #36 enamelled wire. As in the previous case, make the two windings bifilar.

5 Mixers

Mixer circuits are used extensively in radio frequency electronics. Applications include frequency translators (including in radio receivers), demodulators, limiters, attenuators, phase detectors and frequency doublers. There are a number of different approaches to mixer design. Each of these approaches has advantages and disadvantages, and these factors are critical to the selection process.

Linear-vs-non-linear mixers

The word 'mixer' is used to denote both linear and non-linear circuits. This situation is unfortunate because only the non-linear is appropriate for the RF mixer applications listed above.

So what's the difference? The basic linear mixer is actually a *summer* circuit, as shown in Fig. 5.1A (the schematic symbol is in Fig. 5.1B). Some sort of combiner is needed. In the case shown, the combiner is a resistor network. There is no interaction between the two input signals, *F*1 and *F*2. They will share the same pathway at the output, but otherwise do not affect each other. This is the action one expects of microphone and other audio

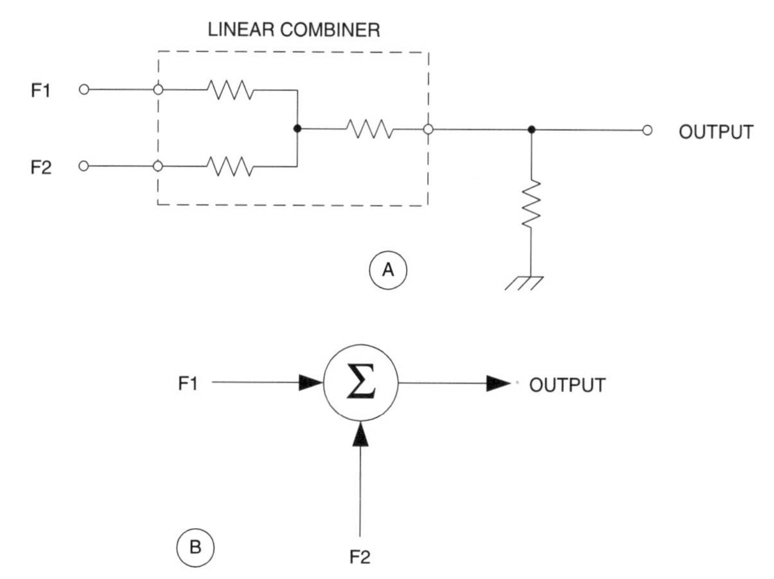

Figure 5.1 Linear combiner (adder) circuit and symbol.

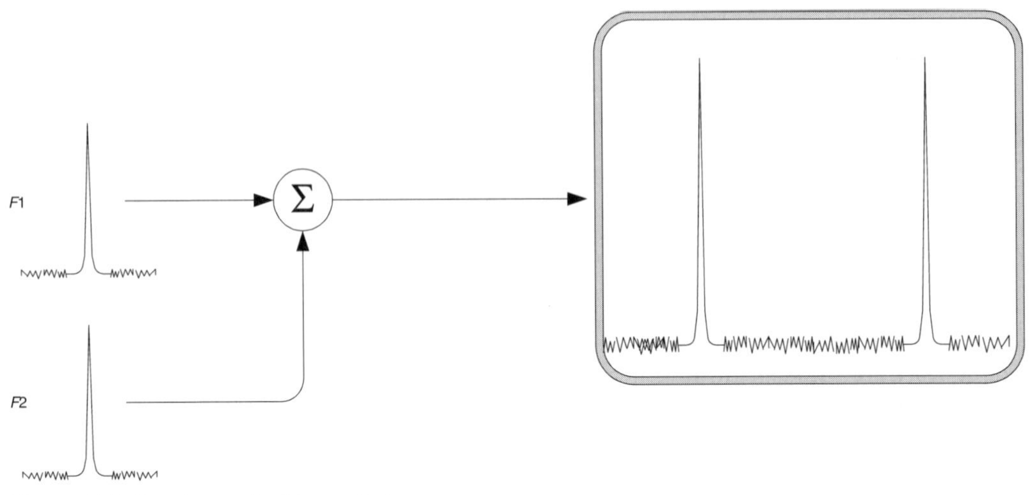

Figure 5.2 Spectrum of adder output.

mixers. If you examine the output of the summer on a spectrum analyser (Fig. 5.2), then you will see the spikes representing the two frequencies, and nothing else other than noise.

The non-linear mixer is shown in Fig. 5.3A, and the circuit symbol in Fig. 5.3B. While the linear mixer is a summer, the non-linear mixer is a *multiplier*. In this particular case, the non-linear element is a simple diode, such as a 1N4148 or similar devices. Mixing action occurs when the non-linear device, such as diode D1, exhibits impedance changes over cyclic excursions of the input signals.

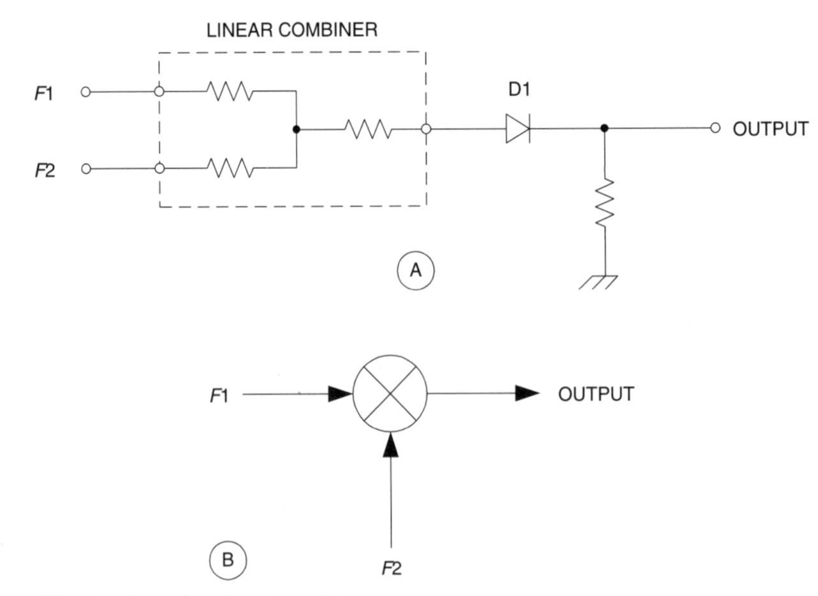

Figure 5.3 Diode mixer circuit and symbol.

Whenever a non-linear element is added to the signal path a number of new frequencies will be generated. If only one frequency is present, then we would still expect to see its harmonics; for example, $F1$ and $nF1$ where n is an integer. But when two or more frequencies are present, a number of other products are also present. The output frequency spectrum from a non-linear mixer is:

$$\pm F_o = mF1 \pm nF2 \tag{5.1}$$

Where:
F_o is the output frequency for a specific (m, n) pair
$F1$ and $F2$ are the applied frequencies
m and n are integers or zero (0, 1, 2, 3, ...)

There will be a unique set of frequencies generated for each (m, n) ordered pair. These new frequencies are called *mixer products* or *intermodulation products*. Figure 5.4 shows how the output would look on a spectrum analyser. The original signals ($F1$ and $F2$) are present, along with an array of mixer products arrayed at frequencies away from $F1$ and $F2$.

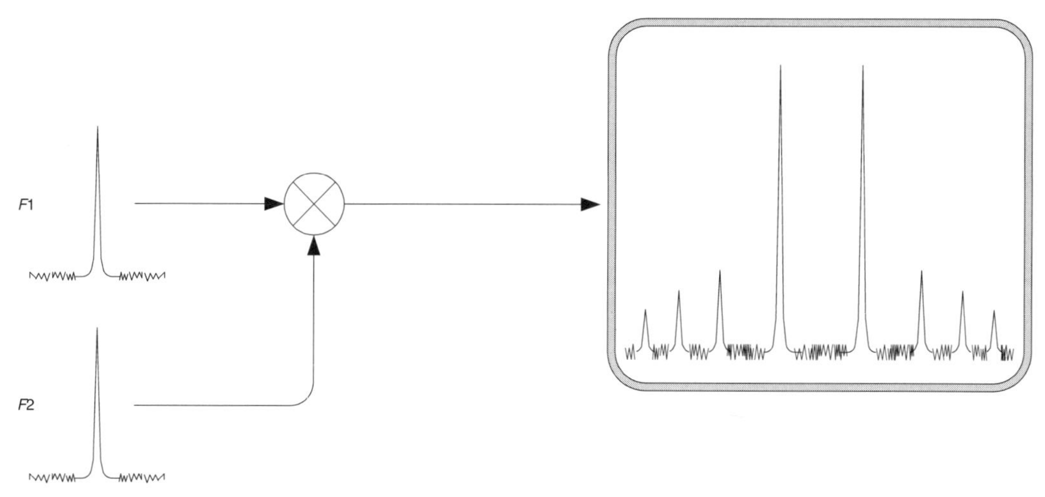

Figure 5.4 Spectrum of mixer output.

The implication of Eq. (5.1) is that there will be a large number of (m, n) frequency products in the output spectrum. Not all of them will be useful for any specific purpose, and some may well cause adverse effects.

So why do we need mixers? There are other ways to generate various frequencies, so why a frequency translator such as a heterodyne mixer? The principal answer is that the mixer will translate the frequency, and in the process transfer the modulation of the original signal. So, when an AM signal is received, and then translated to a different frequency in the receiver, the modulation characteristics of the AM signal convey to the new frequency essentially undistorted (those who know that there are no 'distortionless' circuits please refrain from snickering). Perhaps the most common use for mixers, in this regard, is in radio receivers.

Terminology. In the remainder of this chapter *F*1 and *F*2 will be expressed much of the time as F_{RF} and F_{LO} in view of the receiver being the most common use for mixer devices.

Simple diode mixer

Figure 5.5 shows a block diagram circuit for a simple form of mixer. Although not terribly practical in most cases, the circuit has been popular in a number of receivers in the high UHF and microwave regions since World War II. The two input signals are the RF and LO. The LO signal is at a very much higher level than the RF signal, and is used to switch the diode in and out of conduction, providing the non-linearity that mixer action requires.

There are three filters shown in this circuit. The RF and LO filters are used for limiting the frequencies that can be applied to the mixer. In the case of the RF port it is other radio signals on the band that are being suppressed, while in the case of the LO it is LO noise and harmonics that are suppressed. The RF filter also serves to reduce any LO energy that may be transmitted back towards the RF input.

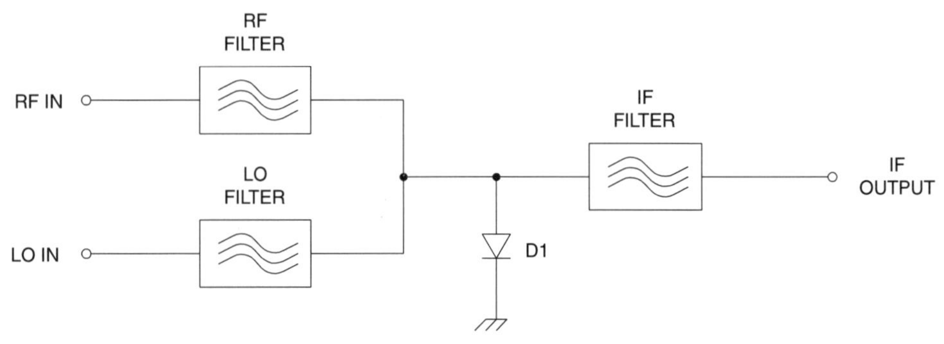

Figure 5.5 Mixer circuit.

There are a number of stories from World War II of receiver LO radiation back through the antenna circuit being responsible for an enemy detecting the location of the receiver, so this effect is rather important. One such story is from British airborne radar history. According to one source of doubtful authority, German submarines sailing on the surface learned to listen for *Beaufighter* centimetric radars using a receiver that was poorly suppressed. The aircrews then learned that they could locate the submarine with just the radar's receiver tuned to listen for the submarine receiver's LO[1].

[1] A similar story is told by R. V. Jones in his book *Most Secret War* (Coronet Books, Hodder & Stoughton, 1978). On page 411 he reports that the Germans thought their local oscillators were being picked up, but this was not in fact the case. They had been misled by a British POW.

The question of 'balance'

One of the ways of classifying mixers is whether or not they are *unbalanced, single balanced* or *double balanced*. Although there are interesting aspects of each of these categories, we are presently interested in how they affect the output spectrum.

Unbalanced mixers

Both F_{RF} and F_{LO} appear in the output spectrum, and there may be poor LO–RF and RF–LO port isolation. Their principal attraction is low cost.

Single balanced mixers

Either F_{RF} or F_{LO} is suppressed in the output spectrum, but not both. The single balanced mixer will also suppress even-order LO harmonics ($2F_{LO}$, $4F_{LO}$, $6F_{LO}$, etc.). High LO–RF isolation is provided, but external filtering must provide LO–IF isolation.

Double balanced mixers

Both F_{RF} and F_{LO} are suppressed in the output. The double balanced mixer will also suppress even-order LO and RF harmonics ($2F_{LO}$, $2F_{RF}$, $4F_{LO}$, $4F_{RF}$, $6F_{LO}$, $6F_{RF}$, etc.). High port-to-port isolation is provided.

Spurious responses

A receiver mixer will use one of the second-order products in order to convert F_{RF} to F_{IF}. Ideally, the receiver would only respond to one RF frequency. Unfortunately, reality sometimes rudely intervenes, and certain spurious responses might be noted.

A *spurious response* in a superheterodyne receiver is any response to any frequency other than the desired F_{RF}, which is strong enough to be heard in the receiver input. Most of these 'spurs' are actually mixer responses, although overloading the RF amplifier can cause some responses as well. The mixer responses may or may not be affected by premixer filtering of the RF signal. Candidate spur frequencies include any that satisfy Eq. (5.2):

$$F_{Spur} = \frac{nF_{LO} \pm F_{iJ}}{m} \tag{5.2}$$

Image

The image response of a mixer is due to the fact that two frequencies satisfy the criteria for F_{FI}. Figure 5.6 shows how the image response works. The frequency that satisfies the image criteria depends on whether the LO is *high-side injected* ($F_{LO} > F_{RF}$) or *low-side injected* ($F_{LO} < F_{RF}$). In the high-side injection case $[(m, n) = (1, -1)]$ shown in Fig. 5.6, the image appears at $F_{RF} + 2F_{IF}$. If low-side injection $[(m, n) = (-1, 1)$ is used, then the image is at $F_{RF} - 2F_{IF}$. The image always appears on the *opposite side of the LO from the RF.*

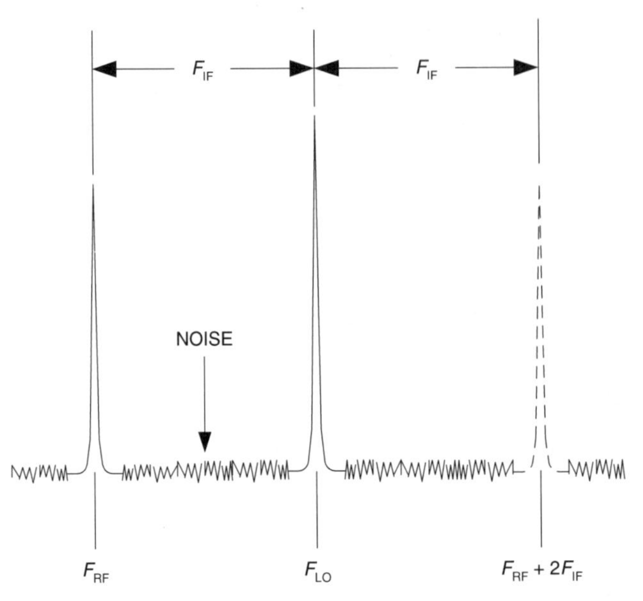

Figure 5.6 Spectrum of mixer circuit.

Let's consider an actual example based on an AM broadcast band (BCB) receiver. The IF is 455 kHz, and the receiver is tuned to F_{RF} = 1000 kHz. The usual procedure on AM BCB receivers is high-side injection, so $F_{LO} = F_{RF} + F_{IF}$ = 1000 kHz + 455 kHz = 1455 kHz. The image frequency appears at $F_{RF} + 2F_{IF}$ = 1000 kHz + (2 × 455 kHz) = 1910 kHz. Any signal on or near 1910 kHz that makes it to the mixer RF input port will be converted to 455 kHz along with the desired signals.

The problem is complicated by the fact that it is not just actual signals present at the image frequency, but noise as well. The noise applied to the mixer input is essentially doubled if the receiver has any significant response at the image frequency. Premixer filtering is needed to reduce the noise. Receiver designers also specify high IF frequencies in order to move the image out of the passband of the RF prefilter.

Half IF

Another set of images occurs when (m, n) is (2, –2) for low side or (–2, 2) for high side. This image is called the *half-IF image*, and is illustrated in Fig. 5.7. An interesting aspect of the half-IF image is that it is created by internally generated harmonics of both F_{RF} and F_{LO}. For our AM BCB receiver where F_{RF} = 1000 kHz, F_{LO} = 1450 kHz and F_{IF} = 455 kHz, then the half-IF frequency is 1000 + (455/2) = 1227.5 kHz.

IF feedthrough

If a signal from outside passes through the mixer to the IF amplifier, and happens to be on a frequency equal to F_{IF}, then it will be accepted as a valid input signal by the IF amplifier. The mixer RF–IF port isolation is critical in this respect.

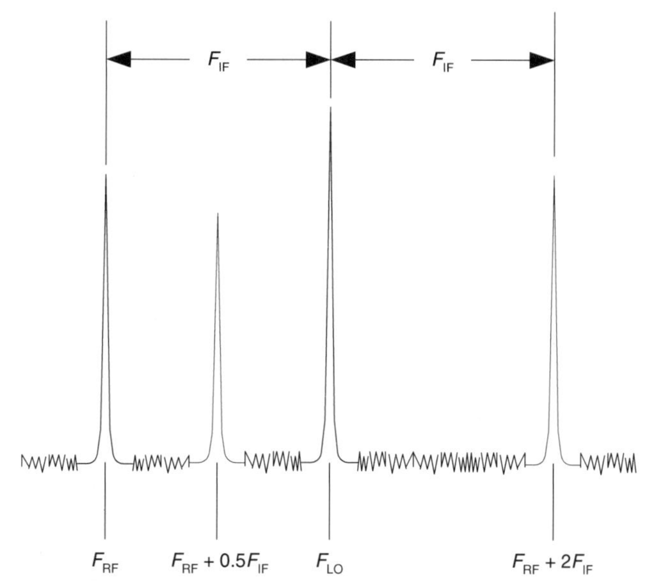

Figure 5.7 Spectrum of mixer circuit with half-IF.

High-order spurs

Thus far we have considered only the case where a single RF frequency is applied to the mixer. But what happens when two RF frequencies (F_{RF1} and F_{RF2}) are applied simultaneously. This is the actual situation in most practical receivers. There are a large number of higher-order responses (i.e. where m and n are both greater than 1) defined by $mF_{RF1} \pm nF_{RF2}$.

Although any of the spurs may prove difficult to handle in some extreme cases, the principal problems occur with the third-order difference products of two RF signals

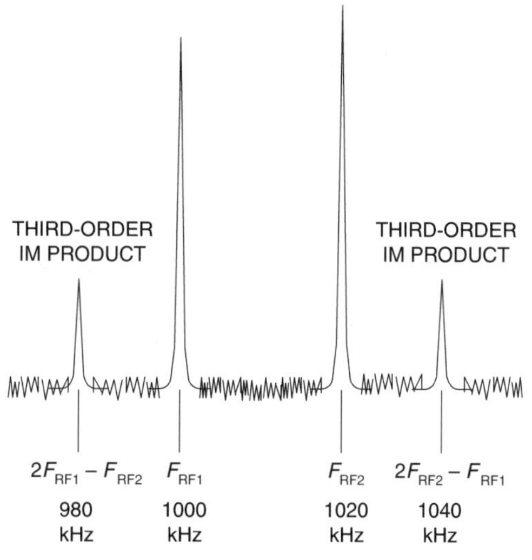

Figure 5.8 Third-order IMD products.

applied to the RF port of the mixer ($2F_{RF1} - F_{RF2}$ and $2F_{RF2} - F_{RF1}$) because they fall close to F_{RF1} and F_{RF2} and may be within the device passband. Figure 5.8 illustrates this effect for our AM BCB receiver. Suppose two signals appear at the mixer input: $F_{RF1} = 1000$ kHz and $F_{RF2} = 1020$ kHz (this combination is highly likely in the crowded AM BCB!). The third-order products of these two signals hitting the mixer are 980 kHz and 1040 kHz, and appear close to F_{RF1} and F_{RF2}. If the premixer filter selectivity is not sufficiently narrow to suppress the unwanted RF frequency, then the receiver may respond to the third-order products as well as the desired signal.

LO harmonic spurs

If the harmonics of the local oscillator are strong enough to drive mixer action, then signal clustered at $\pm F_{IF}$ from each significant harmonic will also cause mixing. Figure 5.9 shows this effect. The passband of the premixer filter is shown as dashed line curves at $F_{LO} \pm F_{IF}$, $2F_{LO} \pm F_{IF}$ and $3F_{LO} \pm F_{IF}$.

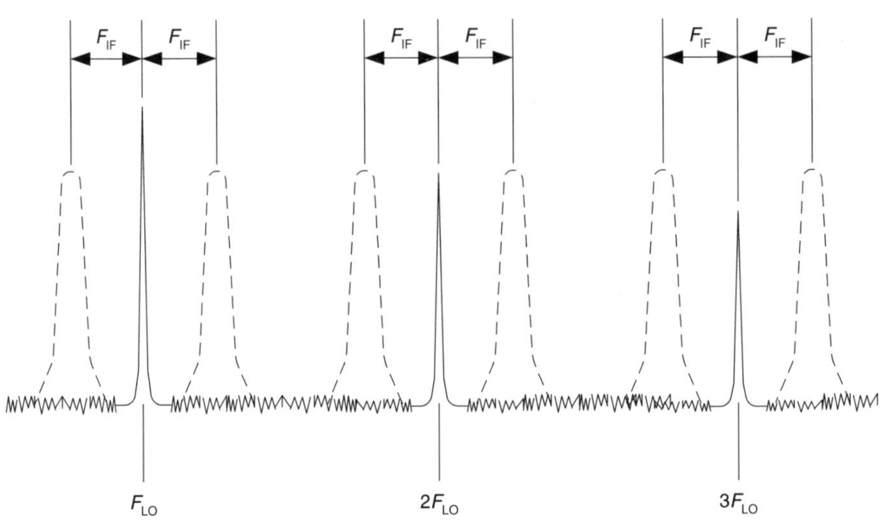

Figure 5.9 Spectrum showing harmonics of local oscillator.

LO noise spurs

All oscillators have noise close to the LO frequency. The noise may be due to power supply noise modulating the LO, or it may be random phase noise about the LO. In either case, the noise close to the LO, and within the limits imposed by the IF filter, will be passed through the mixer to the IF amplifier.

Mixer distortion products

Because mixers are non-linear, they will produce both harmonic distortion products and intermodulation products. Intermodulation was discussed in Chapter 3 in the context of a radio receiver, and the concept of intercept point was introduced. The same ideas can be applied to a mixer on its own.

Third-order intercept point

It can be claimed that the third-order intercept point (TOIP) is the single most important specification of a mixer's dynamic performance because it predicts the performance as regards intermodulation, cross-modulation and blocking desensitization.

When a mixer is used in a receiver, the third-order (and higher) intermodulation products (IP) are normally very weak, and don't exceed the receiver noise floor when the mixer and any preamplifiers are operating in the linear region. As input signal levels increase, forcing the front-end of the receiver toward the saturated non-linear region, the IP emerge from the noise (Fig. 5.10) and begin to cause problems. When this happens, new spurious signals appear on the band and self-generated interference arises.

Figure 5.11 shows a plot of the output signal-vs-fundamental input signal. Note the output compression effect that occurs as the system begins to saturate. The dashed gain line continuing above the saturation region shows the theoretical output that would be produced if the gain did not clip.

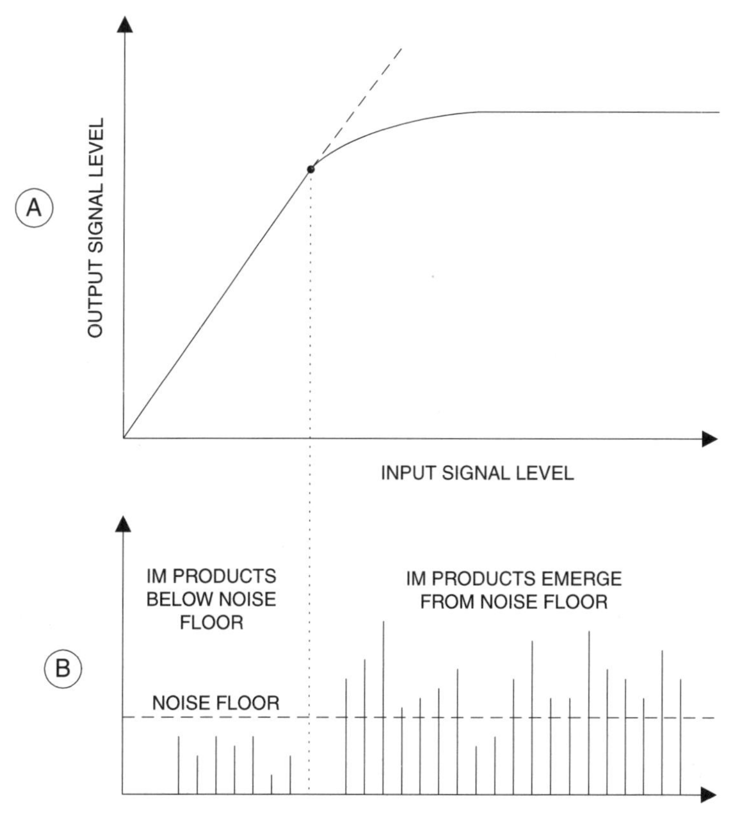

Figure 5.10 Noise floor increases dramatically above TOIP.

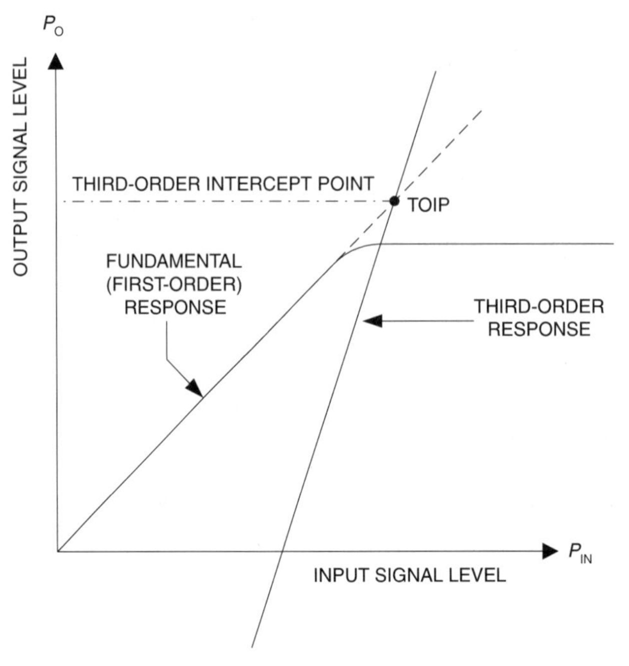

Figure 5.11 Third-order intercept point.

It is the nature of third-order products in the output signal to emerge from the noise at a certain input level, and increase as the cube of the input level. Thus, the third-order line increases 3 dB for every 1 dB increase in the response to the fundamental signal. Although the output response of the third-order line saturates similarly to that of the fundamental signal, the gain line can be continued to a point where it intersects the gain line of the fundamental signal. This point is the *third-order intercept point* (TOIP).

Notice in Fig. 5.11 that the gain (P_O/P_{IN}) begins to decrease in the vicinity of the TOIP. The measure of this tendency to saturation is called the *–1 dB compression point*, i.e. the point where the gain slope decreases by 1 dB.

Interestingly enough, one tactic that can help reduce IP levels back down under the noise is the use of an attenuator ahead of the mixer. Even a few dB of input attenuation is often sufficient to drop the IPs back into the noise, while affecting the desired signals only a small amount. Many modern receivers provide a switchable attenuator ahead of the mixer. This practice must be evaluated closely, however, if low level signals are to be handled. The usual resistive attenuator pad will increase the thermal noise level appearing at the input of the mixer by an amount proportional to its looking back resistance.

The IP performance of the mixer selected for a receiver design can profoundly affect the performance of the receiver. For example, the second-order intercept point affects the half-IF spur rejection, while the third-order intercept point will affect the intermodulation distortion (IMD) performance.

Calculating intercept points

Calculating the nth order intercept point can be done using a two-tone test scheme. A test system is created in which two equal amplitude signals (F_A and F_B) are applied simultaneously to the mixer RF input. These signals are set to a standard level (typically –20 dBm to –10 dBm), and the power of the nth intermodulation product ($P_{\mathrm{IM}n}$) is measured (using a spectrum analyser or, if the spectrum analyser is tied up elsewhere, a receiver with a calibrated S-meter). The nth intercept point is:

$$IP_n = \frac{NP_A - P_{\mathrm{IM}n}}{N - 1} \tag{5.3}$$

Where:

IP_n is the intermod product of order N
N is the order of the intermod product
P_A is the input power level (in dBm) of one of the input signals
$P_{\mathrm{IM}n}$ is the power level (in dBm) of the nth IM product (often specified in terms of the receiver's minimum discernible signal specification)

Once the P_A and P_{IM} points are found any IP can be calculated using Eq. (5.3).

Mixer losses

Depending on its design a mixer may show either loss or gain. The principal loss is conversion loss, which is made up of three elements: *mismatch loss*, *parasitic loss* and *junction loss* (assuming a diode mixer). The conversion loss is simply the ratio of the RF input signal level and the signal level appearing at the IF output ($P_{\mathrm{IF}}/P_{\mathrm{RF}}$). In some cases, it may be a gain, but for many – perhaps most – mixers there is a loss. Conversion loss (L_{C}) is:

$$L_{\mathrm{C}} = L_{\mathrm{M}} + L_{\mathrm{P}} + L_{\mathrm{J}} \tag{5.4}$$

Where:

L_{C} is conversion loss
L_{M} is the mismatch loss
L_{P} is the parasitic loss
L_{J} is the junction loss

Mismatch loss is a function of the impedance match at the RF and IF ports. If the mixer port impedance (Z_{P}) and the source impedance (Z_{S}) are not matched, then a VSWR will result that is equal to the ratio of the higher impedance to the lower impedance ($VSWR = Z_{\mathrm{P}}/Z_{\mathrm{S}}$ or $VSWR = Z_{\mathrm{S}}/Z_{\mathrm{P}}$, depending on which ratio is ≥ 1). The mismatch loss is the sum of RF and IF port mismatch losses. Or expressed in terms of VSWR:

$$L_{\mathrm{M}} = 10 \times \left[\log_{10} \left[\frac{(VSWR_{\mathrm{RF}} + 1)^2}{4VSWR_{\mathrm{RF}}} \right] + \log_{10} \left[\frac{(VSWR_{\mathrm{IF}} + 1)^2}{4VSWR_{\mathrm{IF}}} \right] \right] \tag{5.5}$$

Parasitic loss is due to action of the diode's parasitic elements, i.e. series resistance (R_S) and junction capacitance (C_J). Junction loss is a function of the diode's *I-vs-V* curve. The latter two elements are controlled by careful selection of the diode used for the mixer.

Noise figure

Radio reception is largely an issue of *signal-to-noise ratio* (SNR). In order to recover and demodulate weak signals the noise figure (NF) of the receiver is an essential characteristic. The mixer can be a large contributor to the overall noise performance of the receiver. Indeed, the noise performance of the receiver is seemingly affected far out of proportion to the actual noise performance of the mixer. But a study of signals and noise will show (through Friis' equation) that the noise performance of a receiver or cascade chain of amplifiers is dominated by the first two stages, with the first stage being so much more important than the second stage.

Because of the importance of mixer noise performance, a low noise mixer must be designed or procured. In general, the noise figure of the receiver equipped with a diode mixer first stage (i.e. no RF amplifier, as is common in microwave receivers) is:

$$NF = L_C + IF_{NF} \tag{5.6}$$

Where:
NF is the overall noise figure
L_C is the conversion loss
IF_{NF} is the noise figure of the first IF amplifier stage

To obtain the best overall performance from the perspective of the mixer, the following should be observed:

1 Select a mixer diode with a low-noise figure (this will address the junction and parasitic losses).
2 Ensure the impedance match of all mixer ports.
3 Adjust the LO power level for minimum conversion loss (LO power is typically higher than maximum RF power level).

Noise balance

There is noise associated with the LO signal, and that noise can be transferred to the IF in the mixing process. The tendency of the mixer to transfer AM noise to the IF is called its *noise balance*. In some cases, this transferred noise results in loss that is more profound than the simple conversion loss, so should be evaluated when selecting a mixer.

The total noise picture (Fig. 5.12) includes not simply the AM noise sidebands around the LO frequency, but also the noise sidebands around the LO harmonics. The latter can be eliminated by imposing a filter between the LO output and the mixer's LO input. The noise sidebands around the LO itself, however, are not easily suppressed by filtering because they are close in frequency to F_{LO}. The use of a balanced mixer, however, can

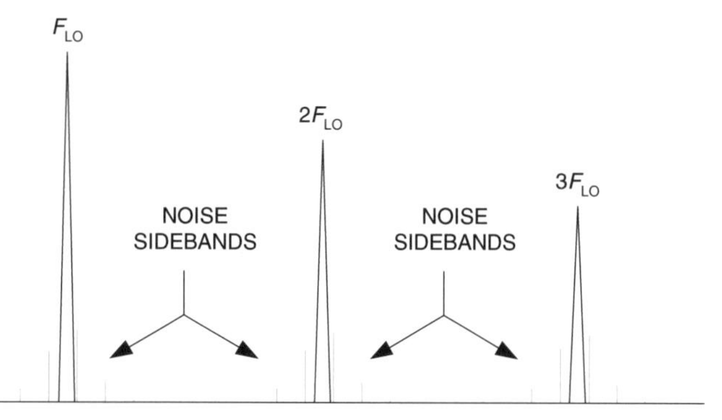

Figure 5.12 Noise sidebands.

suppress all of the LO signal in the output, and that includes the noise sidebands. In the usual way noise balance is specified as the higher the number (in dB) the more suppression of LO AM noise.

Single-ended active mixer circuits

Thus far the only mixer circuit that has been explicitly discussed is the diode mixer. The diode is a general category called a *switching mixer* because the LO switches the diode in and out of conduction. Now let's turn our attention to active single-ended unbalanced mixers.

Figure 5.13 shows the circuit of a simple single-ended unbalanced mixer based on a junction field effect transistor (JFET) such as the MPF-102 or 2N5486. The RF signal is applied to the gate, while the LO signal is applied to the source. If the LO signal has sufficient amplitude to cause non-linear action, then it will permit the JFET to perform as a mixer.

Note that both the RF and LO ports are fitted with bandpass filters to limit the frequencies that can be applied to the mixer. Because these mixers tend to have rather poor LO–RF and RF–LO isolation, these tuned filters will help improve the port isolation by preventing the LO from appearing in the RF output, and the RF from being fed to the output of the LO source.

In many practical cases, the LO filter may be eliminated because it is difficult to make a filter that will track a variable LO frequency. In some cases, the receiver designer will use an untuned bandpass filter, while in others the output of the LO is applied directly to the source of the JFET through either a coupling capacitor or an untuned RF transformer.

The output of the unbalanced mixer contains the full spectrum of $mF_{RF} \pm nF_{LO}$ products, so a tuned filter is needed here also. The drain terminal of the JFET is the IF port in this circuit. The usual case is to use either a double-tuned L–C transformer (T1) as in Fig. 5.13, or some other sort of filter. Typical non-L–C filters used in receivers include ceramic and quartz crystal filters, and mechanical filters.

A MOSFET version of the same type of circuit is shown in Fig. 5.14. In this circuit, a dual-gate MOSFET (e.g. 40673) is the active element. The RF is applied to gate-1, and the LO is applied to gate-2, with the LO signal level being sufficient to drive Q1 into non-

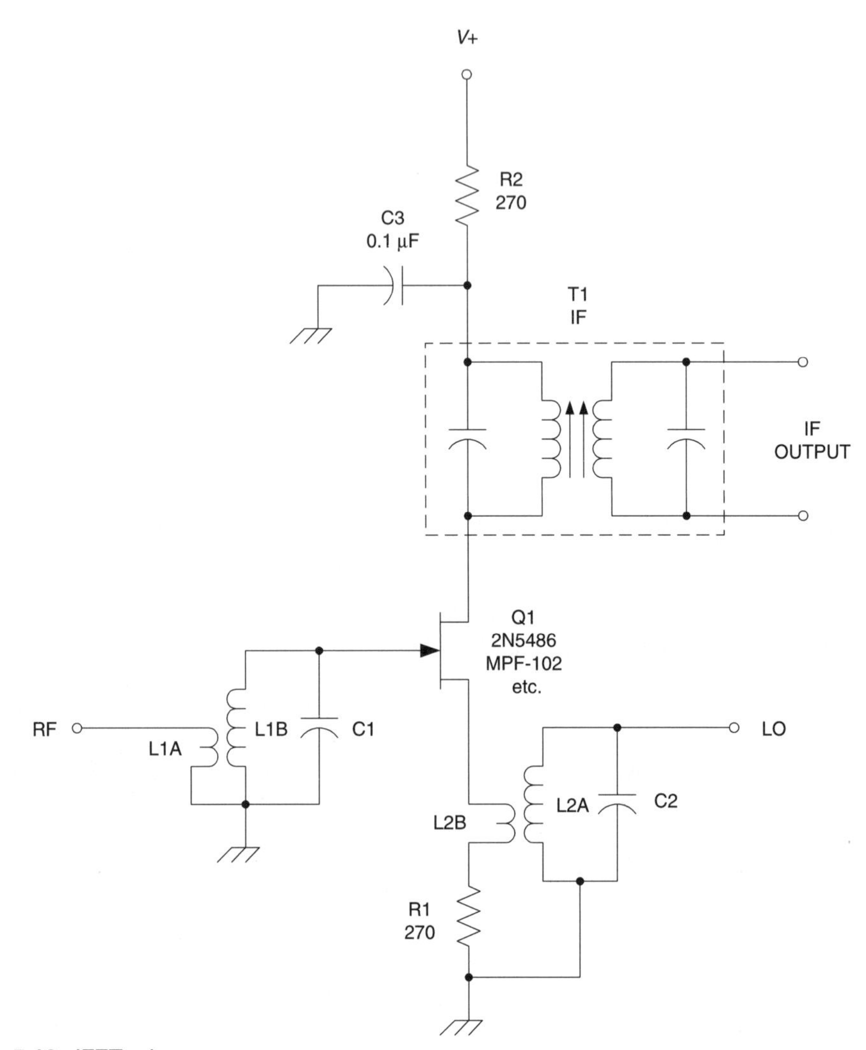

Figure 5.13 JFET mixer.

linear operation. A resistor voltage divider (R3/R4) is used to provide a DC bias level to gate-2. The source terminal is bypassed to ground for RF, and is the common terminal for the mixer.

In this particular case the LO input is broadband, and is coupled to the LO source through a capacitor (C3). The RF input is tuned by a resonant bandpass filter (L1B/C1).

Balanced active mixers

There are a number of balanced active mixers that can be selected. Many of these forms are now available in integrated circuit (IC) form. Because of the intense activity being seen

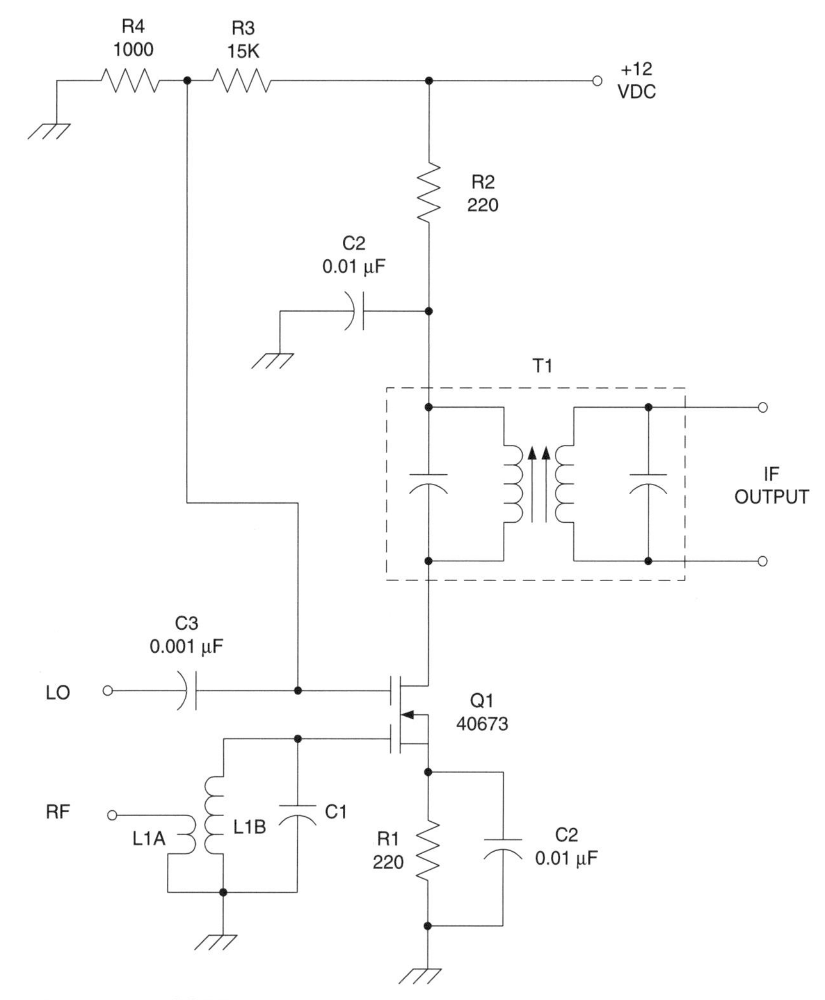

Figure 5.14 Dual-gate MOSFET mixer.

in the development of telecommunications equipment (cellular, PCs and other types), there is a lot of IC development being done in this arena.

One of the earliest types of RF IC on the market was the differential amplifier. Figure 5.15 shows the use of one of these ICs as a mixer stage. Two transistors (Q1 and Q2) are differentially connected by having their emitter terminals connected together to a common current source (Q3). The RF signal is applied to the bases of Q1 and Q2 differentially through transformer T1. The LO signal is used to drive the base terminal of the current source transistor (Q3). The collectors of Q1 and Q2 are differentially connected through a second transformer, T2, which forms the IF port.

Figure 5.16 shows one rendition of the double-balanced mixer used in some HF receivers. It offers a noise figure of about 3 dB. The mixer features a push-pull pair of high pinch-off voltage JFETs (Q1 and Q2) connected in a common source configuration. The LO signal is applied to the common source in a manner similar to Fig. 5.14.

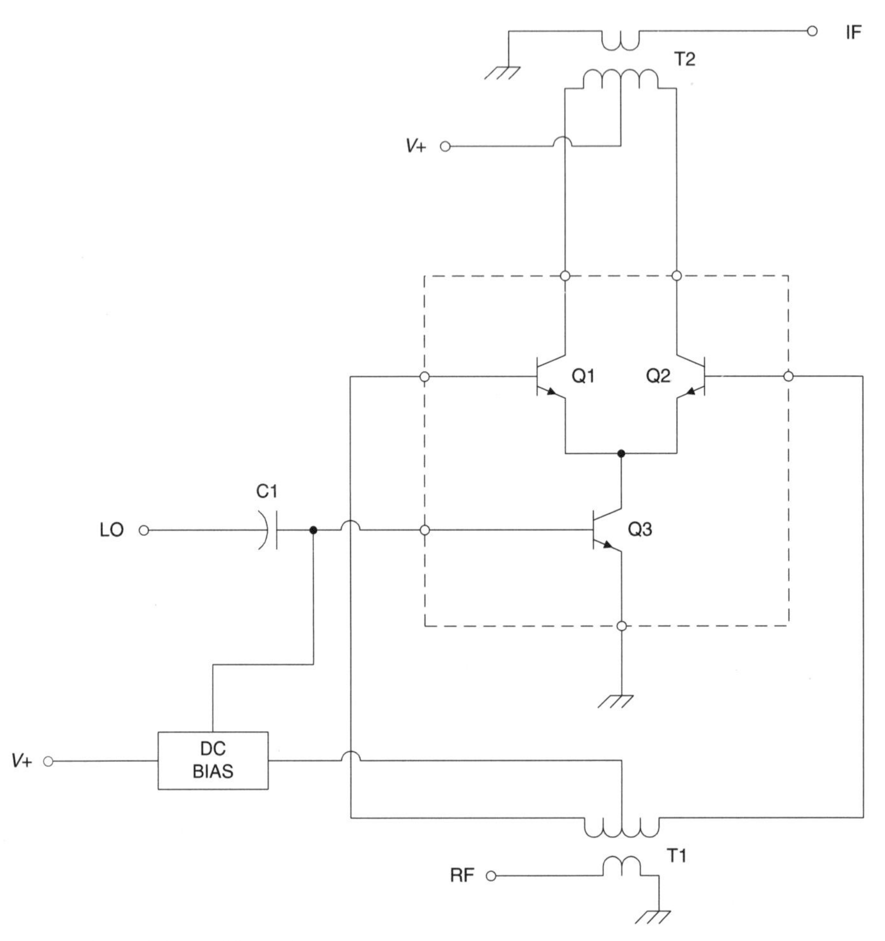

Figure 5.15 Differential amplifier mixer.

The gate circuits of Q1 and Q2 are driven from a balanced transformer, T1. This transformer is trifilar wound, usually on either a toroid or binocular balun core. The dots on the transformer windings indicate the phase sense of the winding. Note that the gate of Q1 is fed from a dotted winding end, while that of Q2 is fed from a non-dotted end. This arrangement ensures that the signals will be 180 degrees out of phase, resulting in the required push-pull action. Some input filtering and impedance matching the 1.5 kohm JFET input impedance to a 50 or 75 ohm system impedance (as need) is provided by L1/C1.

The IF output is similar to the RF input. A second trifilar transformer (T2) is connected such that one drain is to a dotted winding end and the other is to a non-dotted end of T2. Compare the sense of the windings of T1 and T2 in order to avoid signal cancellation due to phasing problems. IF filtering and impedance matching is provided by C5/L2. The tap on L2 is adjusted to match the 5.5 kohm impedance of the JFETs to system impedance.

A MOSFET balanced active mixer is shown in Fig. 5.17. This circuit was discussed in DeMaw and Williams (1981). The dual-gate MOSFET is ideally suited to this type of

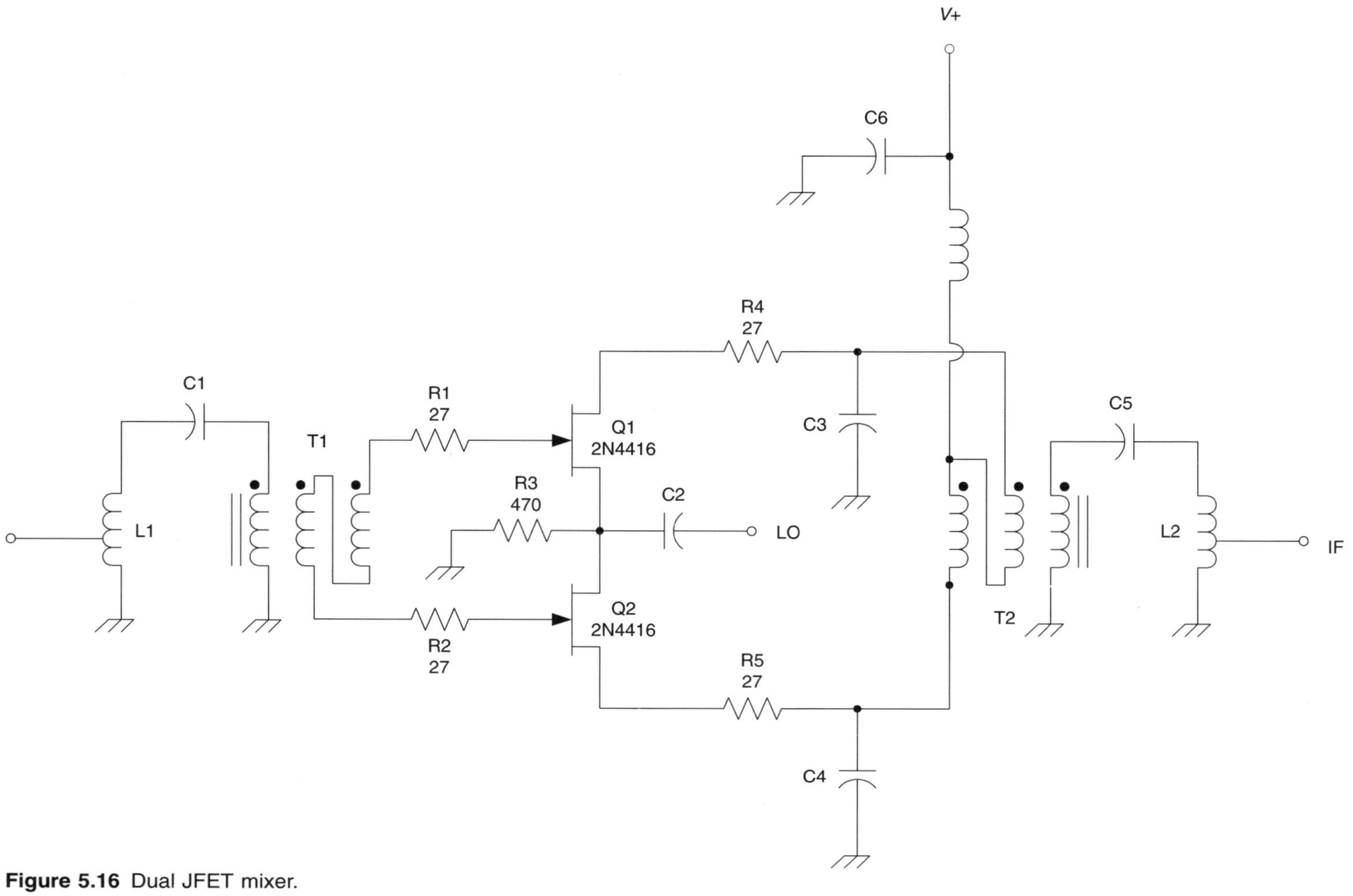

Figure 5.16 Dual JFET mixer.

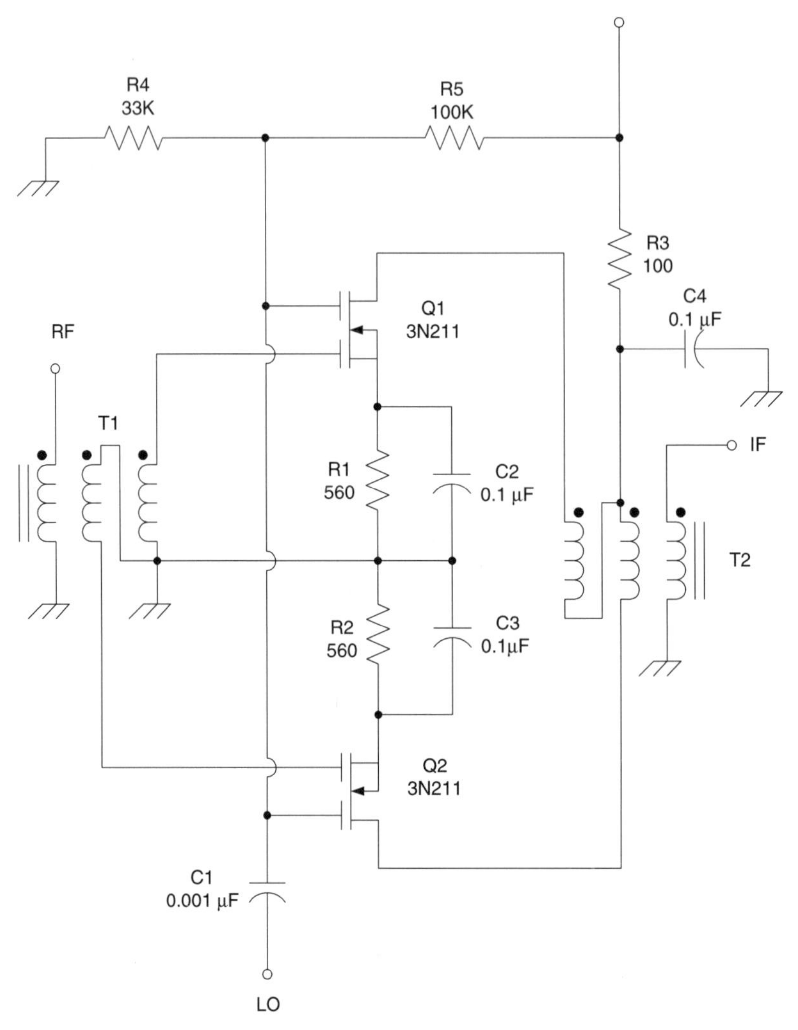

Figure 5.17 Dual MOSFET mixer.

application, but one must be cautious regarding selection. This circuit using 3N211 devices, or their equivalent, will produce low conversion gain (e.g. about 10 dB), but good overall performance. With an LO injection of 8 volts peak-to-peak, and a 10 dBm RF input signal, this circuit will exhibit a respectable third-order intercept point of +17 dBm.

An active double-balanced mixer based on NPN bipolar transistors is shown in Fig. 5.18 (Rohde 1994b). This circuit is usable to frequencies around 500 MHz. Normally, the use of non-IC transistors in a circuit such as this requires matching of the transistors for best performance. That need is overcome by using a bit of degenerative feedback for Q1 and Q2 in the form of unbypassed emitter resistors (R3 and R4).

The base circuits are driven with the LO signal from a balun transformer (T1) in a manner similar to the earlier JFET circuit. The output transformer, however, is rather

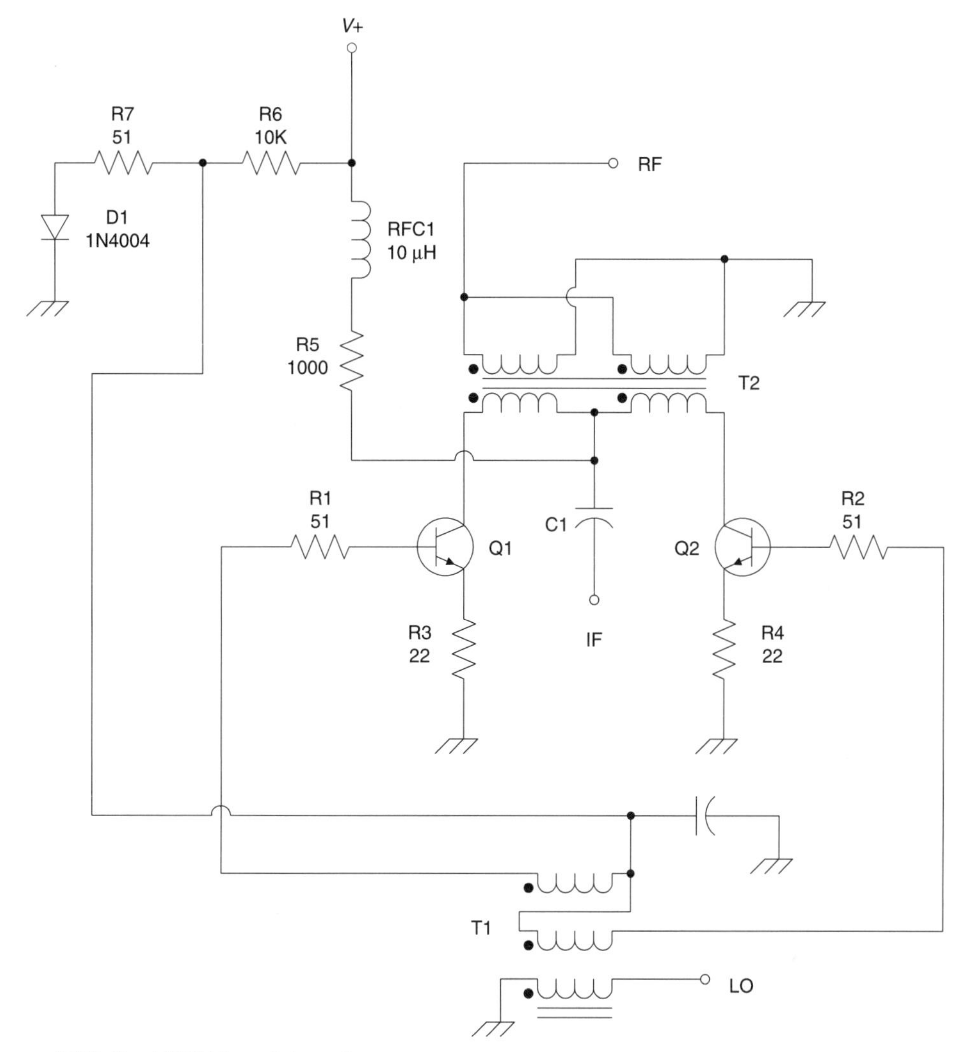

Figure 5.18 Dual NPN transistor mixer.

interesting. It consists of four windings, correctly phased, with the IF being taken from the junction of two of the windings. The RF signal is applied to the remaining two windings of the transformer. This mixer exhibits a third-order intercept of +33 dBm, with a conversion loss of 6 dB, and only 15 to 17 dBm of LO drive power.

Although the use of bipolar transistors can result in an active double-balanced mixer with a high TOIP, there is a distinct trend today towards the use of JFET and MOSFET devices. Typical designs use four active devices. This approach is made easier by the fact that many IC makers are producing RF MOSFET and JFET products that include four matched devices in the same package.

Figure 5.19 shows a mixer circuit based on the use of four JFET devices (Q1–Q4). These transistors are arrayed such that the source terminals of Q1–Q2 are tied together, as are

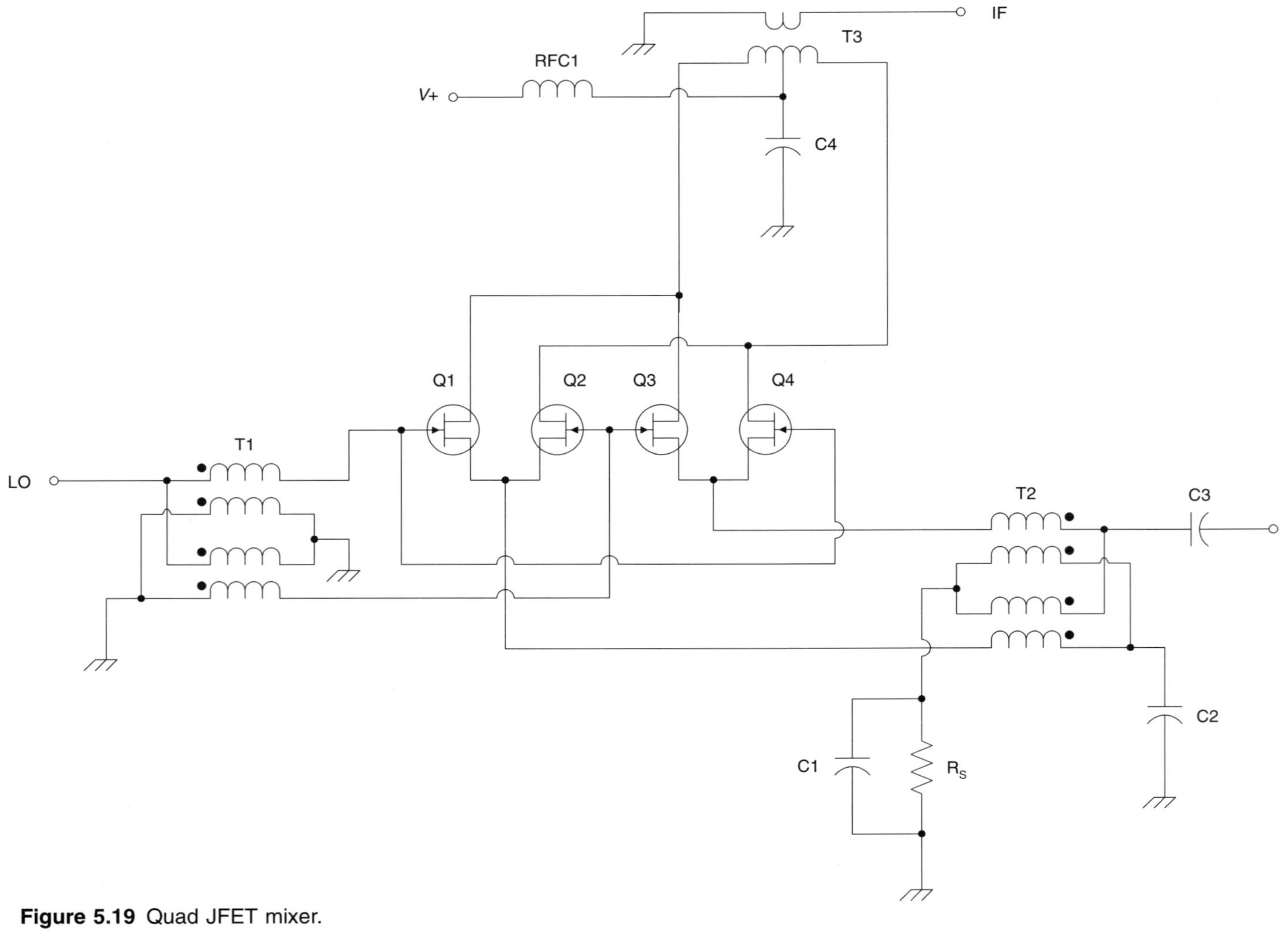

Figure 5.19 Quad JFET mixer.

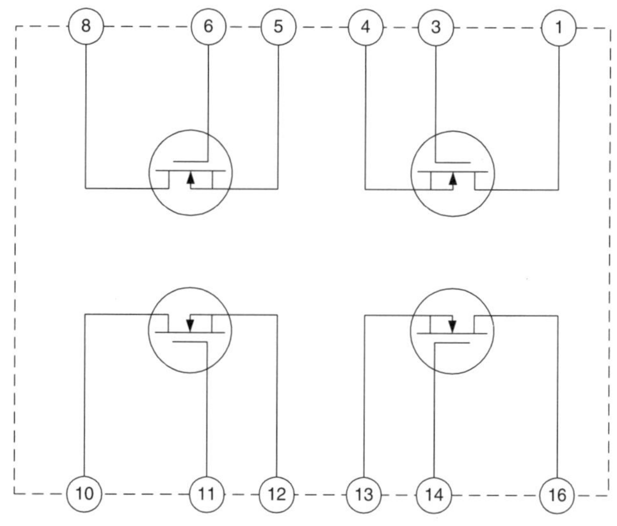

Figure 5.20 Quad MOSFET.

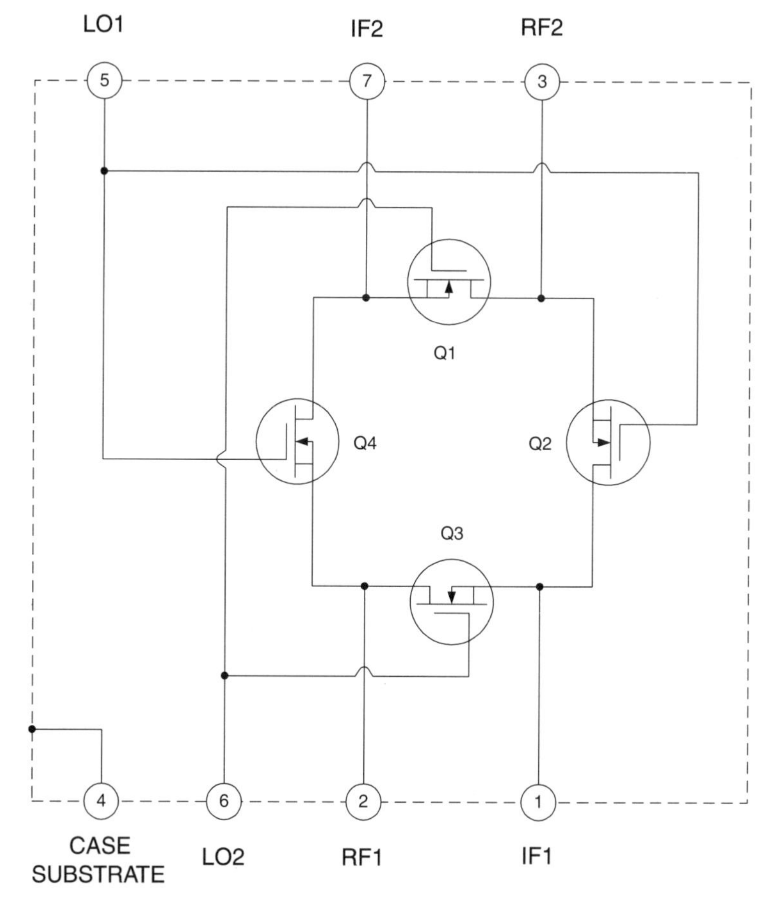

Figure 5.21 8901 device.

the source terminals of Q3–Q4. These source-pair terminals receive the RF input signal from transformer T2. The gates of these transistors are connected such that Q1–Q4 and Q2–Q3 are paired. The LO signal is applied differentially to these gates through transformer T1. The Q1–Q3 and Q2–Q4 drains are tied to the IF output transformer T3.

There are several quad FET ICs on the market that have found favour as mixers in radio receivers. The Siliconix SD5000 DMOS Quad FET is shown in Fig. 5.20. This device contains four DMOS MOSFETs that can be used independently. When connected as a ring, they will form a mixer. Calogic carries the theme a little further in their SD8901 DMOS quad FET mixer IC (Fig. 5.21). The FETs (Q1–Q4) are connected in a ring such that opposite gates are connected together to form two LO ports (LO1 and LO2). The RF signals are applied differentially across drain-source nodes Q1–Q2 and Q3–Q4. Similarly, the IF output is taken from the opposite pair of nodes: Q1–Q4 and Q2–Q3. The SD8901 comes in an eight-pin metal can package.

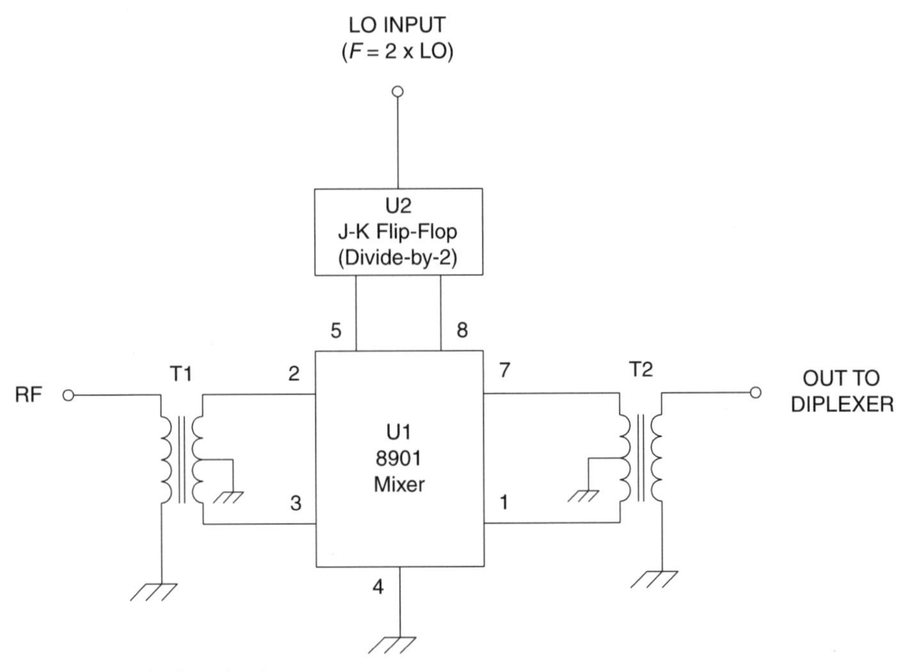

Figure 5.22 8901 device circuit.

The circuit for using the SD8901 (Fig. 5.22) is representative of this class of mixers. The RF output and IF input are connected through transformers T1 and T2, respectively. The LO signal is applied directly to the LO1 and LO2 ports, but requires a J–K flip-flop divide-by -2 circuit. Note that this makes the LO signal a square wave rather than a sine wave. An implication of this circuit is that the LO injection frequency must be twice the expected LO frequency.

Gilbert cell mixers

The *Gilbert transconductance cell* (Fig. 5.23) is the basis for a number of IC mixer (e.g. the NE-602 shown in Fig. 5.24) and analog multiplier (e.g. LM-1496) devices. The circuit consists of two cross-connected NPN pairs fed from a common current source. The RF signal is differentially applied to the transistors that control the apportioning of the current source between the two differential pairs. The LO signal is used to drive the base connections of the differential pairs.

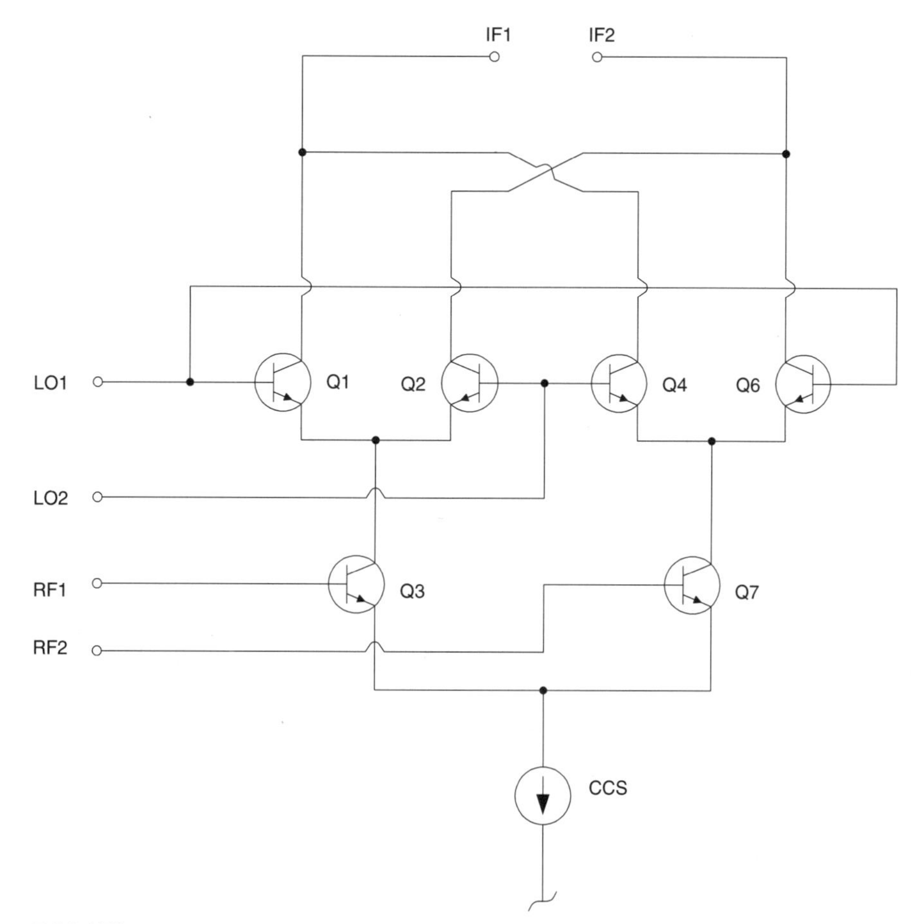

Figure 5.23 Wilson transconductance cell.

IC Gilbert cell devices such as the Philips/Signetics NE-602 are used extensively in low-cost radio receivers. The Gilbert cell is capable of operating to 500 MHz. An on-board oscillator can be used to 200 MHz. One problem seen on such devices is that they often trade off dynamic range for higher sensitivity.

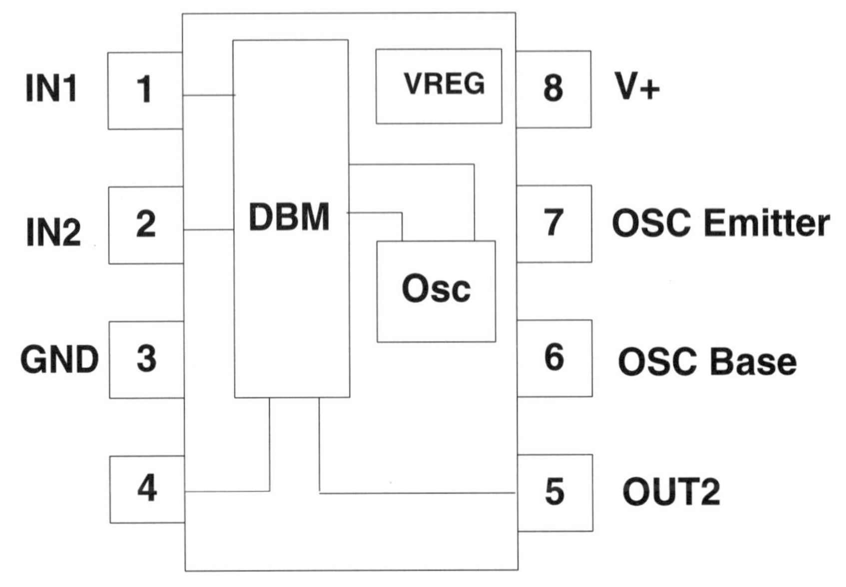

Figure 5.24 NE-602/NE-612 block diagram.

Passive double-balanced mixers

The diode double-balanced mixer (Fig. 5.25) is one of the more popular approaches to DBM design. It has the obvious advantage over active mixers of not requiring a DC power source. This circuit uses a diode ring (D1–D4) to perform the switching action. In the circuit shown in Fig. 5.25 only one diode per arm is shown, but some commercial DBMs use two or more diodes per arm. It is capable of 30 to 60 dB of port-to-port isolation, and is easy to use in practical applications.

With proper design, it is easy to build passive diode DBMs with frequency responses from 1 to 500 MHz, although commercial models are easily obtained into the microwave region. The IF outputs of the typical DBM can be DC to about 500 MHz.

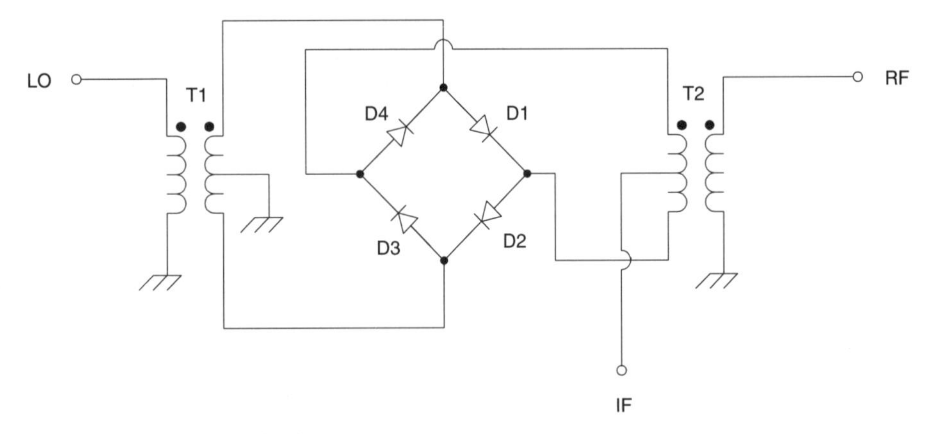

Figure 5.25 Double-balanced mixer.

The diodes used in the ring can be ordinary silicon small-signal diodes such as 1N914 and 1N4148, but these are not as good as hot carrier Schottky diodes (e.g. 1N5820, 1N5821, 1N5822). Whichever diodes are selected, however, they should be matched for use in the circuit because diode differences can deteriorate mixer performance. The usual approach is to match the diode forward voltage drop at some specified standard current such as 5 to 10 mA, depending on the normal forward current rating of the diode. Also of importance is matching the junction capacitance of the diodes.

The DBM in Fig. 5.25 uses two balun transformers, T1 and T2, to couple to the diode ring. The double-balanced nature of this circuit depends on these transformers, and as a result the LO and RF components are suppressed in the IF output.

Diode DBMs are characterized according to their drive level requirements, which is a function of the number of diodes in each arm of the ring. Typical values of drive required for proper mixing action are 0 dBm, +3 dBm, +7 dBm, +10 dBm, +13 dBm, +17 dBm, +23 dBm and +27 dBm.

Figure 5.26 shows the internal circuitry for a commercially available passive DBM made by Mini-Circuits (PO Box 166, Brooklyn, NY, 11235, USA: Phone 714–934–4500; Web site http://www.minicircuits.com). These type no. SBL-x and SRA-x devices are available in a number of different characteristics.

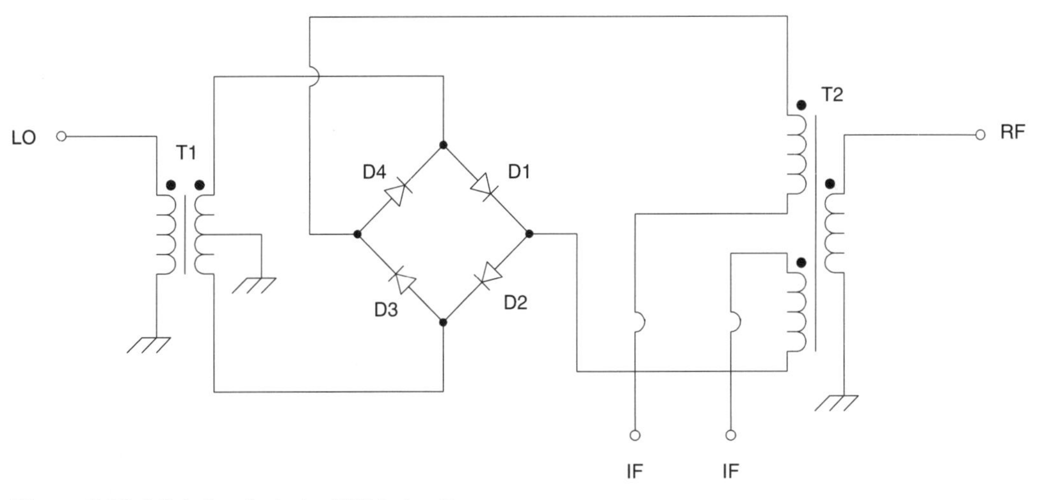

Figure 5.26 Mini-circuits Labs DBM circuit.

The standard package for the SRA/SBL devices is shown in Fig. 5.27. The pins are symmetrical on 5.06 mm (0.20 inch) centres. Pin no. 1 is indicated by a blue dot insulator (the other insulated pins have green insulation). The non-insulated pins are grounded to the case.

The regular SRA/SBL devices use an LO drive level of +7 dBm, and can accommodate RF input levels up to +1 dBm. The devices will work at lower LO drive levels, but performance deteriorates rapidly, so it is not recommended.

Note that the IF output port is split into two pins (3 and 4). Some models tie the ports together, but for others an external connection must be provided for the device to work.

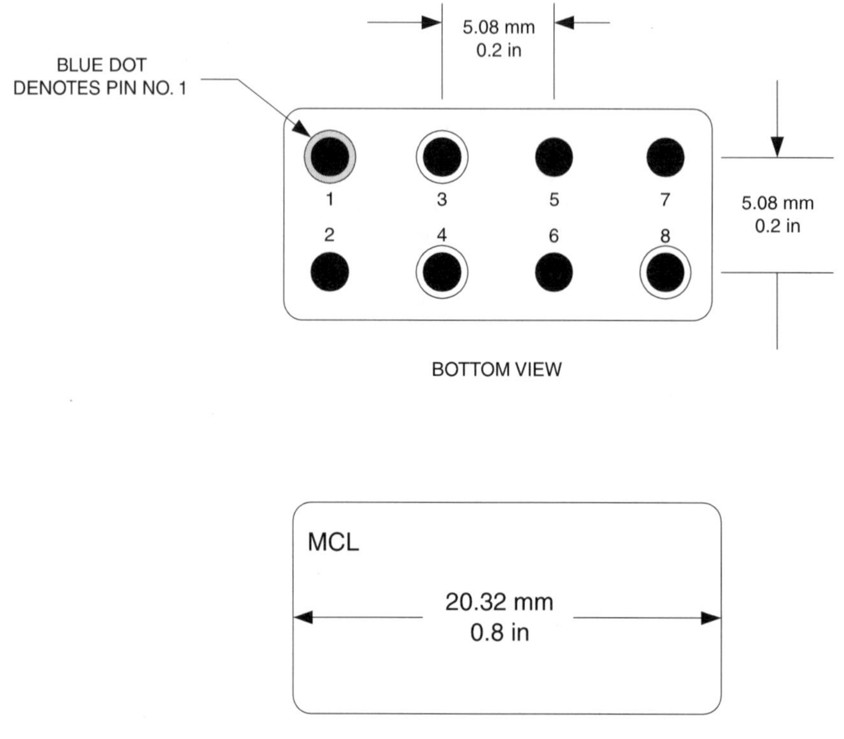

Figure 5.27 SRA/SBL package pin-outs.

The nice thing about this type of commercially available mixer is that the system impedances are already set to 50 ohms. Otherwise, impedance matching would be necessary for them to be used in typical RF circuits. Note, however, that if a circuit or system impedance is other than 50 ohms, then a mismatch loss will be seen unless steps are taken to effect an impedance match.

The mismatch problem becomes considerably greater when the mismatch occurs at the IF port of the mixer. The mixer works properly only when it is connected to a matched resistive load. Reactive loads and mismatched resistive loads deteriorate performance. Figure 5.28 shows a circuit using a passive diode DBM. The diplexer is a critical component to this type of circuit.

Diplexers

The diplexer is a passive RF circuit that provides frequency selectivity at the output, while looking like a constant resistive impedance at its input terminal. Figure 5.29 shows a generalization of the diplexer. It consists of a *high-pass filter* and a *low-pass filter* that share a common input line, and are balanced to present a constant input impedance. With appropriate design, the diplexer will not exhibit any reactance reflected back to the input terminal (which eliminates the reflections and VSWR problem). Yet, at the same time it

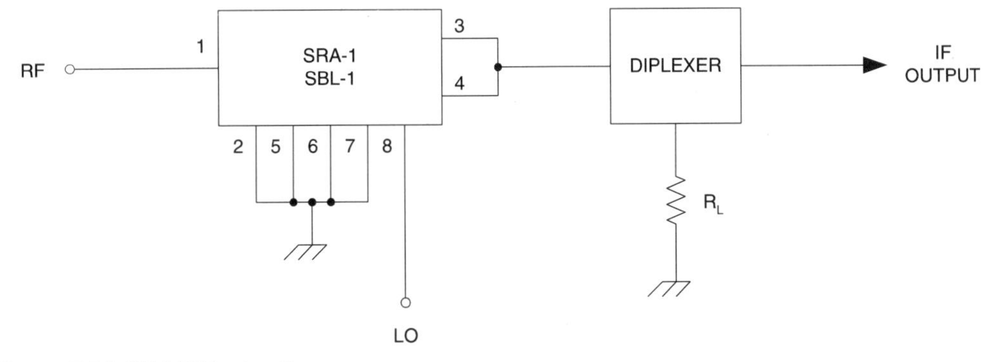

Figure 5.28 SRA/SBL circuit.

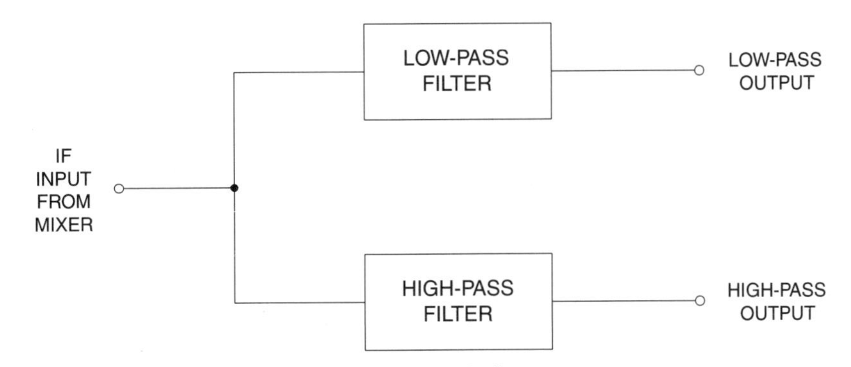

Figure 5.29 High-pass/low-pass diplexer circuit block diagram.

will separate the high and low frequency components into two separate signal channels. The idea is to forward the desired frequency to the output and absorb the unwanted frequency in a dummy load.

Figure 5.30 shows the two cases. In each case, a mixer non-linearly combines two frequencies, $F1$ and $F2$, to produce an output spectrum of $mF1 \pm nF2$, where m and n are integers representing the fundamental and harmonics of $F1$ and $F2$. In some cases, we are interested only in the difference frequency, so will want to use the low-pass output (LPO) of the diplexer (Fig. 5.30A). The high-pass output (HPO) is terminated in a matched load so that signal transmitted through the high-pass filter is fully absorbed in the load.

The exact opposite situation is shown in Fig. 5.30B. Here we are interested in the sum frequency, so use the HPO port of the diplexer, and terminate the LPO port in a resistive load. In this case, the load will absorb the signal passed through the low-pass filter section.

Bandpass diplexers

Figures 5.31 and 5.32 show two different bandpass diplexer circuits commonly used at the outputs of mixers. These circuits use a bandpass filter approach, rather than two separate

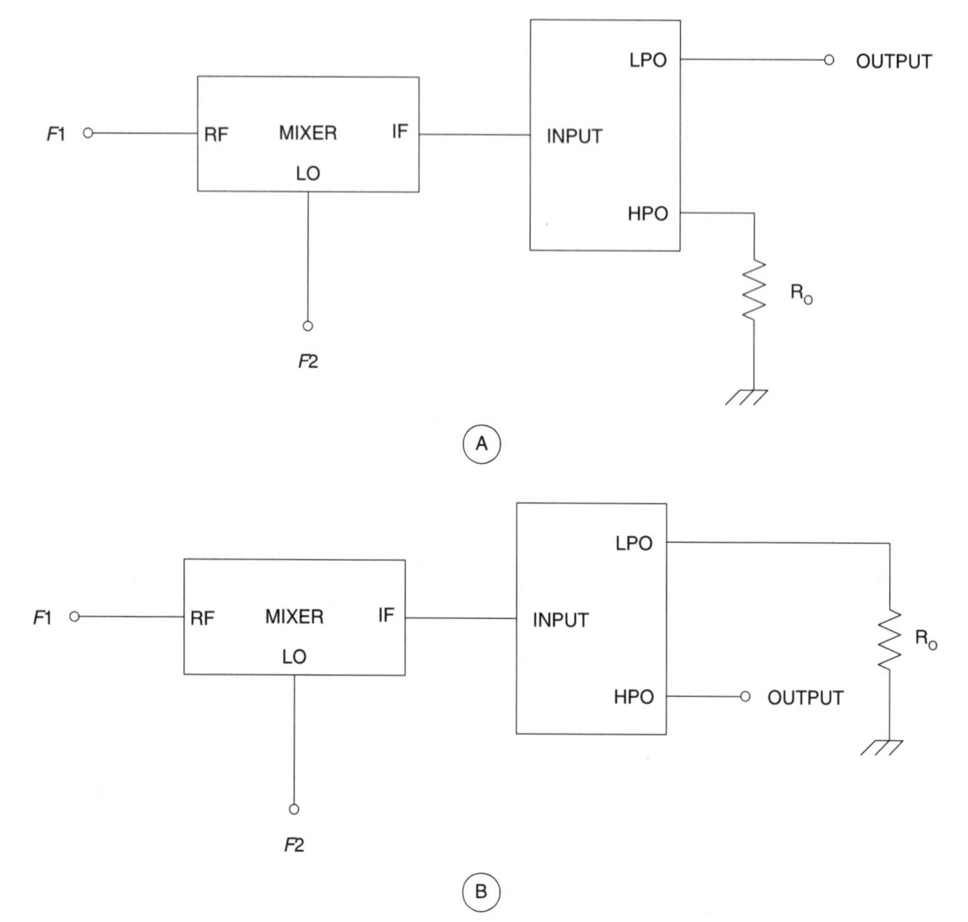

Figure 5.30 Connection of diplexer.

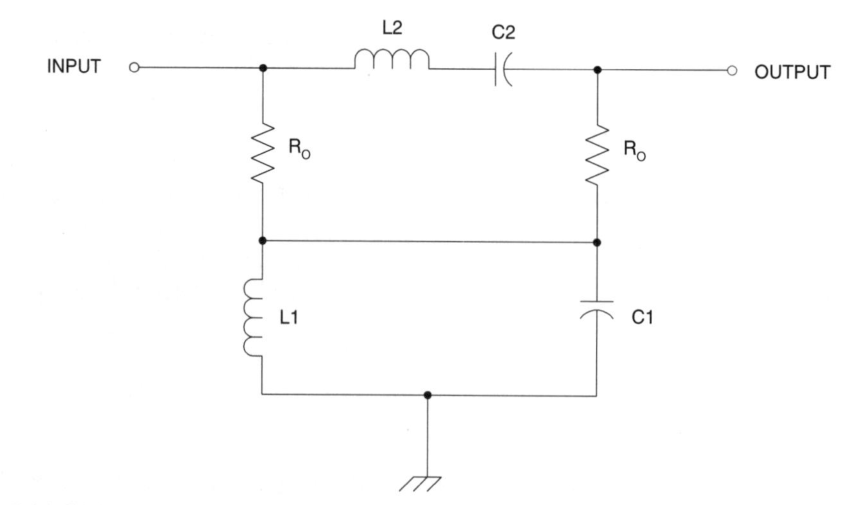

Figure 5.31 Diplexer circuit.

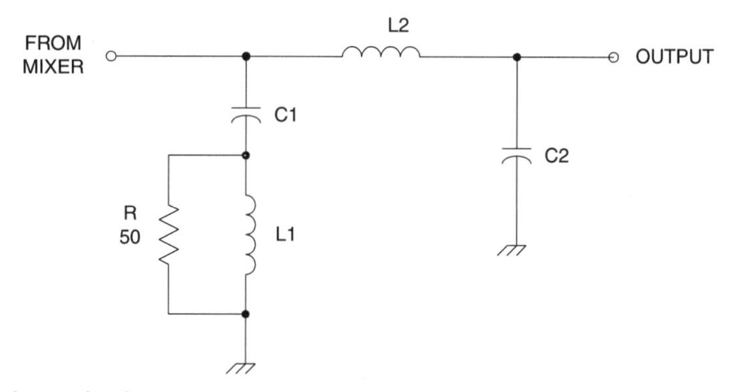

Figure 5.32 Diplexer circuit.

filters. Figure 5.31 is a π-network approach, while the version in Fig. 5.32 is an L-network. In both cases:

$$Q = \frac{f_o}{BW_{3dB}} \tag{5.7}$$

and,

$$\omega = 2\pi f_o \tag{5.8}$$

Where:
f_o is the centre frequency of the pass band in hertz (Hz)
BW_{3dB} is the desired bandwidth in hertz (Hz)
Q is the relative bandwidth

For the circuit of Fig. 5.31:

$$L2 = \frac{R_o Q}{\omega} \tag{5.9}$$

$$L1 = \frac{R_o}{\omega Q} \tag{5.10}$$

$$C2 = \frac{1}{R_o Q \omega} \tag{5.11}$$

$$C1 = \frac{Q}{\omega R_o} \tag{5.12}$$

For the circuit of Fig. 5.32:

$$L2 = \frac{R_o Q}{\omega} \tag{5.13}$$

$$L1 = \frac{R_o}{\omega Q} \tag{5.14}$$

$$C1 = \frac{1}{L1\omega^2} \tag{5.15}$$

$$C2 = \frac{1}{L2\omega^2} \tag{5.16}$$

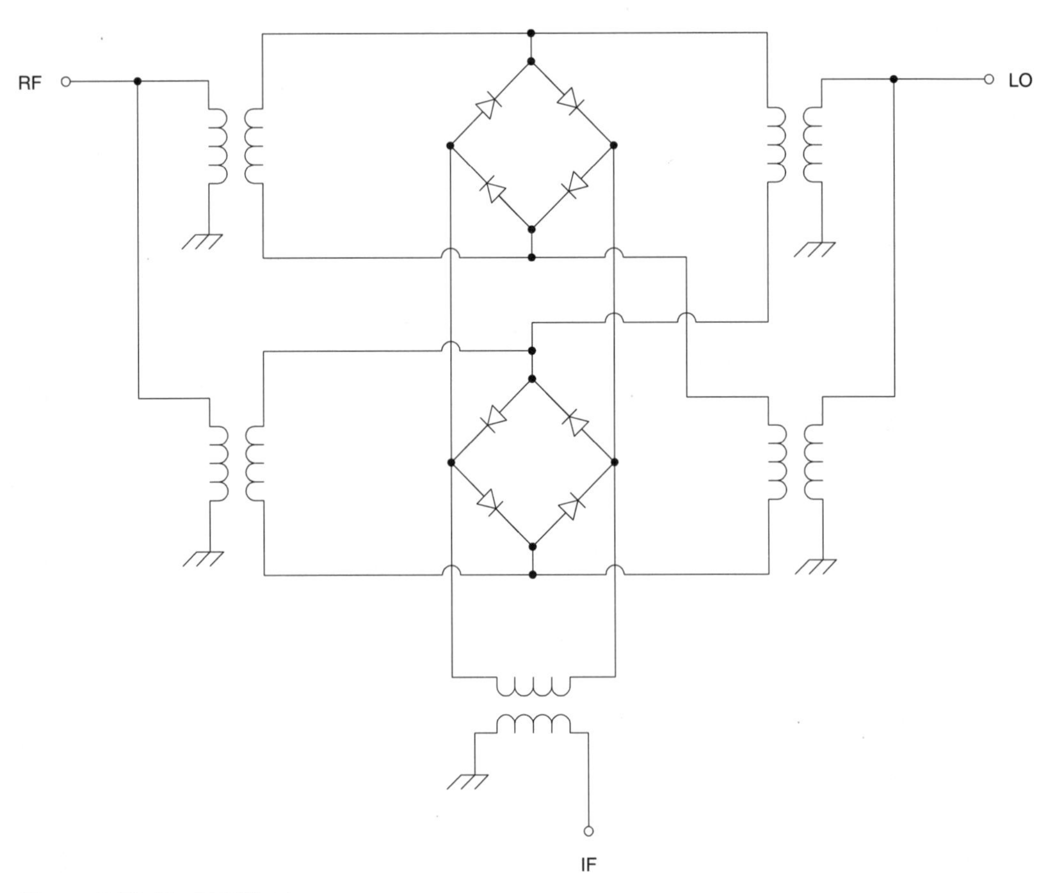

Figure 5.33 Dual DBM mixer.

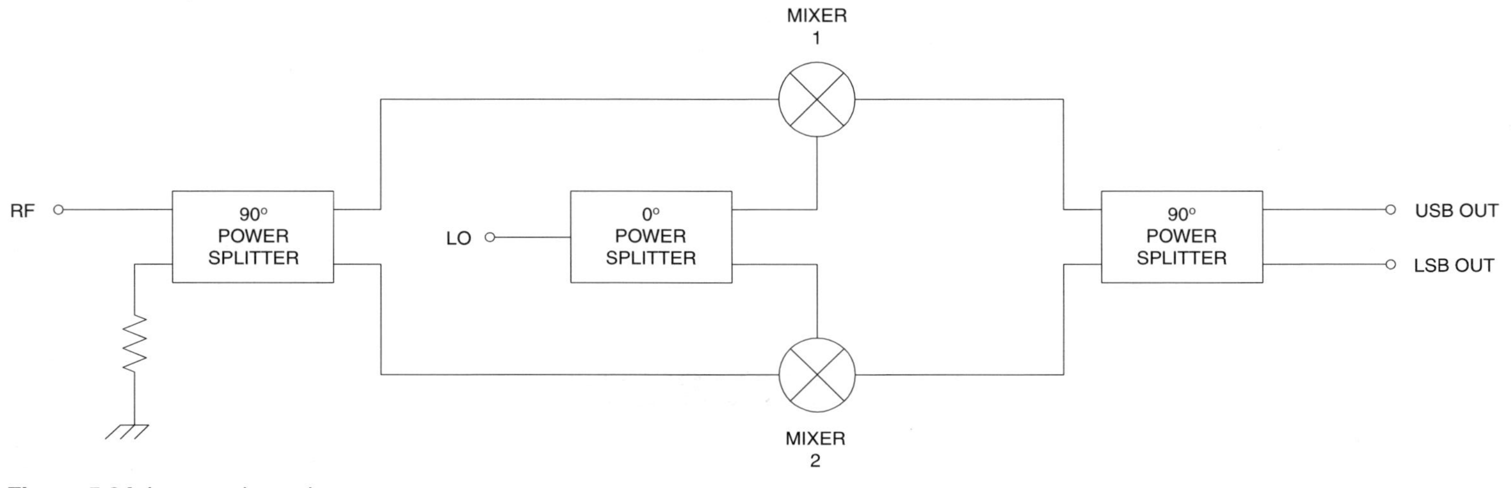

Figure 5.34 Image reject mixer.

Double DBM

The normal passive diode DBM provides relatively high TOIP and –1 dB compression points. It also provides a high degree of port-to-port isolation because the switching action of the diodes in the ring are shut off at the instances where they would feed through the other ports. Where an even higher degree of performance is needed designers sometimes opt for the double DBM as shown in Fig. 5.33.

Image reject mixers

In cases where very good image rejection performance is needed in a receiver, a circuit such as Fig. 5.34 can be used. This circuit uses a pair of passive DBMs, a 0 degree power splitter and two 90 degree power splitters to form an image reject mixer. The LO ports of Mixer-1 and Mixer-2 are driven in-phase from a master LO source. The RF input, however,

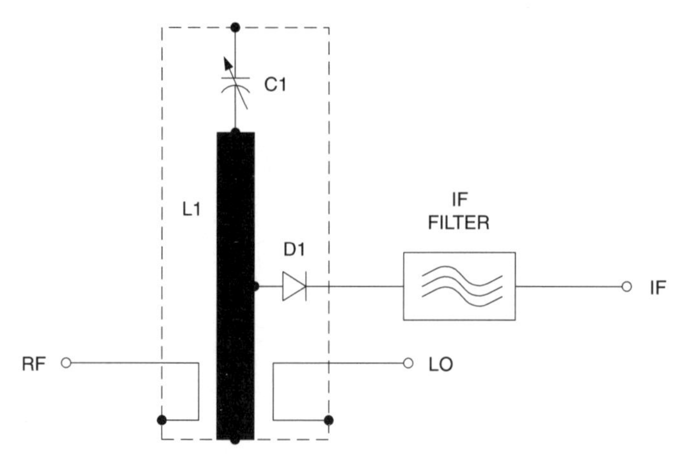

Figure 5.35 Helix-tuned mixer for microwaves.

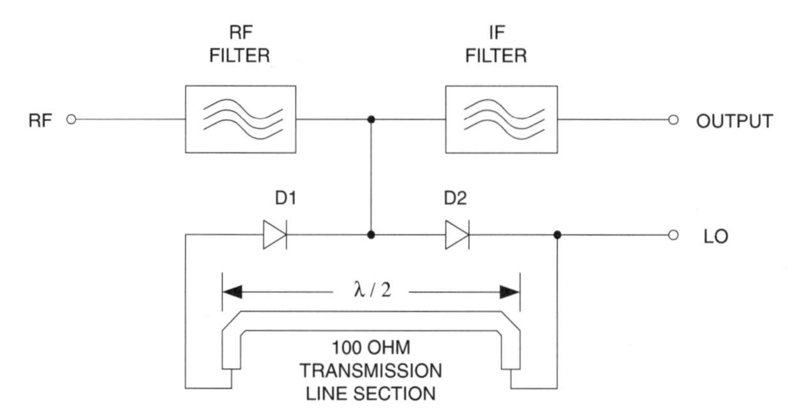

Figure 5.36 Transmission line mixer for microwaves.

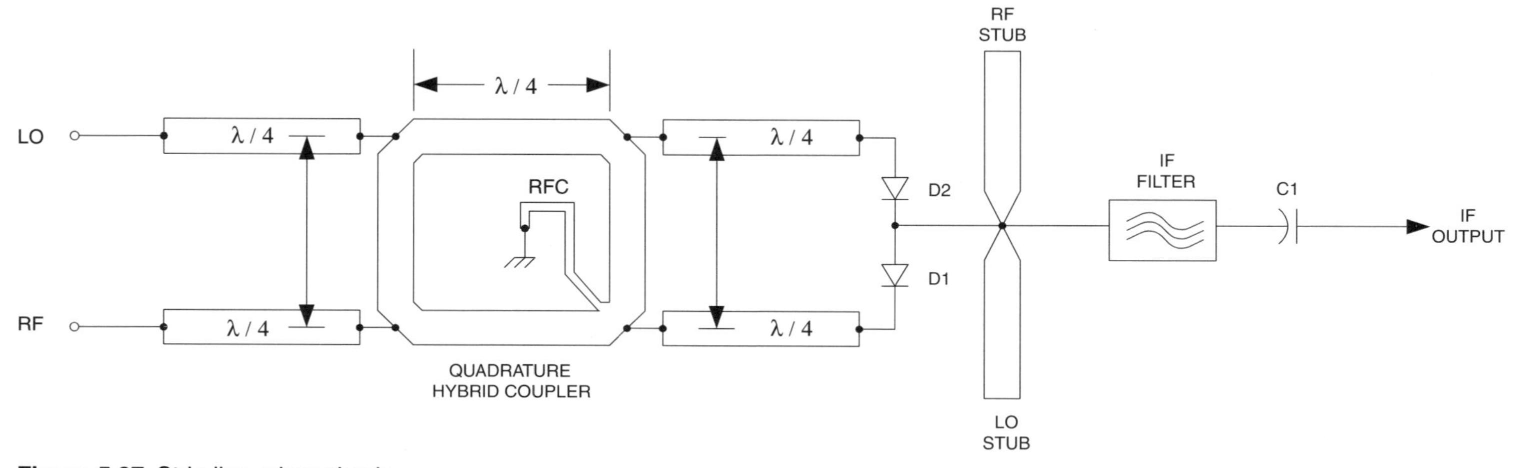

Figure 5.37 Strip-line mixer circuit.

is divided into quadrature signals and applied to the respective RF inputs of the two mixers. The IF outputs of the mixers are then recombined in another quadrature splitter, to form separate USB and LSB IF outputs.

VHF/UHF microwave mixer circuits

When the frequencies used for LO, RF and IF begin to reach into the VHF region and above, design approaches change a bit. Figure 5.35 shows a simple single diode unbalanced mixer. Variants of this circuit have been used in UHF television and other types of receivers. The circuit is enclosed in a shielded space in which a strip inductor (L1) and a variable capacitor (C1) form a resonant circuit. The LO and RF signals are applied to the mixer through coupling loops to L1. A UHF signal diode is connected to L1 at a point that matches its impedance. An IF filter is used to select the mixer product desired for the IF section of the particular receiver.

A single-balanced mixer is shown in Fig. 5.36. This mixer circuit uses two diodes, D1 and D2, connected together and also to the ends of a half-wavelength 100 ohm transmission line. The LO signal is applied to D2 and one end of the transmission line. IF and RF filters are used to couple to the RF and IF ports.

These mixers suffer from RF and LO components appearing in the output. Figure 5.37 shows an improved version that will solve the problem. It is used in the UHF and microwave regions. The LO and RF input signals are applied to two separate ports of a quadrature hybrid coupler. The input and output filtering are made using printed circuit board transmission lines. Each of these is quarter wavelength, although the actual physical lengths must be shortened by the velocity factor of the printed circuit board being used. A printed circuit RF choke ('RFC') is used to provide a return connection for the diodes.

Note the RF and LO stubs at the output of the mixer, prior to the input of the IF filter. These stubs are used to suppress RL and LO components that pass through the mixer.

6 Oscillators

Radio frequency (RF) oscillators can be built using a number of different types of frequency selective resonator. Common types include inductor–capacitor (L–C) networks and quartz crystal resonators. The crystal resonator has by far the best accuracy and stability, but can only be adjusted over a narrow range of frequency.

Feedback oscillators

A feedback oscillator (Fig. 6.1) consists of an amplifier (A1) with an open-loop gain of A_{vol} and a feedback network with a gain (or transfer function) β. It is called a 'feedback oscillator' because the output signal of the amplifier is fed back to the amplifier's own input by way of the feedback network. That it bears more than a superficial resemblance to a feedback amplifier is no coincidence. Indeed, as anyone who has misdesigned or misconstructed an amplifier knows all too well, a feedback oscillator is an amplifier in which special conditions prevail. These conditions are called Barkhausen's criteria for oscillation:

1 Feedback voltage must be in-phase (360 degrees) with the input voltage.
2 The loop gain βA_{vol} must be unity (1).

The first of these criteria means that the total phase shift from the input of the amplifier, to the output of the amplifier, around the loop back to the input, must be 360 degrees (2π radians) or an integer (N) multiple of 360 degrees (i.e. $N2\pi$ radians).

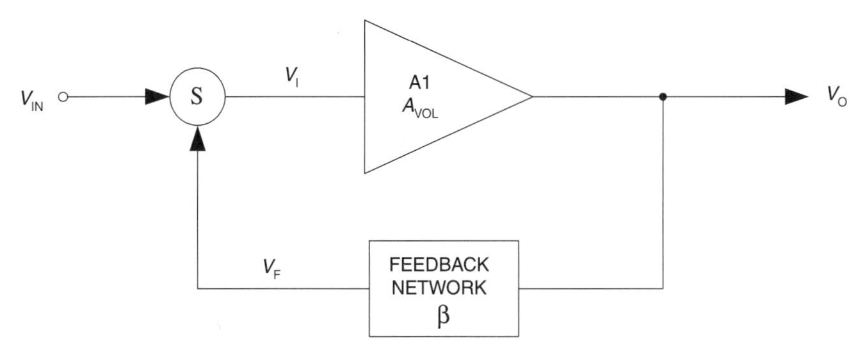

Figure 6.1 Feedback amplifier block diagram.

The amplifier can be any of many different devices. In some circuits it will be a common-emitter bipolar transistor (NPN or PNP devices). In others it will be a junction field effect transistor (JFET) or metal oxide semiconductor field effect transistor (MOSFET). In older equipment it was a vacuum tube. In modern circuits the active device will probably be either an integrated circuit operational amplifier, or some other form of linear IC amplifier.

The amplifier is most frequently an inverting type, so the output is out of phase with the input by 180 degrees. As a result, in order to obtain the required 360 degrees phase shift, an additional phase shift of 180 degrees must be provided in the feedback network at the frequency of oscillation. If the network is designed to produce this phase shift *at only one frequency*, then the oscillator will produce a sine wave output on that frequency.

Looking at it another way, the standard equation for closed loop gain in a feedback amplifier is:

$$A_v = \frac{A_{vol}}{1 - \beta A_{vol}} \tag{6.1}$$

In the special case of an oscillator $V_{in} = 0$ so $A_v \rightarrow \infty$. Implied, therefore, is that the denominator of Eq. (6.1) must also be zero:

$$1 - \beta A_{vol} = 0 \tag{6.2}$$

Therefore, for the case of the feedback oscillator:

$$\beta A_{vol} = 1 \tag{6.3}$$

The term βA_{vol} is the loop gain of the amplifier and feedback network, so Eq. (6.3) meets Barkhausen's second criterion. Thus, when these conditions are met the circuit will oscillate. Hopefully, what we intended to design was an oscillator, and not an amplifier.

In an oscillator amplified noise in the circuit at start-up initiates the oscillation, but it is the feedback voltage that is used to continuously re-excite the crystal or L–C tank circuit to keep it oscillating.

General types of RF oscillator circuits

There are several different configurations for RF oscillators, but the fundamental forms are *Colpitts* and *Hartley*. Figure 6.2 shows the basic difference between these two oscillators. Keep in mind that these are block diagrams, not circuit diagrams, so the apparent 'short' through the coil from the output to ground is not a problem here (there is no DC).

The Colpitts oscillator is shown in Fig. 6.2A. The oscillator is tuned by the resonance between inductor L1 and the combined capacitance of C1 and C2 in series. In actual oscillators there will also be a tuning capacitor in parallel with L1, and the total capacitance used in resonance will include the tuning capacitance, plus C1 and C2 in

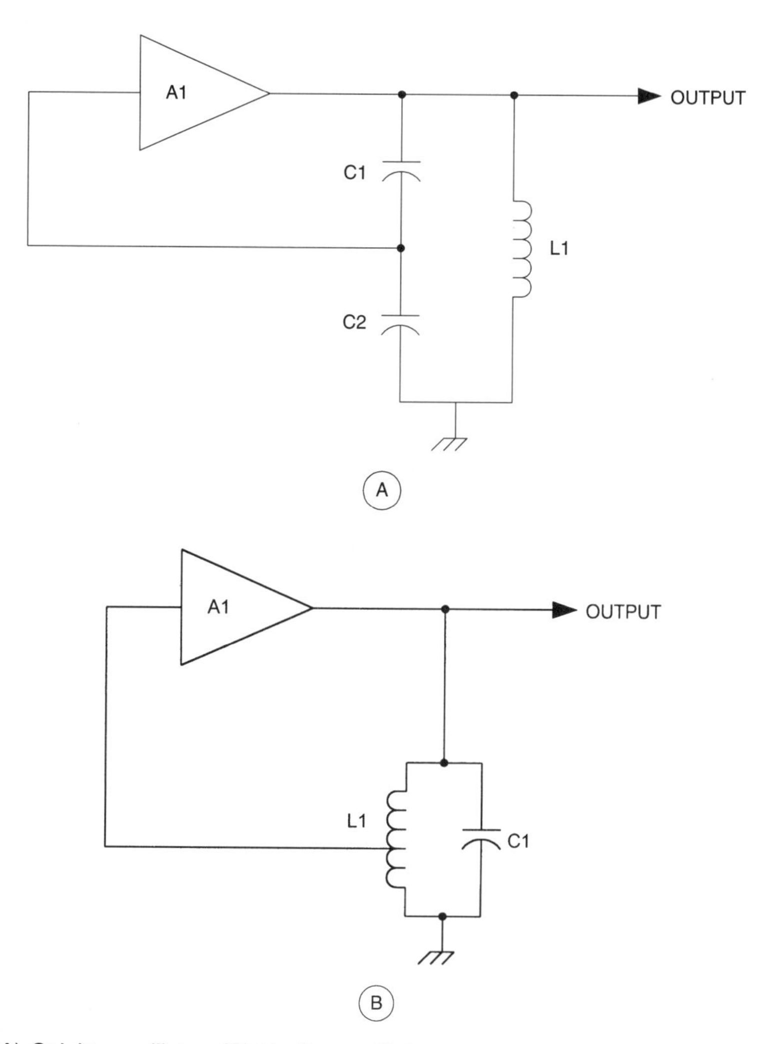

Figure 6.2 (A) Colpitts oscillator; (B) Hartley oscillator.

series. The distinguishing characteristic that identifies the Colpitts oscillator is that the feedback network consists of a tapped capacitive voltage divider (C1/C2). The output of this voltage divider is fed back to the input of the amplifier (A1).

A special variation on the Colpitts theme is the Clapp oscillator. The difference is that the Colpitts uses parallel resonant tuning, while the Clapp uses series resonant tuning. Otherwise, they are both identical (both use the capacitive voltage divider).

The Hartley oscillator is shown in Fig. 6.2B. The tuning is done by an L–C network consisting of L1 and C1. The Hartley oscillator is identified by the fact that the feedback voltage is derived by tapping the tuning inductor L1. There are variations on the Hartley theme that use a tapped coil as part of the feedback network, but a crystal to actually set the frequency of oscillation.

In the past, L–C *variable frequency oscillators* (VFO) were widely used in receivers and transmitters but frequency synthesizers have now largely replaced them, at least in commercial gear. A good place to look for VFO circuits is amateur radio books and journals, such as *QST* (by ARRL in the US) or *Radio Communication* (by RSGB in the UK), particularly if you can get hold of back issues from the 1970s or 1980s.

Piezoelectric crystals

Certain naturally occurring and man-made materials exhibit the property of *piezoelectricity*: Rochelle salts, quartz and tourmaline are examples. Rochelle salts crystals are not used for RF oscillators, although at one time they were used extensively for phonograph

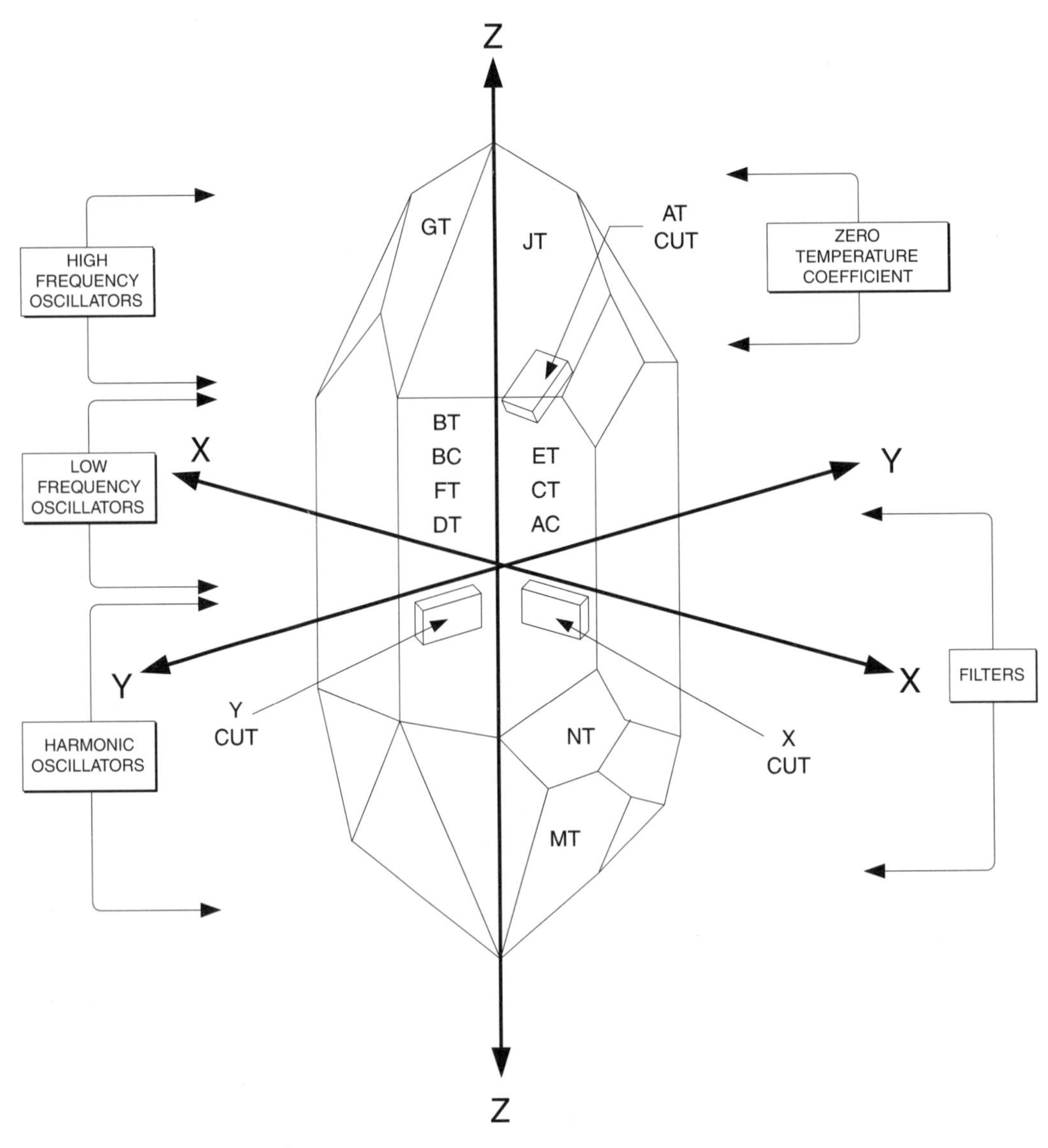

Figure 6.3 Quartz crystal structure.

pick-up cartridges. Tourmaline crystals can be used for some RF applications, but are not often used due to high cost. Tourmaline is considered a semiprecious stone, so tourmaline crystals are more likely to wind up as gemstones in jewellery than radio circuits. That leaves quartz as the preferred material for radio crystals.

Figure 6.3 shows a typical natural quartz crystal. Actual crystals rarely have all of the planes and facets shown. There are three *optical axes* (X, Y and Z) in the crystal used to establish the geometry and locations of various cuts. The actual crystal segments used in RF circuits are sliced out of the main crystal. Some slices are taken along the optical axes, so are called Y-cut, X-cut and Z-cut slabs. Others are taken from various sections, and are given letter designations such as BT, BC, FT, AT and so forth.

Piezoelectricity

All materials contain electrons and protons, but in most materials their alignment is random. This produces a net electrical potential in any one direction of zero. But in crystalline materials the atoms are lined up, so can form electrical potentials. *Piezoelectricity* refers to the *generation of electrical potentials due to mechanical deformation of the crystal*.

Figure 6.4 shows the piezoelectric effect. A zero-centre voltmeter is connected across a crystal slab. At Fig. 6.4A, the slab is at rest, so the potential across the surfaces is zero. In Fig. 6.4B, the crystal slab is deformed in the upward direction, and a positive potential is seen across the slab. When the crystal slab is deformed in the opposite direction, a negative voltage is noted.

If the crystal is mechanically 'pinged' once it will vibrate back and forth, producing an oscillating potential across its terminals, at its resonant frequency. Due to losses the oscillation will die out in short order. But if the crystal is repetitively pinged, then it will generate a sustained oscillation on its resonant frequency.

It is not, however, terribly practical to stand there with a tiny little hammer pinging the crystal all the while the oscillator is running. Fortunately, piezoelectricity also works in the reverse mode: if an electrical potential is applied across the slab it will deform. Thus, if we amplify the output of the crystal, and then feed back some of the amplified output to electrically 're-ping' the crystal, then it will sustain oscillation on its resonant frequency.

Equivalent circuit

Figure 6.5A shows the equivalent R–L–C circuit of a crystal resonator, and Fig. 6.5B shows the impedance-vs-frequency plot for the crystal. There are four basic components of the equivalent circuit: series inductance (L_s), series resistance (R_s), series capacitance (C_s) and parallel capacitance (C_p). Because there are two capacitances, there are two resonances: *series* and *parallel*. The series resonance point is where the impedance curve crosses the zero line, while parallel resonance occurs a bit higher on the curve.

Crystal packaging

Over the years a number of different packages have been used for crystals. Even today there are different styles. Figure 6.6A shows a representation of the largest class of

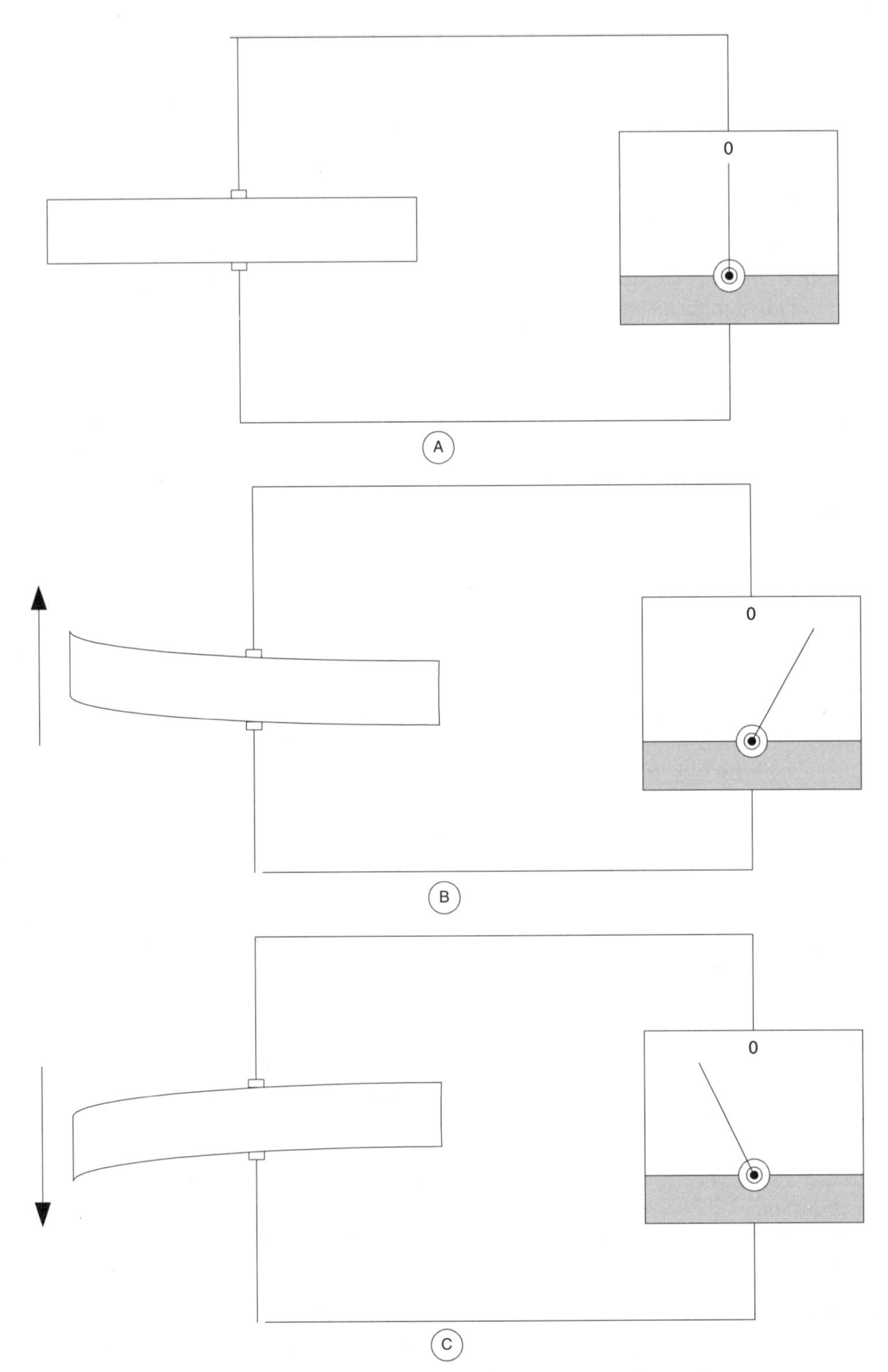

Figure 6.4 Deflection of the crystal causes voltage changes across the surfaces.

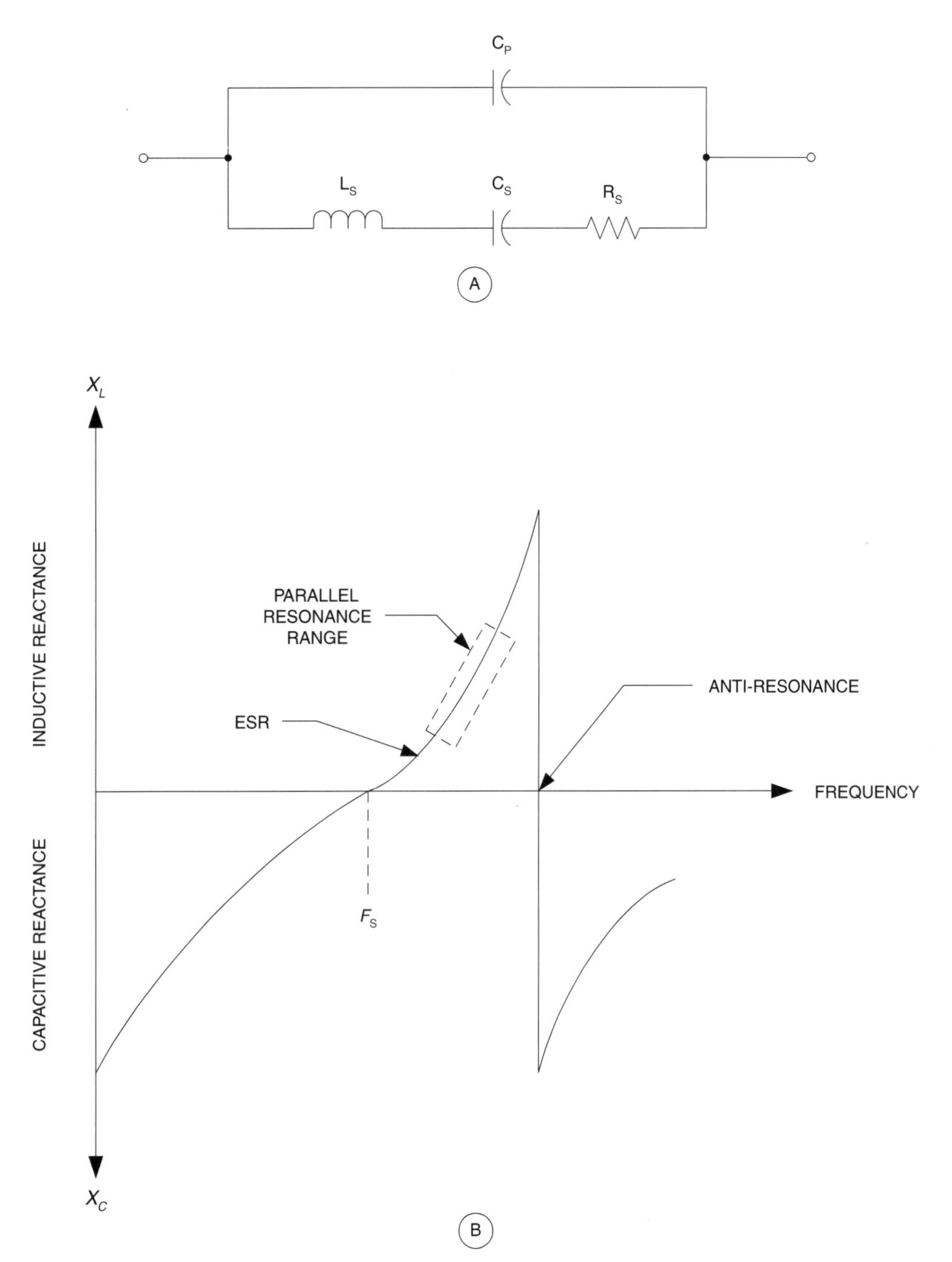

Figure 6.5 (A) Equivalent circuit; (B) impedance relationship.

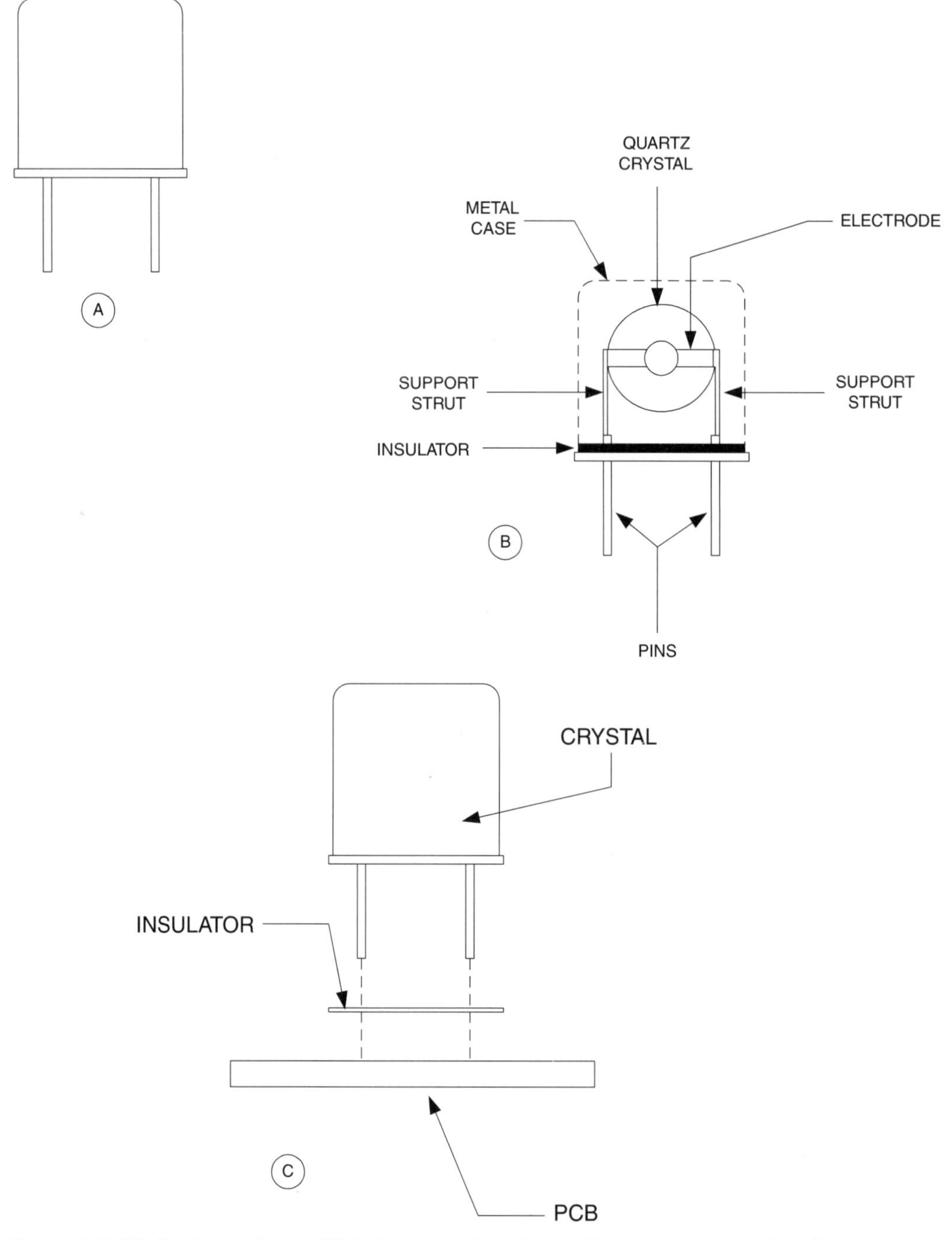

Figure 6.6 (A) Crystal package; (B) inside crystal package, (C) mounting crystal package on dual-sided PWB.

packages. It is a hermetically sealed small metal package, in various sizes. The actual quartz crystal slab is mounted on support struts inside the package (Fig. 6.6B), which are in turn mounted to either a wire header or pins.

Some crystals use pins for the electrical connections, and are typically mounted in sockets. The pin type of package can be soldered directly to a printed circuit board, but care must be taken to keep from fracturing the crystal with heat. Not all pins are easily soldered, although it can help if the pins are scraped to reveal fresh metal before soldering. Normally, however, if the crystal is soldered into the circuit a wire-lead package is used.

Some crystals may short circuit if installed on a printed circuit board with either through via holes or a ground plane on the top side of the board. In those cases, the usual practice is to insert a thin insulator between the PCB and the crystal (Fig. 6.6C).

Temperature performance

There are three basic categories of crystal oscillator: *room temperature crystal oscillators* (RTXO), *temperature-compensated crystal oscillators* (TCXO) and *oven-controlled crystal oscillators* (OCXO). Let's take a look at each of these groups.

Room temperature crystal oscillators

The RTXO takes no special precautions about frequency drift. But with proper selection of crystal cut, and reasonable attention to construction, stability on the order of 2.5 parts per million (ppm), 2.5×10^{-6}, over the temperature range 0°C to 50°C is possible.

Temperature-compensated crystal oscillators

The TCXO circuit also works over the 0°C to 50°C temperature range, but is designed for much better stability. The temperature coefficients of certain components of the TCXO are designed to counter the drift of the crystal, so the overall stability is improved to 0.5 ppm (5×10^{-7}). The cost of TCXOs has decreased markedly over the years to the point where relatively low-cost packaged oscillators can have a rather respectable stability specification.

Oven-controlled crystal oscillators

The best stability is achieved from the OCXO time base. These oscillators place the resonating crystal inside of a heated oven that keeps its operating temperature constant, usually near 70°C or 80°C.

There are two forms of crystal oven used in OCXO designs: *on/off* and *proportional control*. The on/off type is similar to the simple furnace control in houses. It has a snap action that turns the oven heater on when the internal chamber temperature drops below a certain point, and off when it rises to a certain maximum point. The proportional control type operates the heating circuit continuously, and supplies an amount of heating that is proportional to the actual temperature difference between the chamber and the set point. The on/off form of oven is capable of 0.1 ppm (10^{-7}).

OCXOs that use a proportional control oven can reach a stability of 0.0002 ppm (2×10^{-10}) with a 20 minute warm-up and 0.0001 ppm (1.4×10^{-10}) after 24 hours. It is common practice to design equipment to leave the OCTX turned on even when the remainder of the circuit is off. Some portable frequency counters, such as those used in two-way radio servicing, have a battery back-up to keep the OCXO turned on while the counter is in transit.

The variation described above is referred to as the *temperature stability* of the oscillator. We must also consider *short-term stability* and *long-term stability* (aging).

Short-term stability

The short-term stability is the random frequency and phase variation due to noise that occurs in any oscillator circuit. It is sometimes also called either *time domain stability* or *fractional frequency deviation*. In practice, the short-term stability has to be a type of rms value averaged over one second. The short-term stability measure is given as $\sigma(\Delta f/f)(t)$. Typical values of short-term stability are given below for the different forms of oscillator.

RTXO	2×10^{-9} rms	0.002 ppm
TCXO	1×10^{-9} rms	0.001 ppm
OCXO (on/off)	5×10^{-10} rms	0.0005 ppm
OCXO (prop.)	1×10^{-11} rms	0.00001 ppm

Long-term stability

The long-term stability of the oscillator is due largely to crystal aging. The nature of the crystal, the quality of the crystal, and the plane from which the particular resonator was cut from the original quartz crystal are determining factors in defining aging. This figure is usually given in terms of frequency units per month.

RTXO	3×10^{-7}/month	0.3 ppm
TXCO	1×10^{-7}/month	0.1 ppm
OCXO (on/off)	1×10^{-7}/month	0.1 ppm
OCXO (prop.)	1.5×10^{-8}/month	0.015 ppm
OXCO (prop.)	5×10^{-10}/day	0.0005 ppm

Miller oscillators

Miller oscillators are analogous to the tuned input/tuned output variable frequency oscillator because they have a crystal at the input of the active device, and an L–C tuned circuit at the output. Figure 6.7 shows a basic Miller circuit built with a junction field effect transistor (JFET). Any common RF device can be used for Q1 (e.g. the MPF-102). DC bias is provided by R2, which places the source terminal at a potential above ground due to the channel current flowing in Q1. The source must be kept at ground potential for AC, so a bypass capacitor (C4) is provided. The reactance of this capacitor must be less than one-tenth the value of R2 at the lowest intended frequency of operation.

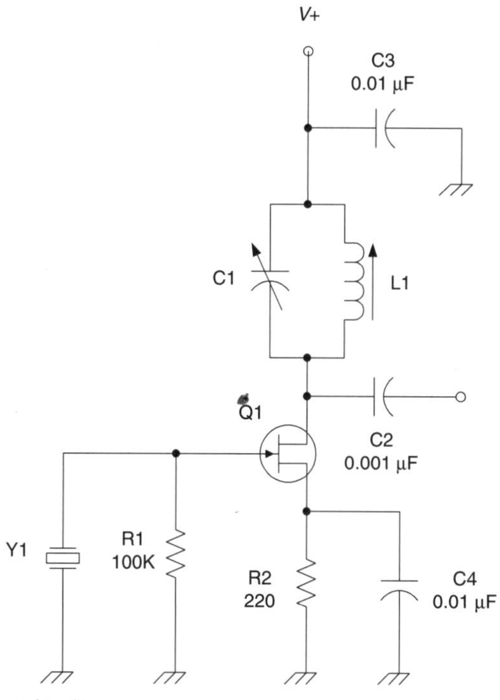

Figure 6.7 Miller oscillator circuit.

The output of the oscillator is tuned by a parallel resonant L–C tank circuit L1/C1. The tuned circuit must be adjusted to the resonant frequency of the oscillator, although best performance usually occurs at a frequency slightly removed from the crystal frequency. If you monitor the output signal level while adjusting either C1 or L1 you will note a distinct difference between the high side and low side of the crystal frequency. Best operation usually occurs at the low side. Whichever is selected, however, care must be taken that the oscillator will start reliably. The output signal can be taken either from capacitor C2 as shown, or through a link coupling winding on L1.

The Miller oscillator circuit of Fig. 6.7 has the advantage of being simple to build, but suffers from some problems as well. One problem is that the feedback is highly variable from one transistor to the next because it is created by the gate-drain capacitance of Q1. There are also output level variations noted, as well as frequency pulling, under output load impedance variations. These are not good attributes for an oscillator! Also, there is a large difference in starting ability between JFETs of the same type number, and different crystals of the same type number from the same manufacturer. I have also noted problems with this circuit when either the JFET or crystal ages. In the case of the JFET, I've seen oscillators that worked well, and then failed. When the JFET was replaced, it started working again. What surprised me was that the JFET tested good!

An improved Miller oscillator is shown in Fig. 6.8. This circuit uses a dual-gate MOSFET transistor such as the 40673 as the active element. It is a fundamental mode oscillator that uses the parallel resonant frequency of the crystal. The crystal circuit is

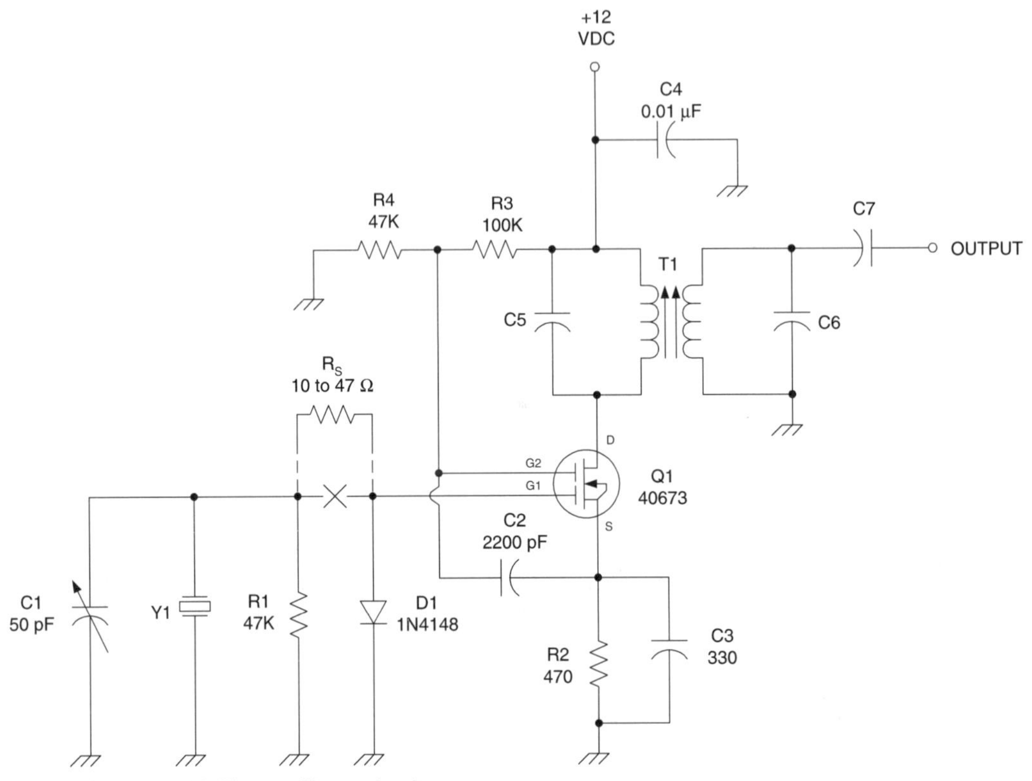

Figure 6.8 Tuned Miller oscillator circuit.

connected to gate-1, while gate-2 is biased to a DC level. This circuit can provide a stability of 15 to 20 ppm if AT-cut or BT-cut crystals are used.

A problem that might be seen with this circuit is parasitic oscillation at VHF frequencies. The MOSFETs used typically have substantial gain at VHF, so could oscillate at any frequency where Barkhausen's criteria are met. There are two approaches to solving this problem. One approach is to insert a ferrite bead on the lead of gate-1 of the MOSFET. The ferrite bead acts like a VHF/UHF RF choke. The second approach, shown in Fig. 6.8, is to insert a snubber resistor R_s between the crystal and gate-1 of the MOSFET. Usually, some value between 10 and 47 ohms will provide the necessary protection. Use the highest value that permits sure starting of the oscillator.

One interesting aspect of the Miller oscillator of Fig. 6.8 is that it can be used as a frequency multiplier (not to be confused with an overtone oscillator) if the tuned network in the drain circuit of Q1 is tuned to an integer multiple of the crystal frequency.

Pierce oscillators

The Pierce oscillator is characterized by the crystal being connected between the output and input of the active device.

An example using a bipolar NPN transistor is shown in Fig. 6.9. This circuit includes a capacitor (C1) for pulling the crystal a small amount in order to precisely tune the frequency.

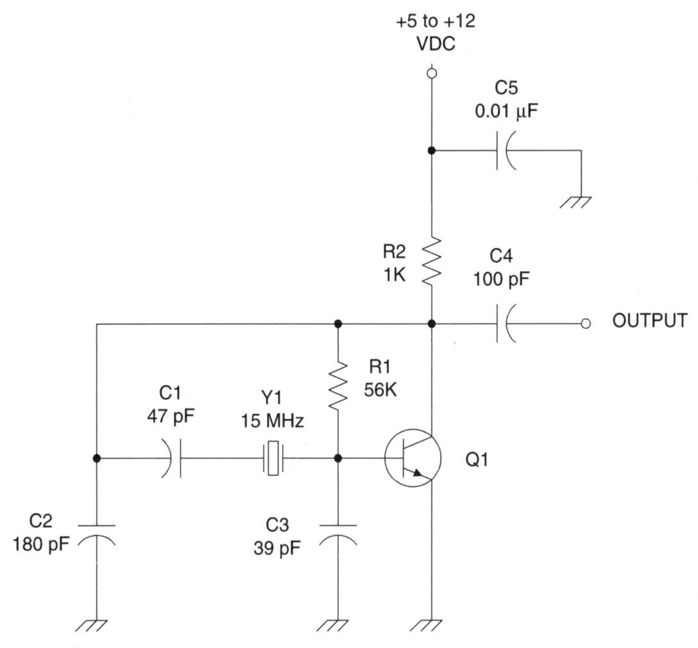

Figure 6.9 Pierce oscillator.

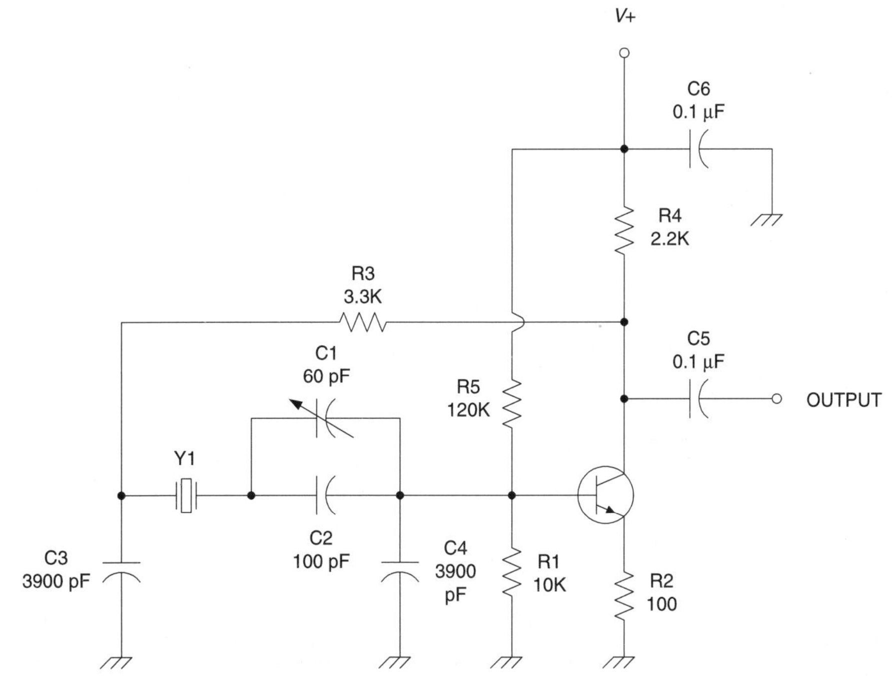

Figure 6.10 LF Pierce oscillator.

This circuit is designed for frequencies between 10 MHz and 20 MHz with the capacitance values shown. If the output is lightly loaded (by keeping C4 small), then the oscillator will provide a reasonable output stability at a level of near 0 dBm. Figure 6.10 is a variation on the theme that will work in the 50 kHz to 500 kHz region. This circuit is very similar to Fig. 6.9 except for increased capacitance values to account for the lower frequency. In both circuits ordinary NPN devices such as the 2N2222 can be used successfully.

Butler oscillators

The Butler oscillator looks superficially like the Colpitts in some manifestations (Fig. 6.11). The difference is that the crystal is connected between the tap on the feedback network and the emitter of the transistor. This particular circuit is a series mode oscillator. The value of R1 should be whatever value between 100 and 1000 ohms will result in reliable oscillation and starting, while minimizing crystal dissipation. A table of capacitance values for feedback network C1/C2 is provided. For the 3 to 10 MHz range, use C1 = 47 pF and C2 = 390 pF; for 10 to 20 MHz select C1 = 22 pF and C2 = 220 pF.

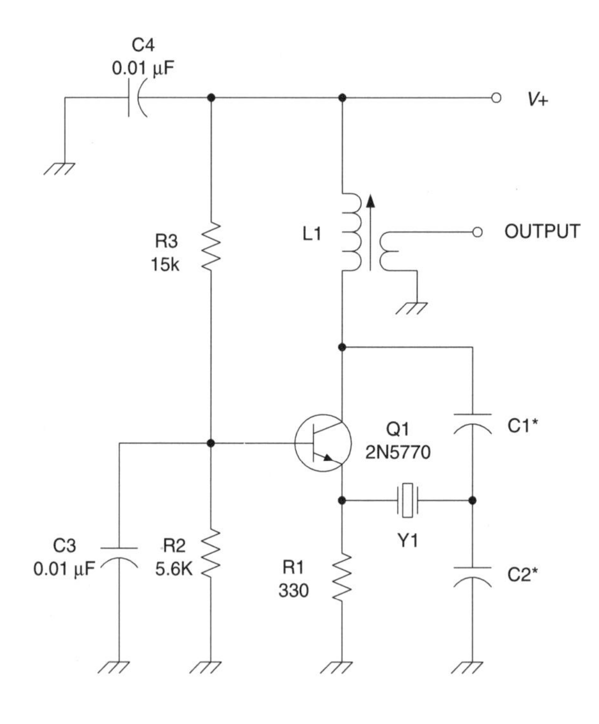

FREQ (MHz)	C1 (pF)	C2 (pF)
3–10	47	390
10–20	22	220

Figure 6.11 Butler oscillator.

The collector circuit is tuned by the combination of C1 and L1. This circuit may well oscillate with the crystal shorted, and care must be taken to ensure that the 'free' oscillation and the crystal oscillation frequencies are the same. The crystal should take over oscillation when it is in the circuit.

The Butler oscillator of Fig. 6.11 is capable of 10 to 20 ppm stability if a buffer amplifier with good isolation is provided at the output. Otherwise, some frequency pulling with load variations might be noted.

The output signal is taken from a coupling winding over L1. This winding is typically only a few turns of wire on one end of L1. Alternatively, a tap on L1 might be provided, and the tap connected to a low-value capacitor. That approach might change some resonances unless care is taken. Another output scheme is to connect a small value capacitor to the collector of Q1. Keep the value low so as to reduce loading, and also to reduce the effects of the output capacitor on the resonance of L1/C1.

A somewhat more complex Butler oscillator is shown in Fig. 6.12. This circuit is sometimes called an aperiodic oscillator circuit. It uses two additional transistors to provide buffering; one of these also serves as part of the feedback circuit. This circuit will operate in the frequency range from about 300 kHz to 10 MHz, although some care must be taken in the selection of the transistor.

Many low frequency crystals exhibit a lower equivalent series resistance (ESR) in one of the higher frequency overtone modes of oscillation than in the fundamental mode, so you might find this circuit oscillating at the wrong frequency. The key to preventing this problem is to use a transistor with a low gain-bandwidth product (e.g. 2N3565 or equivalent).

At this point it might be wise to point out a fact about 'universal' replacement lines of transistors. Because crystal oscillators may operate in an unwanted overtone mode (i.e. at a higher frequency), or because of stray L–C components they may parasitically oscillate on a VHF or UHF frequency, you will want to keep the gain-bandwidth (GBW) product of the active device low. But many replacement lines use a single high frequency transistor with similar gain, collector current and power dissipation ratings as a 'one-size fits all' replacement for transistors with lower GBW products.

I've seen that situation in service replacements on older equipment. The original component may not be available, so a universal 'service shop replacement' line device is selected. It is then discovered that there are parasitic oscillations and other problems because the new replacement has a GBW of 200 MHz, where the old device was a 50 MHz transistor. This problem can show up especially severely in RF amplifiers and low frequency oscillators where L–C components naturally exist, or any circuit where the stray and distributed L–C elements provide the required phase shift on some frequency above the unity GBW point.

The circuit of Fig. 6.12 produces a sine wave output, but not without relatively strong harmonic output. The second and third harmonics are particularly evident. However, if harmonics are desired (as when the oscillator is used in a frequency multiplier circuit), then strong harmonics up to 30 MHz can be generated from a 100 kHz crystal if R5 is reduced to about 1000 ohms.

The output of this oscillator is taken through an emitter follower buffer stage. This circuit can be used as a general buffer for a number of oscillator circuits. It is generally a good practice to use a buffer amplifier with any oscillator in order to reduce loading and smooth out load impedance variations.

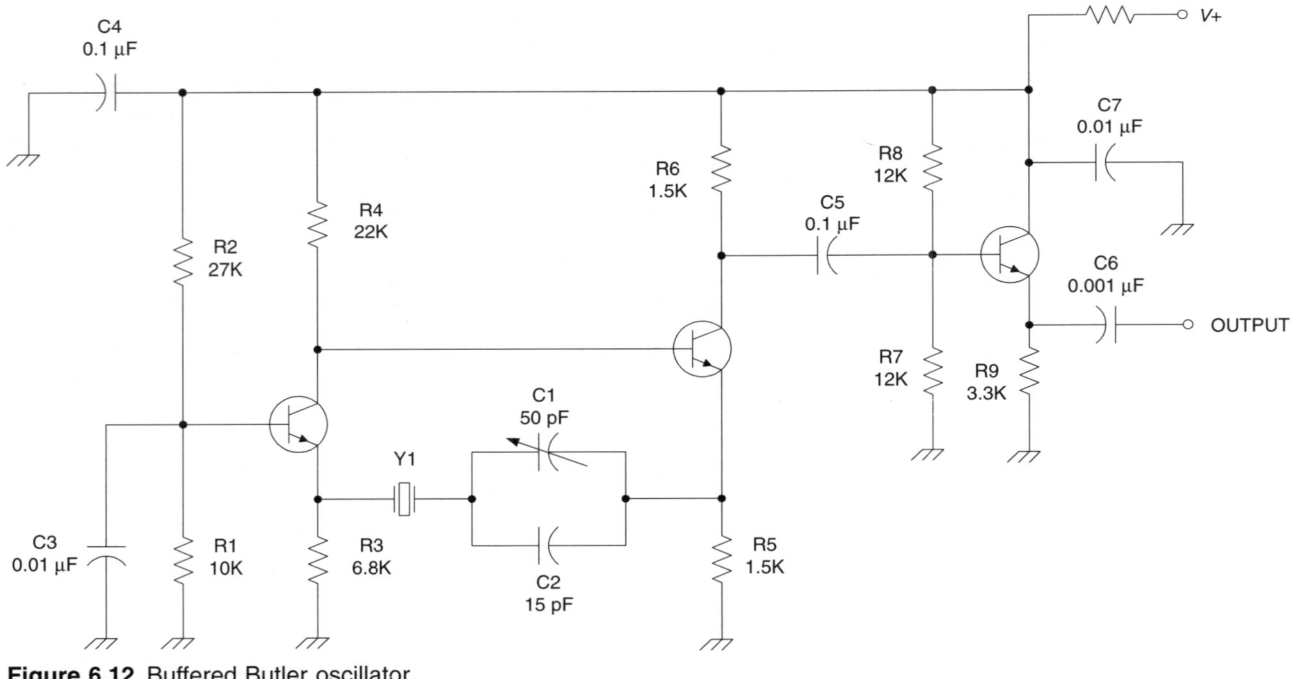

Figure 6.12 Buffered Butler oscillator.

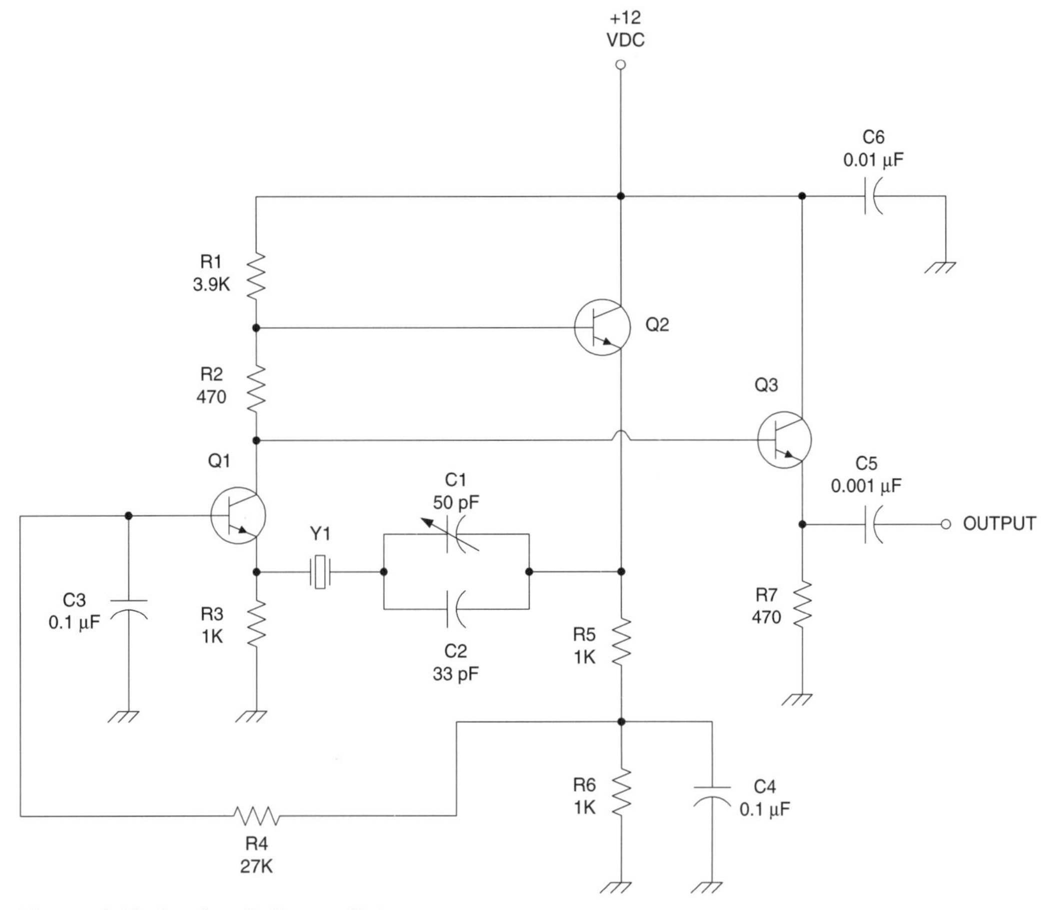

Figure 6.13 Another Butler oscillator.

Another variation on the Butler theme is shown in Fig. 6.13. This circuit is similar to Fig. 6.12, but is a bit less sensitive to frequency pulling due to DC power supply voltage variations. However, it is good engineering practice to use a separate voltage regulator for all oscillator circuits in order to prevent such variation. The availability of low cost three-terminal integrated circuit voltage regulators makes it quite easy to make this happen.

An improved Butler oscillator is shown in Fig. 6.14. This circuit is based on Fig. 6.12. Both circuits can be used at frequencies from LF up to the mid-HF region (about 12 to 15 MHz) if appropriate values of R3 and R5 are used. The improvement of Fig. 6.14 over Fig. 6.12 is the limiting diodes (D1 and D2) provided between the two oscillator transistors (Q1 and Q2). These diodes can be ordinary 1N4148 small-signal diodes. They reduce crystal dissipation, which leads to improved stability and more sure cold starting.

The Butler oscillators above are series mode circuits, but because of the series capacitors are able to use parallel mode crystals. For a strictly series mode circuit eliminate the capacitors in series with the crystal and replace them with a short circuit.

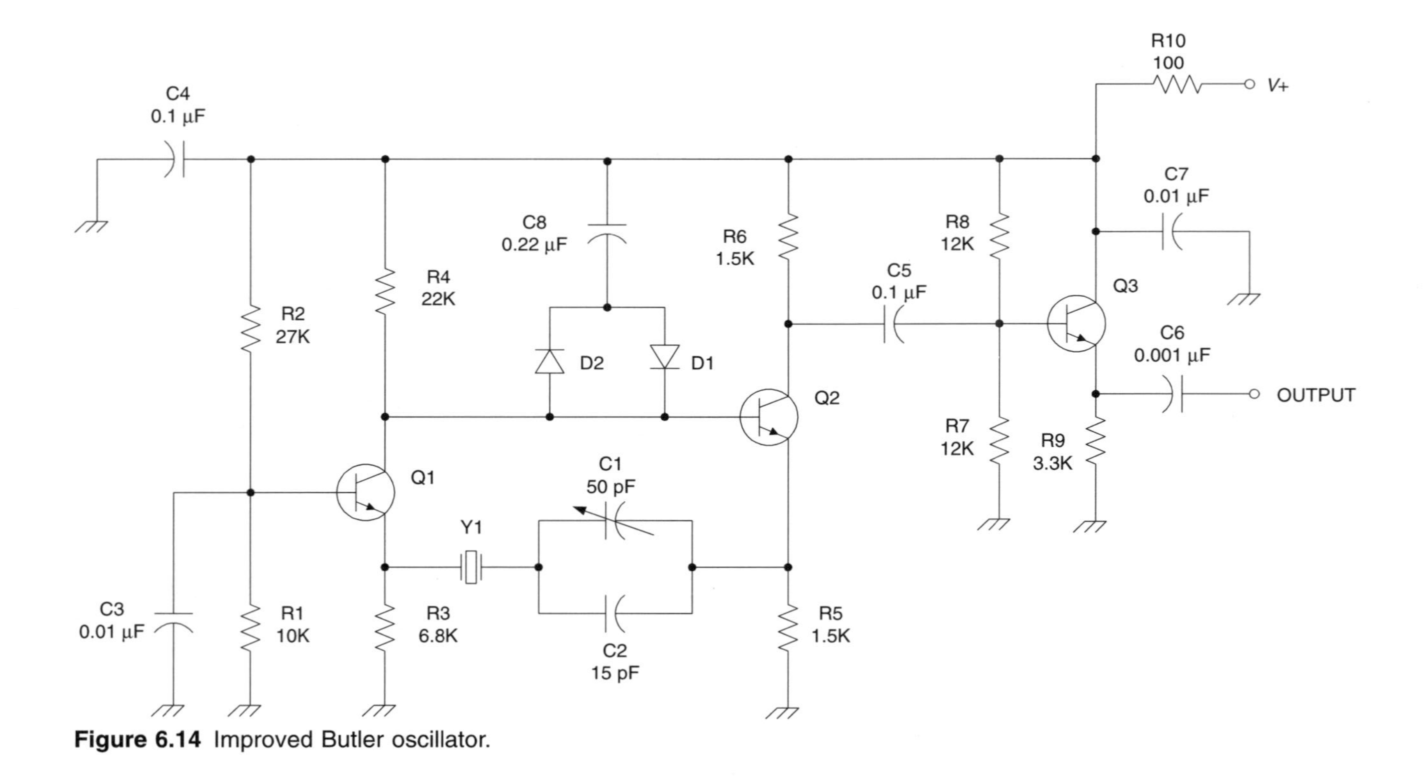

Figure 6.14 Improved Butler oscillator.

Colpitts oscillators

The Colpitts oscillator is characterized by a feedback network consisting of a tapped capacitive voltage divider. In Fig. 6.15 the feedback is provided by C1 and C2, although the situation is somewhat modified by the gate capacitances of Q1. This circuit can be used with parallel mode crystals from about 3 to 20 MHz with proper values of C1 and C2 (see table provided in Fig. 6.15). Frequency trimming of the oscillator can be done by shunting a small value trimmer capacitor across the crystal. The trimmer can also be placed in series with the crystal.

If the oscillator tends to oscillate parasitically in the VHF region, then try using the snubber resistor method (R4 in Fig. 6.15). This could occur because the JFET used at Q1 will have sufficient gain at VHF to permit Barkhausen to have his due at some frequency where strays and distributed L–C elements produce the correct phase shift. A value between 10 and 47 ohms will usually eliminate the problem. Alternatively, a small ferrite bead can be slipped over the gate terminal of Q1 to act as a small value VHF/UHF RF choke.

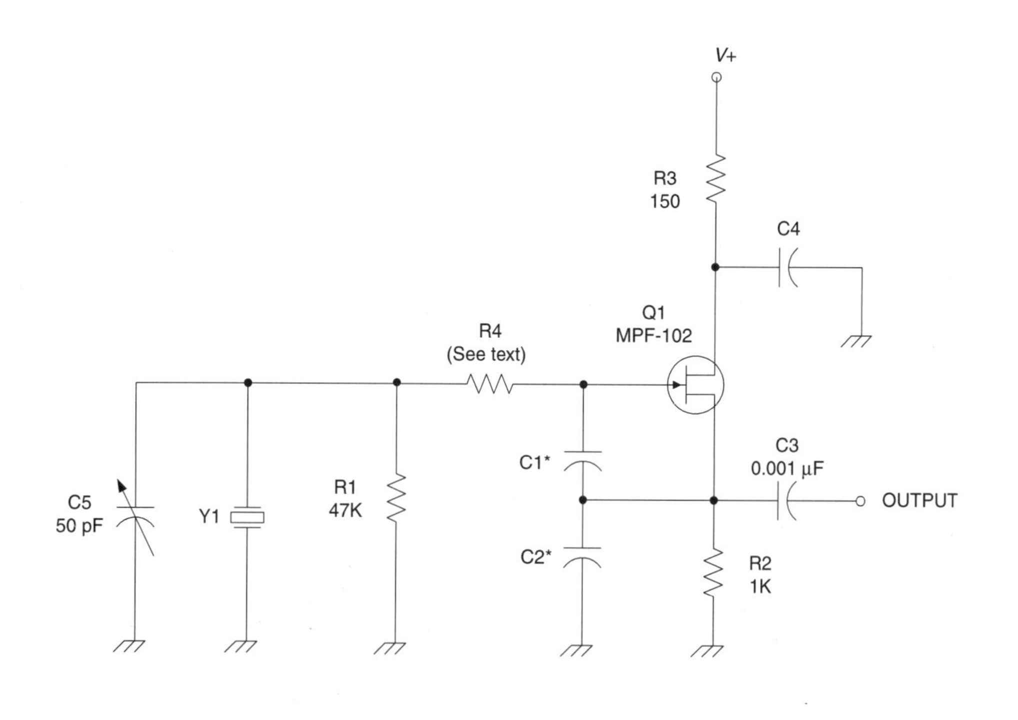

FREQ (MHz)	C1 (pF)	C2 (pF)
3–10	27	68
10–20	10	27

Figure 6.15 JFET Colpitts crystal oscillator.

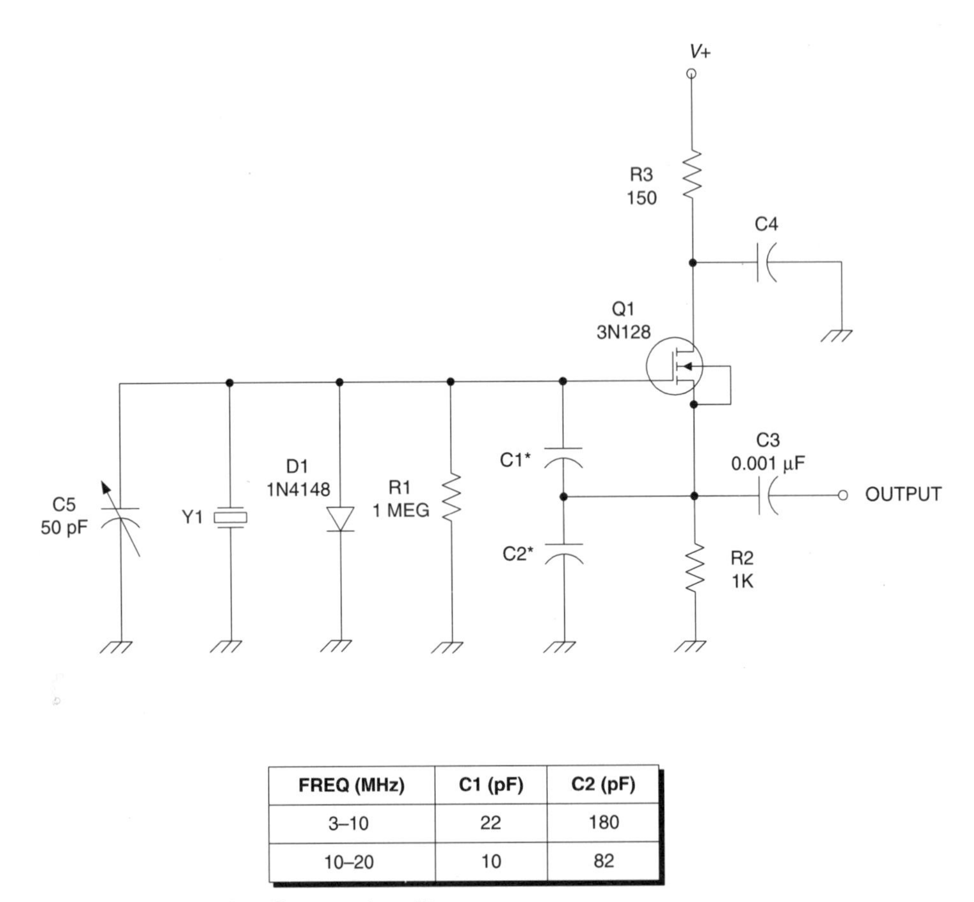

FREQ (MHz)	C1 (pF)	C2 (pF)
3–10	22	180
10–20	10	82

Figure 6.16 MOSFET Colpitts crystal oscillator.

Figure 6.16 is very similar to Fig. 6.15, except for two features. First, the active device is an n-channel MOSFET rather than a JFET. Any of the single-gate devices (e.g. 3N128) can be used. One must, however, be mindful of the possibility of electrostatic damage (ESD) when the MOSFET is used.

The other difference is that a 1N4148 small-signal diode shunts the gate-source path to provide a small amount of automatic gain control action. When the signal appearing across the crystal and feedback network is sufficiently large the diode will rectify the signal and produce a DC bias on the gate that counters the source bias provided by R2. This diode helps smooth out amplitude variations, especially when more than one crystal is switched in and out of the circuit.

Another variation on the Colpitts theme is the impedance inverting oscillator circuit of Fig. 6.17. It will provide stability of 10 ppm over a wide temperature range (0°C to 60°C) if a wise selection of components is made (C1–C3 and L1 are particularly troublesome). The circuit will also remain within ±0.001 per cent over a DC power supply variation of 2:1 (provided the crystal dissipation is not exceeded). Harmonic output of the circuit is typically low.

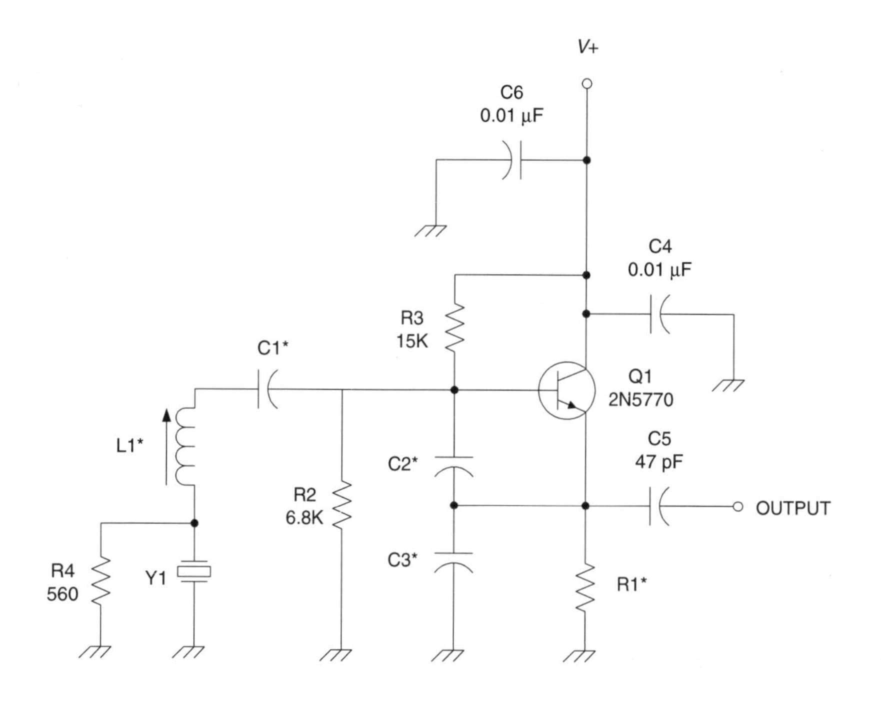

FREQ (MHz)	C1 (pF)	C2 (pF)	C3 (pF)	R1	L1
3–10	1000	270	270	1.5K	
10–15	100	220	220	680	15 turns
15–20	100	100	100	680	10 turns

2–4 MHz	60 turns
4–6 MHz	40 turns
6–10 MHz	25 turns

Figure 6.17 Impedance inverting Colpitts crystal oscillator.

The oscillation frequency is set by adjusting inductor L1. The turns counts shown in the table in Fig. 6.17 presume a 6.5 mm slug-tuned coil form designed for use in the frequency range 3 to 20 MHz. Some experimentation is needed depending on the particular former used. The idea is to set the resonant frequency of the coil and C1–C3 combined to something near the crystal frequency.

It is sometimes appealing to add a tuned circuit to the output circuit of oscillators. The harmonics of the oscillator are suppressed when this is done. But in this case, a transistor equivalent of the old-fashioned TGTP oscillator will result because of the action of the output-tuned circuit and the L1/C1–C3 combination. Don't do it!

Overtone oscillators

Thus far only the fundamental oscillating mode has been discussed. But crystals oscillate at more than one frequency. The oscillations of a crystal slab are in the form of *bulk acoustic waves* (BAWs), and can occur at any wave frequency that produces an odd half-wavelength of the crystal's physical dimensions (e.g. $1\lambda/2$, $3\lambda/2$, $5\lambda/2$, $7\lambda/2$, $9\lambda/2$, where the fundamental mode is $1\lambda/2$). Note that these frequencies are not harmonics of the fundamental mode, but are actually valid oscillation modes for the crystal slab. The frequencies fall close to, but not directly on, some of the harmonics of the fundamental (which probably accounts for the confusion). The overtone frequency will be marked on the crystal, rather than the fundamental. It is rare to find fundamental mode crystals above 20 MHz or so, because their thinness makes them more likely to fracture at low values of power dissipation.

The problem to solve in an overtone oscillator is encouraging oscillation on the correct overtone, while squelching oscillations at the fundamental and undesired overtones. Crystal manufacturers can help with correct methods, but there is still a responsibility on the part of the oscillator designer. Figure 6.18 shows a third-overtone Butler oscillator that will operate at frequencies between 15 MHz and 65 MHz. The inductor (L1) is set to resonate close to the crystal frequency, and is used in part to ensure overtone mode oscillation. If moderate DC supply voltages are used (e.g. 9 to 12 volts in most cases), the harmonic content is low (–40 dB), and stability is at least as good as a similar fundamental mode Butler oscillator.

Figure 6.19 is a third-overtone impedance inverting Colpitts style oscillator that will operate over the 15 MHz to 65 MHz range. As in similar circuits, inductor L1 is tuned to the overtone, and is resonated with C1 (combined with the capacitances of C2 and C3).

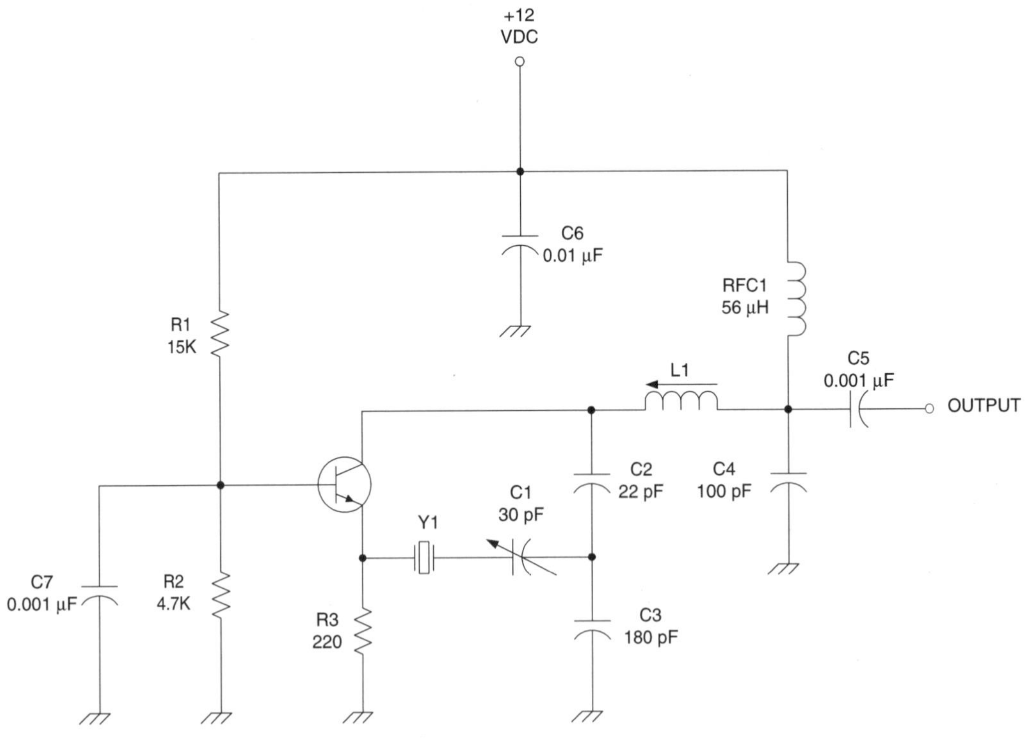

Figure 6.18 Butler overtone crystal oscillator.

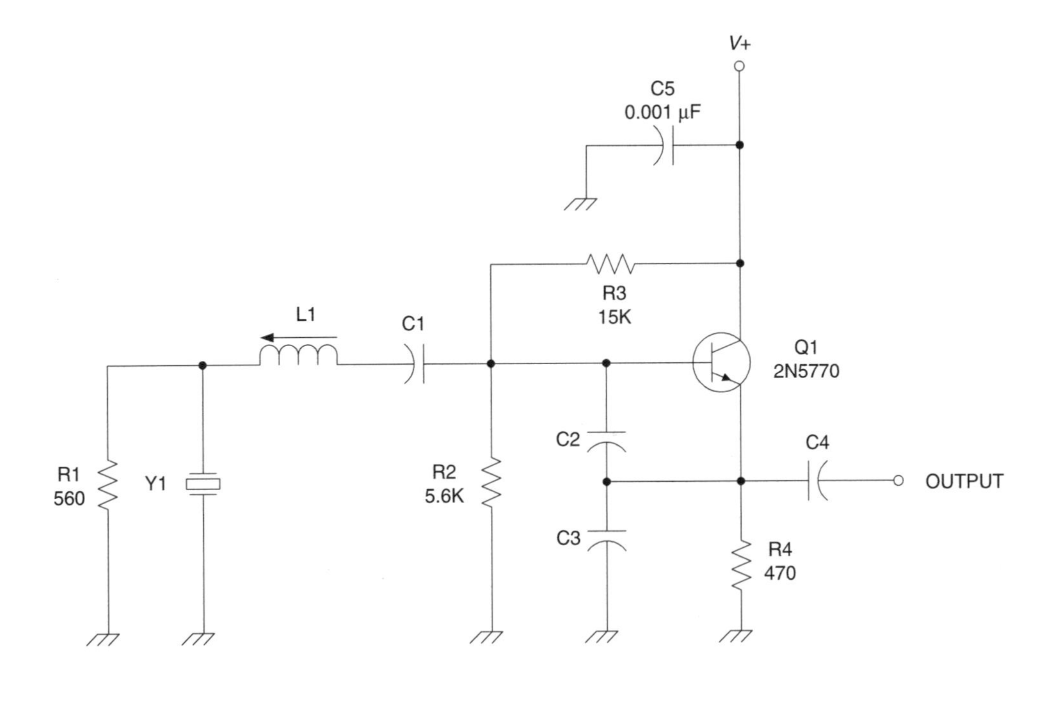

FREQ (MHz)	C1 (pF)	C2 (pF)	C3 (pF)	C4 (pF)	L1 (0.25 inch form)
15–25	100	100	68	33	12 t, #30, CW
25–55	100	68	47	33	8 t, #30, CW
50–65	68	33	15	22	6 t, #22, CW

Figure 6.19 Third overtone Colpitts oscillator.

Values for C1 through C3, and winding instructions for a 6.5 mm low-band VHF coil former are shown in the table.

Note the resistor across crystal Y1. This resistor tends to snub out oscillations in modes other than the overtone, including the fundamental. Care must be taken to not make L1 too large, otherwise it will resonate at a lower frequency (with C1–C3), forming an oscillator on a frequency not related to either the crystal's fundamental or overtones. The oscillator may well be perfectly happy to think of itself as a series-tuned Clapp oscillator! Operation of this circuit to 110 MHz, with fifth- or seventh-overtone crystals, can be accomplished by modifying it to the form shown in Fig. 6.20.

Frequency stability

An ideal sine wave RF oscillator ideally produces a nice, clean harmonic- and noise-free output on a frequency that is stable, but real RF oscillators tend to have certain problems that deteriorate from the quality of their output signals. Load impedance variation and

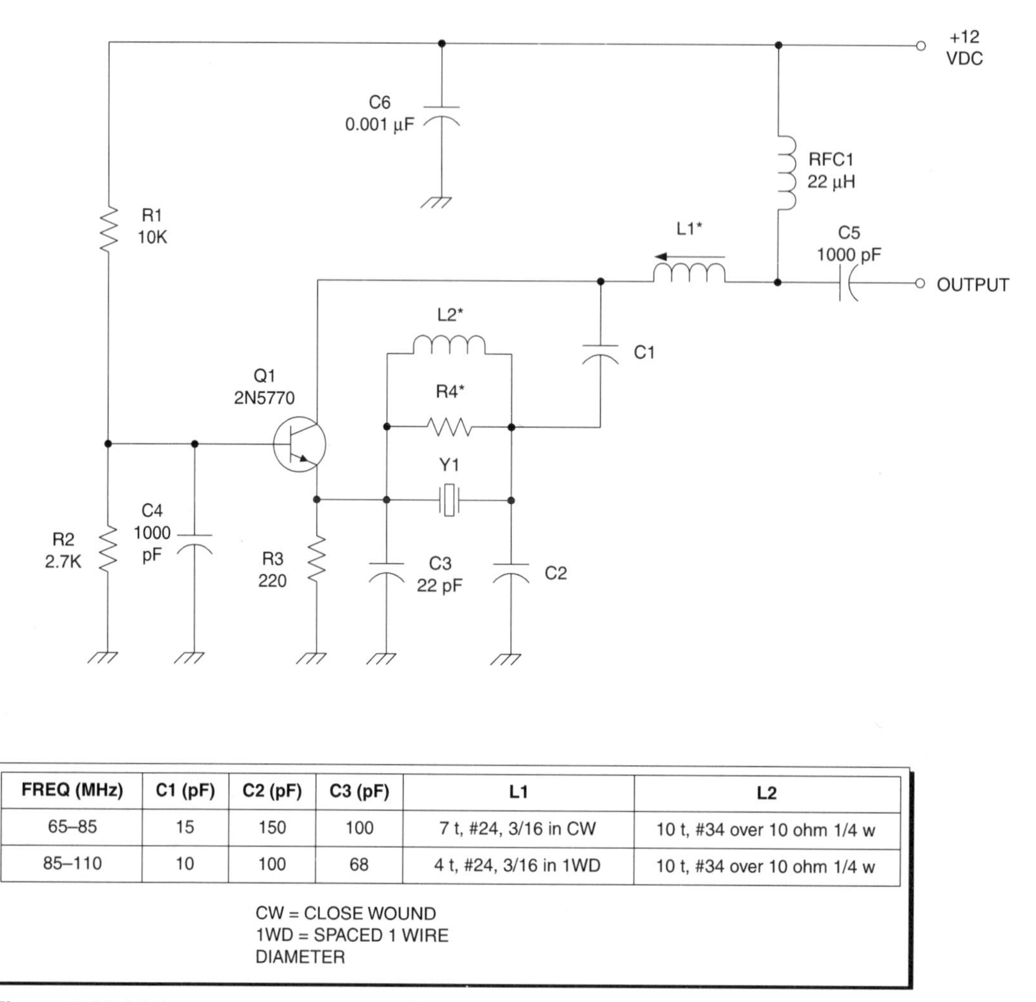

FREQ (MHz)	C1 (pF)	C2 (pF)	C3 (pF)	L1	L2
65–85	15	150	100	7 t, #24, 3/16 in CW	10 t, #34 over 10 ohm 1/4 w
85–110	10	100	68	4 t, #24, 3/16 in 1WD	10 t, #34 over 10 ohm 1/4 w

CW = CLOSE WOUND
1WD = SPACED 1 WIRE DIAMETER

Figure 6.20 Higher overtone crystal oscillator.

DC power supply variation can both cause the oscillator frequency to shift. Temperature changes also cause frequency change problems.

An oscillator that changes frequency without any help from the operator is said to *drift*. Frequency stability refers generally to freedom from frequency changes over a relatively short period of time (e.g. few seconds to dozens of minutes). This problem is different from *aging*, which refers to frequency change over relatively long periods of time (i.e. hertz/year) caused by aging of the components (some electronic components tend to change value with long use). Temperature changes are a large contributor to drift in two forms: warm-up drift (i.e. drift in the first 15 minutes) and ambient temperature change drift.

If certain guidelines are followed, then it is possible to build a very stable oscillator. For the most part, the comments below apply to both crystal oscillators and L–C tuned oscillators (e.g. variable frequency oscillators, VFOs), although in some cases one or the other is indicated by the text.

Temperature

Temperature variation has a tremendous effect on oscillator stability. Avoid locating the oscillator circuit near any source of heat. In other words, keep it away from power transistors or IC devices, voltage regulators, rectifiers, lamps or other sources of heat.

Thermal isolation

One approach is to thermally isolate the oscillator circuits. Figure 6.21 shows one such method. The oscillator is built inside a metal shielded cabinet as usual, but has styrofoam insulation applied to the sides. There is a type of poster board which has a backing of styrofoam to give it substance enough for self-support. It is really easy to cut using hobby or razor knives. Cut the pieces to size, and then glue them to the metal surface of the shielded cabinet.

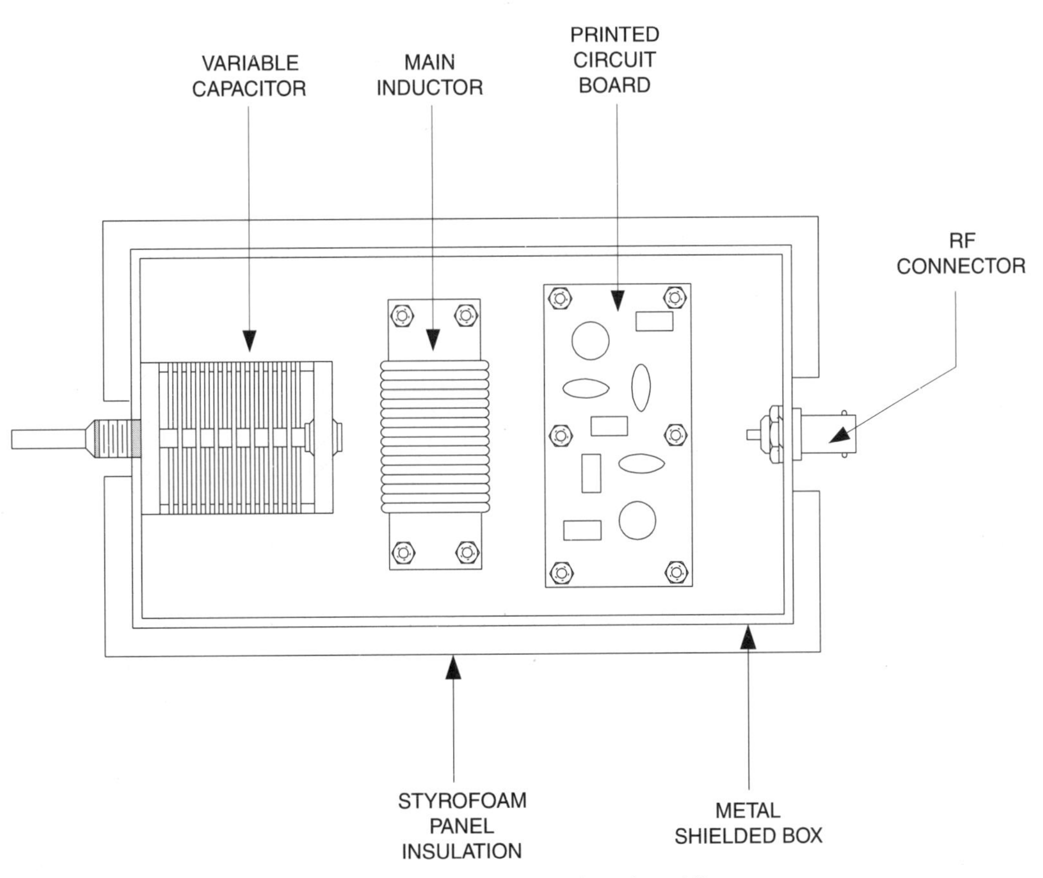

Figure 6.21 Variable frequency oscillator well shielded against drift.

Avoid self-heating

Operate the oscillator at as low a power level as can be tolerated so as to prevent self-heating of the active device and associated frequency determining components. It is generally agreed that a power level on the order of 1 to 10 mW is sufficient. If a higher power is needed, then a *buffer amplifier* can be used. The buffer amplifier also serves to isolate the oscillator from load variations.

Other stability criteria

Use low frequencies

In general, L–C controlled VFOs should not be operated at frequencies above about 12 MHz. For higher frequencies, it is better to use a lower frequency VFO and heterodyne it against a crystal oscillator to produce the higher frequency. For example, one common combination uses a 5 to 5.5 MHz main VFO for all HF bands in SSB transceivers.

Feedback level

Use only as much feedback in the oscillator as is needed to ensure that the oscillator starts quickly when turned on, stays operational, and does not 'pull' in frequency when the load impedance changes. In some cases, there is a small value capacitor between the L–C resonant circuit and the gate or base of the active device. This reduces drift by only lightly loading the tuned circuit. The most common means for doing this job is to use a 3 to 12 pF NP0 disk ceramic capacitor. Adjust its value for the minimum level to ensure good starting and freedom from frequency changes under varying load conditions.

Output isolation

A *buffer amplifier,* even if it is a unity gain emitter follower, is also highly recommended. It will permit building up the oscillator signal power, if that is needed, without loading the oscillator. The principal use of the buffer is to isolate the oscillator from variations in the output load conditions.

Figures 6.22 and 6.23 show typical buffer amplifier circuits. The circuit in Fig. 6.22 is a standard emitter follower (i.e. common collector) circuit. It produces near unity voltage gain, but some power gain. Figure 6.23 is a feedback amplifier using a 4:1 balun-style toroidal core transformer in the collector circuit. This circuit exhibits an input impedance near 50 ohms, so should only be used if the oscillator circuit can drive 50 ohms. In some cases, a higher impedance circuit is needed.

DC power supply

Power supply voltage variations have a tendency to frequency modulate the oscillator signal. It is a good idea to use a voltage regulated DC power supply on the oscillator.

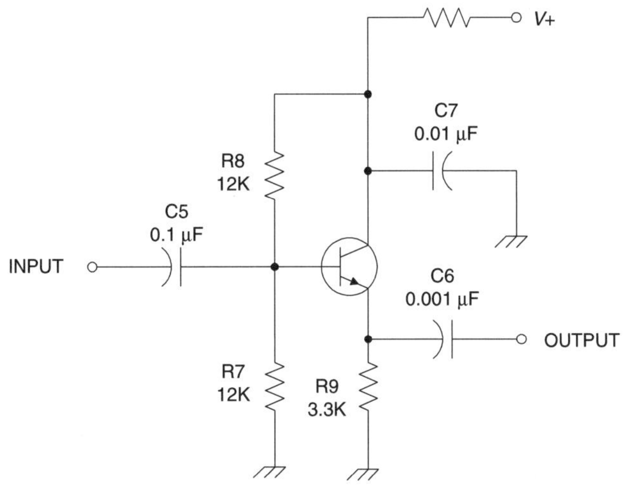

Figure 6.22 NPN bipolar buffer amplifier.

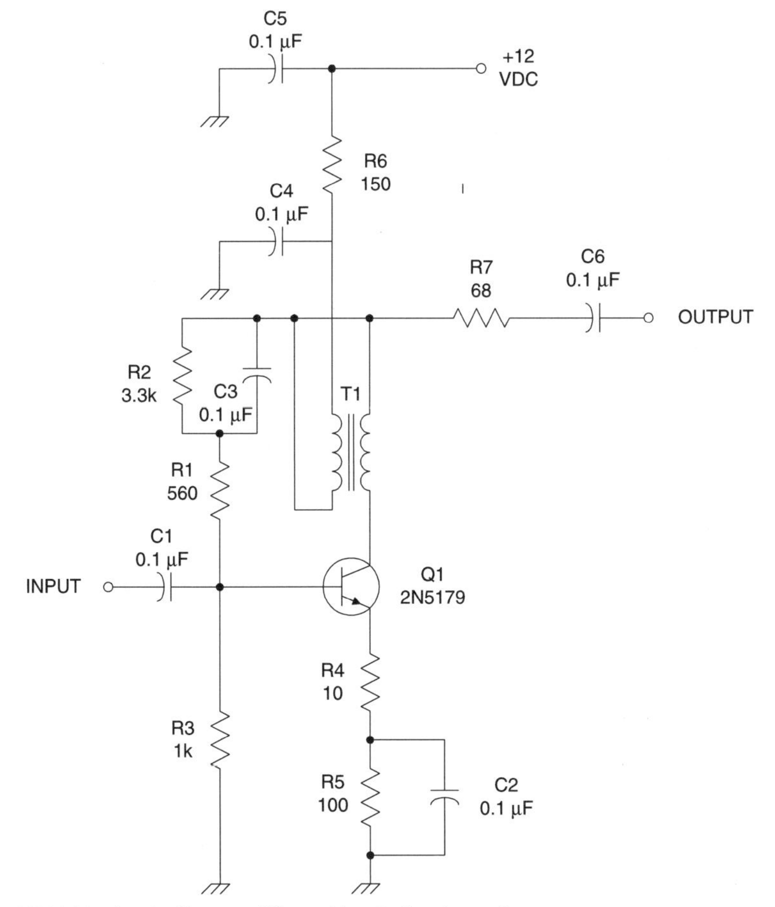

Figure 6.23 NPN bipolar buffer amplifier with 50 ohm impedance.

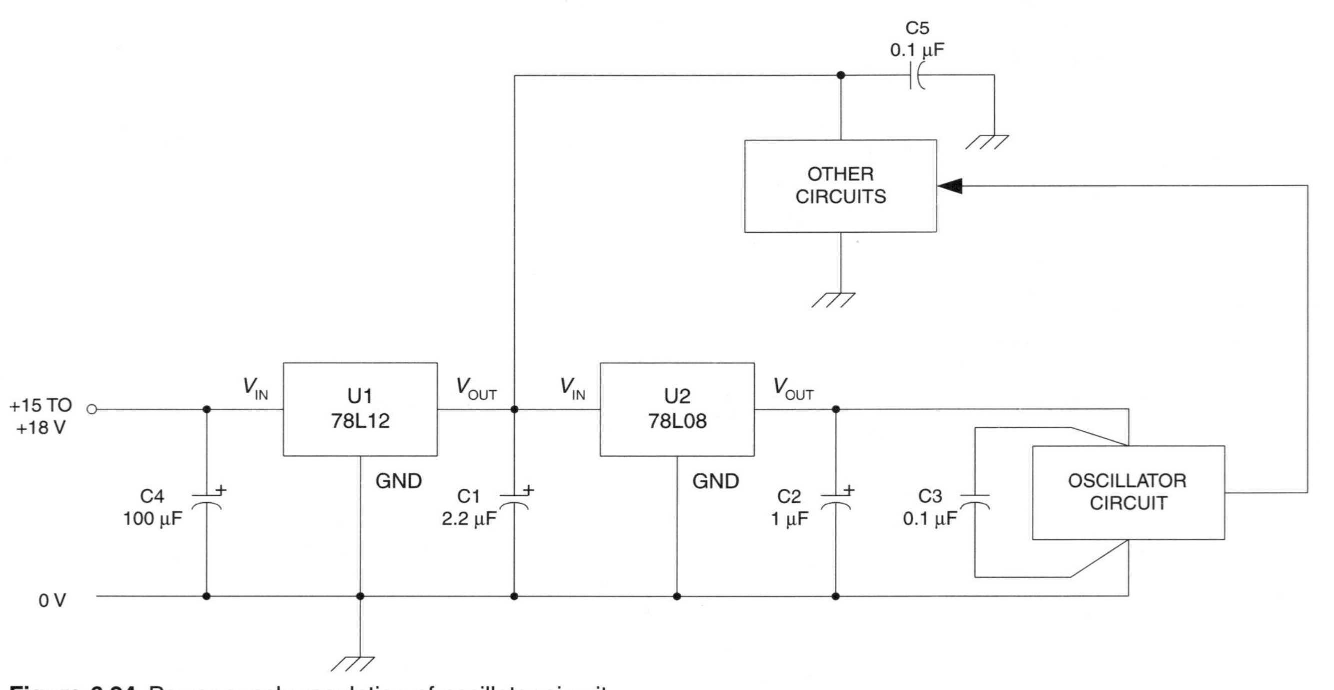

Figure 6.24 Power supply regulation of oscillator circuit.

It is worth using a voltage regulator to serve the oscillator alone, even if another voltage regulator is used to regulate the voltage applied to other circuits (see Fig. 6.24). This double regulation approach is reasonable given the low cost of three-terminal IC voltage regulators and the advantage gained. For most low powered oscillators, a simple low power 'L-series' (e.g. 78L06) three-terminal integrated circuit voltage regulator is sufficient (see U2 in Fig. 6.24). The L-series devices provide up to 100 mA of current at the specified voltage, and this is enough for most oscillator circuits.

Capacitors on both input and output sides of the voltage regulator (C1 and C2 in Fig. 6.24) add further protection from noise and transients. The values of these capacitors are selected according to the amount of current drawn. The idea is to have a local supply of stored current to temporarily handle sudden demand changes, allowing time for a transient to pass, or the regulator to 'catch up' with the changed situation.

Vibration isolation

The frequency setting components of the oscillator can also affect the stability performance. The inductor should be mounted so as to prevent vibration. While this requirement means different things to different styles of coil, it is nonetheless important. Some people prefer to rigidly mount coils, while others will put them on a vibration isolating shock absorber material. I've seen cases where a receiver that required a very low SSB noise sideband local oscillator was not able to meet specifications until some rubber shock absorbing material was placed underneath the oscillator printed wiring board.

Coil core selection

Air core coils are generally considered superior to those with either ferrite or powdered iron cores because the magnetic properties of the cores are affected by temperature variation. Of those coils that do use cores, slug tuned are said to be best because they can be operated with only a small amount of the tuning core actually inside the windings of the coil, reducing the vulnerability to temperature effects. Still, toroidal cores have a certain endearing charm, and can be used wherever the ambient temperature is relatively constant. The Type SF material is said to be the best in this regard, and it is easily available Type-6 material. For example, you could wind a T-50-6 core and expect relatively good frequency stability.

Coil-core processing

One source recommends tightly winding the coil wire onto the toroidal core, and then heat annealing the assembly. This means placing it in boiling water for several minutes, and then removing and allowing it to cool in ambient room air while sitting on an insulated pad. I haven't personally tried it, but the source reported remarkable freedom from inductor-caused thermal drift.

For most applications, especially where the temperature is relatively stable, the coil with a magnetic core can be wound from enamelled wire (#20 to #32 AWG are usually specified), but for best stability it is recommended that *Litz* wire be used. Although a bit hard to get in small quantities, it offers superior performance over relatively wide changes in temperature. Be aware that this nickel-based wire is difficult to solder properly, so be prepared for a bit of frustration.

Air core coils

For air core coils, use #18 SWG or larger bare solid wire, wound on low temperature coefficient of expansion formers. Figure 6.25 shows how the coil can be mounted in a project. Stand-off insulators provide adequate clearance for the coil, and hold it to the chassis. The mount shown has shock absorbers to prevent movement.

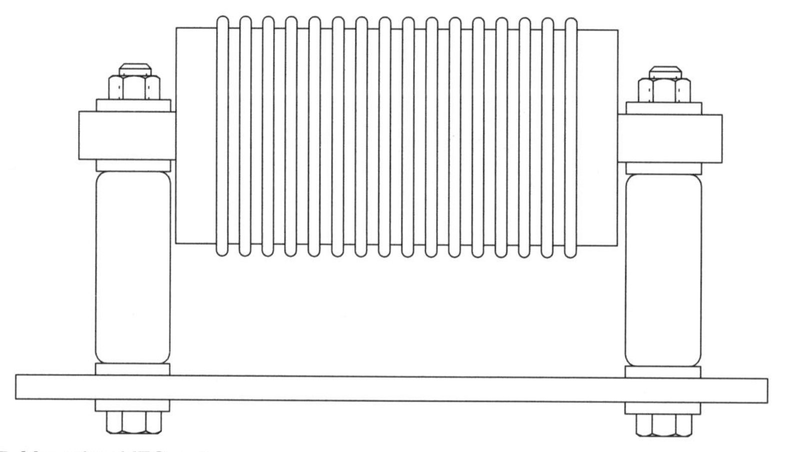

Figure 6.25 Mounting VFO coils.

Capacitor selection

The trimmer capacitors used in the oscillator circuit should be air dielectric types, rather than ceramic or mica dielectric trimmers, because of lower temperature coefficient.

The small fixed capacitors used in the oscillator should be either NPO disk ceramics (i.e. zero temperature coefficient), silvered mica or polystyrene types. Some people dislike the silvered mica types because they tend to be a bit quirky with respect to temperature coefficient. Even out of the same batch they can have widely differing temperature coefficients on either side of zero.

Sometimes, you will find fixed capacitors with non-zero temperature coefficient in an oscillator frequency determining circuit. They are there to make temperature compensated oscillators. The temperature coefficients of certain critical capacitors are selected to cancel out the natural drift of the circuit.

The main tuning air variable capacitor should be an old-fashioned double bearing (i.e. bearing surface on each end-plate) type. For best results, it should be made with either brass or iron stator and rotor plates (not aluminium).

Tempco circuit

Temperature-compensated crystal oscillators can be built by using the temperature coefficient of some of the capacitors in the circuit to cancel drift in the opposite direction. The circuit in Fig. 6.26 was used in an amateur radio transceiver at one time. It is a standard Colpitts crystal oscillator, but with the addition of a temperature coefficient cancellation circuit (C8A, C8B, C9 and C10).

The key to this circuit is C8, which is a *differential variable capacitor*, i.e. it contains two sections connected so that as the shaft turns one is increasing capacitance as the other is decreasing capacitance. In most cases, C9 and C10 are of equal value, but one is NPO and the other N750 or N1500. If C8 is differential, and C9 = C10, then the net capacitance across crystal Y1 is constant with C8 shaft rotation, but the net temperature coefficient changes. The idea is to crank in the amount of temperature coefficient that exactly cancels the drift in the circuit's other components.

The problem with this circuit is that differential capacitors are quite expensive. It may be useful, however, to connect the circuit as shown, and find the correct setting. Once the setting is determined, disconnect C8–C10 without losing the setting, and measure the capacitance of the two sections. With this information you can calculate the capacitances required for each temperature coefficient and replace the network with appropriately selected fixed capacitors.

In general, when selecting temperature compensating capacitance, keep in mind the following relationship for the temperature coefficient of frequency (TCF):

$$TCF = \frac{-\left[T_{CL} + \dfrac{T_{C1} \times C1}{C_{total}} + \dfrac{T_{C2} \times C1}{C_{total}}\right]}{2} \qquad (6.4)$$

Where:

TCF is the temperature coefficient of frequency
T_{CL} is the temperature coefficient of the inductor
T_{C1} is the temperature coefficient of C1
T_{C2} is the temperature coefficient of C2
C_{total} is the sum of all capacitances in parallel with L1, including C1, C2 and all strays and other capacitors

This equation assumes that two capacitors, C1 and C2, are shunted across an inductor in a VFO circuit.

Varactors

Voltage variable capacitance diodes ('varactors') are often used today as a replacement for the main tuning capacitor. If this is done, then it becomes critical to temperature control the environment of the oscillator. Temperature variations will result in changes in diode pn junction capacitance, and that contributes much to thermal drift. Varactor temperature coefficients of 450 ppm are not unusual.

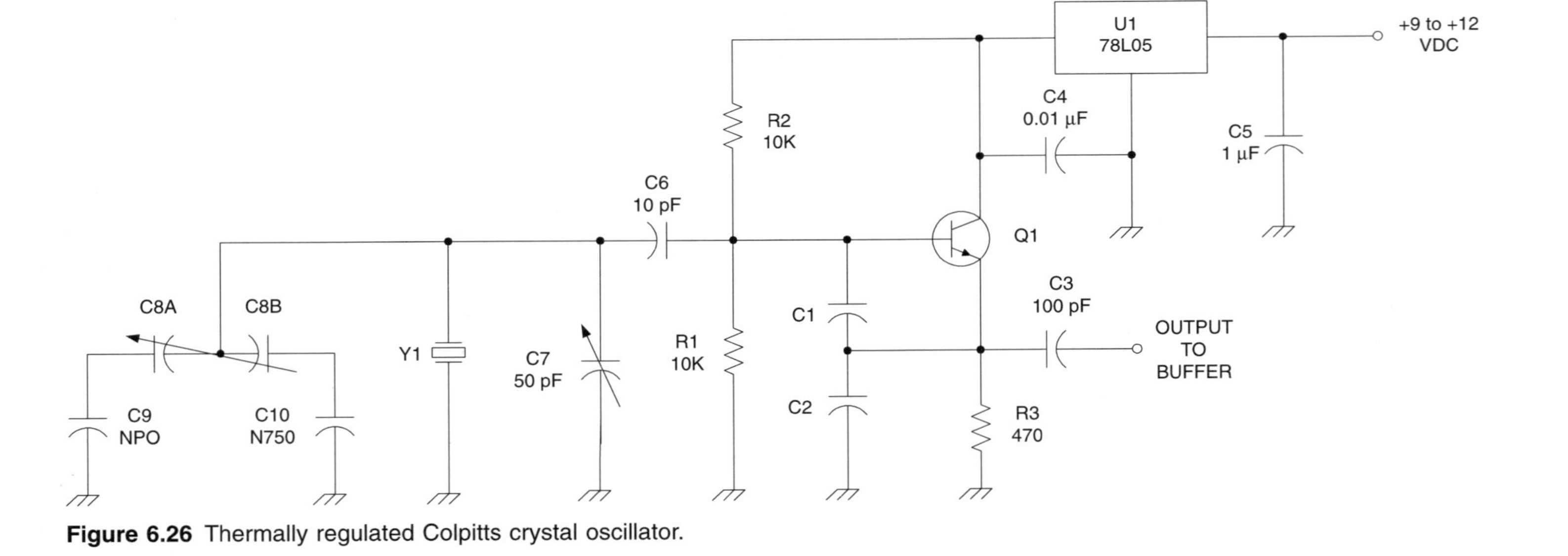

Figure 6.26 Thermally regulated Colpitts crystal oscillator.

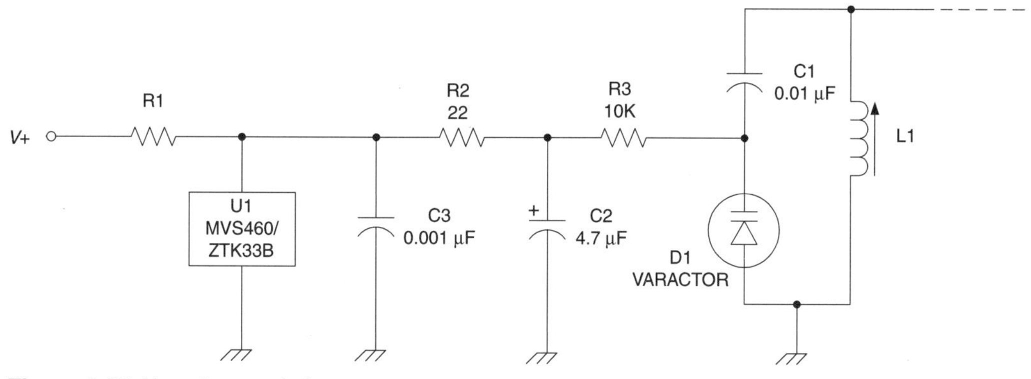

Figure 6.27 Varactor regulation.

A good way to stabilize these circuits is to supply a voltage regulator that has a temperature coefficient that is opposite the direction of the varactor drift. By matching these two, it is possible to cancel the drift of the varactor.

Figure 6.27 shows a basic circuit that will accomplish this job. Diode D1, in series with DC blocking capacitor C1, is in parallel with the inductor (L1). These components are connected into an oscillator circuit (not shown for sake of simplicity).

Normally, resistor R3 (10 kohms to 100 kohms, typically) is used to isolate the diode from the tuning voltage source, and will be connected to the wiper of a potentiometer that varies the voltage. In this circuit, however, an MVS460 (USA type number) or ZTK33B (European type number) varicap voltage stabilizer is present. This device is similar to a zener diode, but has the required –2.3 mV/°C temperature coefficient. It operates at a current of 5 mA (0.005 A) at a supply voltage of 34 to >40 volts DC. The value of resistor R1 is found from:

$$R1 = \frac{(V+) - 33 \text{ volts}}{0.005 \text{ amps}} \tag{6.5}$$

For example, when *V*+ is 40 volts, then the resistor should be 1400 ohms (use 1.5 kohms).

Frequency synthesizers

A *synthesizer* architecture is shown in Fig. 6.28. This type of oscillator is more modern than other types, and is capable of producing very accurate, high quality signals. There are three main sections to this signal source: *reference section*, *frequency synthesizer*, and *output section*. In Fig. 6.28 each section is broken down into further components for sake of easy analysis.

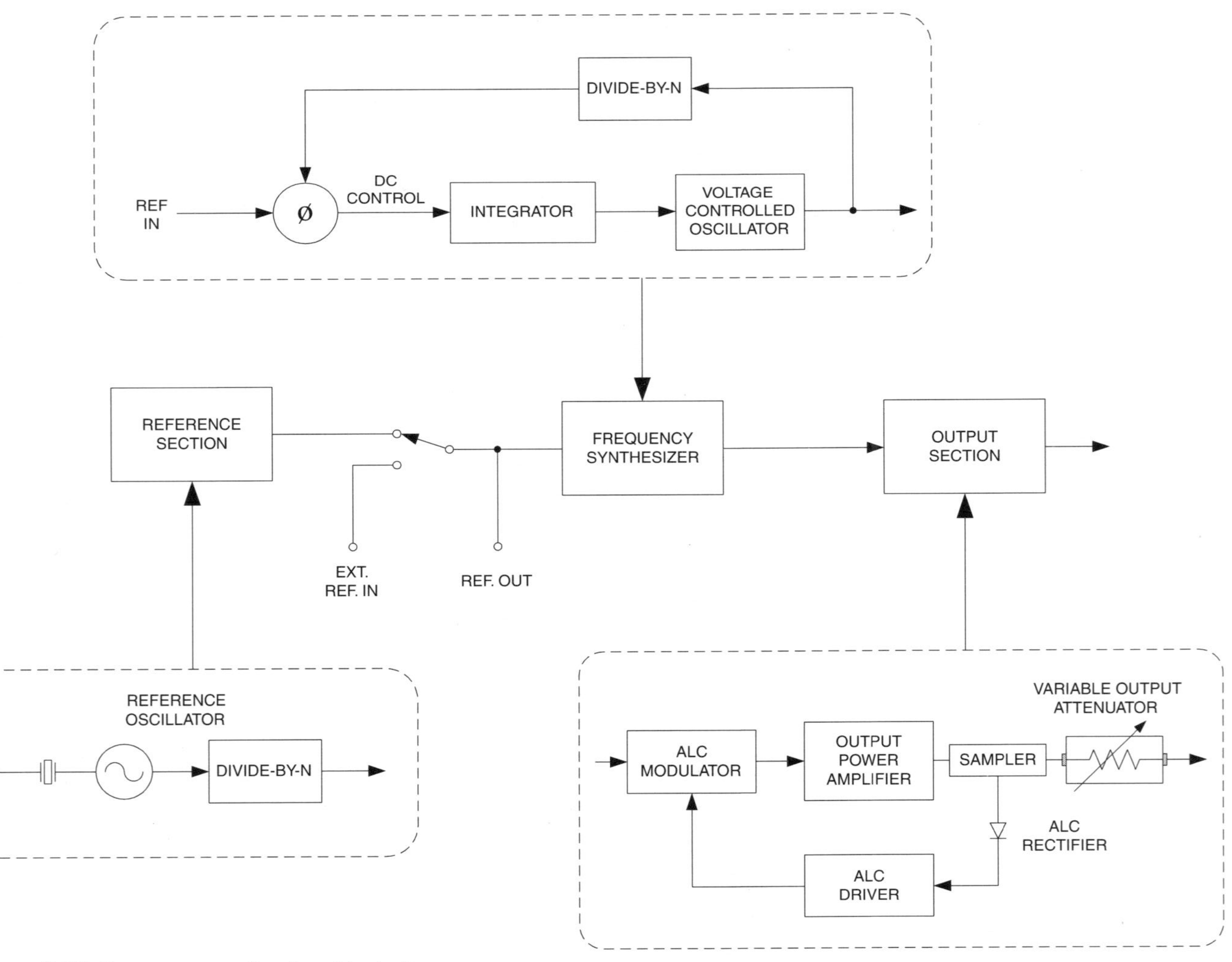

Figure 6.28 Frequency synthesizer block diagram.

Reference section

The reference section is at the very core of the signal generation process. It is an accurate, stable fixed frequency source such as a crystal oscillator. The frequency of the reference section must be precisely adjustable over a small range so that it can be compared to a higher-order standard, such as a Cesium beam oscillator or WWVB comparator receiver, for purposes of calibration.

Because it controls the frequency synthesizer, the stability of the reference section determines the overall stability of the oscillator. Typically, either *temperature-compensated crystal oscillators* (TCXO) or *oven-controlled crystal oscillators* (OCXO) are used for the reference section. The TCXO will typically exhibit crystal aging of better then ±2 ppm/year, and a temperature aging of ±1 ppm/year. The OCXO is capable of 0.1 ppm/year for aging and 0.01 ppm/year for temperature.

The crystal oscillator is usually operated at a frequency such as 5 MHz, but lower frequencies are often needed. To generate these lower frequencies a divide-by-N digital counter is provided. This circuit will divide the output frequency by some integer, N, to produce a much lower reference frequency.

Frequency synthesizer section

The actual signal is produced in the frequency synthesizer section. It is generated by a *voltage-controlled oscillator* (VCO) whose output is compared to the reference signal. Voltage variable capacitance diodes ('varactors') can be used for the VCO, as can *surface acoustical wave* (SAW) oscillators (which are used at higher frequencies and in the microwave bands).

The frequency of the VCO is set by a DC control voltage applied to its tuning input line. This control voltage is generated by integrating the output of a phase detector or phase comparator that receives the reference frequency and a divide-by-N version of the VCO frequency as inputs. When the two frequencies are equal, then the output of the phase detector is zero, so the VCO tuning voltage is at some quiescent value.

If the frequency of the VCO drifts, the phase detector output becomes non-zero. The integrator output ramps up in the direction that will cancel the frequency drift of the VCO. The VCO frequency is continuously held in check by corrections from the integrated output of the phase detector. This type of circuit is called a *phase-locked loop* (PLL).

Suppose, for example, a signal source has a reference of 5 MHz, and it is divided by 20 to produce a 250 kHz reference. If the frequency synthesizer divide-by-N stage is set for, say, $N = 511$, then the VCO output frequency will be $0.25\,\text{MHz} \times 511 = 127.75\,\text{MHz}$. Band switching, operating frequency and frequency resolution are controlled by manipulating the reference frequency divide-by-N and VCO divide-by-N settings. In some cases, the frequency is entered by keypad, and this sets these values. Alternatively, 'tunable' equipment may have a digital encoder shaft connected to a front panel control.

Output section

The output section performs three basic functions: it boosts power output to a specified maximum level, it provides precision control over the actual output level, and keeps the output level constant as frequency is changed.

The power amplifier is a wideband amplifier that produces an output level of some value in excess of the required maximum output level (e.g. +13 dBm). An attenuator can then be used to set the actual output level to any lower value required (e.g. +10 dBm).

The accuracy of the output power setting is dependent on keeping the RF power applied to the attenuator input constant, despite the fact that oscillators (including VCOs) tend to exhibit output signal amplitude changes as frequency is changed. An *automatic level control* (ALC) circuit accomplishes this job.

Automatic level control (ALC)

The ALC modulator is essentially an amplitude modulator that is controlled by a DC voltage developed by rectifying and filtering a sample of the RF output level. The ALC driver compares the actual output level with a preset value, and adjusts the control signal to the ALC modulator in a direction that cancels the error.

7 IF amplifiers and filters

The subject of IF amplifiers and filters is very interesting. The IF amplifier provides most of the gain and selectivity of a superheterodyne receiver.

IF filters: general filter theory

There are a number of IF filters used in radio receivers today. In addition to the L–C filters, there are various types of crystal filter, monolithic ceramic filters, and mechanical filters to consider. In this chapter, we will look at the various types of IF filter, their characteristics and their applications.

Before delving into the topic, however, let's look at some general filter theory as applied to IF pass band filters, and how these filters are used. Figure 7.1A shows the Butterworth pass band characteristic. It is characterized by the fact that the pass band is relatively flat. The Chebyshev filter is shown in Fig. 7.1B. It has a rippled pass band, but steeper slopes than the Butterworth design.

The *bandwidth* of the filter is the bandwidth between the –3 dB points (Fig. 7.1A). The Q of the filter is the ratio of centre frequency to bandwidth, or:

$$Q = \frac{F_O}{B} \tag{7.1}$$

Where:
Q is the quality factor of the filter
F_O is the centre frequency of the filter
B is the –3 dB bandwidth of the filter
(Note: F_O and B are in the same units)

The *shape factor* of the filter is defined as the ratio of the –60 dB bandwidth to the –6 dB bandwidth. This is an indication of how well the filter will reject out of band interference. The lower the shape factor the better (shape factors of 1.2:1 are achievable).

Figure 7.1C shows a generic IF amplifier with the filters in place. The IF filters provide most of the selectivity in a receiver, and have the narrowest bandpass of all the filters in the receiver.

The IF amplifier may or may not use two (or more) filters, depending upon the type of design. In cases where only one filter is used, then the filter will usually be placed at the input of the amplifier in order to eliminate the mixer products that can affect the IF amplifier performance. The noise produced by the IF amplifier can be significant, which means that an output IF filter is indicated in order to eliminate that noise.

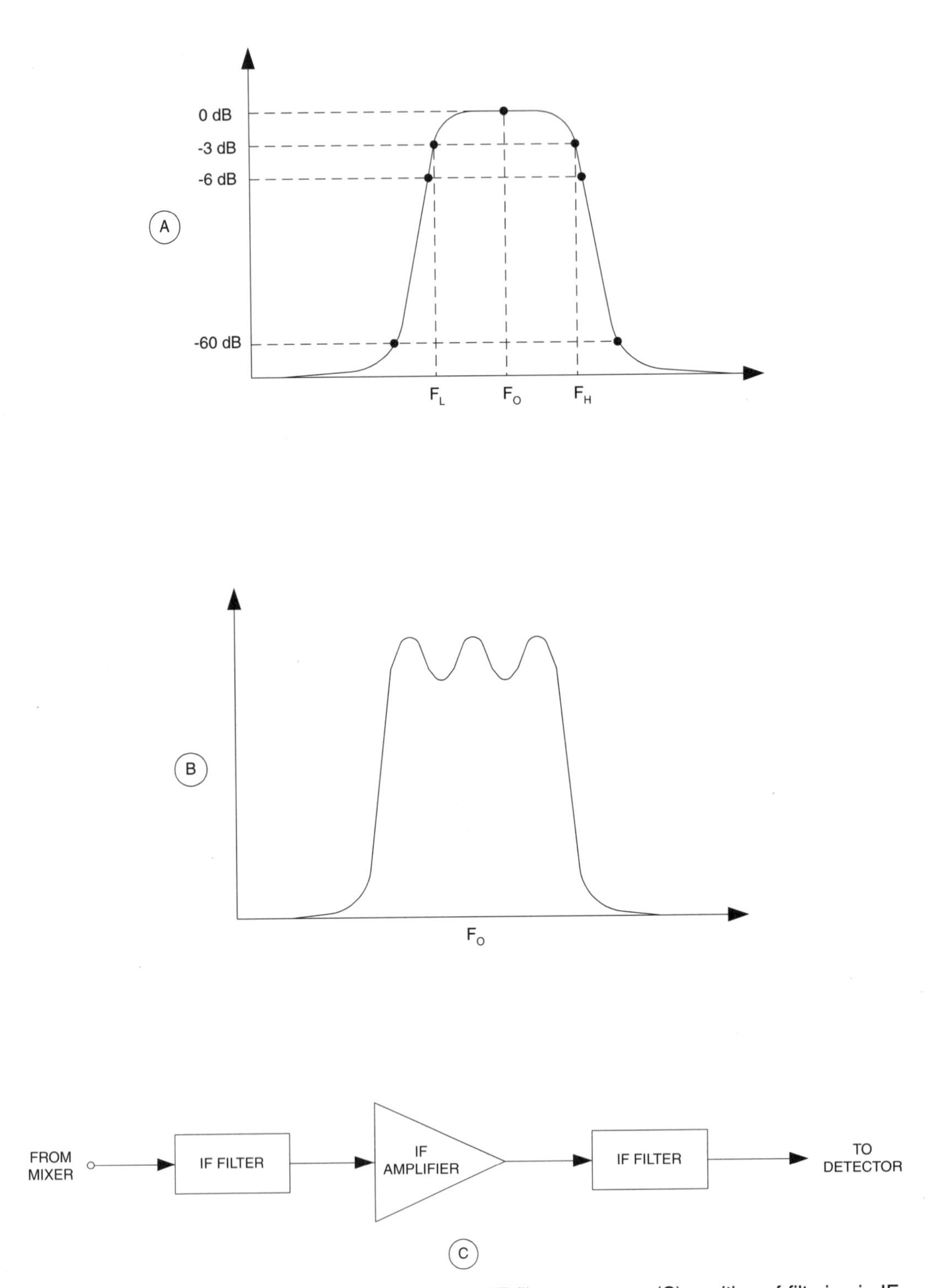

Figure 7.1 (A) Ideal IF filter response; (B) practical IF filter response; (C) position of filtering in IF amplifier.

L–C IF filters

The basic type of filter, and once the most common, is the *L–C filter,* which comes in various types (Fig. 7.2). The type shown in Fig. 7.2A contains two parallel-tuned L–C sections. Although it is not apparent here, the input and output sides of the L–C network need not be the same impedance, although that is usually the case. Once the most common form of IF amplifier filter, this type has largely been eclipsed by other types except in certain IC amplifiers.

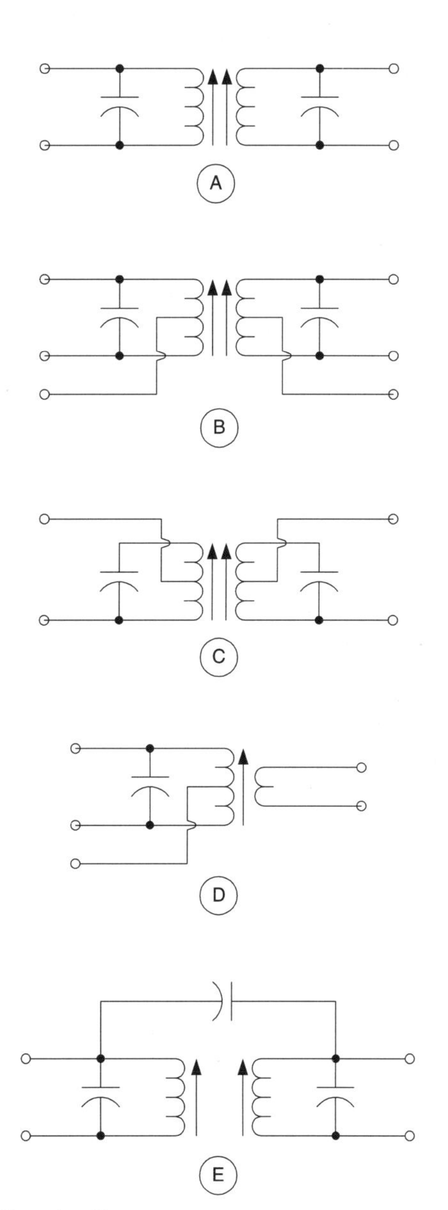

Figure 7.2 Various L–C IF filter circuits.

A more common form today is shown in Fig. 7.2B. This form has a low impedance tap for transistor or IC applications, with the high impedance portions still available. In Fig. 7.2C we see a common form of IF filter in which the low impedance tap is available to both input and output sides, but one side of the high impedance portions of the transformer are not. In Fig. 7.2D we see a single-tuned IF filter. It has a standard IF filter input side, but it has only a low impedance link on the output side. The IF filtering is performed by the tuned L–C circuit, and the low impedance link is for impedance matching.

In Fig. 7.2E we see a double-tuned IF filter that is not magnetically coupled (so it is not a transformer), but rather is coupled through a common impedance. In this case a small value capacitor is used as the common impedance, but inductors can also be used.

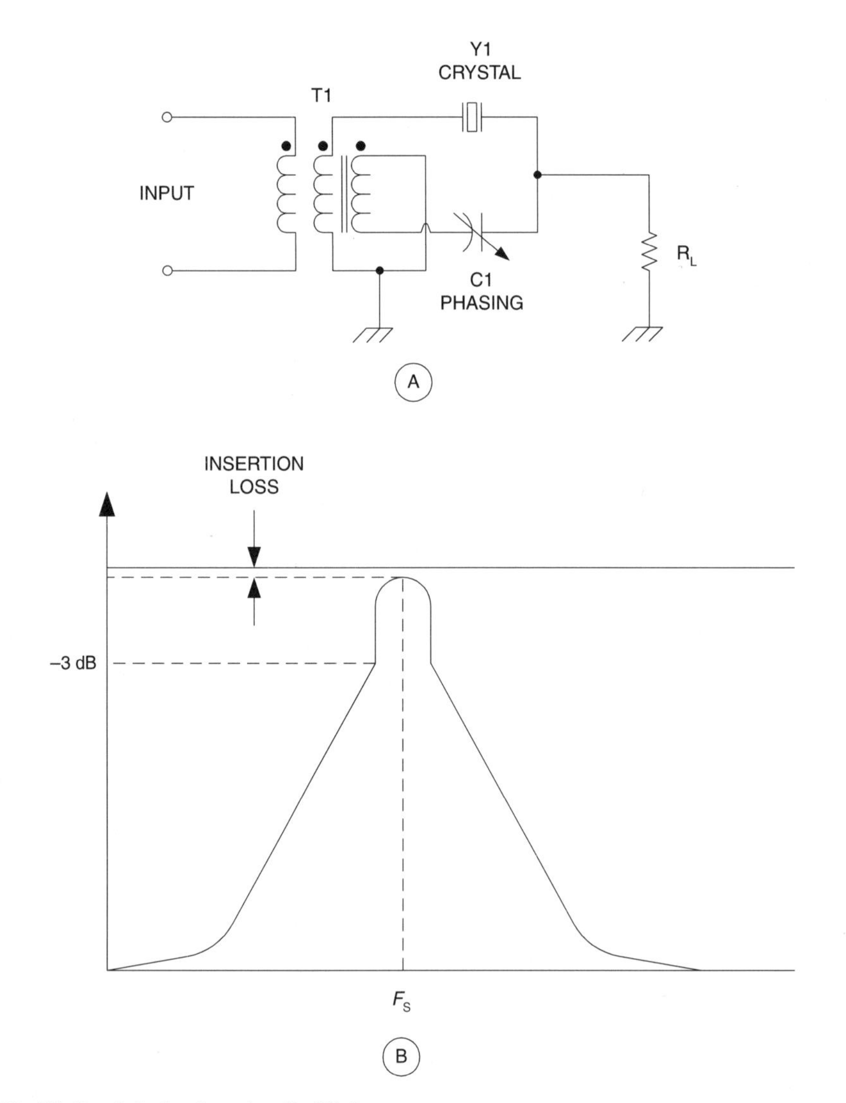

Figure 7.3 (A) Crystal phasing circuit; (B) frequency response.

Crystal filters

The quartz piezoelectric crystal resonator is ideal for IF filtering because it offers high Q (narrow bandwidth) and behaves as an L–C circuit. Because of this feature, it can be used for high quality receiver design as well as single sideband transmitters (filter type).

Figure 7.3A shows a simple crystal filter that has been around since the 1930s in one form or another. Figure 7.3B shows the attenuation graph for this filter. There is a 'crystal phasing' capacitor, adjustable from the front panel, that cancels the parallel capacitance. This cancels the parallel resonance, leaving the series resonance of the crystal.

Figure 7.4A shows the circuit for a *half-lattice crystal* filter, while Fig. 7.4B shows the attenuation curve. This type of crystal filter is used in lower-cost radios. Like the simple

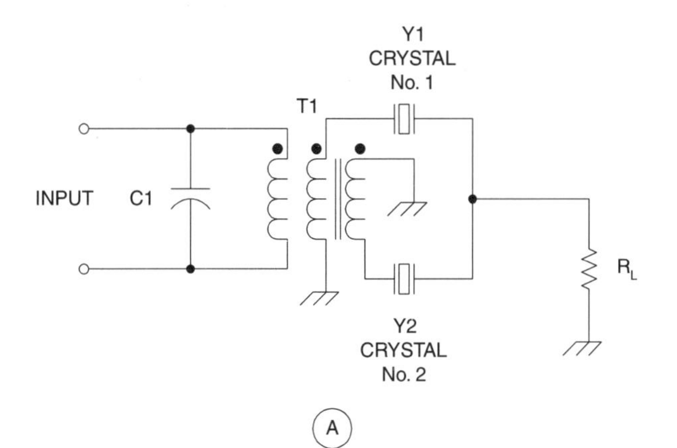

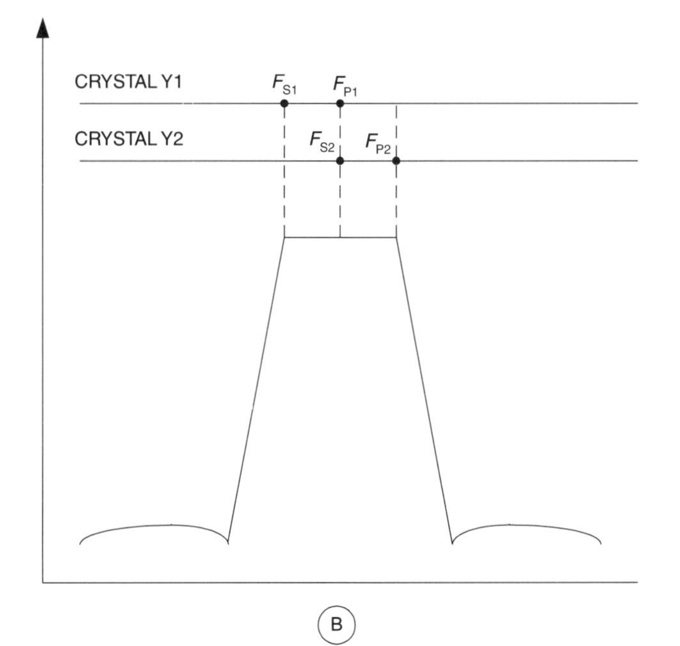

Figure 7.4 (A) Dual crystal filter; (B) frequency response.

crystal filter described above, this version uses a trifilar coil for T1. Instead of the phasing capacitor there is a second crystal in the circuit. The frequency relationship between the two crystals is shown in Fig. 7.4B. They have overlapping parallel and series resonance points such that the parallel resonance of crystal no. 1 is the same as the series resonance of crystal no. 2.

We can use the half-lattice filter to build a cascade half-lattice filter (Fig. 7.5) and a full lattice crystal filter (Fig. 7.6). The cascade half-lattice filter has increased skirt selectivity and fewer spurious responses compared with the same pass band in the half-lattice type of filter. It is a back-to-back arrangement on a bifilar transformer (T1). In practice, close matching is needed to make the cascaded half-lattice filter work properly.

The full lattice crystal filter (Fig. 7.6) uses four crystals like the cascade half-lattice, but the circuit is built on a different basis than the latter type. It uses two tuned transformers (T1 and T2), with the two pairs of crystals that are cross-connected across the tuned sections of the transformers. Crystals Y1 and Y3 are of one frequency, while Y2 and Y4 are the other frequency in the pair.

A different sort of filter is shown in Fig. 7.7A, with its asymmetrical attenuation curve shown in Fig. 7.7B. This filter has a more gradual fall-off on one side than on the other. The filter has the advantage that the frequencies of crystals Y1 and Y2 are the same. There is

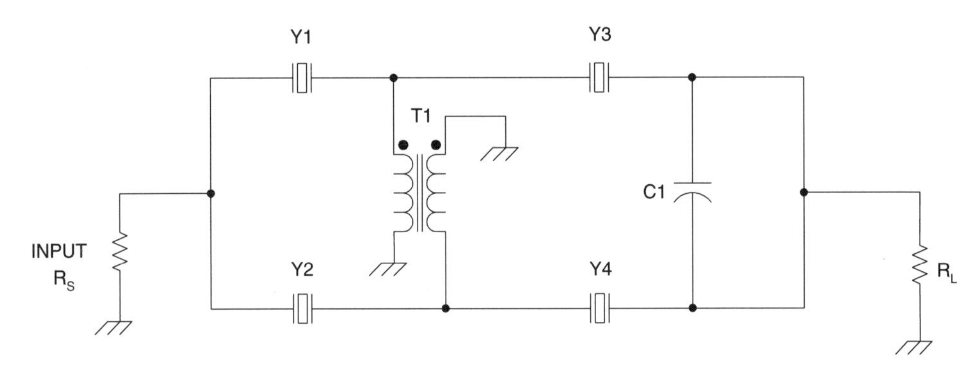

Figure 7.5 Four crystal filter.

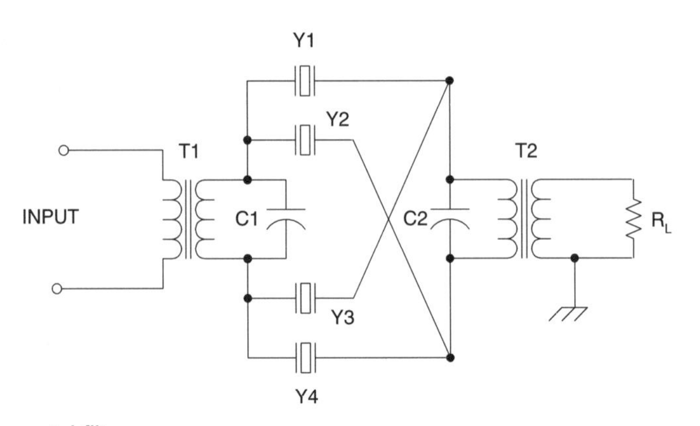

Figure 7.6 Four crystal filter.

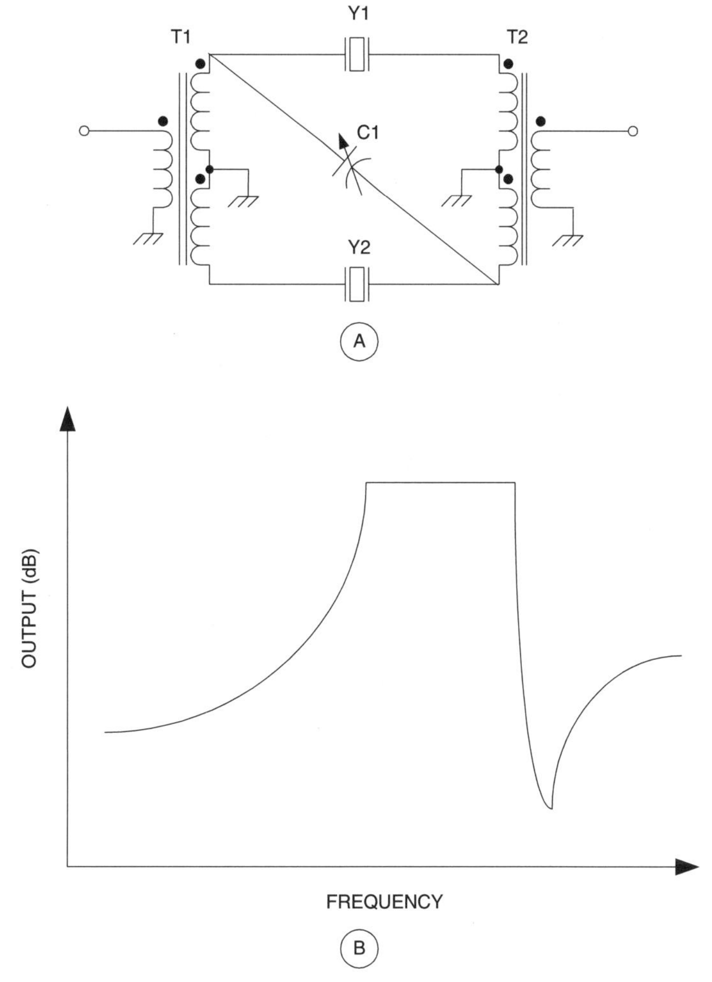

Figure 7.7 Two-crystal phasing network.

a capacitor (C1) in the circuit to tune the desired pass band. The bandwidth of this filter is only half what is expected from the half-lattice crystal filter above.

Crystal ladder filters

Figure 7.8 shows a crystal ladder filter. This filter has several advantages over the other types:

1 All crystals are the same frequency (no matching is required).
2 Filters may be constructed using an odd or even number of crystal.
3 Spurious responses are not harmful (especially for filters over four or more sections).
4 Insertion loss is very low.

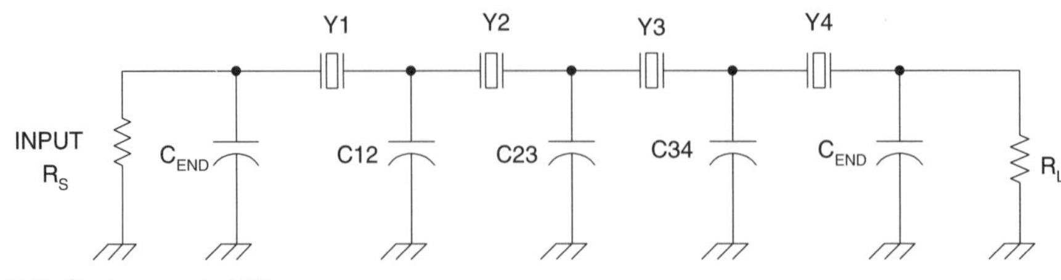

Figure 7.8 Series crystal filter.

Both Butterworth and the equi-ripple or Chebyshev responses can be created using this design. Ideally, in the Chebyshev designs the number of positive peaks should be the same as the number of crystals, and be of equal amplitude over the pass band of the filter. In reality, fewer peaks than that are found, some being merged with each other. In addition, the amplitude of the ripple increases towards the edges of the band.

The design of this filter can be simplified by using a test fixture to dope out the problem first. The value of the end capacitors is:

$$\mathrm{C_{END}} = \left[\frac{1.59 \times 10^5}{R_S F_O}\right] \times \left[\sqrt{\frac{R_S}{R_{END}} - 1}\right] - 5 \tag{7.2}$$

The value of the coupling capacitors is:

$$\mathrm{C_{JK}} = 1326\left[\frac{\Delta f}{B k_{JK} F_O}\right] - 10 \tag{7.3}$$

And the value of R_{END} is:

$$R_{END} = \left[\frac{120B}{q\Delta f}\right] - R_S \tag{7.4}$$

Where:

B is the bandwidth in hertz (Hz)
$\mathrm{C_{END}}$ is the end capacitance in picofarads (pF)
$\mathrm{C_{JK}}$ are the shunt capacitors in picofarads (pF)
F_O is the crystal centre frequency
Δf is the bandwidth measured in a test fixture
k_{JK} is the normalized values given in Tables 7.1 and 7.2
q is normalized end section Q given in Tables 7.1 and 7.2
R_S is the end termination of the filter ($R_S > R_{END}$)
R_{END} is the end termination to be used without matching capacitors

A special version of the crystal ladder filter is the *Cohn filter* or 'minimum loss' filter of Fig. 7.9. This filter rotates the end capacitors, and makes the shunt capacitors equal value. It preserves a reasonable shape factor, while minimizing loss when built with practical resonators. Like the crystal ladder filter, the Cohn filter uses the same frequency crystals

Table 7.1 Normalized *k* and *q* values (Butterworth)

N	*q*	K_{12}	K_{23}	K_{34}	K_{45}
2	1.414	0.7071	XX	XX	XX
3	1.0	0.7071	0.7071	XX	XX
4	0.785	0.8409	0.4512	0.8409	XX
5	0.618	1.000	0.5559	0.5559	1.0

Table 7.2 Normalized *k* and *q* values (Chebyshev)

N	*q*	K_{12}	K_{23}	K_{34}	K_{45}
2	1.638	1.6362	0.7016	XX	XX
3	1.433	0.6618	0.6618	XX	XX
4	1.345	0.685	0.5421	0.685	XX
5	1.301	0.7028	0.5355	0.5355	0.7028

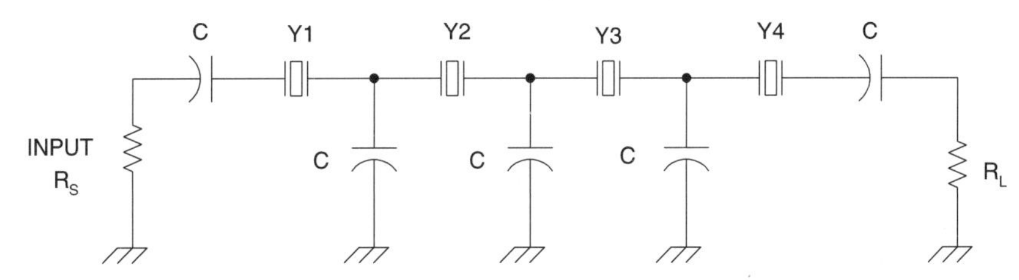

Figure 7.9 Alternate series crystal filter.

throughout. The error in frequency between the crystals (ΔF_O) should be less than 10 per cent of the desired bandwidth of the filter.

The design procedure given by Hayward (1987) is simplified:

1 Pick a crystal frequency (2 to 12 MHz). Hayward used 3.579 MHz colour burst crystals.
2 Pick a capacitance for the filter (200 pF is a good start – higher capacitance yields narrower bandwidth but higher insertion loss).
3 Vary the end termination impedance to obtain a ripple-free pass band while providing sufficient stop band attenuation.

Table 7.3 gives various Cohn filter bandwidths, termination impedances and capacitor values for a three crystal filter.

Table 7.3 Cohn three-crystal filter

Bandwidth	C (pF)	R_{END} (ohms)
380	200	150
600	130	238
1000	70	431
1800	30	1500
2500	17	3300

Monolithic ceramic crystal filters

Figure 7.10 shows a monolithic crystal filter. They are often made with synthetic piezoelectric resonators, rather than quartz. These filters are made in small packages, with some being made in crystal packages and some being made in special packages.

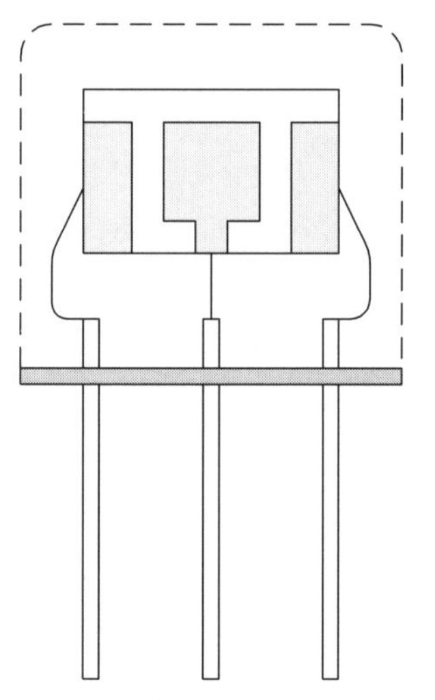

Figure 7.10 Monolithic crystal filter.

Mechanical filters

Considerable improvement in filter action is possible with the use of the *mechanical filter*. These filters were once used in Collins high-end radio receivers and SSB transmitters, but are now more widespread (although the Rockwell/Collins company still makes the filters).

The basic principle of operation is the phenomenon of *magnetostriction*, that is the *length or circumference of a piece of material will change when it is magnetized*. Nickel has this material, although the effect is one part in about 20 000. Other materials, such as the

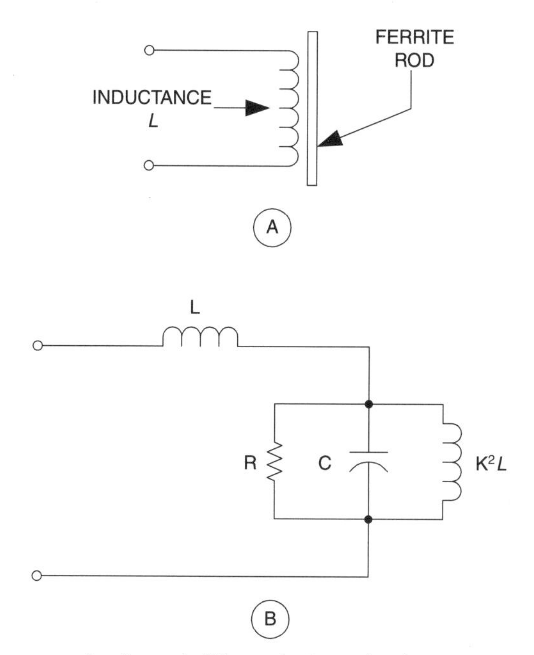

Figure 7.11 (A) Inductance on ferrite rod; (B) equivalent circuit.

ferrites or powdered iron type (61, Q1 or 4C4), provide much stronger magnetostriction effects. In addition, these materials have a high electrical resistivity, so eddy current losses are minimized, and they have a mechanical *Q* on the order of several thousand. The *Q* is determined by the proportions of oxides in the ferrite material.

Figure 7.11A shows a magnetostrictive resonator, while Fig. 7.11B shows the equivalent circuit. The ferrite rod is wound with a coil such that it is a slip fit. It will be biased magnetically with either a permanent magnet or a DC component to the electrical signal applied to the coil (L). When alternative current flows in the coil, L, it adds to or subtracts from the magnetic field of the bias, causing the ferrite length to oscillate.

Figure 7.12 shows a mechanical filter built using toroidal resonators. Various mechanical filters are available in frequencies between 60 kHz and 600 kHz. Magnetostrictive transducers are located at either end of the filter to translate electrical energy to mechanical energy, and vice versa. The resonators supply a sharp shape factor of up to 1.2:1 (60 to 6 dB), with *Q*s of 8000 to 12 000 (this is up to 150 times the *Q* of crystal filters). Small rods (not shown) couple the tranducers together. Over a temperature range of –25°C to +85°C the change of resonant frequency is as little as 1.5 parts per million. In one test, the frequency shift for 8 months was one part per million.

SAW filters

A *surface acoustic wave* (SAW) filter consists of a thin substrate sliced from a single crystal on which are vapour deposited two sets of aluminum comb-shaped electrodes. When a high frequency is applied to one of these electrodes it is converted to mechanical energy,

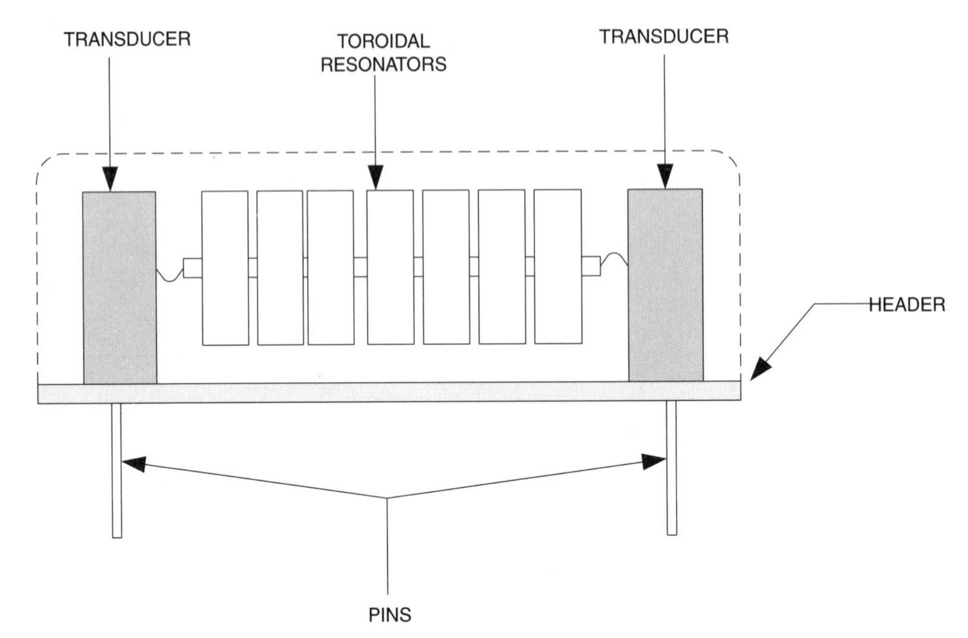

Figure 7.12 Mechanical filter.

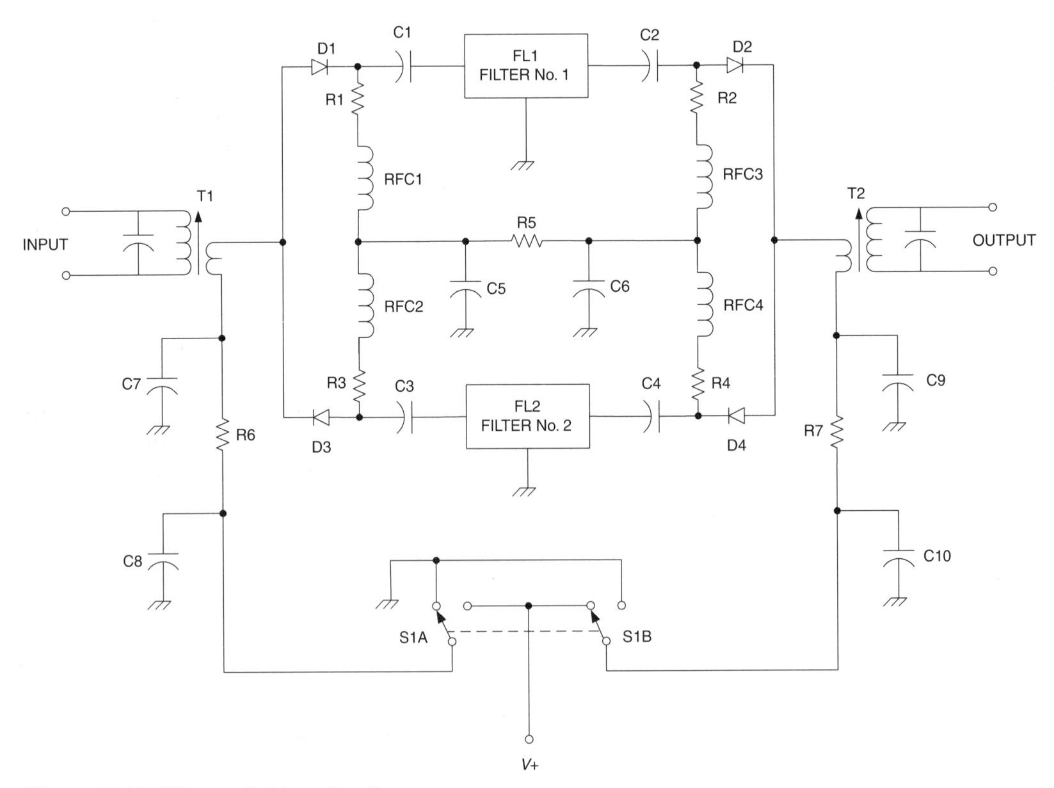

Figure 7.13 Filter switching circuit.

transmitted across the crystal, and then converted back to electrical energy at the other end. The most common substrate materials include quartz, lithium niobate, lithium tantalate crystals, langasite, with some lithium tetraborate.

SAW filters tend to be used for higher frequency IFs, typically for VHF/UHF/ microwave receivers. SAW filters can be had from about 21 MHz and up, so easily fill the bill for 30 MHz, 50 MHz or 70 MHz IFs. They are also widely used in TV receivers.

Filter switching in IF amplifiers

Switching filters is necessary to accommodate different modes of transmission. AM requires 4 or 6 kHz on short wave, and 8 kHz on the AM broadcast band (BCB). Single sideband requires 2.8 kHz, RTTY/RATT requires 1.8 kHz and CW requires either 270 Hz or 500 Hz. Similarly, FM might require 150 kHz for the FM BCB, and as little as 5 kHz on land mobile communications equipment. These various modes of transmission require different filters, and those filters have to be switched in and out of the circuit.

The switching could be done directly, but that requires either a coaxial cable between the switch and the filter, or placing the switch at the site of the filters. A better solution is found in Fig. 7.13: diode switching. A diode has the ability to pass a small AC signal on top of the DC bias. Switch S1 is used to apply the proper polarity voltage to the diodes in the circuit. In one sense of S1, the positive voltage is applied to D4 through the primary winding of T2, through RFC4 and RFC2, to D3 and then through the secondary winding of T1 to ground. The response is to turn on filter FL2. This same current flow reverse biases diodes D1 and D2, disconnecting FL1. The converse happens when S1 is turned to the other position.

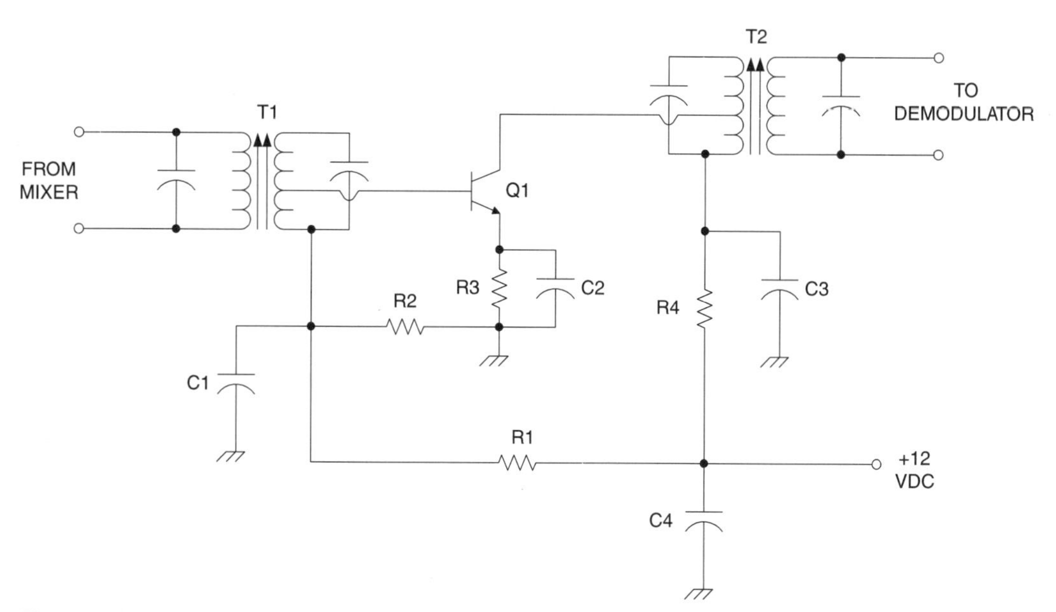

Figure 7.14 NPN IF amplifier.

Amplifier circuits

A simple IF amplifier is shown in Fig. 7.14. A simple AM band radio may have one such stage, while FM receivers, shortwave receivers and other types of communications receiver may have two to four stages like this. Transformer T1 has a low impedance tap on its secondary connected to the base of a transistor (Q1). Similarly with T2, but in this case the primary winding is tapped for the collector of the transistor.

Resistors R1 and R2 provide the bias for the transistor. Capacitor C1 is used to place the cold end of the T1 secondary at ground potential for AC signals. Resistor R3 is used to provide a bit of stability to the circuit. Capacitor C2 is used to keep the emitter of the transistor at ground potential for AC signals, while keeping it at a potential of $I_E R3$ for DC. The reactance of capacitor C2 should be $<R3/10$. Resistor R4 forms part of the collector load for transistor Q1. Capacitors C3 and C4 are used to bypass and decouple the circuit.

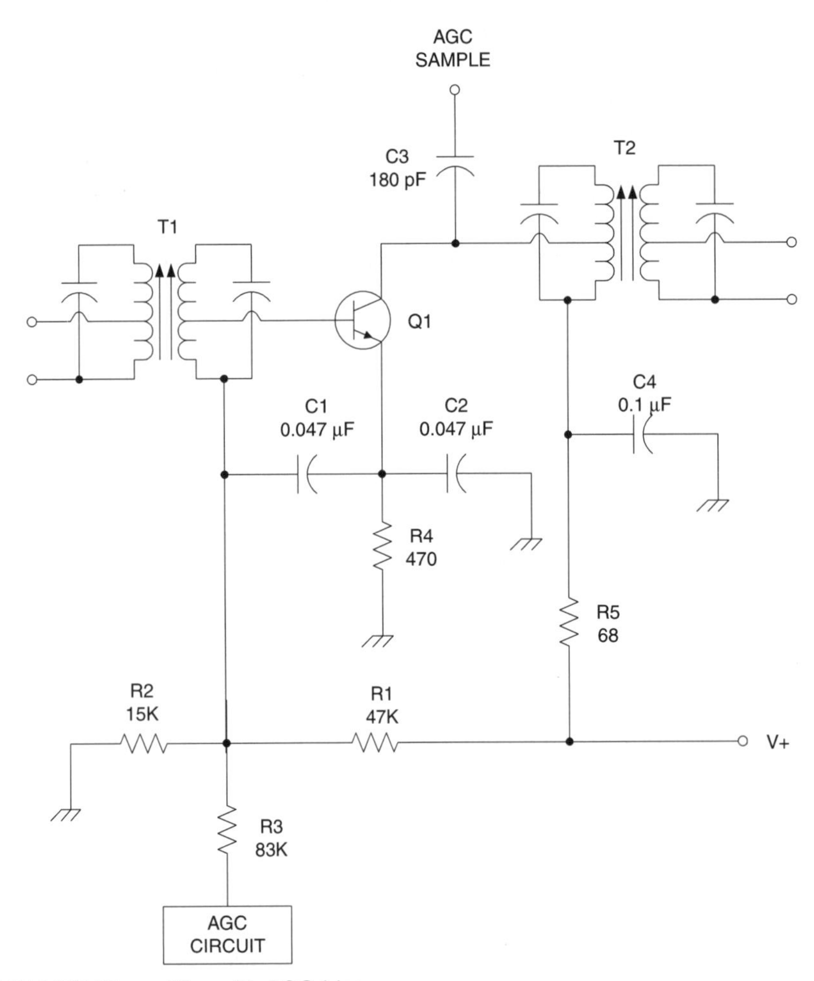

Figure 7.15 NPN IF amplifier with AGC bias.

Figure 7.15 shows a gain-controlled version of Fig. 7.14. This particular circuit is designed for a low frequency (240 kHz through 500 kHz), although with certain component value changes it could be used for higher frequencies as well. Another difference is that this circuit has double-tapped transformers for T1 and T2.

The major change is the provision for automatic gain control. A capacitor is used to sample the signal for some sort of AGC circuit. Furthermore, the circuit has an additional resistor (R3) to the DC control voltage from the AGC circuit.

Cascode pair amplifier

A cascode pair amplifier is shown in Fig. 7.16. This amplifier uses two transistors (both JFETs) in an arrangement that puts Q1 in the common source configuration and Q2 in the common gate configuration. The two transistors are direct coupled. Input and output tuning is accomplished by a pair of L–C filters (L2C1 and L3C2). To keep this circuit from oscillating at the IF frequency a neutralization capacitor (C3) is provided. This capacitor is connected from the output L–C filter on Q2 to the input of Q1.

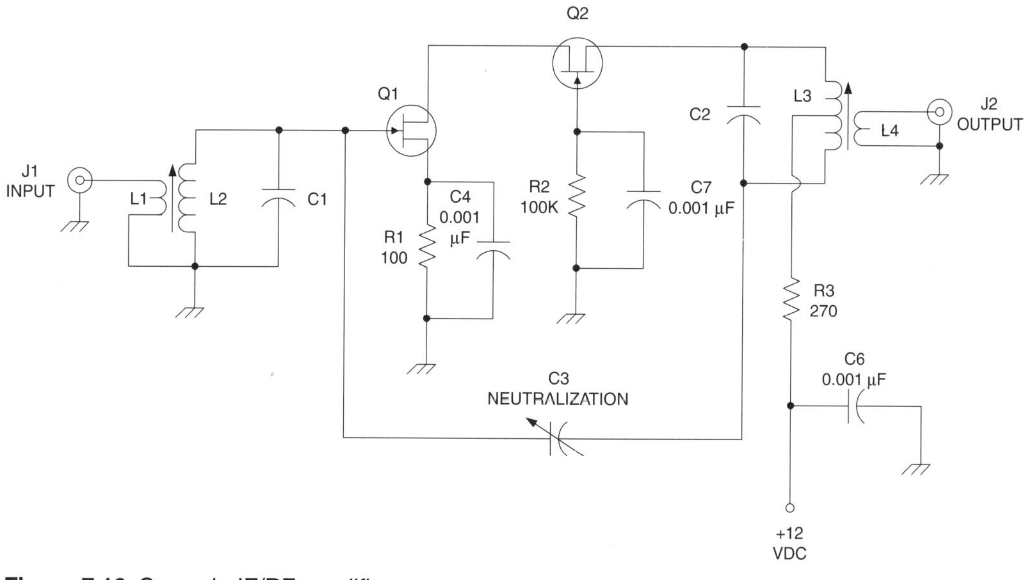

Figure 7.16 Cascode IF/RF amplifier.

'Universal' IF amplifier

The IF amplifier in Fig. 7.17 is based on the popular MC-1350P integrated circuit. This chip is easily available through any of the major mail order parts houses, and many small ones. It is basically a variation on the LM-1490 and LM-1590 type of circuit, but is a little easier to apply.

If you have difficulty locating MC-1350P devices, the exact same chip is available in the service replacement lines such as ECG and NTE. These parts lines are sold at local electronics parts distributors, and are intended for the service repair shop trade. I used

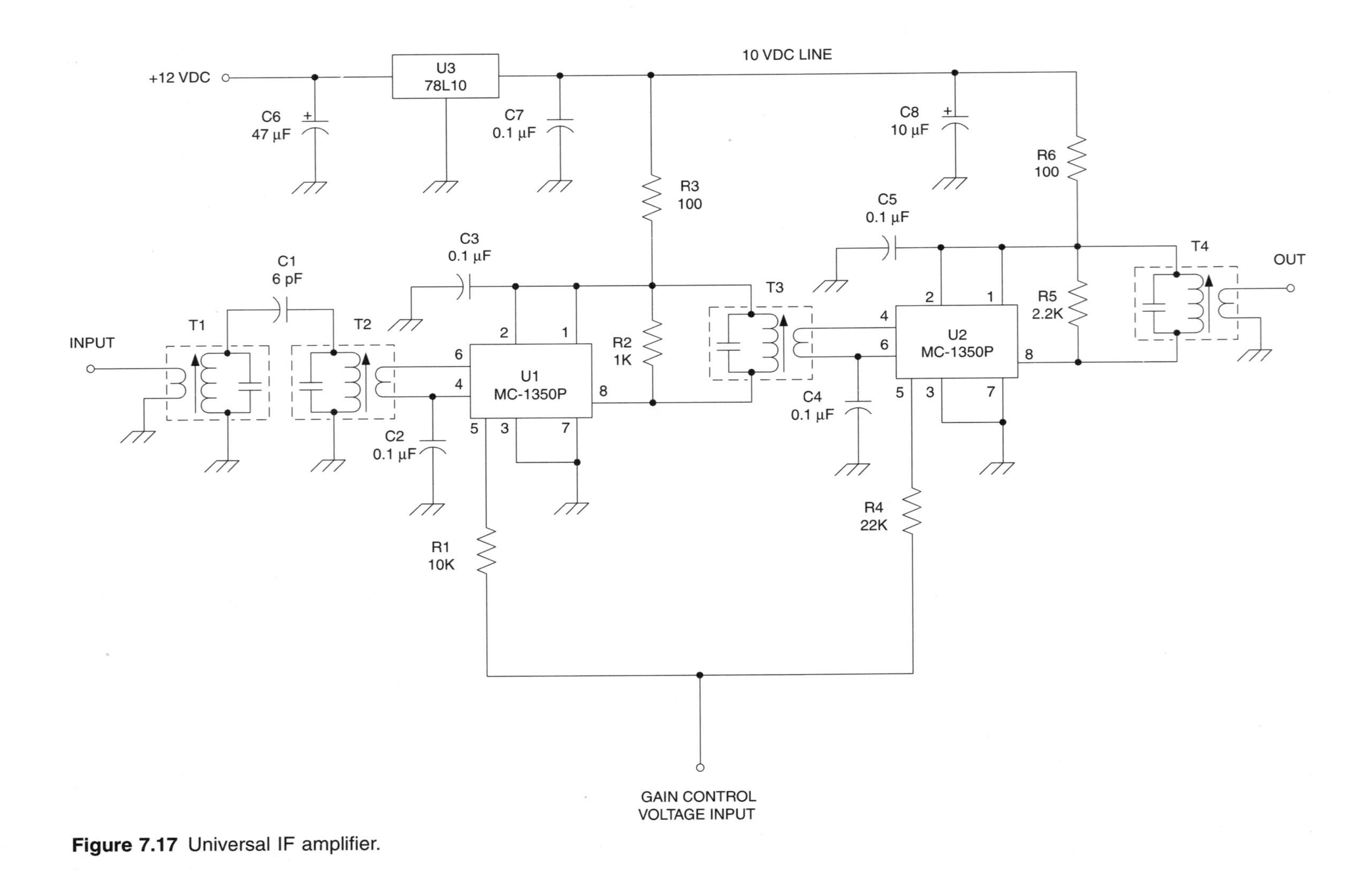

Figure 7.17 Universal IF amplifier.

actual MC-1350P chips in one version, and NTE-746 (same as ECG-746) chips in the other, without any difference in performance. The NTE and ECG chips are actually purchased from the sources of the original devices, and then renumbered.

Two MC-1350P devices in cascade are used. Each device has a differential input (pins 4 and 6). These pins are connected to the link windings on IF transformers T2/T3. In both cases, one of the input pins is grounded for AC (i.e. RF and IF) signals through a bypass capacitor (C2 and C4).

In the past I've had difficulties applying the MC-1350P devices when two were used in cascade. The problem is that these are high gain chips, and any coupling at all will cause oscillation. I've built several really good MC-1350P oscillators . . . the problem is that I was building IF *amplifiers*. The problem is basically solved by two tactics that I'd ignored in the past. This time I reversed the connections to the input terminals on the two devices. Note that pin no. 4 is bypassed to ground on U1, while on U2 it is pin no. 6. The other tactic is to use different value resistors at pin no. 5.

Pin no. 5 on the MC-1350P device is the gain control pin. It is used to provide either manual gain control (MGC) or automatic gain control (AGC). The voltage applied to this pin should be between +3 volts and +9 volts, with the highest gain being at +3 volts and nearly zero gain at +9 volts.

The outputs of the MC-1350P are connected to the primaries of T3 and T4. Each output circuit has a resistor (R2 and R5) across the transformer winding. The transformers used can be standard 'transistor radio' IF transformers provided that the impedance matching requirements are met.

The DC power is applied to the MC-1350P devices through pins 1 and 2. Bypass capacitors C3 and C5 are used to decouple the DC power lines, and thereby prevent oscillation. All of the bypass capacitors (C2, C3, C4 and C5) should be mounted as close to the bodies of U1 and U2 as possible. They can be disk ceramic devices, or some of the newer dielectric capacitors, provided of course that they are rated for operation at the frequency you select. Most capacitors will work to 10.7 MHz, but if you go to 50 MHz or so, some capacitor types might show too much reactance (disk ceramic works fine at those frequencies, however).

The DC power supply should be regulated at some voltage between +9 VDC and +15 VDC. More gain can be obtained at +15 VDC, but I used +10 VDC with good results. In each power line there is a 100 ohm resistor (R3 and R6), which helps provide some isolation between the two devices. Feedback via the power line is one source of oscillation in high frequency circuits.

There are several ways to make this circuit work at other frequencies. If you want to use a standard IF frequency up to 45 MHz or so, then select IFTs with the configuration shown in Fig. 7–17.

If you want to make the circuit operate in the HF band on a frequency other than 10.7 MHz, then it's possible to use a 10.7 MHz transformer. If the desired frequency is less than 10.7 MHz, then add a small value fixed or trimmer capacitor in parallel with the tuned winding. It will add to the built-in capacitance, reducing the resonant frequency. I don't know how low you can go, but I've had good results at the 40 metre amateur band (7–7.3 MHz) using additional capacitance across a 10.7 MHz IF transformer.

On frequencies higher than 10.7 MHz you must take some more drastic action. Take one of the transformers, and turn it over so that you can see the pins. In the middle of the bottom header, between the two rows of pins, there will be an indentation containing the tuning capacitor. It is a small tubular ceramic capacitor (you may need a magnifying glass to see it well if your eyes are like mine). If it is colour coded, then you can obtain the value using your knowledge of the standard colour code. Take a small screwdriver and crush the capacitor. Clean out all of the debris to prevent shorts at a later time. You now have an untuned transformer with an inductance probably somewhere around 2 μH. You can use the standard L–C resonance formula to calculate a new trial capacitance value.

If the original capacitor was marked as to value or colour coded, then you can calculate the approximate capacitance needed by taking the ratio of the old frequency to the new frequency, and then square it. The square of the frequency ratio is the inverse of the capacitance ratio. For example, suppose a 110 pF capacitor is used for 10.7 MHz, and you want to make a 20.5 MHz coil. The ratio is $(10.7\,\text{MHz}/20.5\,\text{MHz})^2 = 0.272$. The new capacitance will be about $0.272 \times 110\,\text{pF} = 30\,\text{pF}$. For other frequencies, you might consider using homebrew toroid inductors.

A variation on the theme is to make the circuit wideband. This can be done for a wide portion of the HF spectrum by removing the capacitors from the transformers, and not replacing them with some other capacitor. In that case, IF filtering is done at the input (between the IF amplifier and the mixer circuit).

Coupling to block filters

Some filters may require special coupling methods.

Figure 7.18 shows an approach that accommodates mechanical filters as well as crystal filters. The particular circuit shown is for very high frequency IF amplifiers (e.g. 50 MHz), but with changes to the values of the components, this IF amplifier could be used from VLF through VHF regions. The input circuit is tuned by L1 and C1, but care must be taken to get good matching otherwise noise or IMD performance could be affected. The amplifier is a MOSFET transistor device connected in the common source configuration. Gate G1 is used for the signal, and G2 is used for DC bias and gain control.

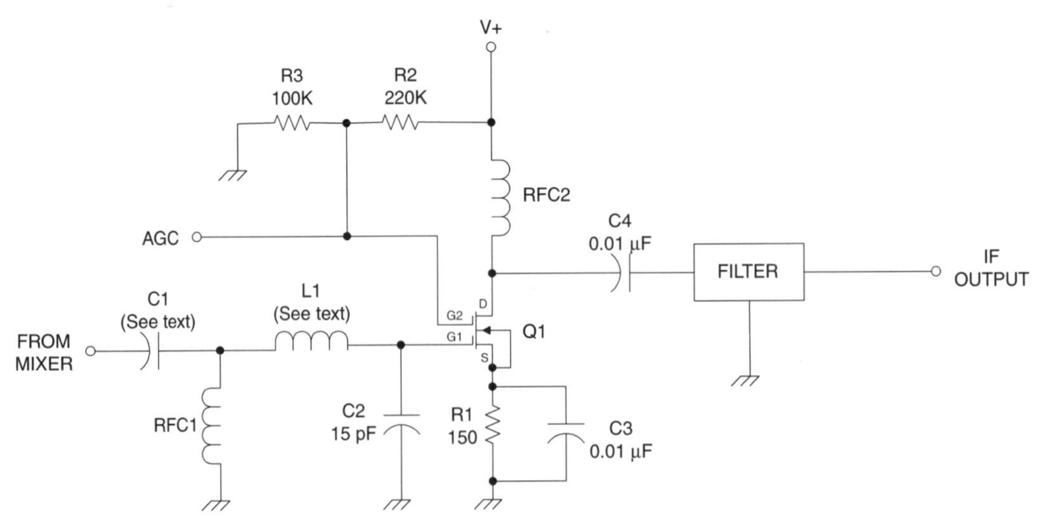

Figure 7.18 AGC-controlled IF amplifier.

The filter is connected to the amplifier through a capacitor to block the DC at the drain of the MOSFET device (similar capacitors would be used in bipolar circuits as well). The output of the filter may or may not be capacitor coupled depending upon the design of the circuits to follow this one.

More IC IF amplifiers

MC-1590 circuit

Figure 7.19 shows an amplifier based on the 1490 or 1590 chips. This particular circuit works well in the VHF region (30 to 80 MHz). Input signal is coupled to the IC through capacitor C1. Tuning is accomplished by C2 and L1, which forms a parallel resonant tank circuit. Capacitor C3 sets the unused differential input of the 1590 chip to ground potential, while retaining its DC level.

Output tuning in Fig. 7.19 reflects the fact that the 1590 chip has differential output as well as differential input. The L–C tuned circuit, consisting of the primary of T1 and capacitor C6, is parallel resonant and is connected between pins 5 and 6. A resistor across the tank circuit reduces its loaded *Q*, which has the effect of broadening the response of the circuit.

V+ power is applied to the chip both through the *V*+ terminal and pin no. 6 through the coil L2. Pin no. 2 is used as an AGC gain control terminal.

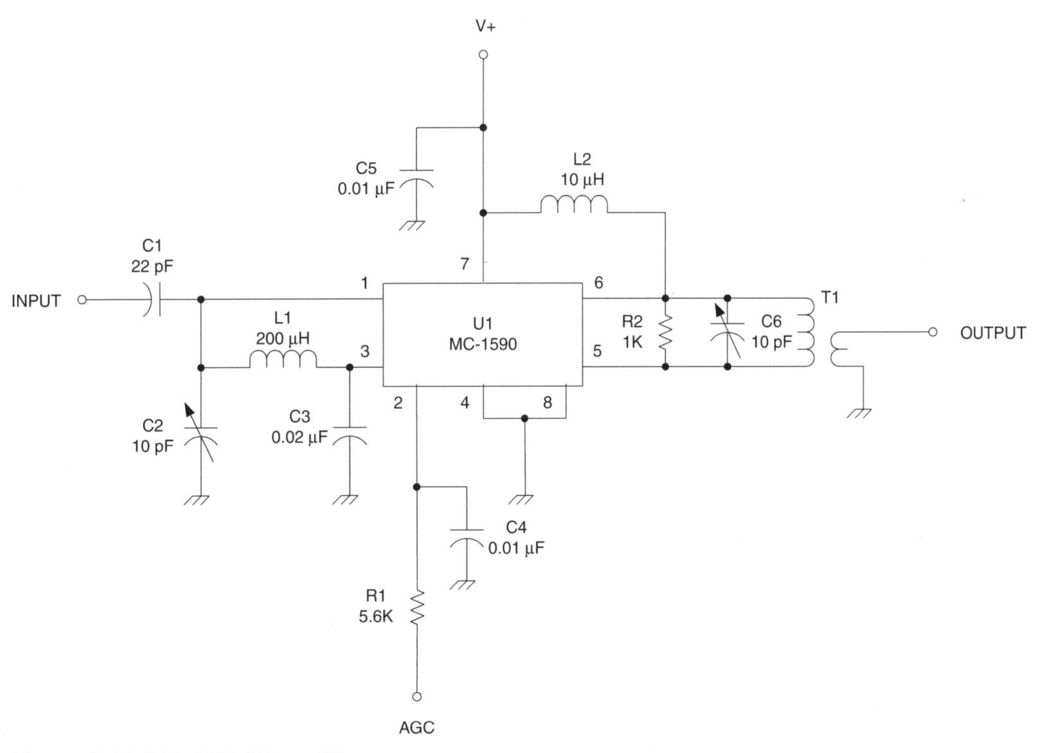

Figure 7.19 MC-1590 IF amplifier.

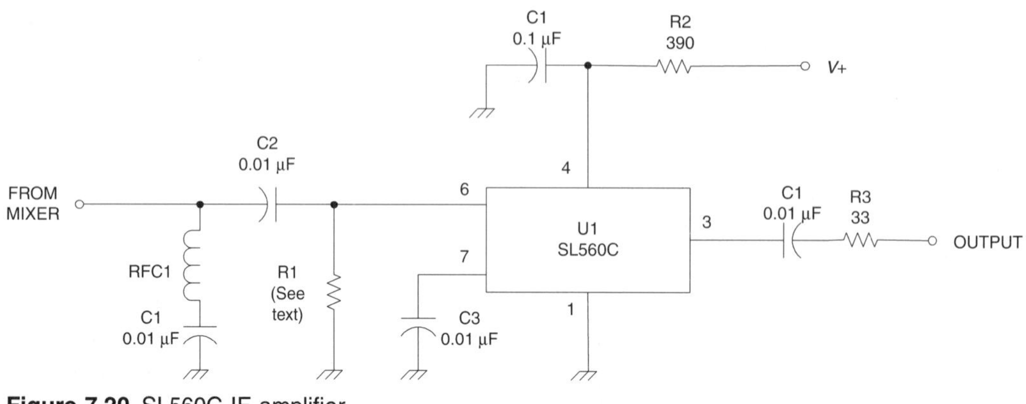

Figure 7.20 SL560C IF amplifier.

SL560C circuits

The SL560C is basically a gain block that can be used at RF and IF frequencies. Figure 7.20 shows a circuit based on the SL560C. The input of the SL560C is differential, but this is a single-ended circuit. That requires the unused input to be bypassed to ground through capacitor C3. Because this is a wideband circuit, there is no tuning associated with the input or output circuitry. The input circuitry consists of a 0.02 μF coupling capacitor, and an RF choke (RFC1).

A tuned circuit version of the circuit is shown in Fig. 7.21. This circuit replaces the input circuitry with a tuned circuit (T1), and places a transformer (T2) in the output circuit. Also different is that the V+ circuit in this case uses a zener diode to regulate the DC voltage.

FM IF amplifier

There are several forms of IC used for FM processing. The CA3189E (Fig. 7.22) is one of several that are used in broadcast and communications receivers. The input circuitry consists of a filter, although L–C-tuned circuits could be used as well. In this version, the filter circuit is coupled via a pair of capacitors (C1 and C2) to the CA3189E. The input impedance is set by resistor R1, and should reflect the needs of the filter rather than the IC (filters don't produce the same response when mismatched).

The CA3189E is an IF gain block and FM demodulator circuit, all in one IC. Coil L1 and C6 are used for the quadrature detector, and their value depends on the frequency used. There are three outputs used on the CA3189E. The audio output is derived from the demodulator, as is the automatic frequency control (AFC) output. There is also a signal strength output that can be used to drive an S-meter (M1), or left blank.

Successive detection logarithmic amplifiers

Where signal level information is required, or where instantaneous outputs are required over a wide range of input signal levels, a logarithmic amplifier might be used. Radar receivers frequently use log amps in the IF amplifier stages.

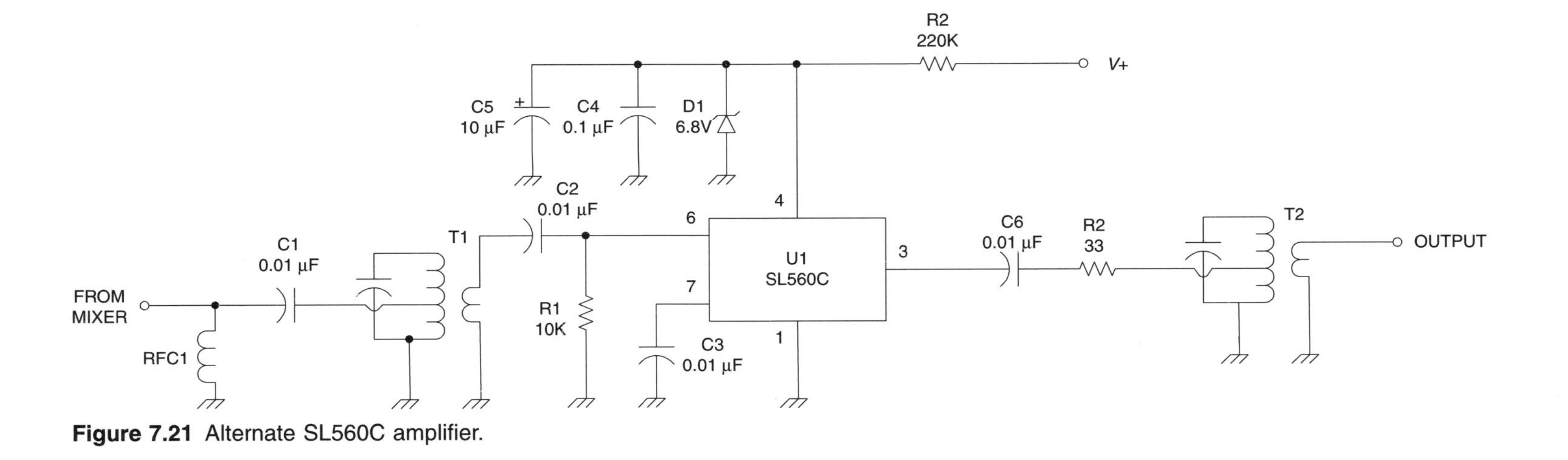

Figure 7.21 Alternate SL560C amplifier.

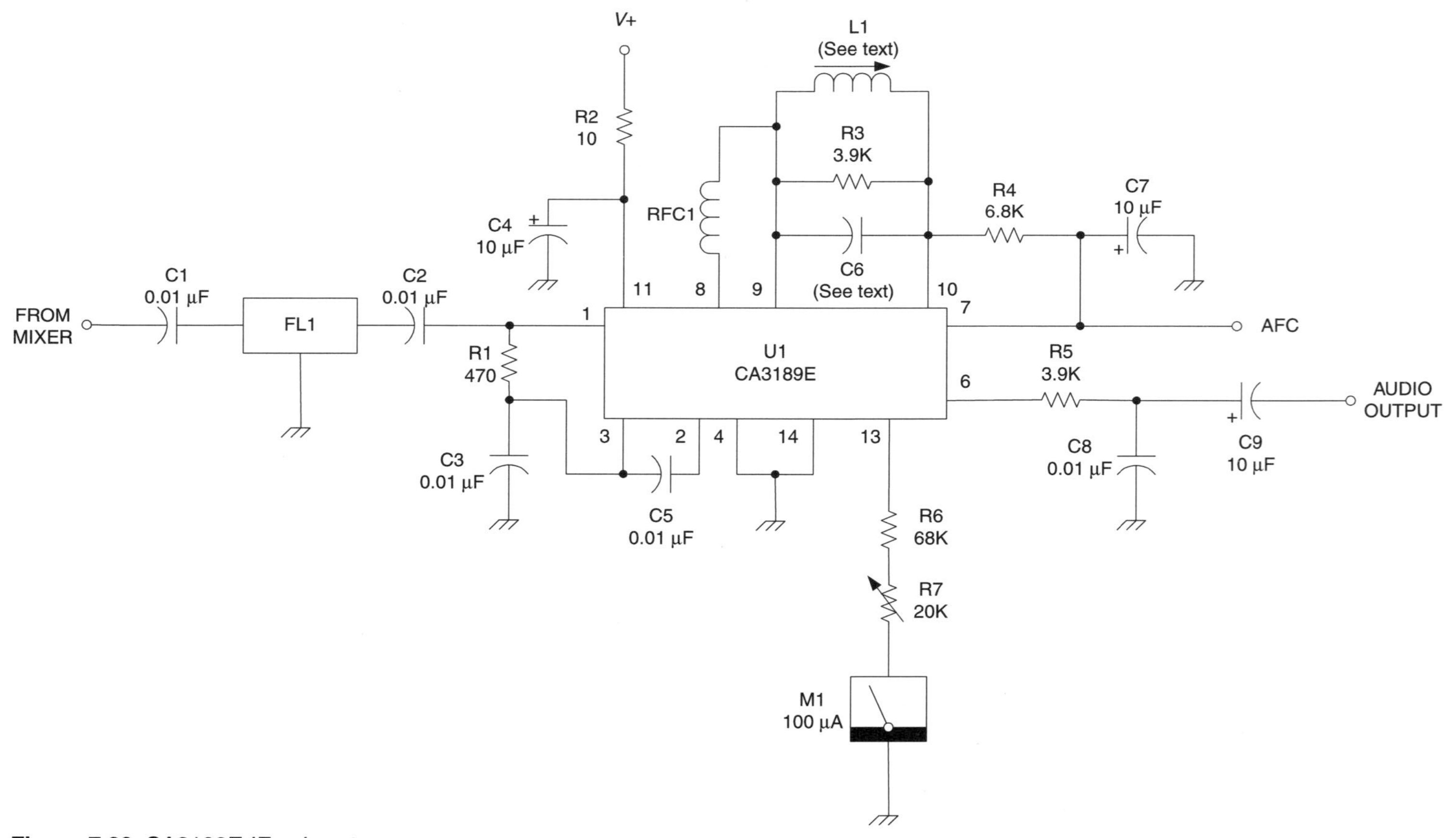

Figure 7.22 CA3189E IF subsystem.

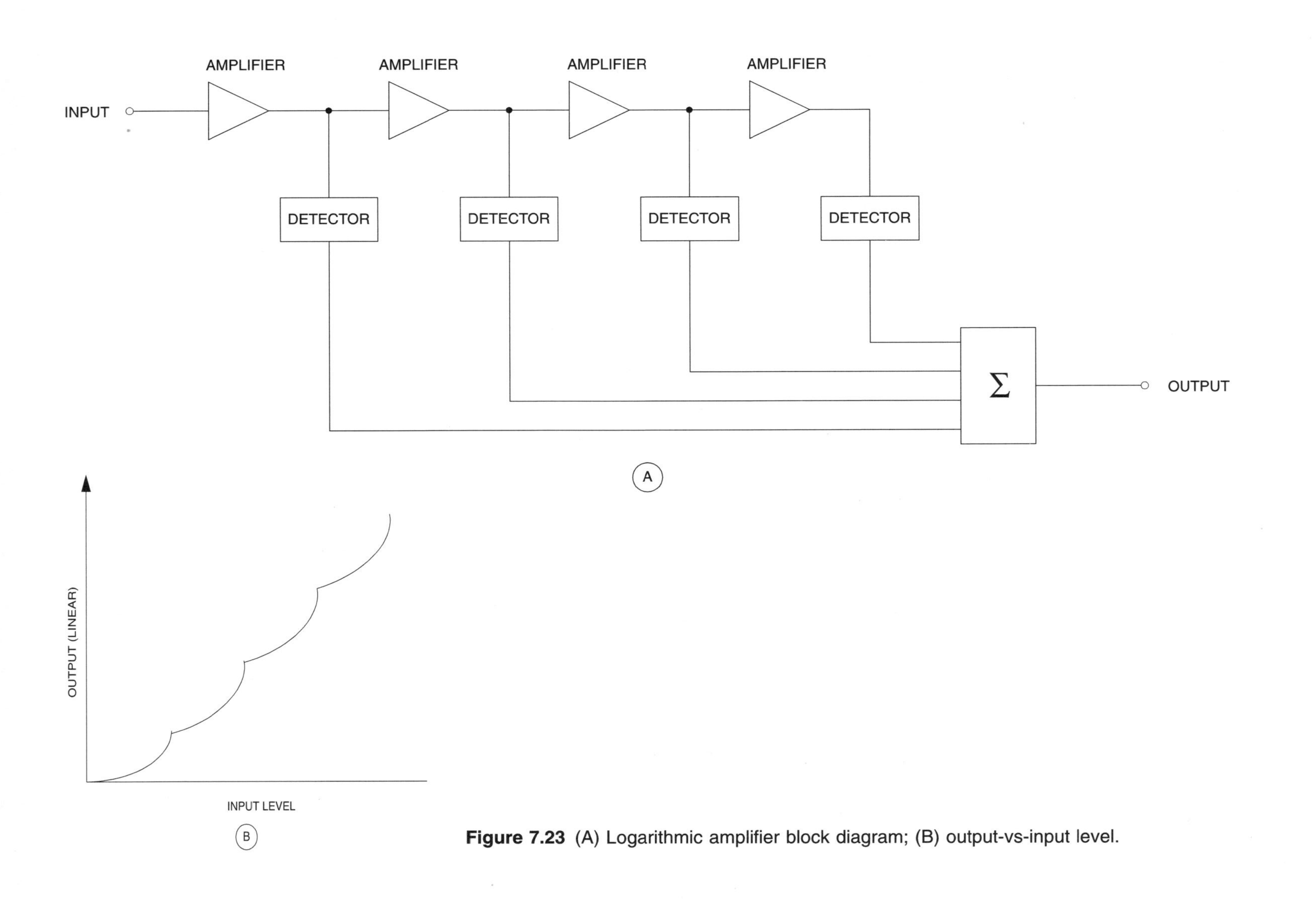

Figure 7.23 (A) Logarithmic amplifier block diagram; (B) output-vs-input level.

The successive detection method is used because it is difficult to produce high gain logarithmic amplifiers. The successive detection method uses several log amps, and then detects all outputs and sums these outputs together. Each amplifier has an output voltage equal to:

$$V_o = k \log V_{\text{in}} \tag{7.5}$$

Where:
V_o is the output voltage
V_{in} is the input voltage
k is a constant

Figure 7.23A shows the block diagram of such an amplifier, while Fig. 7.23B shows a hypothetical output-vs-input characteristic. The circuit uses four or more stages of non-linear amplification so signals are amplitude compressed. Because of this, the circuit is sometimes called a *compression amplifier.*

8 Demodulators

The purpose of a detector or demodulator circuit is to recover the intelligence impressed on the radio carrier wave at the transmitter. They are also sometimes called *second detectors* in superheterodyne receivers.

In a superheterodyne receiver the detector or demodulator circuit is placed between the IF amplifier and the audio amplifier (Fig. 8.1). This position is the same in AM, FM, pulse modulation and digital receivers (although in digital receivers the demodulator might be in a circuit called a modem).

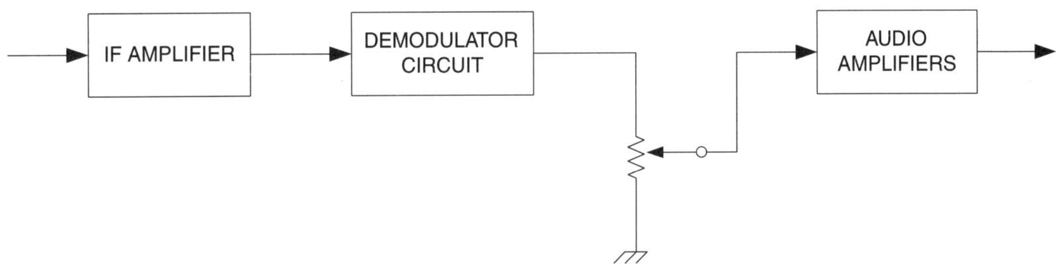

Figure 8.1 Position of demodulator in circuit.

AM envelope detectors

An AM signal consists of an RF carrier wave whose amplitude varies with the waveform of a lower frequency audio signal. Amplitude modulation is essentially a mixing process in which the RF carrier and AF signals are multiplied together, producing sum (RF + AF) and difference (RF–AF) signals in addition to the inputs. Because of the selectivity of the transmitter circuits, only the RF carrier and the sum and difference signals appear in the output. The sum signal is known as the *upper sideband* (USB), while the difference signal

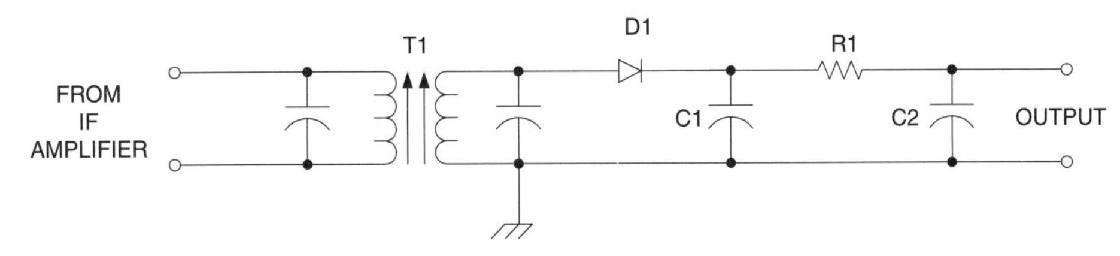

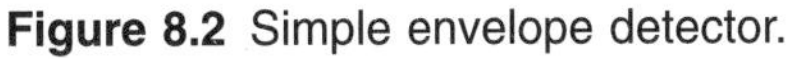

Figure 8.2 Simple envelope detector.

is known as the *lower sideband* (LSB). Hence, the bandwidth of the AM signal is determined by the highest audio frequency transmitted, and is equal to twice that frequency.

Figure 8.2 shows a simple AM *envelope detector* circuit, while Fig. 8.3 shows the waveforms associated with this circuit. The circuit consists of a signal diode rectifier connected to the output of an IF amplifier. There is a capacitor (C1) and resistive load connected to the rectifier. When the input signal is received (Fig. 8.3A), it is rectified (Fig. 8.3B), producing an average current output that translates to the voltage waveform of Fig. 8.3C. Low-pass filter R1C2 takes out the residual RF/IF signal.

The important attribute of a demodulator is for it to have a non-linear response, preferably with a sharp cut-off. Figure 8.4 shows the input and output characteristics compared with the diode's *I-vs-V* curve. At low signal levels the diode acts like a *square law detector*, but at higher signal levels the operation is somewhat linear.

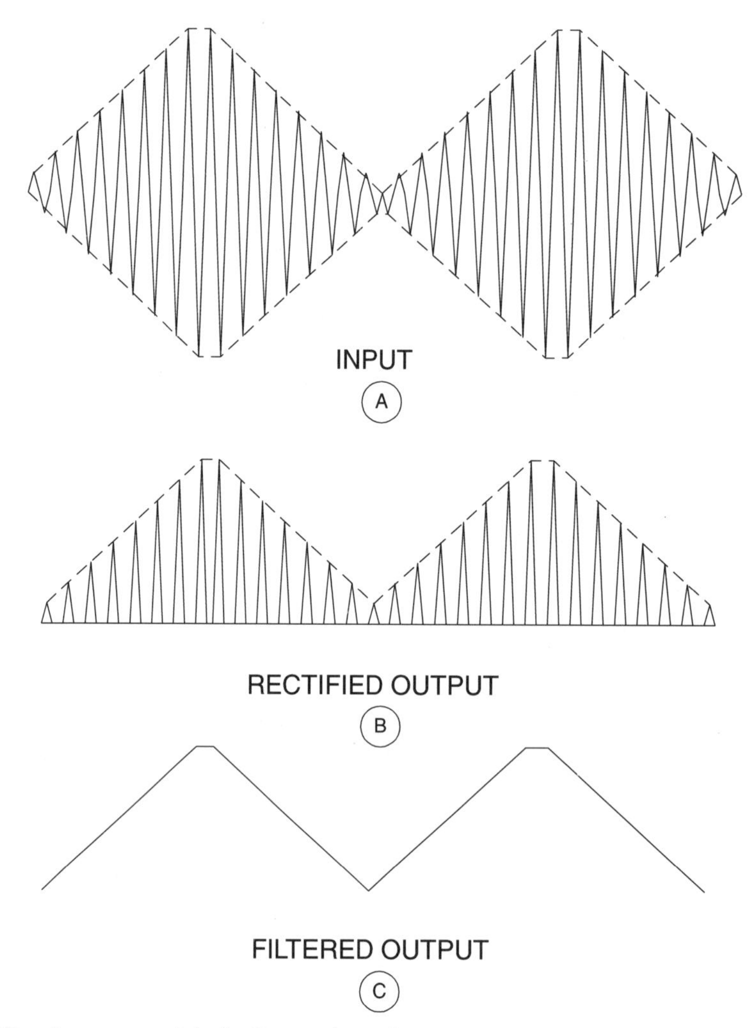

Figure 8.3 Waveforms associated with envelope detector.

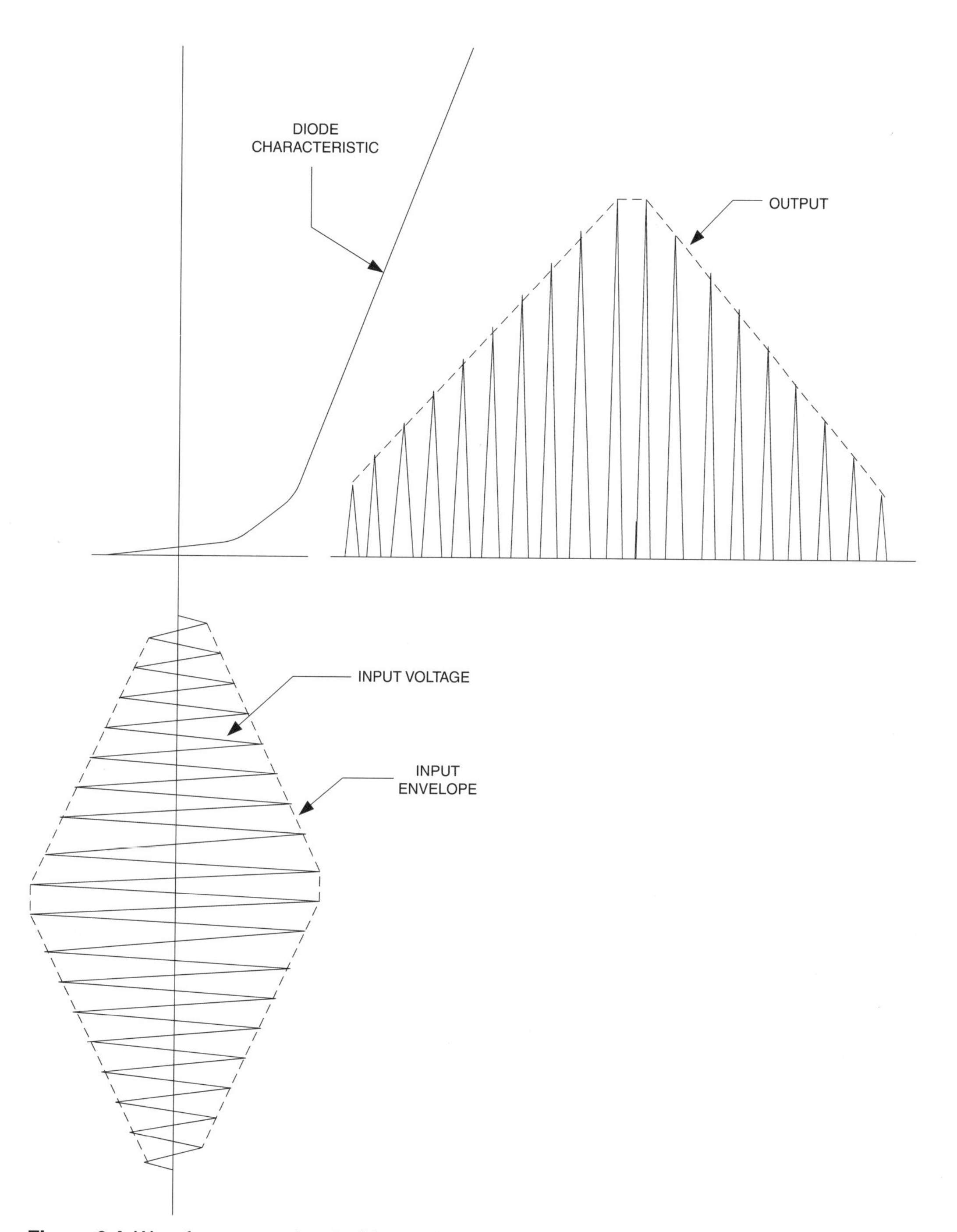

Figure 8.4 Waveforms associated with envelope detector superimposed on diode characteristic.

Consider a diode with an *I-vs-V* characteristic of:

$$I_d = a_0 + a_1 V_d + a_2 V_d^2 \tag{8.1}$$

Let:

$$V_d = A[1 + ms(t)]\cos(2\pi f_c t) \tag{8.2}$$

We can then write:

$$I_d = a_0 + a_1 A[1 + ms(t)]\cos(2\pi f_c t) + a_2 A^2[1 + ms(t)]^2\cos^2(2\pi f_c t) \tag{8.3}$$

$$I_d = a_0 + a_1\ A[1 + ms(t)]\cos(2\pi f_c t) + \frac{a_2 A^2}{2} + \ldots$$

$$\ldots + \left[a_2 A^2 ms(t) + \frac{a_2 A^2}{2} m_2 S^2(t)\right] + \left[\frac{a_2 A^2}{2}[1 + ms(t)]^2\cos(4\pi f_c t)\right] \tag{8.4}$$

Where:
m is the modulation index
f_c is the carrier frequency
A is the peak amplitude
a_0, a_1 and a_2 are constants

The terms in the brackets are the modulation and distortion products. The second-order terms are modulation and distortion, where the higher-order terms are distortion only. What falls out of the equation is that to keep distortion low, modulation index (m) must be low as well.

Figure 8.5 shows what happens at the capacitor. The dashed lines represent the output of the diode, which is a half-wave rectified RF/IF signal; the heavy line represents the capacitor voltage (V_c). The capacitor charges to the peak value, and then the diode cuts off. The voltage across the capacitor drops slightly to the point where its voltage is equal to the input voltage, when it turns on again. The diode may be modelled as a switch with resistance. During the non-conduction period, the switch is open and the capacitor discharges. During the conduction period, the 'switch' conducts and charges the

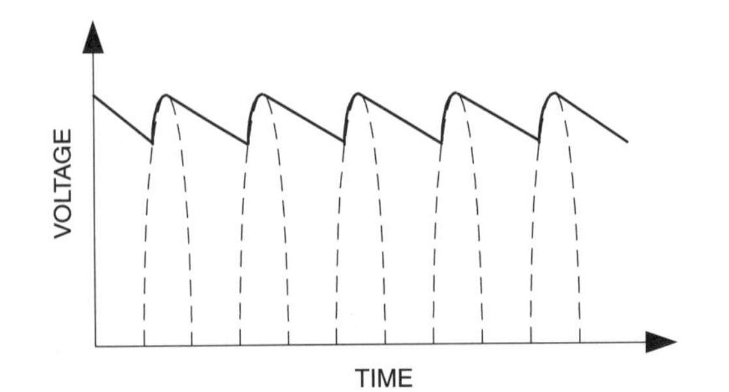

Figure 8.5 Rectified output.

capacitor. The waveform across the capacitor when there is no modulation is close to a sawtooth at the carrier frequency (f_c), and it represents the residual RF. This is reduced by the R–C filter following the envelope detector (R1C2), and the response of the audio amplifiers. These components are typically about 30 dB below the carrier level.

The maximum time constant of the filtering action of C1, plus ($R1 + R$), where R is the resistance to ground (typically a volume control), that can be accommodated depends on the maximum audio frequency that must be processed. A frequency of 3 000 Hz, and a time constant of 10 μs yields:

$$2\pi f_m T_c = 2\pi 3000 \times 10^{-5}\,\text{s} = 0.1884 \qquad (8.5)$$

That produces an output reduction of

$$\sqrt{1 + (2\pi f_m T_c)^2} = 1.018 \qquad (8.6)$$

Which is 0.16 dB. At 10 kHz these values are 0.628 and 1.18, or 1.44 dB.

Figure 8.6 shows a version of the envelope detector that uses a high-pass filter at the output. The time constant R2C2 is set to eliminate low frequency audio. This could be hum and noise from a poor quality transmitter, or LF signals which a small loudspeaker could not handle. The speech requirement in communications receivers is 300 to 3000 Hz, so a 60 Hz hum is easily accommodated.

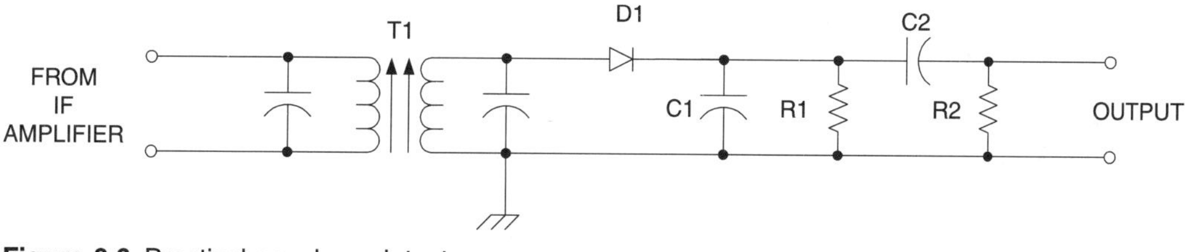

Figure 8.6 Practical envelope detector.

An envelope detector with different AC and DC loads (such as Fig. 8.6) can generate audio distortion if the modulation index is high enough. In the linear mode of operation of the diode, the modulation index for which distortion begins is:

$$m = \frac{(R1 + R2)R_d + R1R2}{(R1 + R2)(R1 + R_d)} \qquad (8.7)$$

The distortion will be small if

$$\frac{|Z_m|}{R1} < m \qquad (8.8)$$

Where:
Z_m is the impedance of the circuit at the modulating frequency

Equation (8.7) reduces to this form when $R_d << R1$, and $Z_m = R1$ in parallel with $R2$.

Another form of distortion occurs when the RF/IF waveform is distorted. Both in-phase and quadrature distortion can occur in the RF/IF waveform, especially if the bandpass filters used in the RF/IF circuit have complex poles and zeros distributed asymmetrically about the filter centre frequency, or when the carrier is not tuned to the exact centre of the filter pass band.

AM noise

All radio reception is basically a game of signal-to-noise ratio (SNR). At low signal levels (Fig. 8.7) the SNR is poor, and the noise dominates the output. As the signal level increases, the noise level increases but at a slower rate than the signal output. The noise level comes to rest (see dashed line) about 3.7 dB over the no-carrier state, but the signal level continues upward.

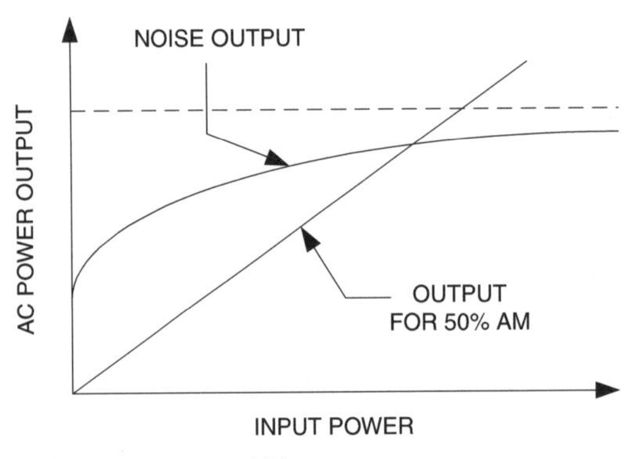

Figure 8.7 Noise-vs-output for 50 per cent AM.

Synchronous AM demodulation

One of the factors that controls the comfort level of listening to demodulated AM transmissions is that the carrier and two sidebands can fade out of phase with each other. The bad effects of this can be reduced with quasi-synchronous demodulation or synchronous demodulation. Both require that the incoming carrier be replaced by a locally generated carrier in the receiver. The difference is that in quasi-synchronous demodulation the reinserted carrier has random phase, whereas in synchronous demodulation it is locked in phase with the original carrier via a phase-locked loop (PLL).

Double sideband (DSBSC) and single sideband (SSBSC) suppressed carrier demodulators

Double and single sideband suppressed carriers are a lot more efficient than straight AM. In straight AM, the carrier contains two-thirds of the RF power, with one-third split between the two sidebands. Interestingly enough, the entire intelligent content of the

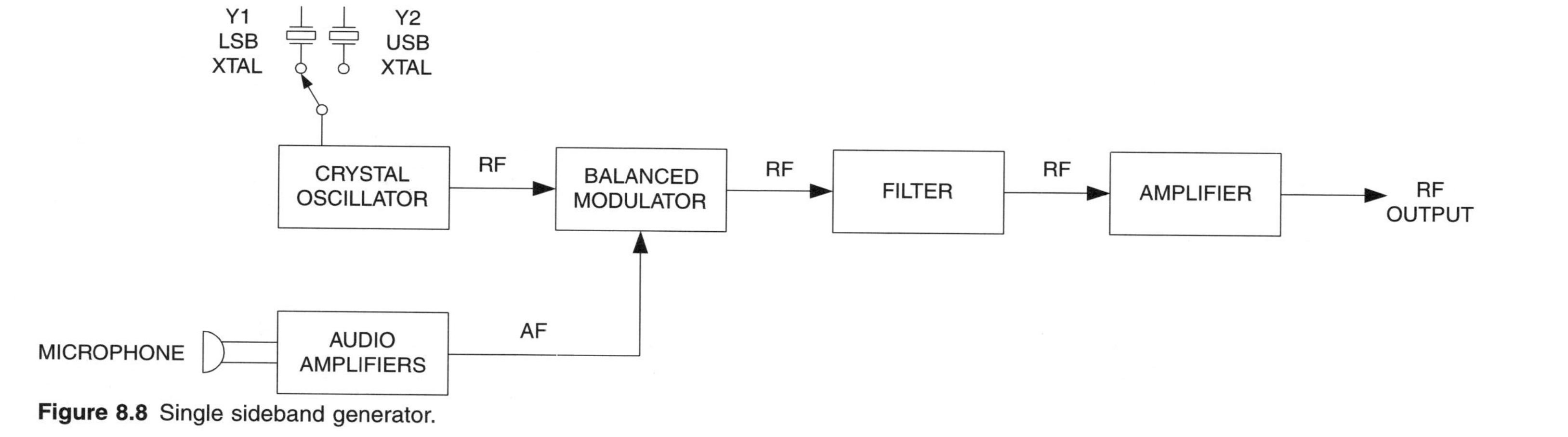

Figure 8.8 Single sideband generator.

speech waveform is fully contained within one sideband, so the other sideband and carrier are superfluous.

Figure 8.8 shows a single sideband suppressed carrier (SSBSC, usually shortened to 'SSB') transmitter. The heart of the circuit is a *balanced modulator* circuit. This circuit is balanced to produce an RF output from the crystal oscillator only when the audio frequency (AF) signal is present. As a result, the carrier is suppressed in the double sideband output.

The next stage is a symmetrical bandpass filter circuit that removes the unwanted sideband, leaving only the desired sideband. The crystal oscillator determines which sideband is generated. By positioning the frequency of the oscillator on the lower skirt of the filter, an *upper sideband* (USB) signal is generated. By positioning the frequency of the oscillator on the upper skirt of the filter, the *lower sideband* (LSB) is generated. This is shown in Fig. 8.9, where F_c is the centre frequency of the filter.

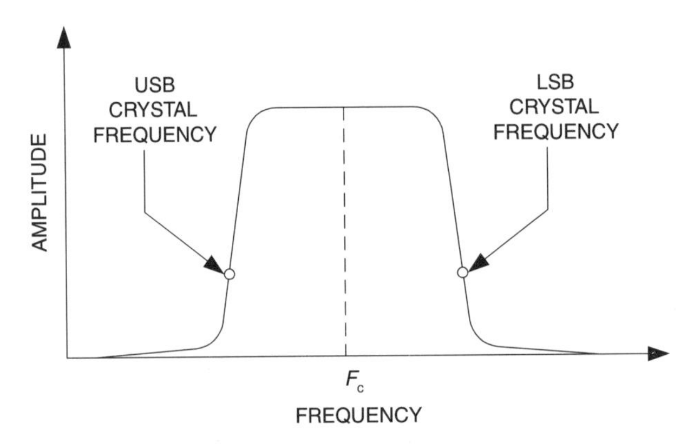

Figure 8.9 Location of USB and LSB crystals.

Following the filter circuit is any amplification or frequency mixing needed to accomplish the purposes of the transmitter. All stages following the balanced modulator are expected to be linear amplifiers, because non-linear stages will distort the envelope of the SSB signal. This means that heterodyning must be used in order to translate the frequencies, rather than multipliers or other non-linear means. It is generally the way of transmitter designers to generate the SSB signal at a fixed frequency, and then heterodyne it to the desired operating frequency. There is also a phasing method of generating an SSB signal. Double sideband transmitters are the same as Fig. 8.8, except that the filter is not present.

The basis for SSB and DSB demodulation is the *product detector* circuit (see Fig. 8.10). In Fig. 8.10A we see the SSB or DSB signal, with the carrier suppressed (it's actually a DSB signal, but an SSB signal would lack the other sideband). This signal is combined with a strong local oscillator signal (called a *beat frequency oscillator* or *BFO*), Fig. 8.10B, to produce the baseband signal (Fig. 8.10C). This is a mixing process, with the baseband signal being the difference signal.

Figure 8.11 shows a simple form of product detector circuit. It is like the envelope detector except for the extra diode (D2) and a carrier regeneration oscillator (also called a

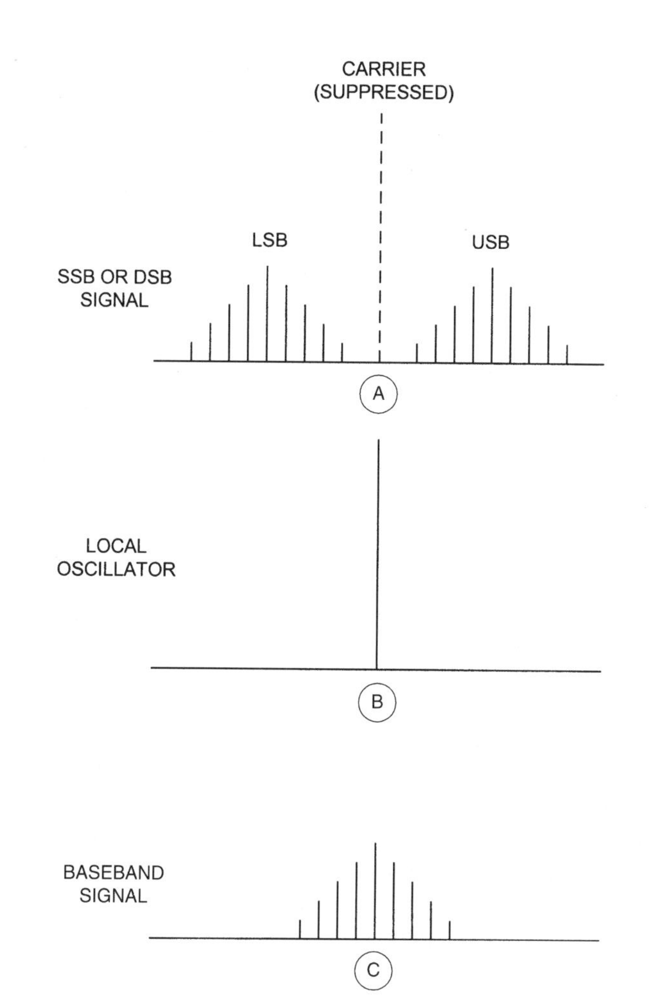

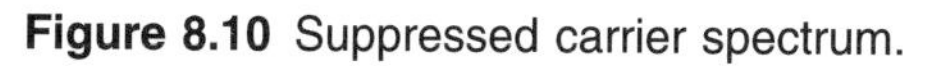

Figure 8.10 Suppressed carrier spectrum.

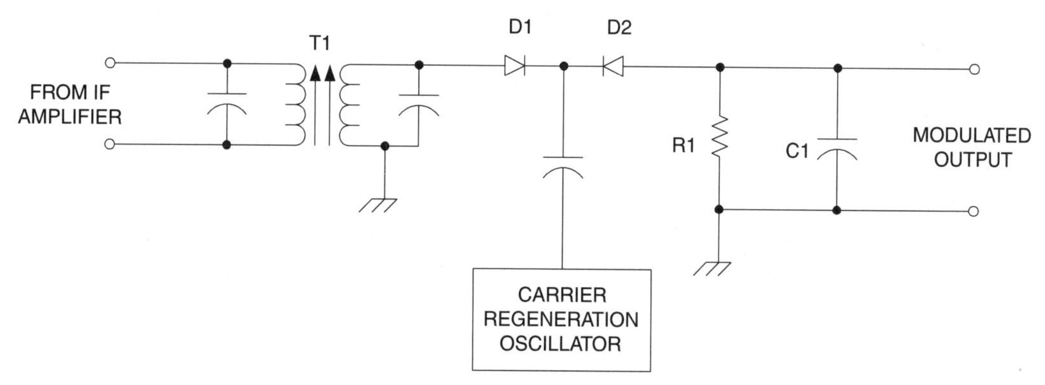

Figure 8.11 Simple diode product detector.

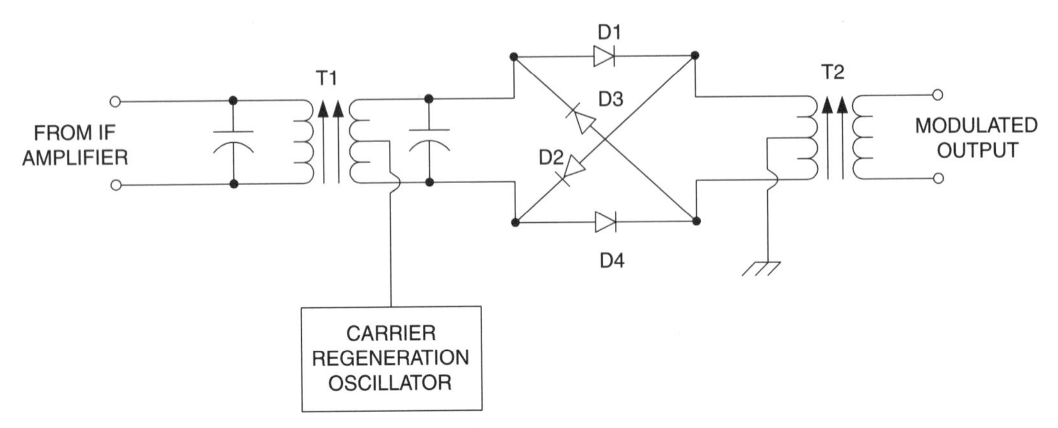

Figure 8.12 More complex diode product detector.

BFO). The circuit works by switching the diodes into and out of conduction with the oscillator signal. When the output of the oscillator is negative, the diodes will conduct, passing signal to the output. On positive excursions of the local oscillator signal the diodes are blocked from conducting by the bias produced by the oscillator signal. The residual RF/IF signal is filtered out by capacitor C1.

A superior circuit is shown in Fig. 8.12. This circuit is a balanced product detector. It consists of a balanced ring demodulator coupled through a pair of centre tapped transformers (T1 and T2). Transformer T1 is the last IF transformer. It has a centre tapped secondary to receive the strong local oscillator signal used for carrier regeneration. The output transformer (T2) is also centre tapped but in that case the centre-tap is grounded.

The circuit works by switching in and out of the circuit pairs of diodes on alternate half cycles of the oscillator waveform. On positive half cycles diodes D1 and D4 conduct, while on negative half cycles D2 and D3 conduct. The result is that there is a signal path through for both cycles, but with opposite polarity due to the balanced transformers; the diodes act like a changeover switch. The oscillator signal itself is nulled in the primary winding of T2 so does not appear in the output.

A differential pair of junction field effect transistors (JFETs) is used in Fig. 8.13 to produce the product detection. The SSB signal is applied to the gate of Q1, while the local oscillator signal is used to disrupt the operation of the circuit from the Q2 side. Keep in mind that the local oscillator signal is very much greater in amplitude than the SSB signal. A low-pass filter tuned to the spectrum that is to be recovered (typically audio) is connected to the common drain circuits, and thence to the output.

A dual-gate MOSFET transistor is the subject of Fig. 8.14. The normal signal input of the MOSFET, gate 1, is used to receive the SSB signal from the IF amplifier. The MOSFET is turned on and off by the local oscillator signal applied to gate 2. Again, an audio low-pass filter is present at the output (drain) circuit to limit the residual IF signal that gets through to the modulated output.

Notice in this circuit that there are two capacitors in the source circuit of the MOSFET transistor (C1 and C2). Typically, one of these will be for RF and the other for AF, although with modern capacitors it might not be strictly necessary.

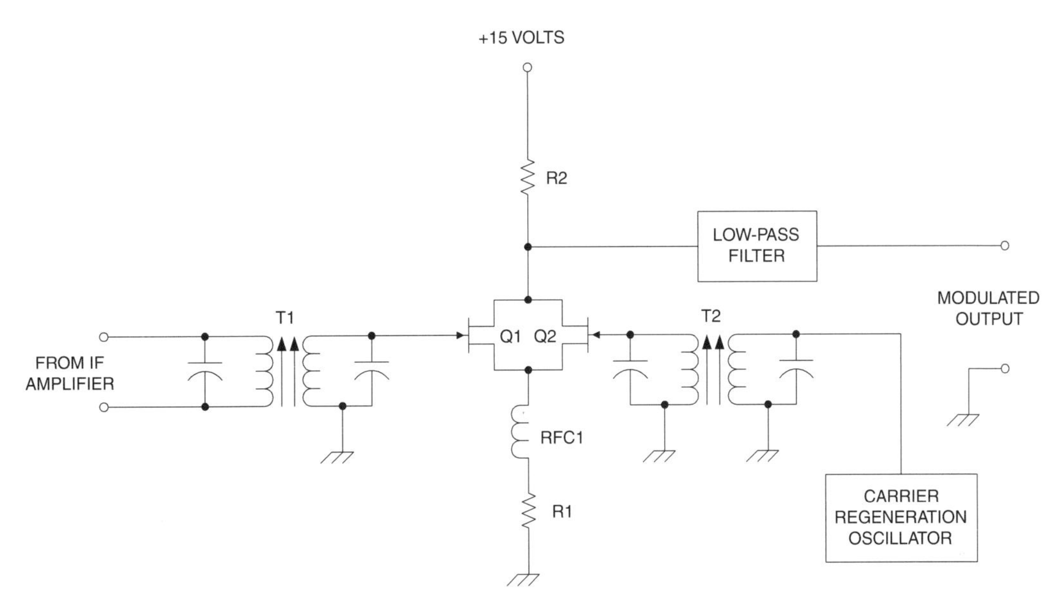

Figure 8.13 JFET differential amplifier product detector.

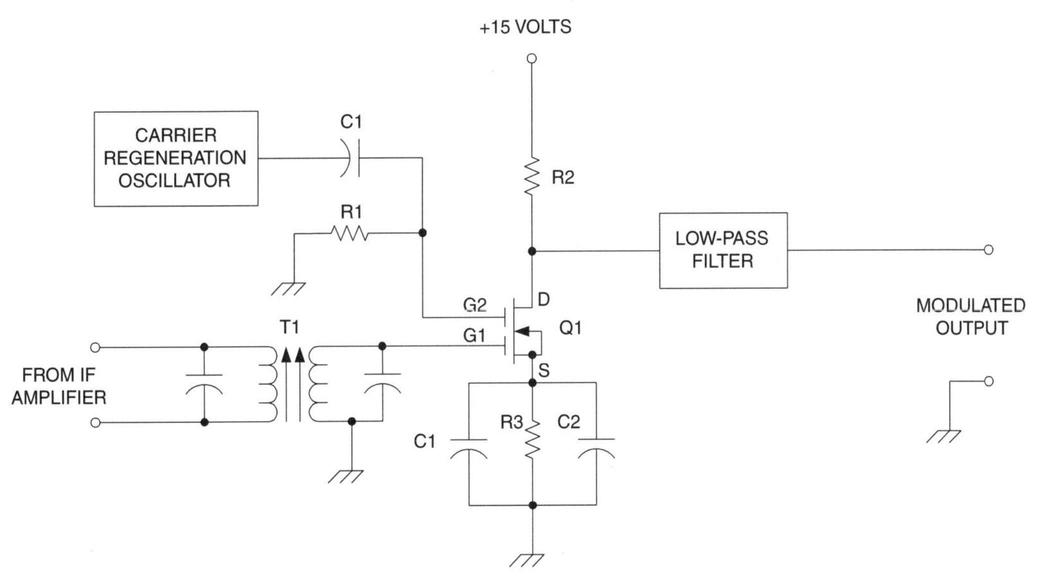

Figure 8.14 Dual-gate MOSFET product detector.

An integrated circuit SSB product detector is shown in Fig. 8.15. This circuit is based on the MC-1496 analogue multiplier chip. It contains a transconductance cell demodulator that is switched on and off by the action of the local oscillator. The SSB IF signal is input through pin no. 1, and the local oscillator through pin no. 10. The alternate pins in each case are biased and not otherwise used.

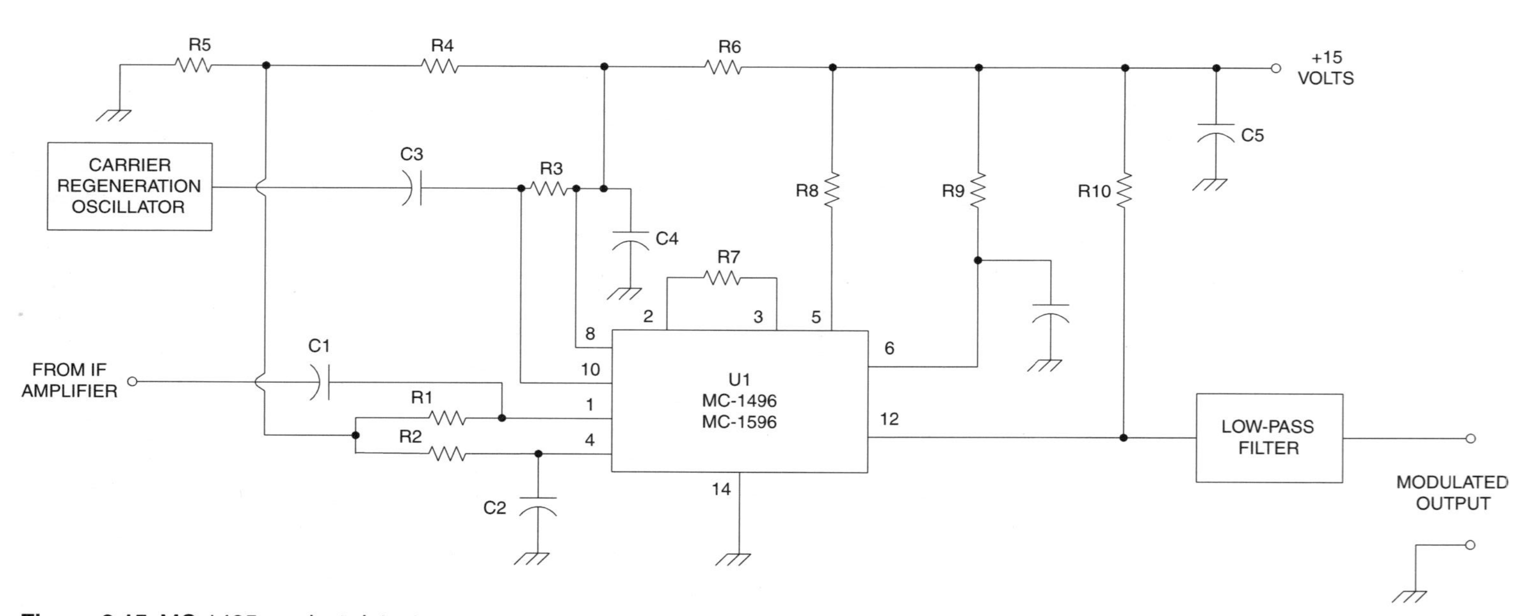

Figure 8.15 MC-1495 product detector.

Phasing method

An SSB signal is usually demodulated after IF filtering to eliminate unwanted signals, which may include the other sideband frequency. There is also a phasing method used in some cases. Figure 8.16 shows this method in block diagram form. There is a similar method for generating SSB without a filter.

The phasing method splits the SSB IF signal into two paths, I and Q. They are mixed with a pair of oscillator signals that are the same except for phasing. The I signal is mixed with the $\cos(2\pi n F_c/F)$ version of the local oscillator signal, while the Q is mixed with the $-\sin(2\pi n F_c/F)$ version of the same signal. The Q channel is passed through a Hilbert transformer, which has the effect of further phase shifting it by 90 degrees. The I channel signal is delayed an amount equal to the delay of the Q channel signal. The two signals are then summed in a linear mixer circuit to produce the output. This method will yield the lower sideband part of the signal. If we subtract the two signals instead of adding them we will yield the upper sideband signal.

FM and PM demodulator circuits

Frequency modulation and *phase modulation* are examples of *angle modulation*. Figure 8.17 shows this action graphically. The audio signal causes the frequency (or phase) to shift plus and minus from the quiescent value, which exists when there is no modulation present. Frequency and phase modulation are different, but similar enough to make the demodulation schemes the same.

The difference between FM and PM transmitters is the location of the *reactance modulator* used to generate the modulated signal (Fig. 8.18). In the FM transmitter the reactance modulator is part of the frequency determining circuitry (Fig. 8.18A), where in the PM transmitter it follows that circuitry (Fig. 8.18B).

FM audio signals sometimes have high frequency pre-emphasis applied to reduce received noise. PM signals generate this pre-emphasis automatically.

Now that we've discussed FM and PM transmitters, let's take a look at the receiver circuits used to demodulate the FM and PM signals.

Foster–Seeley discriminator

The Foster–Seeley discriminator circuit is shown in Fig. 8.19, while the waveforms are shown in Fig. 8.20. The output voltage is the algebraic sum of the voltages developed across the R2 and R3 load resistances. Figure 8.20A shows the relationship of the output voltage and the frequency.

The primary-tuned circuit is in series with both halves of the secondary winding. When the signal is at the centre frequency, the IF voltage across the secondary is 90 degrees out of phase with the primary voltage. This makes the voltages applied to each diode equal (Fig. 8.20B), resulting in zero output. When the frequency deviates (Fig. 8.20C), the voltages applied to the diodes are no longer equal and that creates an output from the

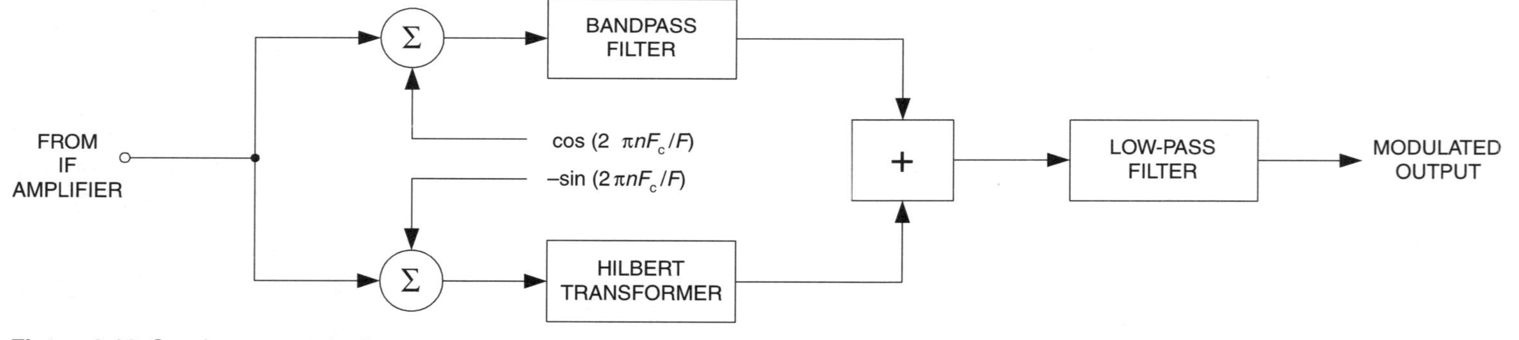

Figure 8.16 Synchronous style detector.

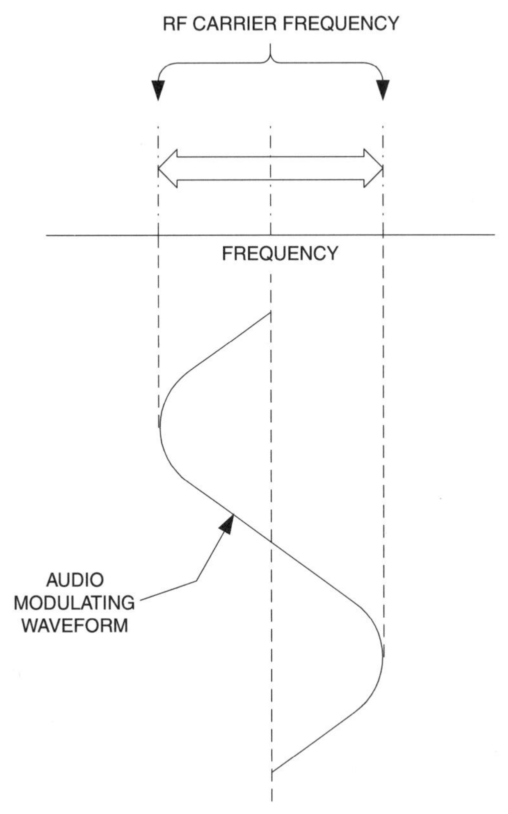

Figure 8.17 Relationship of audio to RF frequency in FM.

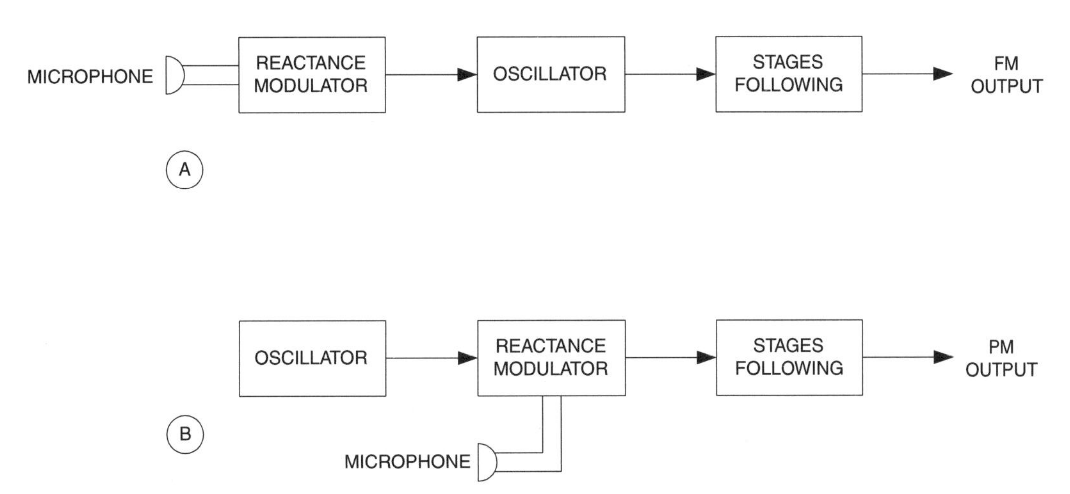

Figure 8.18 (A) Straight FM generation; (B) phase modulation generation.

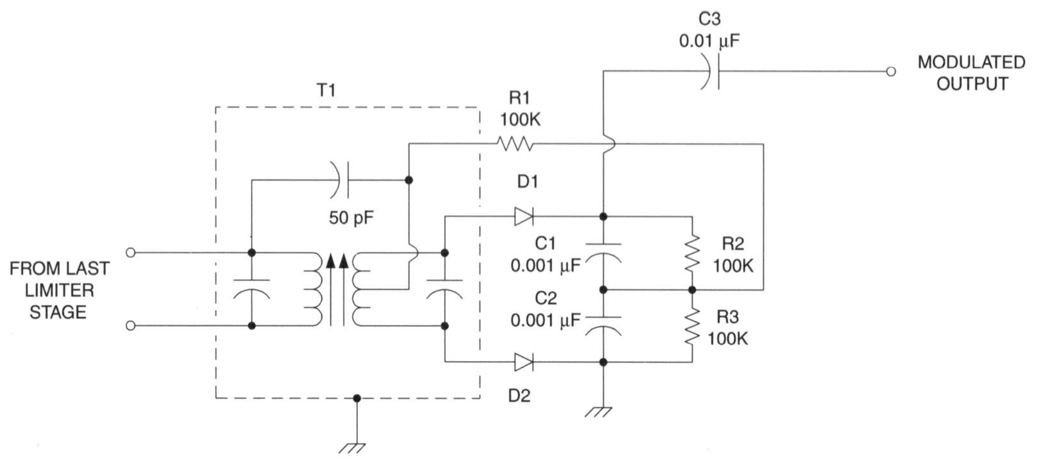

Figure 8.19 FM discriminator.

detector that is frequency sensitive. As the input signal deviates back and forth across the centre frequency of the tuned circuit an audio signal is created equal to the modulation.

In order for the FM/PM transmitter to be received 'noise free' it is necessary to precede the discriminator circuit with a *limiter circuit*. This circuit limits the positive and negative voltage excursions of the IF signal, thus clipping off AM noise.

Ratio detector

The *ratio detector* circuit is shown in Fig. 8.21. This circuit uses a special transformer in which there is a small capacitor between the centre tap on the primary winding and the centre tap on the secondary winding. Note that the diodes are connected to aid each other, rather than buck each other as was the case in the Foster–Seeley discriminator circuit. When the signal is unmodulated, the voltage appearing across R3 is one-half the AGC (automatic gain control) voltage appearing across R2 because the contribution of each diode is the same. However, that situation changes as the input signal is modulated above or below the centre frequency. In that case, the relative contribution of each diode changes. The total output voltage is equal to their ratio, hence the name ratio detector.

There are several advantages of a ratio detector over a Foster–Seeley discriminator. First, there is no need for a limiter amplifier ahead of the ratio detector, as there is with the Foster–Seeley discriminator. Furthermore, the circuit provides an AGC voltage, which can be used to control the gain of preceding RF or IF amplifier stages. However, the ratio detector is sensitive to AM variations of the incoming signal, so the AGC should be used on the stage preceding the ratio detector to limit those AM excursions. The capacitor, C3, also helps eliminate the AM component of the signal, which is noise.

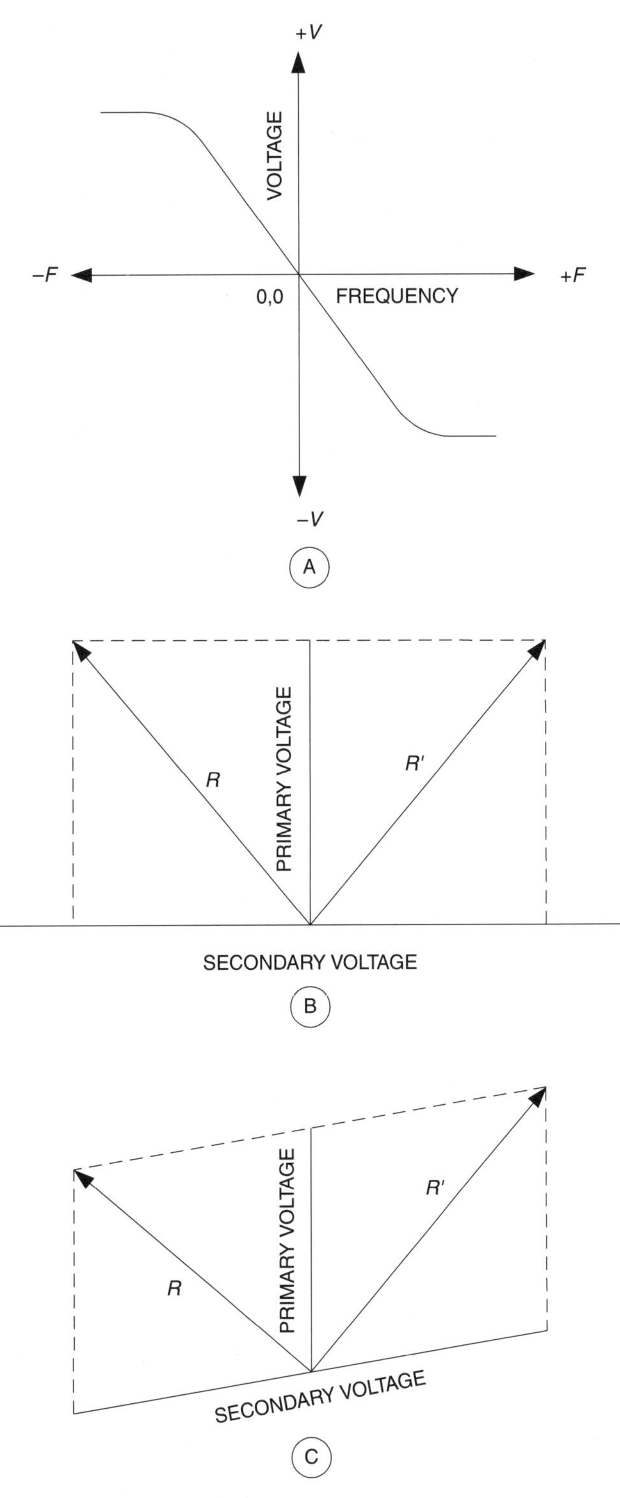

Figure 8.20 Vector relationships in discriminator.

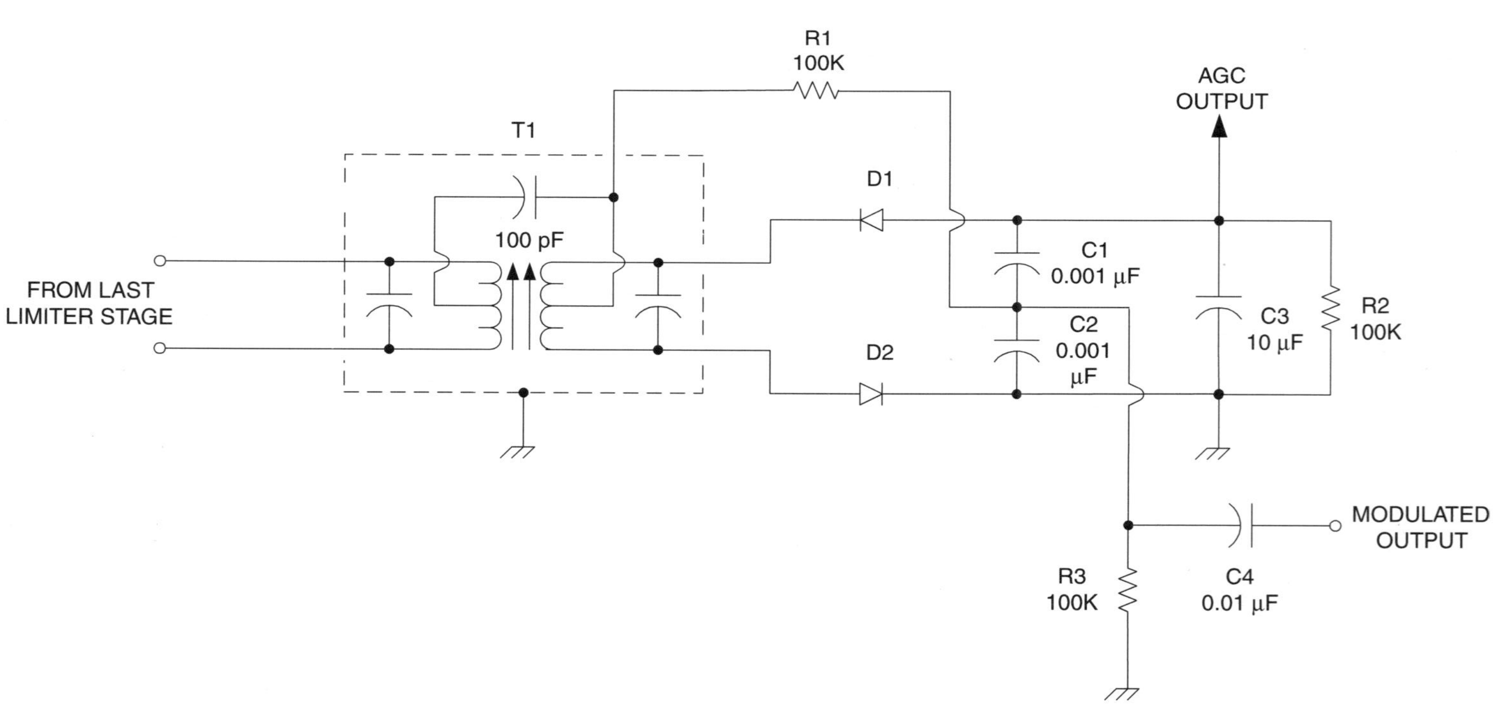

Figure 8.21 Ratio detector.

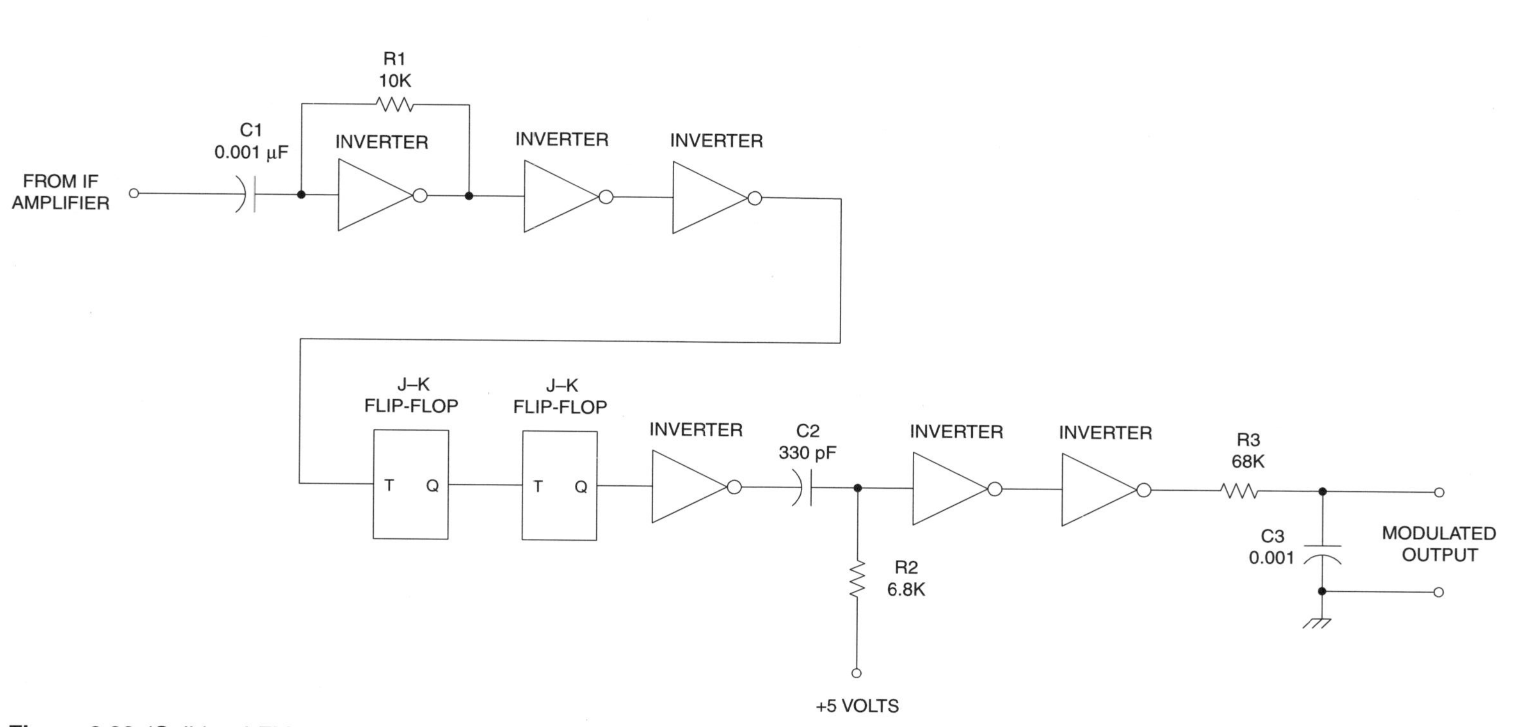

Figure 8.22 'Coil-less' FM detector.

Pulse counting detector

The FM/PM detectors thus far considered have required special transformers to make them work. In this section we are going to look at a species of coil-less FM detector. A pulse counting detector is shown in Fig. 8.22. This example uses two integrated circuits, a hex inverter and a dual J–K flip-flop. The hex inverter has six inverter stages. The first

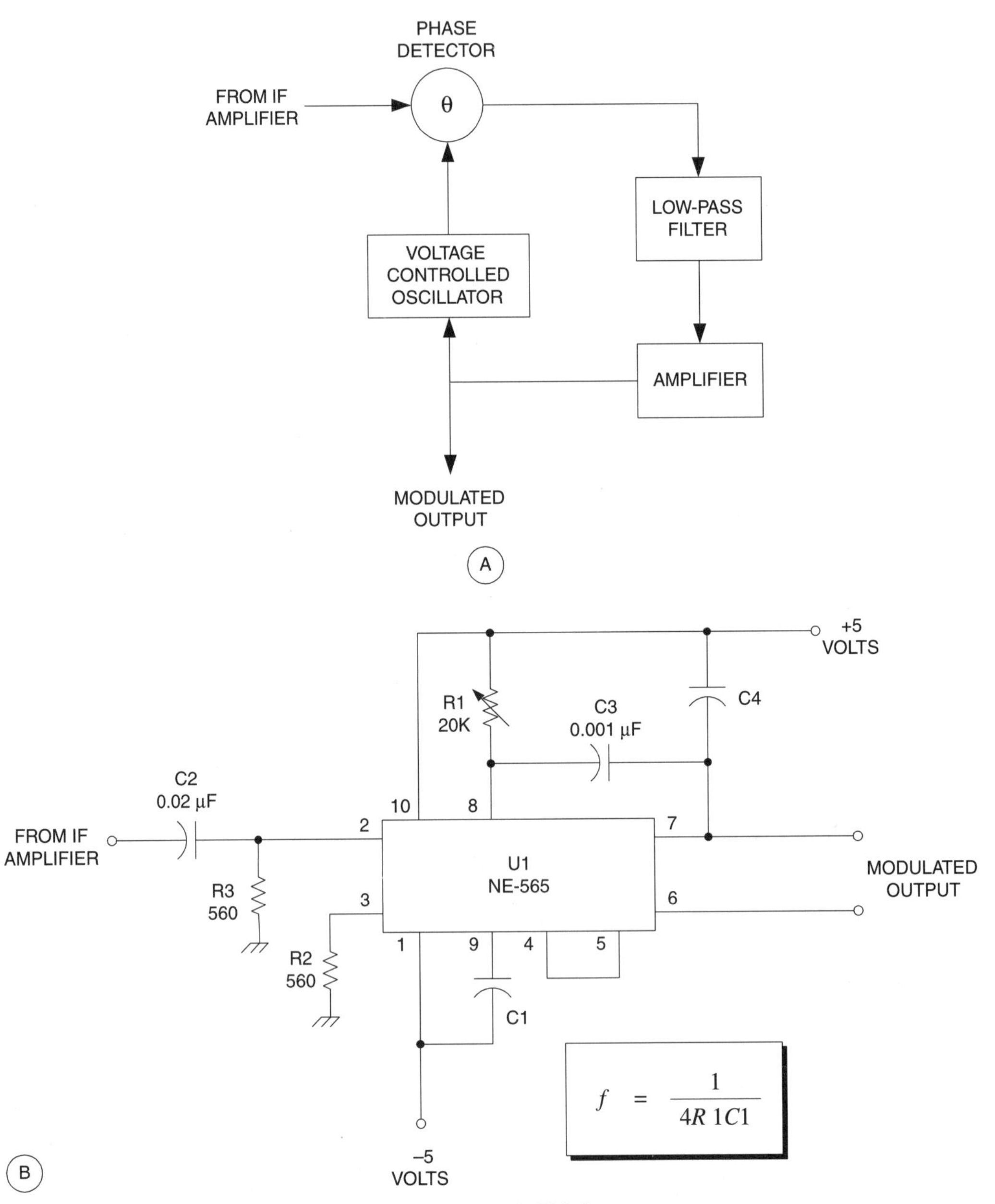

Figure 8.23 (A) PLL block diagram; (B) NE-565 PLL FM detector.

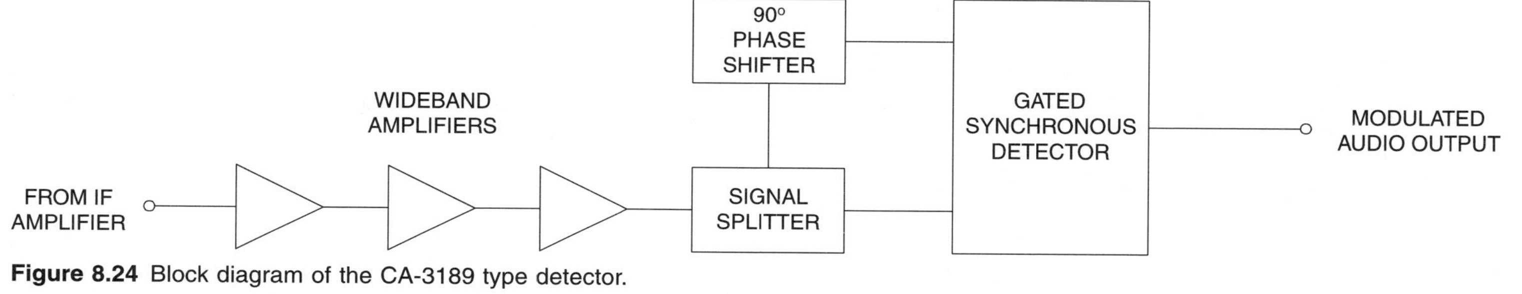

Figure 8.24 Block diagram of the CA-3189 type detector.

stage is used as an amplifier, while the next two are used to produce an output that is free of AM noise (most noise is AM). This is followed by a pair of divide-by-2 (total divide-by-4) stages consisting of a pair of J–K flip-flops. An inverter at the output of the flip-flops is used to drive a circuit which is basically half a multivibrator (but with no feedback), followed by another limiter.

The idea is to generate a sequence of pulses all of the same period (set by C2, R2), but at a rate set by the input frequency. This means that the mark–space ratio, and hence the average voltage, varies with frequency. It is realized as audio in the low-pass filter consisting of R3–C3.

Phase-locked loop FM/PM detectors

The *phase-locked loop* (PLL) circuit can be used as an FM demodulator if its control voltage is monitored. Figure 8.23A shows the basic PLL circuit. It consists of a voltage-controlled oscillator (VCO), phase detector, low-pass filter and an amplifier. The FM signal from the IF amplifier is applied to one port of the phase detector, and the output of the VCO is connected to the other port. When the two frequencies are equal, there is no output from the circuit (or, the value is quiescent). When the FM IF signal deviates above or below the frequency of the VCO, there will be an error term generated. This error signal is processed in the low-pass filter and amplifier to control the VCO. Its purpose is to drive the VCO back on the right frequency. It's this error signal that becomes the modulated output of the PLL FM demodulator circuit.

Figure 8.23B shows a PLL based on the NE-565 PLL integrated circuit. The resonant frequency is set by R1 and C1, which should be the centre frequency of the FM signal. As the signal deviates up and down, the error voltage is monitored, and becomes the modulated output signal.

Quadrature detector

Figure 8.24 shows the *quadrature detector* circuit. This circuit is implemented in integrated circuit form (e.g. MC-1357P, CA-3089), and uses a single phase shifting external coil to accomplish its goals. This is probably the most widely used form of FM demodulator today.

The typical quadrature detector IC uses a series of wideband amplifiers to boost the signal and limit it, eliminating the AM noise modulation that often rides on the signal. This signal is applied to the signal splitter. The two outputs of the signal splitter are applied to a gated synchronous detector, but one is phase shifted 90 degrees. The output of the gated synchronous detector is the modulated audio.

Part 3 Components

9 Capacitors

Capacitors are devices that store electrical energy in an internal electrical field in an insulating dielectric material. They are one of the two components used in RF tuning circuits. Like the inductor, the capacitor is an energy storage device. While the inductor stores electrical energy in a magnetic field, the capacitor stores energy in an *electrical* (or *electrostatic*) field; electrical charge (Q) is stored in the capacitor. But more about that shortly.

The basic capacitor consists of a pair of metallic plates facing each other, and separated by an insulating material called a *dielectric*. This arrangement is shown schematically in Fig. 9.1A and in a more physical sense in Fig. 9.1B. The fixed capacitor shown in Fig. 9.1B consists of a pair of square metal plates separated by a dielectric (i.e. an insulator). Although this type of capacitor is not terribly practical, it was once used quite a bit in radio transmitters. Spark gap transmitters of the 1920s often used a glass and tin-foil capacitor fashioned very much like Fig. 9.1B. Layers of glass and foil are sandwiched together to form a high voltage capacitor. A one-foot square capacitor made of $\frac{1}{8}$ inch thick glass and foil has a capacitance up to about 2000 pF, depending on the specific glass material used.

Units of capacitance

The *capacitance* (C) of the capacitor is a measure of its ability to store current, or more properly electrical charge. The principal unit of capacitance is the *farad* (named after physicist Michael Faraday). *One farad is the capacitance that will store one coulomb of electrical charge (6.28×10^{18} electrons) at an electrical potential of one volt.* Or, in math form:

$$C_{\text{farads}} = \frac{Q_{\text{coulombs}}}{V_{\text{volts}}} \tag{9.1}$$

The farad is far too large for practical RF electronics work, so sub-units are typically used instead. The *microfarad* (μF) is 0.000001 farads ($1\,\text{F} = 10^6\,\mu\text{F}$). The *picofarad* (pF) is 0.000001 μF, which is 0.000000000001 F, or 10^{-12} farads. In older radio texts and schematics the picofarad was called the *micromicrofarad* (μμF or mmF), but never fear: 1 μμF = 1 pF.

The capacitance of the capacitor is directly proportional to the area of the plates (in terms of Fig. 9.1B, $L \times W$), inversely proportional to the thickness (T) of the dielectric (or the spacing between the plates, if you prefer), and directly proportional to the *dielectric constant* (K) of the dielectric.

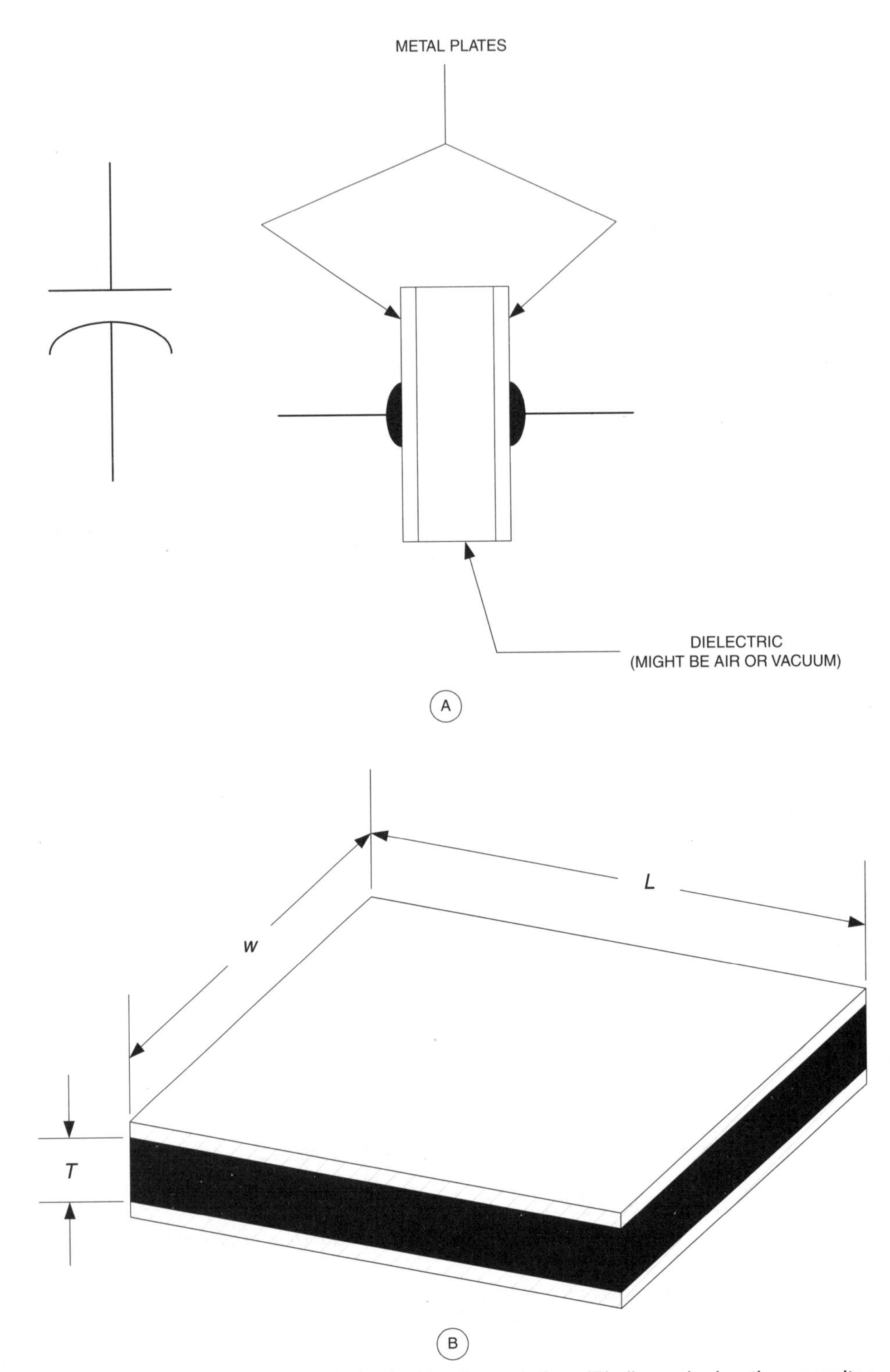

Figure 9.1 (A) Capacitor symbol and physical implementation; (B) dimensioning the capacitor.

Dielectric constant is a property of the insulator material, and a measure of the material's ability to support electric flux; it is thus analogous to the permeability of a magnetic material. The standard of reference for dielectric constant is a perfect vacuum, which by definition has a value of $K = 1.0$. Other materials are compared with the vacuum. The values of K for some common materials are:

Vacuum	1.0000
Dry air	1.0006
Paraffin (wax) paper	3.5
Glass	5 to 10
Mica	3 to 6
Rubber	2.5 to 35
Dry wood	2.5 to 8
Pure (distilled) water	81

The value of capacitance in any given capacitor is found from:

$$C = \frac{0.0224KA(N-1)}{T} \qquad (9.2)$$

Where:
C is the capacitance in picofarads (pF)
K is the dielectric constant
A is the area of one of the plates ($L \times W$), assuming that the two plates are identical, in square inches
N is the number of identical plates
T is the thickness of the dielectric

Breakdown voltage

The capacitor works by supporting an electrical field between two metal plates. When the electrical potential, i.e. the voltage, gets too large, free electrons in the dielectric material (there are a few, but not many, in any insulator) may flow. If a stream of electrons gets started, then the dielectric may break down and allow a current to pass between the plates. The capacitor is shorted. The maximum breakdown voltage of the capacitor must not be exceeded. However, for practical purposes there is a smaller voltage called the *DC working voltage* (WVDC) rating that defines the *maximum safe voltage* that can be applied to the capacitor. Typical values found in common electronic circuits range from 8 WVDC to 1000 WVDC, although multi-kilovolt WVDC ratings are also available.

Circuit symbols for capacitors

The circuit symbols used to designate fixed value capacitors are shown in Fig. 9.2A, and for variable capacitors in Fig. 9.2B. Both types of symbol are common. In certain types of capacitor, the curved plate shown on the left in Fig. 9.2A is usually the outer plate, i.e. the

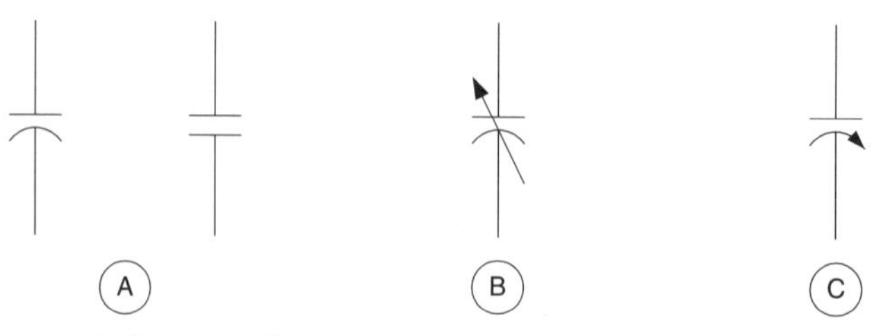

Figure 9.2 Circuit symbols for capacitors.

one closest to the outside package of the capacitor. This end of the capacitor is often indicated with a colour band next to the lead attached to that plate.

The symbols for the variable capacitor are shown in Fig. 9.2B. This symbol is the fixed value symbol with an arrow through the plates. Small trimmer and padder capacitors are often denoted by the symbol of Fig. 9.2C. The variable set of plates is designated by the arrow. Sometimes a trimmer looks like a variable capacitor but with the arrow head replaced by a little bar.

Fixed capacitors

There are several types of fixed capacitor found in typical electronic circuits, and these are classified by dielectric type: *paper, mylar, ceramic, mica, polyester* and others.

Paper dielectric capacitors

The construction of old-fashioned paper capacitors is shown in Fig. 9.3. It consists of two strips of metal foil sandwiched on both sides of a strip of paraffin wax paper. The strip sandwich is then rolled up into a tight cylinder. This rolled-up cylinder is then packaged in either a hard plastic, bakelite, or paper-and-wax case. When the case is cracked, or the wax end plugs are loose, replace the capacitor even though it tests good . . . it won't be for long. Paper capacitors come in values from about 300 pF to about 4 μF. The breakdown voltages will be 100 WVDC to 600 WVDC.

The paper capacitor is used for a number of different applications in older circuits such as bypassing, coupling, and DC blocking. Unfortunately, no component is perfect. The long rolls of foil used in the paper capacitor exhibit a significant amount of stray inductance. As a result, the paper capacitor is not used for high frequencies. Although they are found in some early shortwave receiver circuits, they are not used at all at VHF.

Mylar dielectric capacitors

In modern applications, or when servicing older equipment that used paper capacitors, use a *mylar* dielectric capacitor in place of the paper capacitor. These capacitors use a sheet of stable mylar synthetic material for the dielectric, and are of dipped construction (Fig. 9.4). Select a unit with exactly the same capacitance value, and a WVDC rating that is equal to or greater than the original WVDC rating.

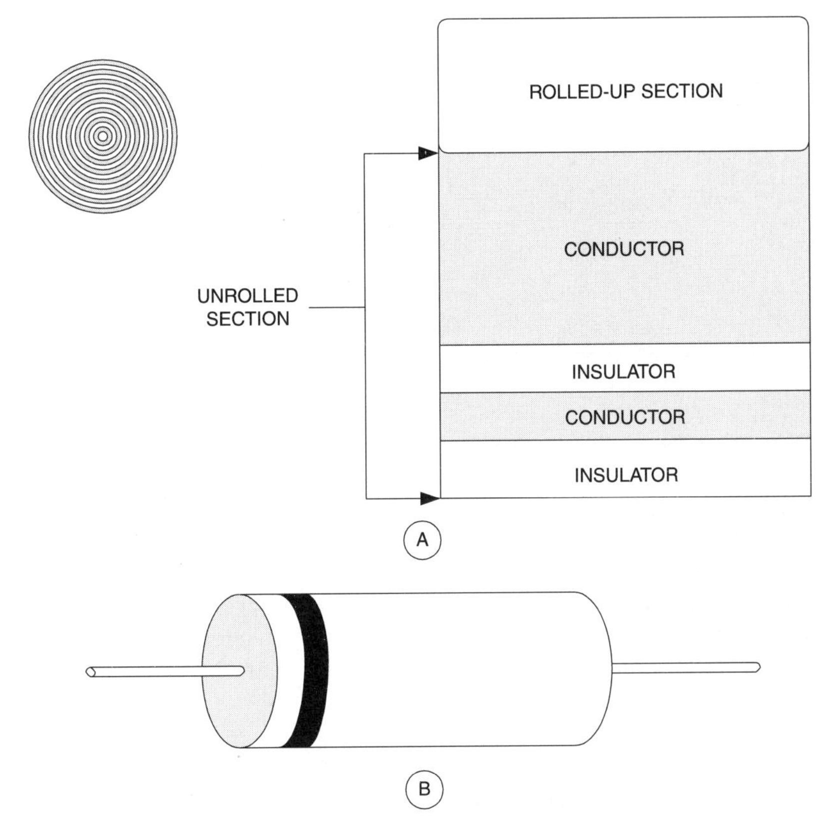

Figure 9.3 (A) Rolled paper capacitor; (B) Paper capacitor.

Ceramic dielectric capacitors

Several different forms of ceramic capacitors are shown in Fig. 9.5. These capacitors come in values from a few picofarads up to 0.5 μF. The working voltages range from 63 WVDC to more than 30 000 WVDC. The tubular ceramic capacitors are typically much smaller in value than disk or flat capacitors, and are used extensively in VHF and UHF circuits for blocking, decoupling, bypassing, coupling, and tuning.

Figure 9.4 Dipped capacitor.

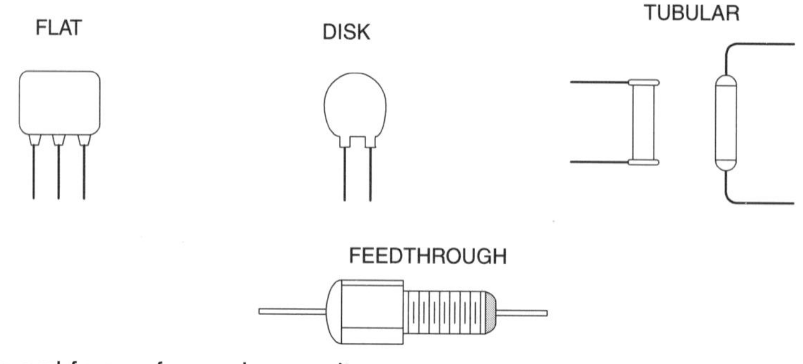

Figure 9.5 Several forms of ceramic capacitor.

The feedthrough type of ceramic capacitor is used to pass DC and low frequency AC lines through a shielded panel. These capacitors are often used to filter or decouple lines that run between circuits that are separated by the shield for purposes of electromagnetic interference (EMI) reduction.

Ceramic capacitors are often rated as to *temperature coefficient*. This specification is the change of capacitance per change of temperature in degrees Celsius. A 'P' prefix indicates a positive temperature coefficient, an 'N' indicates a negative temperature coefficient, and the letters 'NPO' indicate a zero temperature coefficient (NPO stands for 'negative positive zero'). Do not ad-lib on these ratings when servicing a piece of electronic equipment. Use exactly the same temperature coefficient as the original manufacturer used. Non-zero temperature coefficients are often used in oscillator circuits to temperature compensate the oscillator's frequency drift.

Mica dielectric capacitors

Several different types of mica capacitor are shown in Fig. 9.6. The fixed mica capacitor consists of metal plates on either side of a sheet of mica, or a sheet of mica that is silvered with a deposit of metal on both sides. The range of values for mica capacitors tends to be 50 pF to 0.02 μF at voltages in the range of 400 WVDC to 1000 WVDC.

The mica capacitor shown in Fig. 9.6C is called a *silvered mica* capacitor. These capacitors are low temperature coefficient, although for most applications an NPO disk ceramic will serve better than all but the best silvered mica units. Mica capacitors are typically used for tuning and other uses in higher frequency applications.

Other capacitors

Today the equipment designer has a number of different dielectric capacitors available that were not commonly available (or available at all) a few years ago. The *polycarbonate, polyester* and *polyethylene* capacitors are used in a wide variety of applications where the above discussed capacitors once ruled supreme. In digital circuits we find tiny little 100 WVDC capacitors with values of 0.01 μF to 0.1 μF. These are used for decoupling the noise on the +5 VDC power supply line. In circuits such as timers and op-amp Miller

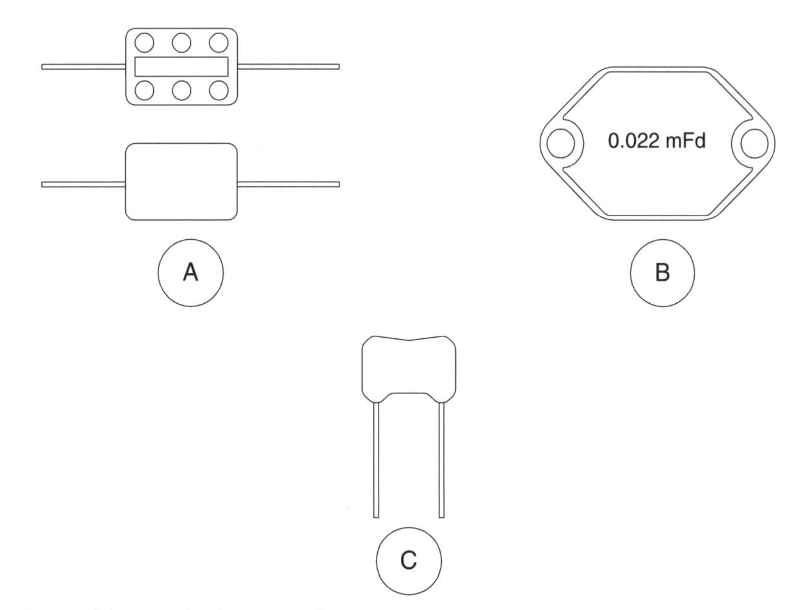

Figure 9.6 Several forms of mica capacitor.

integrators, where the leakage resistance across the capacitor becomes terribly important, we might want to use a polyethylene capacitor. Check current catalogues for various old and new style capacitors – the applications paragraph in the catalogue will tell you in which applications they will serve, and that is a guide to the type of antique capacitor it will replace.

Variable capacitors

Variable capacitors are, like all capacitors, made by placing two sets of metal plates parallel to each other (Figs 9.1B and 9.7A), separated by a dielectric of air, mica, ceramic, or a vacuum. The difference between variable and fixed capacitors is that, in variable capacitors, the plates are constructed in such a way that the capacitance can be changed.

There are two principal ways to vary the capacitance: either the spacing between the plates is varied, or the cross-sectional area of the plates that face each other is varied. Figure 9.7B shows the construction of a typical variable capacitor used for the main tuning control in radio receivers. The capacitor consists of two sets of parallel plates. The *stator* plates are fixed in their position, and are attached to the frame of the capacitor. The *rotor* plates are attached to the shaft that is used to adjust the capacitance.

Another form of variable capacitor found in radio receivers is the *compression capacitor* shown in Fig. 9.7C. This consists of metal plates separated by sheets of mica dielectric. The entire capacitor will be mounted on a ceramic or other form of holder. If mounting screws or holes are provided, then they will be part of the holder assembly.

Still another form of variable capacitor is the *piston capacitor* shown in Fig. 9.7D. This type of capacitor consists of an inner cylinder of metal coaxial to, and inside of, an outer cylinder of metal. An air, vacuum or (as shown) ceramic dielectric separates the two cylinders. The capacitance is increased by inserting the inner cylinder further into the outer cylinder.

The small compression or piston style variable capacitors are sometimes combined with air variable capacitors. Although not exactly correct usage, the smaller capacitor used in conjunction with the larger air variable is called a *trimmer capacitor.* These capacitors are often mounted directly on the air variable frame, or very close by in the circuit. In many radios the 'trimmer' is actually part of the air variable capacitor.

There are actually two uses for small variable capacitors in conjunction with the main tuning capacitor in radios. First, there is the true 'trimmer', i.e. a small-valued variable capacitor in *parallel* with the main capacitor (Fig. 9.8A). These capacitors are used to trim the exact value of the main capacitor. The other form of small capacitor is the *padder*

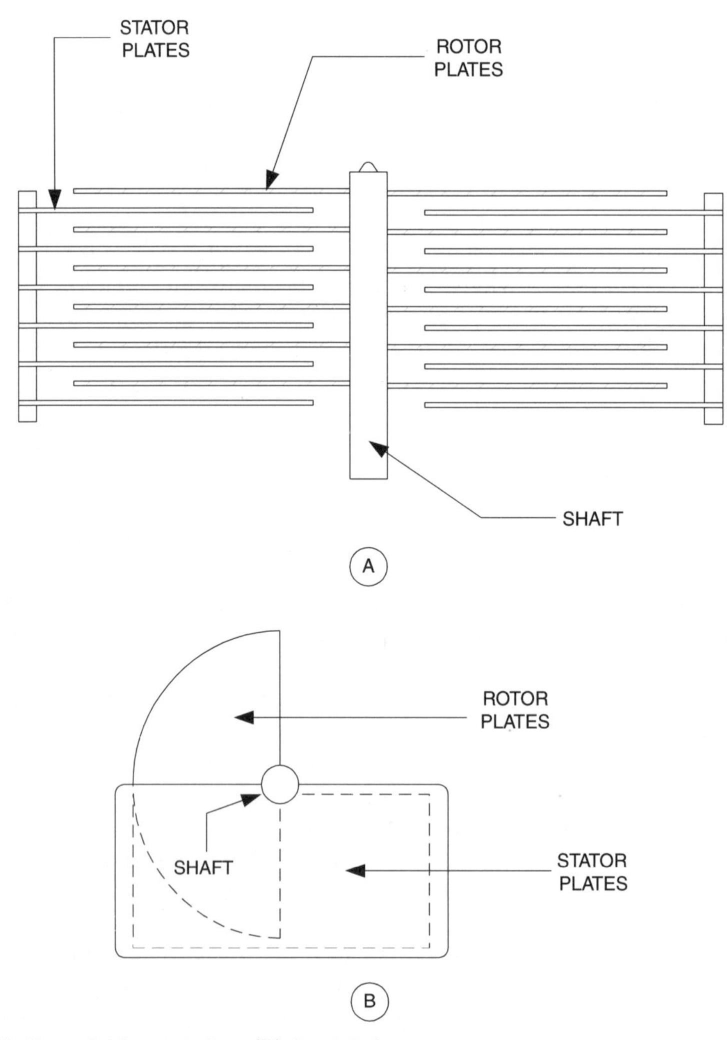

Figure 9.7 (A) Air variable capacitor; (B) frontal view;

capacitor (Fig. 9.8B), which is connected in *series* with the main capacitor. The error in terminology referred to above is calling both series and parallel capacitors 'trimmers', when only the parallel connected capacitor is properly so called.

Air variable main tuning capacitors

The capacitance of an air variable capacitor at any given setting is a function of how much of the rotor plate set is shaded by the stator plates. In Fig. 9.9A, the rotor plates are completely outside of the stator plate area. Because the overlap is zero, the capacitance is minimum. In Fig. 9.9B, however, the rotor plate set has been partly meshed with the stator plate, so some of its area overlaps the stator. The capacitance in this position is at an

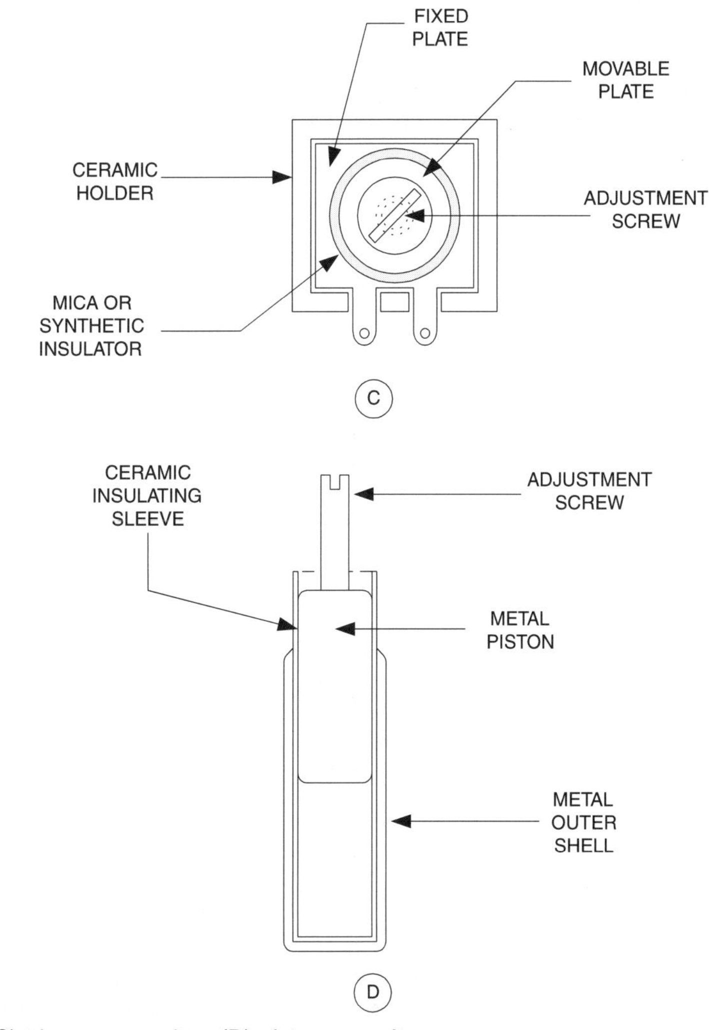

Figure 9.7 (C) trimmer capacitor; (D) piston capacitor.

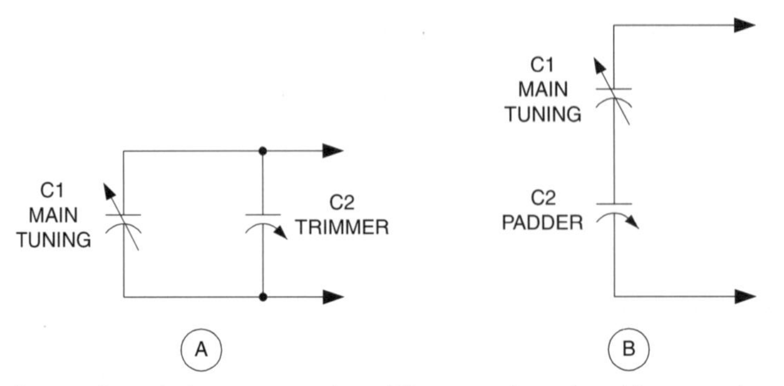

Figure 9.8 (A) Connection of trimmer capacitor; (B) connection of padder capacitor.

intermediate value. Finally, in Fig. 9.9C the rotor is completely meshed with the stator so the cross-sectional area of the rotor that overlaps the stator is maximum. Therefore, the capacitance is also maximum.

Figure 9.10 shows a typical single-section variable capacitor. The stator plates are attached to the frame of the capacitor, which in most radio circuits is grounded. Front and

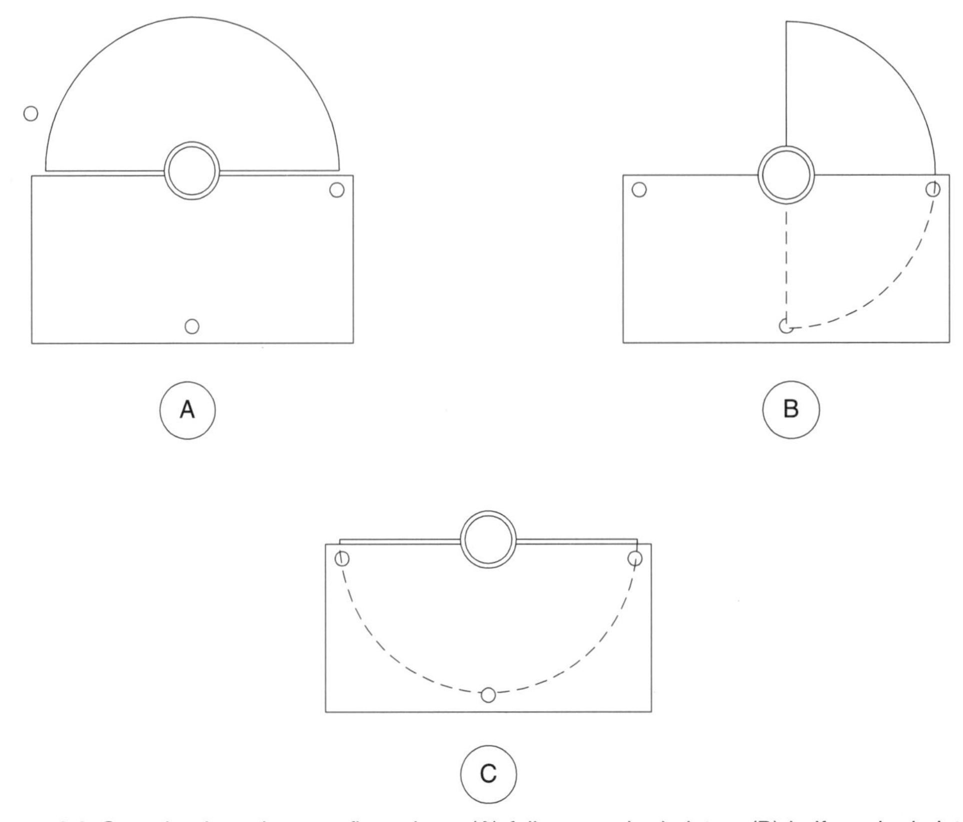

Figure 9.9 Capacitor in various configurations: (A) fully unmeshed plates; (B) half-meshed plates; (C) fully meshed plates.

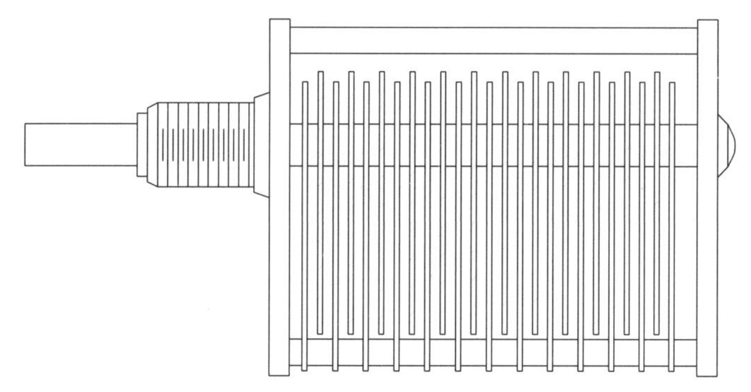

Figure 9.10 Side view of a single section air variable.

rear plates have bearing surfaces to ease the rotor's action. These capacitors were often used in early multi-tuning knob TRF radio receivers (the kind where each RF-tuned circuit had its own selector knob). But that design was not terribly good, so the *ganged variable capacitor* (Fig. 9.11) was invented. These capacitors are basically two or three (as in Fig. 9.11) variable capacitors mechanically ganged on the same rotor shaft.

The sections of a ganged variable capacitor may have the same capacitance, or may be different. If one with identical sections is used in a superheterodyne radio, the section used for the local oscillator (LO) tuning must be padded with a series capacitance in order to reduce the overall capacitance. This trick is done to permit the higher frequency LO to track with the RF amplifiers on the dial.

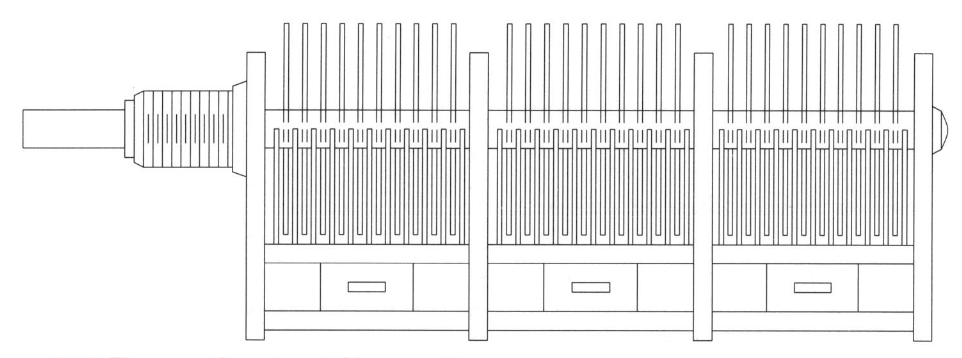

Figure 9.11 Three section air variable capacitor.

In many superheterodyne radios you will find variable tuning capacitors in which one section (usually the front section) has fewer plates than the RF amplifier section (an example is shown in Fig. 9.12). These capacitors are sometimes called *cut-plate capacitors* because the LO section plates are cut to permit tracking of the LO with the RF.

Capacitor tuning laws – SLC-vs-SLF

The variable capacitor shown in Fig. 9.9 has the rotor shaft in the geometric centre of the rotor plate half-circle. The capacitance of this type of variable capacitor varies linearly

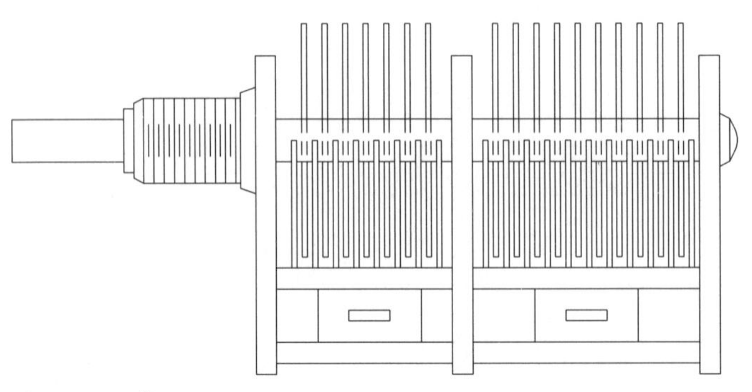

Figure 9.12 Cut-plate capacitor.

with the rotor shaft angle. As a result, this type of capacitor is called a *straight line capacitance* (SLC) model. Unfortunately, the frequency of a tuned circuit based on inductors and capacitors is not a linear (straight line) function of capacitance; it goes like the inverse of the square root of the capacitance. If a straight line capacitance unit is used for the tuner, then the *frequency* units on the dial will be cramped at one end and spread out at the other (you've probably seen such radios).

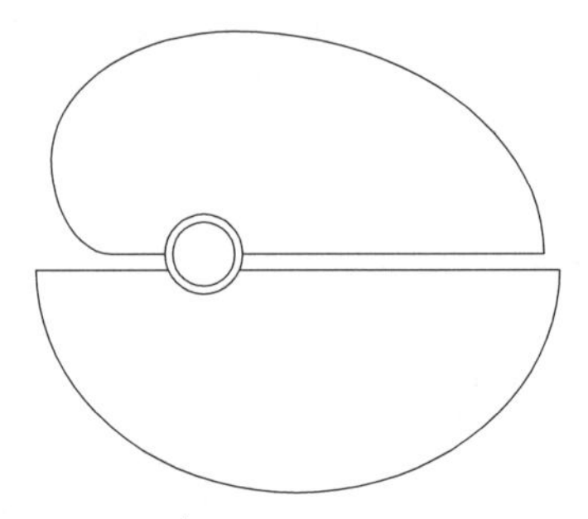

Figure 9.13 Straight line frequency capacitor.

Some capacitors have an offset rotor shaft and funny shaped plates (Fig. 9.13) to compensate for the tuning circuit. They are designed to produce a roughly linear relationship between the shaft angle and the resonant frequency of the tuned circuit in which the capacitor is used. This is a *straight line frequency* (SLF) capacitor. The SLF law works best if the tuning ratio matches what the capacitor designer intended, e.g. about 3:1 to cover the medium wave broadcast band. Some very old radios might have a straight line wavelength (SLW) tuning capacitor.

Special variable capacitors

In the sections above the standard forms of variable capacitor were discussed. These capacitors are largely used for tuning radio receivers, oscillators, signal generators and other variable frequency LC oscillators. In this section we will take a look at some special forms of variable capacitor.

Split stator capacitors

The split stator capacitor is one in which two variable capacitors are mounted on the same shaft. The symbol for the split stator capacitor is shown in Fig. 9.14. The split stator capacitor uses a pair of identical capacitors turned by the same shaft with a common rotor. They are normally used for balanced tuned circuits in tube transmitters or antenna tuning units.

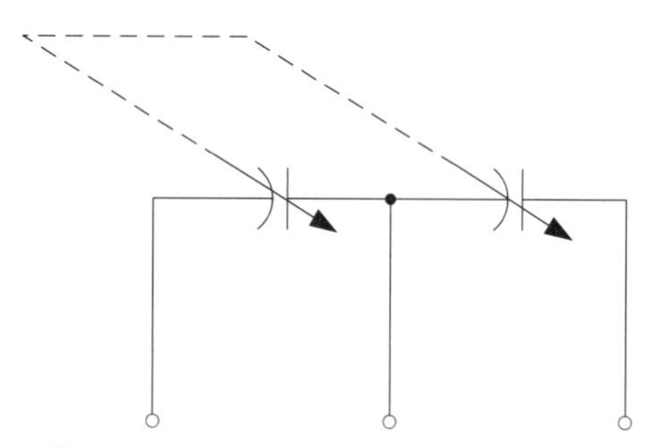

Figure 9.14 Split stator capacitors.

Differential capacitors

Although some differential capacitors are often mistaken for split stator capacitors, they are actually quite different. The split stator capacitor is tuned in tandem, i.e. both capacitor sections have the same value at any given shaft setting. The differential capacitor, on the other hand, is arranged so that one capacitor section increases in capacitance, while the other section decreases by exactly the same amount.

Figure 9.15 shows both the mechanical construction and circuit symbol for a differential capacitor. Note that the rotor plate is set to equally shade both stator-A and stator-B. If the shaft is moved clockwise, it will shade more of stator-B, and less of stator-A, so C_a will decrease and C_b will increase by exactly the same amount. Note: the total capacitance (C_t) is constant no matter what position the rotor shaft takes, only the proportion between C_a and C_b changes.

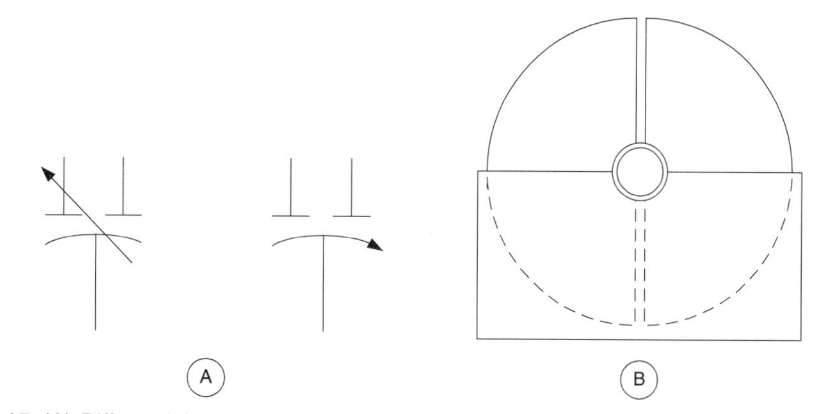

Figure 9.15 (A) Differential capacitors; (B) physical implementation.

Differential capacitors are used in impedance bridges, RF resistance bridges, and other such instruments. If you buy or build a high quality RF impedance bridge for antenna measurements, for example, it is likely that it will have a differential capacitor as the main adjustment control. The two capacitors are used in two arms of a Wheatstone bridge circuit.

'Transmitting' variable capacitors

The one requirement of transmitting variable capacitors (and certain antenna tuner capacitors) is the ability to withstand high voltages reliably. The high power ham radio or AM broadcast transmitter will have a DC potential of 1500 volts to 7500 volts on the RF amplifier anode, depending on the type of tube used. If amplitude modulated, the potential can double. Also, if certain antenna defects arise, then the RF voltages in the circuit can rise quite high. As a result, the variable capacitor used in the final amplifier plate circuit must be able to withstand these potentials.

There are two forms of transmitting variable typically found in RF power amplifiers and antenna tuners. A transmitting *air variable* capacitor is like the other forms of air variable shown in this chapter, except that the plate spacing is wider to account for higher voltages used in transmitters. The other form of transmitting variable is the *vacuum variable*. This type of capacitor is a variation of the piston capacitor, but it has a vacuum dielectric (*K*-factor = 1.0000). The model shown in Fig. 9.16 is a 18 pF to 1000 pF model that is driven from a 12 VDC electric motor. Other vacuum variables are manually driven.

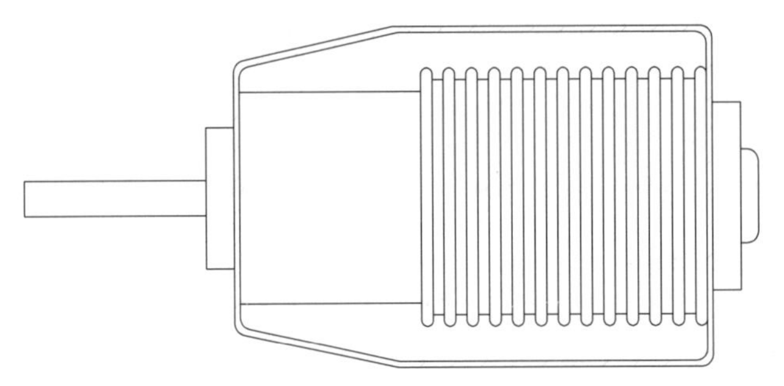

Figure 9.16 Vacuum variable.

Variable capacitor cleaning note

Antique radio buffs often find that the main tuning capacitors in their radios are full of crud, grease and dust. Similarly, ham radio operators working the hamfest circuit looking for linear amplifier and antenna tuner parts often find just what they need, but the thing is full of scum, crud, grease and other stuff. There are several things that can be done about it. First, try using dry compressed air. It will remove dust, but not grease. Aerosol cans of compressed air can be bought from a lot of sources, including automobile parts stores and photography stores.

Another method, if you have the hardware, is to ultrasonically clean the capacitor. The ultrasonic cleaner, however, is expensive.

Still another way is to use a product such as *Birchwood Casey Gun Scrubber.* This product is used to clean firearms, and is available in most gun shops. Firearms become clogged up because gun grease, oil, unburned powder and burned powder residue combine to create a crusty mess that's every bit as hard to remove as capacitor gunk. A related product is the degunking compound used by car mechanics.

At one time, carbon tetrachloride was used for this purpose . . . and you will see it listed in old radio books. However, carbon tet is now well recognized as a health hazard. *Do not use carbon tetrachloride* for cleaning, despite the advice to the contrary found in old radio books.

Using and stabilizing a varactor diode

Have you tried to buy an air variable capacitor for a receiver project recently? They are very rare these days. I've seen them advertised in some amateur radio parts catalogues, in British electronics catalogues, and in antique radio supplies catalogues in the USA, but otherwise it's catch as catch can. So what to do? Well, it seems that commercial radio manufacturers today use *voltage variable capacitance diodes*, commonly called *varactors*, for the radio tuning function. These special semiconductor diodes exhibit a capacitance across the PN junction that is a function of the reverse bias potential (see Fig. 9.17).

The diode representations of Figs 9.17A and 9.17B are in the form of PN junction diode block diagrams. In the N-type region negative charge carriers (electrons) predominate, while in the P-type region positive charge carriers (holes) predominate. When a reverse bias potential is applied, as in Fig. 9.17A, the charge carriers are pulled away from the junction region to form a *depletion zone* that is depleted of charge carriers (hence acts like an insulator or 'dielectric'). The situation is the same as in a charged capacitor: an insulator separating two electrically conductive regions. Thus, a capacitance is formed across the junction that is a function of the width of the depletion zone. Because the size of the depletion zone is a function of applied voltage (compare Figs 9.17A and 9.17B), the *capacitance of the junction is also a function of applied voltage.* A varactor is a diode in which this function is enhanced and stabilized.

Figures 9.18A and 9.18B show two common circuit symbols for a varactor diode. In both cases, the normal diode 'arrow' symbol is somehow combined with a pair of parallel lines representing a capacitor. In some cases, I've seen a variant on Fig. 9.18A in which an arrow is drawn through the parallel plates by extending one side of the arrow symbol. I suppose that's used to indicate the property of 'variableness'.

Several different varactors are listed in Table 9.1. Several of these are also available in the easily available ECG and NTE replacement transistor lines sold by parts houses that normally deal with radio-tv repair shops. Look up the specs for NTE-611 to NTE-618, or ECG-611 to ECG-618 to see if they are appropriate for your application. Alternatively, look up the replacements for those diodes in the table from the ECG or NTE crossover directories.

Varactor tuning circuits

The varactor diode wants to see a voltage that is proportional to the desired capacitance. Several different circuits are used to provide this function, some of which are shown in

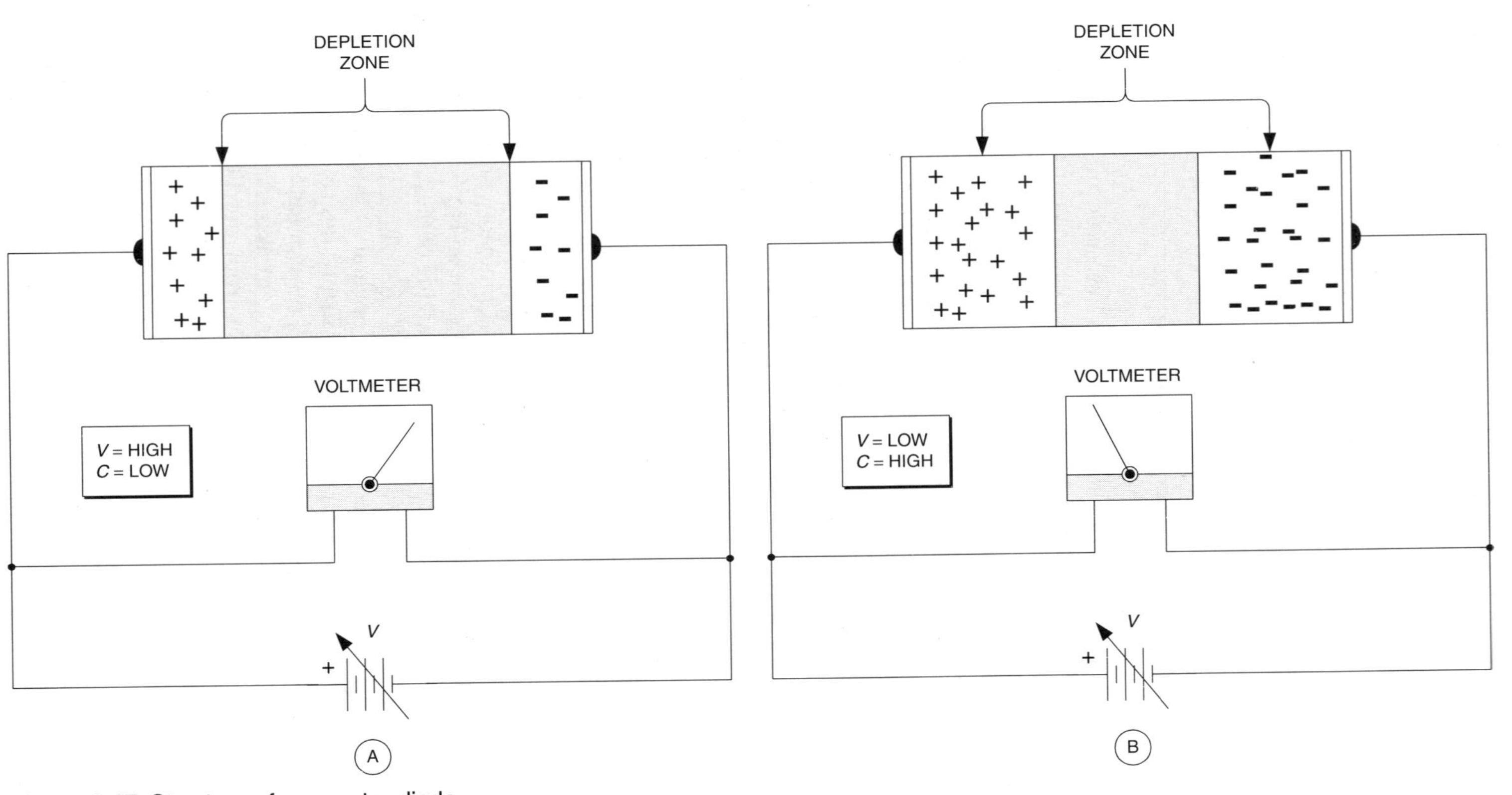

Figure 9.17 Structure of a varactor diode.

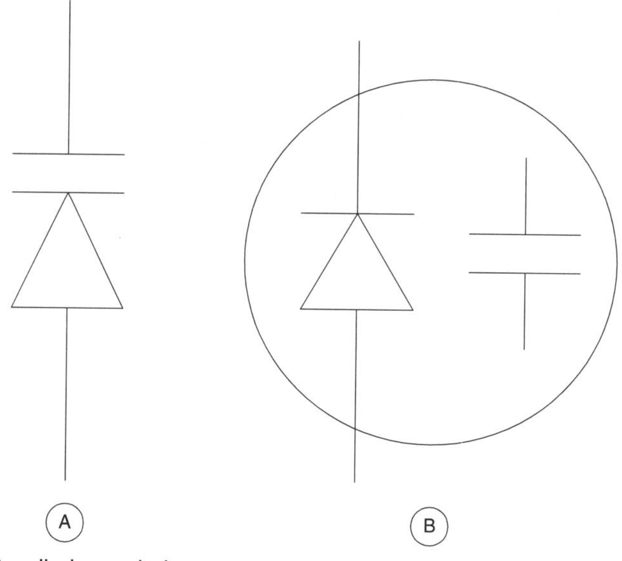

Figure 9.18 Varactor diode symbols.

Table 9.1

Type no.	Capacitance range	Tuning ratio	Frequency ratio
1N5139	6.8–47 pF	2.7–3.4	1.6–1.8
MV2101	6.8–100 pF	1.6–3.3	1.6–1.8
MMBV105G	120–550 pF	10–14	3.2–3.7
MV209	30 pF	5–6.5	2.2–2.5
BB212	10–550 pF	—	7.4

Figs 9.19 and 9.20. In all cases, the tuning voltage must be supplied from a reference voltage source that is very stable. It is normally considered good engineering practice to provide $+V_{REF}$ from a separate voltage regulator that serves only the varactor, even when the maximum value of the voltage is the same as the rest of the circuit (e.g. +12 volts). Therefore, *always use a voltage regulator to provide the tuning voltage source potential.* Most varactors use a maximum voltage around +30 to +40 volts, while many intended for car radio applications are rated only to +12 or +18 volts (check!).

The simplest and probably most popular circuit is shown in Fig. 9.19A. In this circuit, a potentiometer (R1) is connected across the V_{REF} supply, so the tuning voltage (V_T) is a function of the potentiometer wiper position. In many cases, a 0.001 μF to 0.01 μF capacitor is connected from the wiper of the potentiometer to ground in order to snuff any noise pulses so they don't alter the tuning (they are, as far as the diode is concerned, valid tuning voltage signals!). A series current-limiting resistor (R1), usually of a value between 4.7 kohms and 100 kohms, is used to protect the diode in case the voltage gets to the breakover point, and also to isolate C1 capacitance from the tuning circuit. In many cases, a DC blocking capacitor (C2) is needed to prevent the tuning voltage from affecting following circuits, or other circuit voltages from affecting the varactor diode tuning voltage. From the point in Fig. 9.19A marked 'To Circuit' the varactor network acts like a variable capacitor.

A variant circuit is shown in Fig. 9.19B. In this circuit the tuning voltage is only a small portion of the reference voltage. Thus, the tuning voltage is produced by a voltage divider made up of three resistors: R1, R2 and R3. In some cases, one or more of the other resistors will be a trimmer potentiometer to set the 'fine' or 'vernier' frequency of the overall circuit.

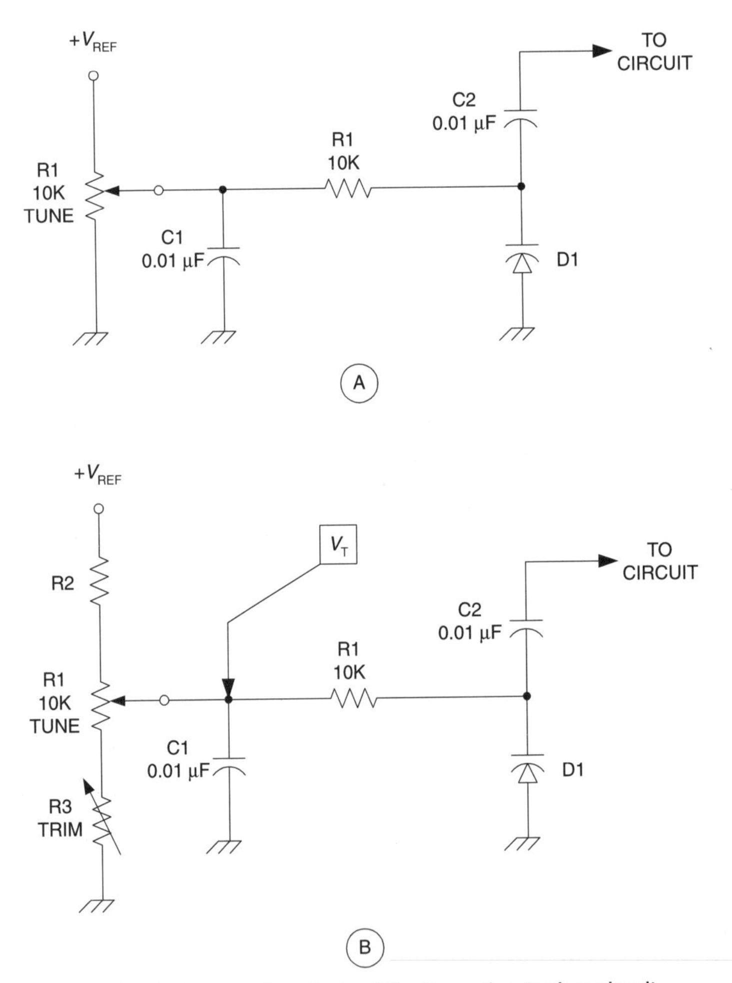

Figure 9.19 (A) Tuning circuit for varactor diode; (B) alternative tuning circuit.

Regardless of which tuning circuit is used, the resistors, including the potentiometer, should be low temperature coefficient types in order to reduce thermal drift. Ordinary carbon composition resistors are probably not suitable for most applications.

If you wish to sweep a band of frequencies, i.e. in a sweep generator or swept receiver (e.g. panadaptor or spectrum analyser), then replace the $+V_{REF}$ potential with a sawtooth

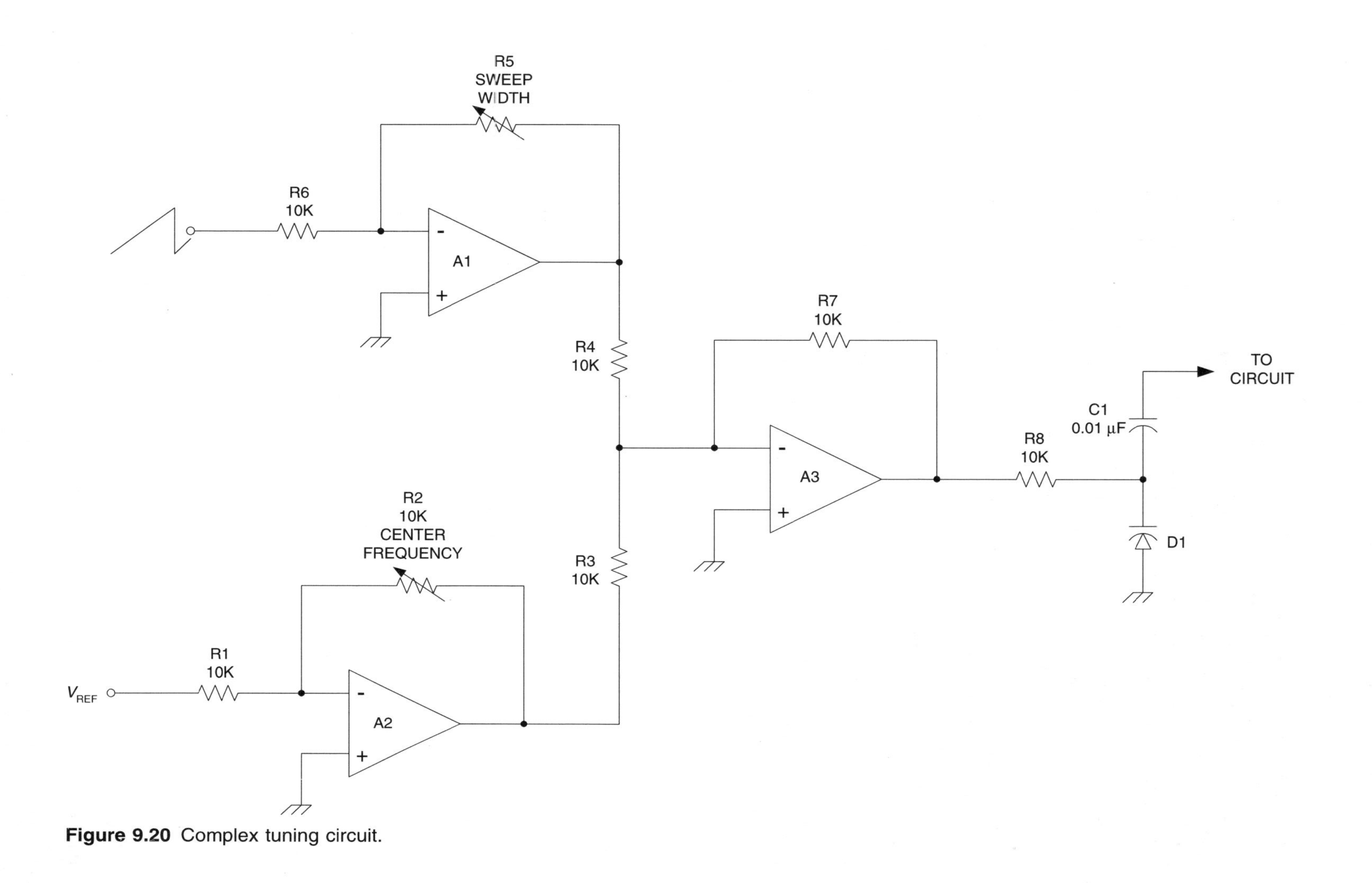

Figure 9.20 Complex tuning circuit.

waveform. The sawtooth waveform is a linear ramp that rises to a specified maximum voltage, and then drops back to zero abruptly. Unfortunately, it is rarely the case that the sawtooth voltage range, the desired swept frequency range, and the varactor voltage characteristic are in sync with each other. For those situations we need to be able to provide a sawtooth of variable amplitude to set the *sweep width* and a DC offset tuning voltage to provide the *centre frequency* function. Figure 9.20 shows how this might be done.

The circuit of Fig. 9.20 uses three operational amplifiers to provide the combination tuning voltage. Op-amp A1 provides a variable amplitude sweep width control to change the sawtooth amplitude. If feedback resistor R5 is made 10 kohms, then the output sawtooth will have the same amplitude as the input sawtooth. If higher or lower amplitude is needed, then adjust the gain of A1 by selecting a different R5 value: Gain = $-R5/R6 = -R5/10\,\text{kohms}$ (the '–' indicates that the circuit is an inverter). For tuning voltages to 18 volts, ordinary 741s can be used for A1 through A3.

Digital frequency control can be accomplished by supplying the reference voltage ($+V_{REF}$) from a digital-to-analogue converter (DAC) that has a voltage output. The binary number applied to the DAC binary inputs will set the tuning voltage, which in turn sets the capacitance of the diode. Those who wish to experiment with low cost components will find that the eight-bit National DAC0800 series devices (available in most local parts stores in the *Jameco Jim-Pak* display) will provide 256 different steps of voltage (hence also of capacitance and frequency). An op-amp is recommended to convert the current output of the DAC080x to a voltage (the *National Linear Data Book* gives example circuits as well as specs for the different devices in the series).

Temperature compensation

There is one nasty little problem with the varactor tuning circuit – the thermal drift can be horrible! According to one source, the temperature coefficient of capacitance (ppm/°C) varied from about 30 ppm/°C at $+V_{REF}$ = 30 volts to 587 ppm/°C at $+V_{REF}$ = 1 volt. Ouch! There are three approaches to this problem: (1) ignore it; (2) use Fig. 9.21, (3) use Fig. 9.22.

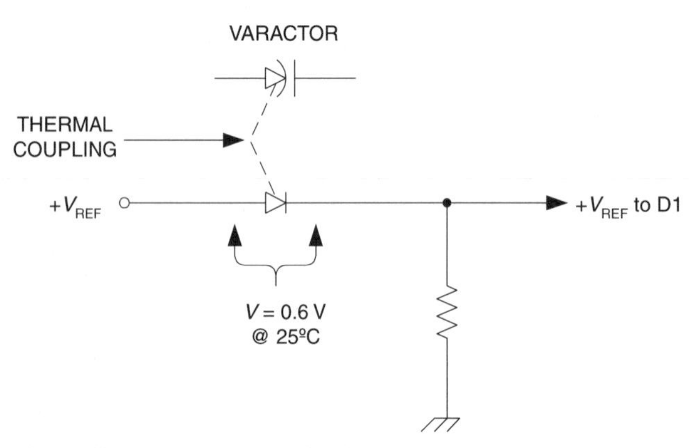

Figure 9.21 Thermal coupling gives thermal feedback.

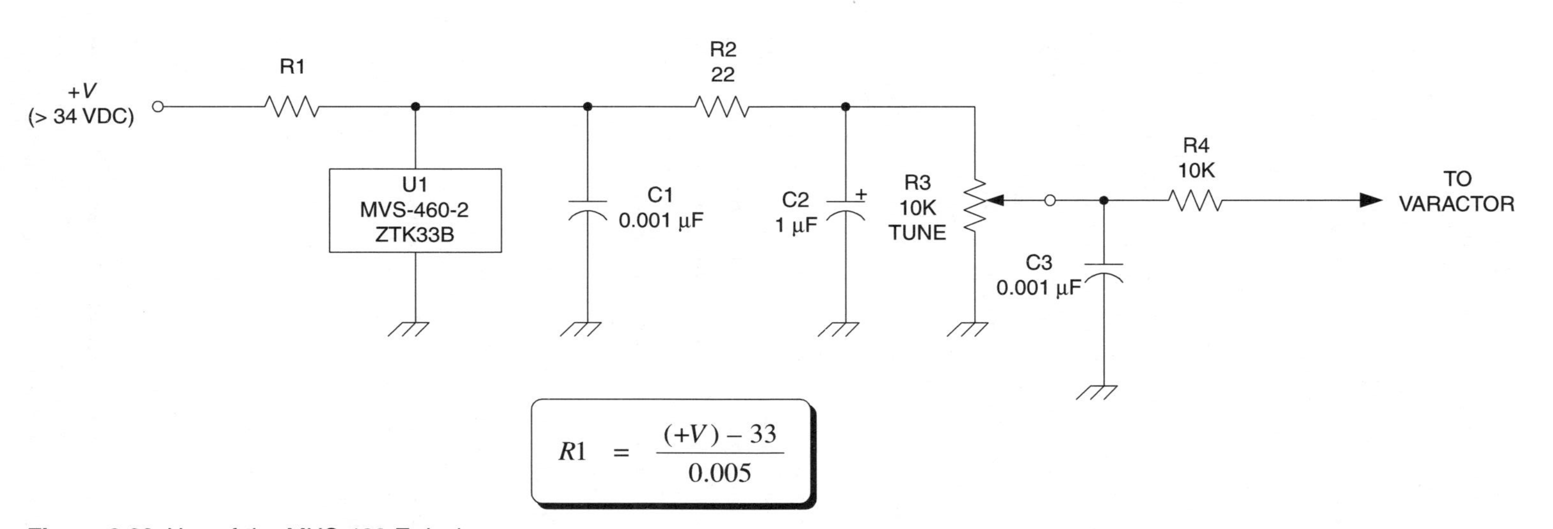

Figure 9.22 Use of the MUS-460-Z device.

The circuit of Fig. 9.21 uses a fixed, regulated voltage for $+V_{REF}$, but passes it through an ordinary silicon diode (D2) that is in close thermal proximity to the varactor diode (so they see the same temperature environment). When resistor R1 is set to draw a current through D2 sufficient to get the voltage drop into the 0.6 volt region, then the output voltage $+V_{REF}$ will track the thermal changes to counteract the change of capacitance. In practice, R1 can be the tuning potentiometer when diodes such as 1N4148 or 1N914 are used.

Figure 9.22 shows a circuit using a special zener diode voltage regulator sold in Europe under both MVS-460-2 and ZTK33B type numbers. It appears to be a +33 volt zener diode that has a −2.3 mV/°C temperature coefficient. It will provide a nominal +33 volt output for all input voltages (V) greater than 34 VDC. Again, the temperature stabilizer (which looks like a diode) is placed in close thermal proximity to the varactor diode being protected. The MVS-460-2 part is in a TO-92-like plastic package, while the ZTK33B is in the normal glass diode package (similar to 1N60 devices).

Unfortunately, the MVS-460-2 and ZTK33B are hard to find in the USA. I bought some from *Maplins Professional Supplies* in England (PO Box 777, Rayleigh, Essex, SS6 8LU, England) for £0.382 (at the time of writing £1 = $1.62, but the rate changes daily so check before sending money order denominated in £ sterling) each in lots of 25 or more. Unfortunately, with a minimum practical order of several pounds sterling, plus a shipping charge of £8 for USA and Canada, means that it is best to order 25 or so. This translates to $27.38 or so, if the price still holds as of publication date. Ordering from the UK is reasonably easy. You can get an international money order denominated in pounds sterling at many banks, but the fee might make you think otherwise (my bank gets $15, which is why I opened a UK checking account). Alternatively, they will accept *Visa*, *Mastercard* or *American Express* cards. The card company will make the currency conversion for you, and they use the rate in effect on the day they make the conversion. I've used all three types of cards to make purchases from UK electronic and old book dealers (my other passion), and experienced no problems. Give them the card number, expiration date and your signature authorizing the charge.

Varactor applications

Varactors are electronically variable capacitors. In other words, they exhibit a variable capacitance that is a function of the applied reverse bias potential. This phenomenon leads to several common applications in which capacitance is a consideration. Figure 9.23 shows a typical varactor-tuned LC tank circuit. The link coupled inductor (L2) is used to input RF signal to the tank when the circuit is used for RF amplifiers, and in many oscillator circuits it serves as the output to take signal energy to other circuits. The principal LC tank circuit consists of a main inductor (L1) and a capacitance made up from the series equivalent of C1 and varactor D1; or:

$$C_t = \frac{C1 C_{D1}}{C1 + C_{D1}}$$

In addition, you must also take into account the stray capacitance (C_s) that exists in all electronic circuits. The blocking capacitor and series resistor functions were discussed previously. Capacitor C2 is used to filter the tuning voltage, V_{in}.

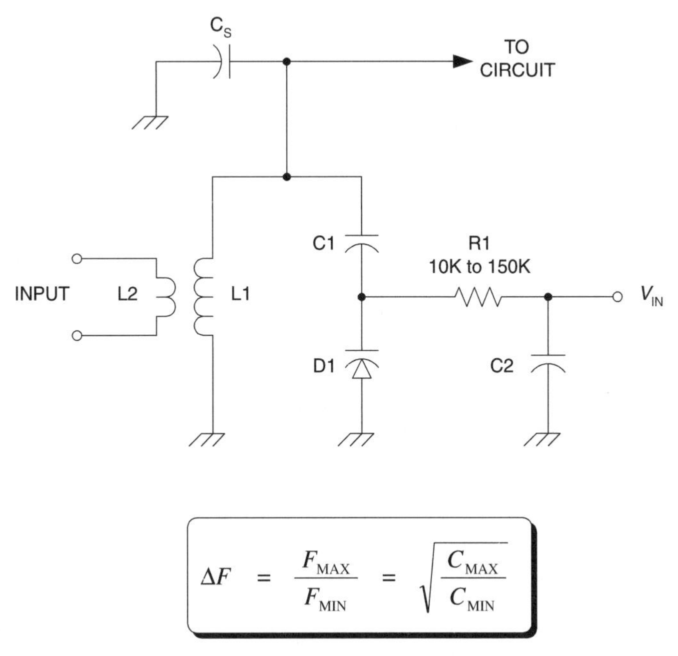

Figure 9.23 Varactor in circuit.

Because the resonant frequency of an LC-tuned tank circuit is a function of the square root of the inductance capacitance product, the maximum/minimum frequency of the varactor-tuned tank circuit varies as the square root of the capacitance ratio of the varactor diode. This value is the ratio of the capacitance at minimum reverse bias over the capacitance at maximum allowable reverse bias. A consequence of this fact is that the tuning characteristic curve (voltage-vs-frequency) is basically a parabolic function.

Well, that's that for varactors. If you want to know more theoretical smoke about the subject, then I recommend Motorola Semiconductor's application note AN847 'Tuning Diode Design Techniques' (Motorola Technical Literature Distribution Center, POB 20912, Phoenix, AZ 85036, USA).

10 Inductors

Inductors form a very large part of RF electronic circuitry with applications ranging from radio tuning, to filters, to RFI/EMI suppression, to impedance matching. In this chapter we will take a look at the subjects of inductance, inductors (coils), and how to build the inductors that you need in practical RF circuits. We will also discuss the related topic of RF transformers.

Inductor circuit symbols

Figure 10.1 shows various circuit symbols used in schematic diagrams to represent inductors. Figures 10.1A and 10.1B represent alternate but equivalent forms of the same thing; i.e. a fixed value, air-core inductor ('coil' in the vernacular). The other forms of inductor symbol shown in Fig. 10.1 are based on Fig. 10.1A, but are just as valid if the 'open-loop' form of Fig. 10.1B is used instead.

The form shown in Fig. 10.1C is a *tapped fixed value air-core inductor.* By providing a tap on the coil, different values of fixed inductance are achieved. The inductance from one end of the coil to the tap is a fraction of the inductance available across the entire coil. By providing one or more taps, several different fixed values of inductance can be selected. Radio receivers and transmitters sometimes use the tap method, along with a *bandswitch*, to select different tuning ranges or 'bands'.

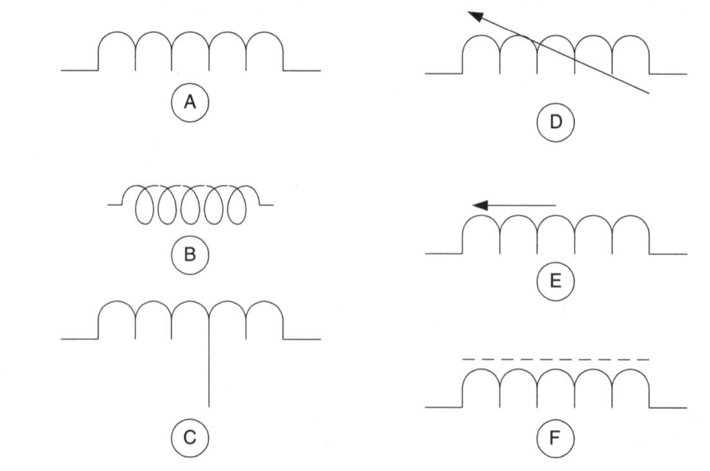

Figure 10.1 Inductor symbols.

Variable inductors are shown in Figs 10.1D and 10.1E. Both forms are used in schematic diagrams, although in some countries Fig. 10.1D implies a form of construction whereby a wiper or sliding electrical contact rides on the uninsulated turns of the coil. Figure 10.1E implies a construction where variable inductance is achieved by moving a magnetic core inside the coil.

Figure 10.1F indicates a fixed value (or tapped, if desired) inductor with a powdered iron, ferrite or non-ferrous (e.g. brass) core. The core will increase (ferrite or powdered iron) or decrease (brass) the inductance value relative to the same number of turns on an air-core coil.

Inductance and inductors

Inductance (L) is a property of electrical circuits that *opposes changes in the flow of current.* Note the word 'changes', for it is very important. Inductance is somewhat analogous to the concept of inertia in mechanics. An inductor *stores energy in a magnetic field* (a fact that we will see is quite important). In order to understand the concept of inductance we must understand two physical facts:

1 When a conductor is in a magnetic field that is changing, an *electromotive force* (EMF or 'voltage') appears across the ends of the conductor.
2 When an electrical current moves in a conductor, a magnetic field is set up around the conductor.

According to *Lenz's law*, the EMF induced into a circuit is 'in a direction that opposes the effect that produced it'. From this fact we can see the following effects:

1 A current induced by a change in a magnetic field always flows in the direction that sets up a magnetic field that opposes the original change.
2 When a current flowing in a conductor changes, the magnetic field change that it generates is in a direction that induces a further current into the conductor that opposes the current change.
3 The EMF generated by a change in current will have a polarity that opposes the potential that created the change.

The unit of inductance (L) is the *henry* (H). A henry is *the inductance that creates an EMF of one volt when the current in the inductor is changing at a rate of one ampere per second,* or mathematically:

$$V = L\left(\frac{\Delta I}{\Delta t}\right) \tag{10.1}$$

Where:
V is the created EMF in volts (V)
L is the inductance in henrys (H)
I is the current in amperes (A)
t is the time in seconds (s)
Δ indicates a 'small change in . . .'

The henry is the appropriate unit for large inductors such as the smoothing filter chokes used in DC power supplies, but is far too large for RF circuits. In those circuits the sub-units of *millihenry* (mH) and *microhenry* (μH) are used. These are related to the henry by: 1 henry = 1000 millihenrys (mH) = 1 000 000 microhenrys (μH). Thus, 1 mH = 10^{-3} H and 1 μH = 10^{-6} H.

The term *inductance* when used alone typically means *self-inductance* (the effect of a change on the conductor which initiated the change), and will be so used in this chapter unless otherwise specified. There is also *mutual inductance,* which is the effect one conductor has on another.

Inductance of a single straight wire

Although it is commonly assumed that 'inductors' are 'coils', and therefore consist of at least one, usually more, turns of wire around a cylindrical form, it is also true that a single, straight piece of wire possesses inductance. The inductance of a wire in which the length is at least 1000 times its diameter is given by:

$$L_{\mu H} = (0.00508)a\left(\ln\left(\frac{4a}{d}\right) - 0.75\right) \tag{10.2}$$

(Ratios less than $l/d > 1000$ are more difficult to calculate.)

The inductance value of representative small wires is very small in absolute numbers, but at higher frequencies becomes a very appreciable portion of the whole inductance needed. Consider a 12 inch length of #30 wire (dia. = 0.010 inches). Plugging these values into Eq. (10.2) yields:

$$L_{\mu H} = (0.00508)(12\,\text{in})\left(\ln\left(\frac{4 \times 12\,\text{in}}{0.010\,\text{in}}\right) - 0.75\right) \tag{10.3}$$

$$L_{\mu H} = 0.471\,\mu\text{H} \tag{10.4}$$

An inductance of 0.471 μH seems terribly small, and at 1 MHz it is small compared with inductances typically used as that frequency. But at 100 MHz, 0.471 μH could easily be more than the entire required circuit inductance. RF circuits have been created in which the inductance of a straight piece of wire is the total inductance, but when the inductance is an unintended consequence of the circuit wiring it can become a disaster at higher frequencies. Such unintended inductance is called *stray inductance,* and can be reduced by using broad, flat conductors. An example is the 'printed circuit' coils wound on cylindrical forms shown in Fig. 10.2.

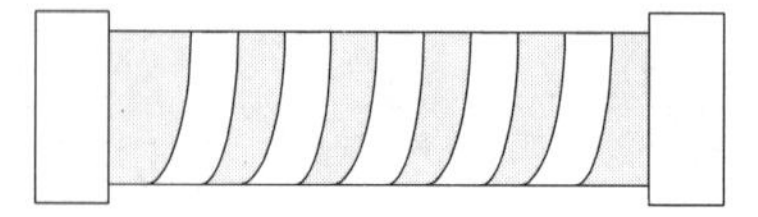

Figure 10.2 Printed circuit circular inductor.

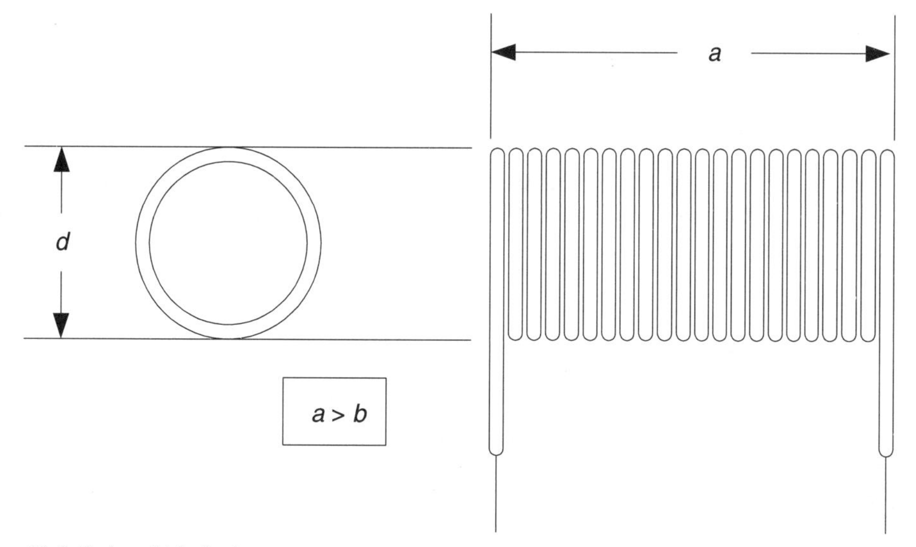

Figure 10.3 Solenoid inductor.

Self-inductance can be increased by forming the conductor into a multi-turn coil (Fig. 10.3); this makes the magnetic field in adjacent turns reinforce each other. The turns of the coil must be insulated from each other.

Several factors affect the inductance of a coil. Perhaps the most obvious are the length, the diameter and the number of turns in the coil. Also affecting the inductance is the nature of the *core* material and its cross-sectional area. In the example of Fig. 10.3 the core is air and the cross-sectional area is directly related to the diameter, but in many radio circuits the core is made of powdered iron or ferrite materials (about which, more later).

Combining two or more inductors

When inductors are connected together in a circuit their inductances combine similar to the resistances of several resistors in parallel or series. For *inductors in which their respective magnetic fields do not interact*:

1 *Series connected inductors*:

$$L_{\mathrm{TOTAL}} = L1 + L2 + L3 + \ldots + L_N \tag{10.5}$$

2 *Parallel connected inductors*:

$$L_{\mathrm{TOTAL}} = \frac{1}{\left[\frac{1}{L1} + \frac{1}{L2} + \frac{1}{L3} + \ldots + \frac{1}{L_N}\right]} \tag{10.6}$$

Or, in the special case of two inductors in parallel:

$$L_{TOTAL} = \frac{L1 \times L2}{L1 + L2} \tag{10.7}$$

If the magnetic fields of the inductors in the circuit interact, then the total inductance becomes somewhat more complicated to express. For the simple case of two inductors in series, the expression would be:

1 *Series inductors*:

$$L_{TOTAL} = L1 + L2 \pm 2M \tag{10.8}$$

Where M is the *mutual inductance* caused by the interaction of the two magnetic fields (note: $+M$ is used when the fields aid each other, and $-M$ is used when the fields are opposing).

2 *Parallel inductors*:

$$L_{TOTAL} = \frac{1}{\left(\frac{1}{L1 \pm M}\right) + \left(\frac{1}{L2 \pm M}\right)} \tag{10.9}$$

Some LC tank circuits use air-core coils in their tuning circuits. Where multiple coils are used, adjacent coils are often aligned at right angles to their neighbour. This helps prevent unintended interaction of the magnetic fields of the coils. In general, for coils in close proximity to each other:

1 Maximum interaction between the coils occurs when the coils' axes are parallel to each other.
2 Minimum interaction between the coils occurs when the coils' axes are at right angles to each other.

For the case where the coils' axes are along the same line, the interaction depends on the distance between the coils.

Air-core inductors

An air-core inductor actually has no core, so might also be called a *coreless coil*. Three different forms of air-core inductor can be recognized. If the length (b) of a cylindrical coil is greater than, or equal to, the diameter (d), then the coil is said to be *solenoid wound*. If the length is much shorter than the diameter, then the coil is said to be *loop wound*.

There is a grey area around the break point between these inductors where the loop wound coil seems to work somewhat like a solenoid wound coil, but in the main most loop wound coils are such that $b << d$. The principal uses of loop wound coils is in making loop antennas for interference nulling and radio direction finding (RDF) applications.

Solenoid wound air-core inductors

An example of the solenoid wound air-core inductor was shown in Fig. 10.3. This form of coil is longer than its own diameter. The inductance of the solenoid wound coil is given by:

$$L_{\mu H} = \frac{a^2N^2}{9a + 10b} \tag{10.10}$$

Where:
$L_{\mu H}$ is the inductance in microhenrys (μH)
a is the coil radius in inches (in)
b is the coil length in inches (in)
N is the number of turns in the coil

The above equation will allow calculation of the inductance of a known coil, but it is usually the case that we need to know the number of turns (N) required to achieve some specific inductance value determined by the application. For this purpose we rearrange the equation in the form:

$$N = \sqrt{\frac{L(9a + 10b)}{a}} \tag{10.11}$$

The air-core coil shown in Fig. 10.4 is typical of those found in older radio transmitters, where it forms the anode tuning inductance. This type of coil is wound of heavy solid wire, and mounted on a plexiglass or Lucite panel. Because several different frequency bands must be accommodated, there are actually several taps on the same form. The required sections are switch selected according to the band of operation.

Adjustable coils

There are several practical problems with the standard fixed coil discussed above. For one thing, the inductance cannot easily be adjusted either to tune the radio or to trim the tuning circuits to account for the tolerances in the circuit.

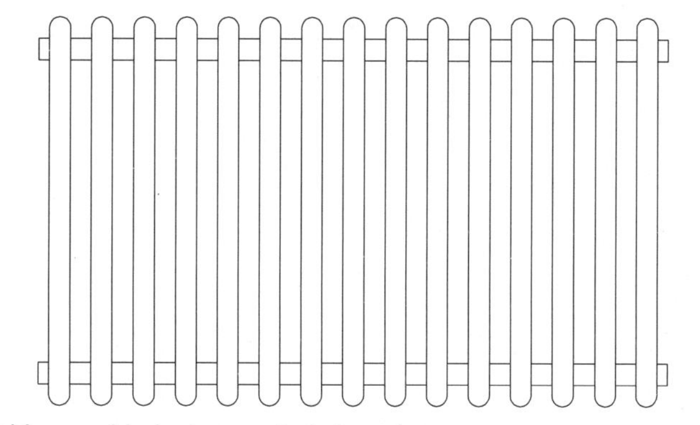

Figure 10.4 Solenoid wound inductor on struts type former.

Air-core coils are difficult to adjust. They can be lengthened or shortened; the number of turns can be changed; or a tap or series of taps can be established on the coil in order to allow an external switch to select the number of turns that are allowed to be effective. None of these methods is terribly elegant, even though all have been used in one application or another.

The solution to the adjustable inductor problem that was developed relatively early in the history of mass produced radios, and still used today, is to insert a powdered iron or ferrite core (or 'slug') inside the coil form (Figs 10.5A and 10.5B). The permeability of the core will increase or decrease the inductance according to how much of the core is inside the coil. If the core is made with either a hexagonal hole or screwdriver slot, then the inductance of the coil can be adjusted by moving the core in or out of the coil. These coils are called *slug-tuned inductors*.

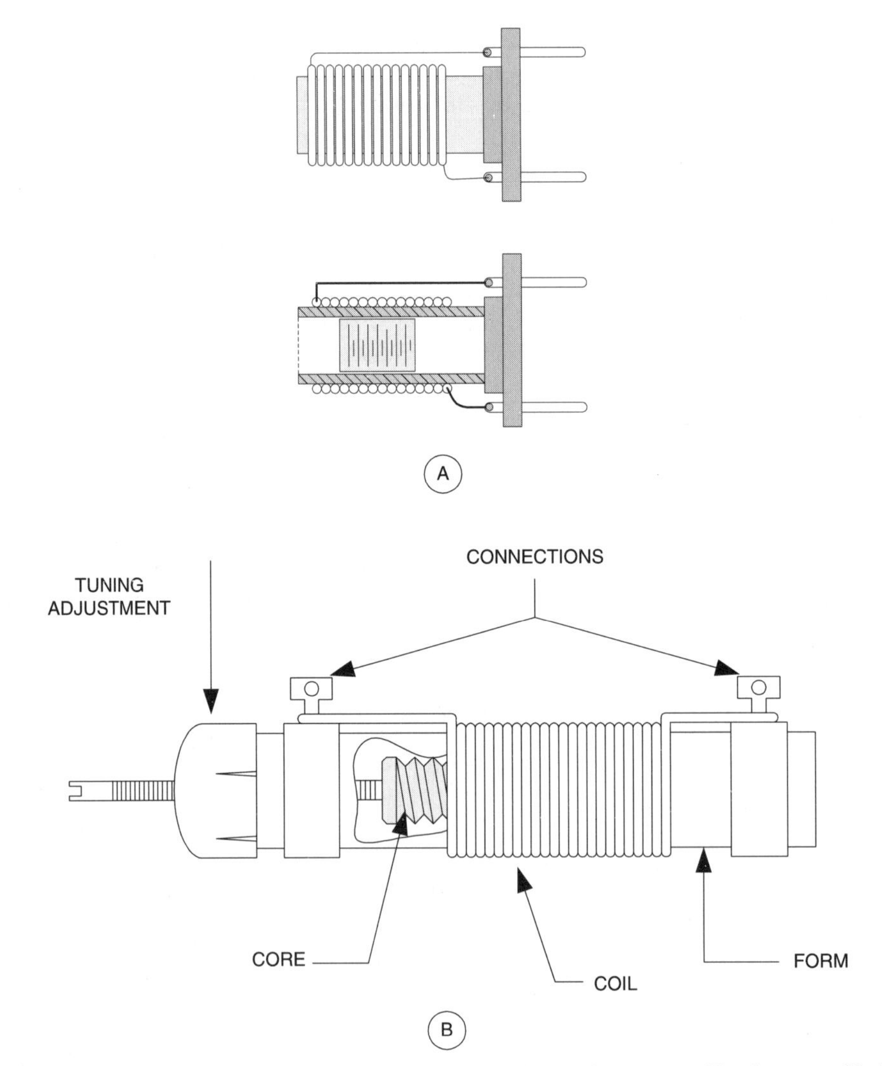

Figure 10.5 (A) Two views of a slug-tuned inductor; (B) different former used in slug-tuned inductor.

Winding your own coils

Inductors (L) and capacitors (C) are the principal components used in RF tuning circuits (also called *resonant circuits* and *LC tank circuits*). The *resonant frequency* of a tank circuit is the frequency to which the LC combination is tuned to, and is found from:

$$f = \frac{1}{2\pi\sqrt{LC}} \tag{10.12}$$

or, if either the inductance (L) or capacitance (C) is either known or preselected, then the other can be found by solving Eq. (10.12) for the unknown:

$$C = \frac{1}{4\pi^2 f^2 L} \tag{10.13}$$

or,

$$L = \frac{1}{4\pi^2 f^2 C} \tag{10.14}$$

In all three equations, L is in henrys, C is in farads, frequency is in hertz (don't forget to convert values to microhenrys and picofarads after calculations are made).

Capacitors are easily obtained in a wide variety of values. But tuning inductors are either unavailable, or are available in other people's ideas of what you need. As a result, it is often difficult to find the kinds of parts that we need. In this section we will take a look at how to make your own slug-tuned adjustable inductors, RF transformers and IF transformers (yes, you *can* build your own IF transformers!).

Tuning inductors can be either air-core or ferrite/powdered iron-core coils. The air-core coils are not usually adjustable unless clumsy taps are provided on the winding of the coil. However, the ferrite and powdered iron-core coils are adjustable if the core is adjustable.

Figure 10.5 showed one form of 'slug-tuned' adjustable coil. The form is made of plastic, phenolic, fibreglass, nylon or ceramic materials, and is internally threaded. The windings of the coil (or coils in the case of RF/IF transformers) are wound onto the form. The *tuning slug* is a ferrite or powdered iron-coil core that mates with the internal threads in the coil form. A screwdriver slot or hex hole in either (or both) ends allows adjustment. The inductance of the coil depends on how much of the core is inside the coil windings.

Amidon Associates coil system

Although blank coil formers were once easily obtained, they fell into disuse. Hobbyists who wanted to build their own RF coils had to use older stocks, or do something else. But *Amidon Associates, Inc.* (2216 East Gladwick, Dominguez Hills, CA, 90220, USA; 310–763–5770 (voice) or 310–763–2250 (fax); http:/www.amidoncorp.com) sells a series of slug-tuned inductor forms that can be used to make any value coil that you are likely to need. Figure 10.6 shows a sectioned view of the *Amidon* forms.

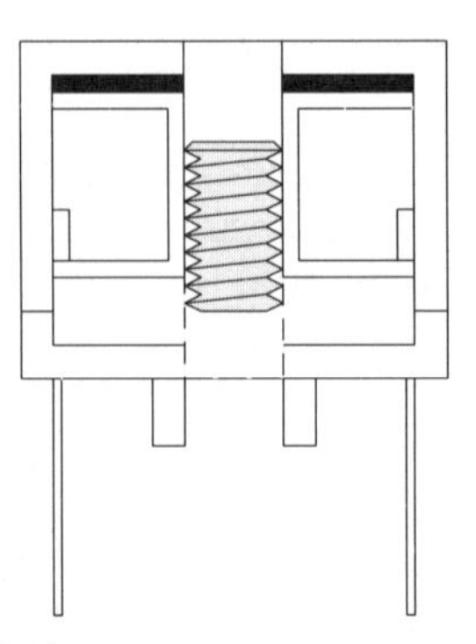

Figure 10.6 Slug-tuned shielded inductor.

Three sizes of coil form are offered. The L-33-X are 0.31 inch square and 0.40 inch high; the L-43-X are 0.44 inch square and 0.50 inch high; and the L-57-X are 0.56 inch square and 0.50 inch high. The 'X' in each type number indicates the type of material, which in turn translates to the operating frequency range. Amidon provides tables giving core data, including A_L. This is a factor which allows you to calculate the number of turns needed for a particular inductance value.

To calculate the number of turns (N) required to make any specific inductance, use the following equation:

$$N = 100\sqrt{\frac{L_{\mu H}}{0.9A_L}} \qquad (10.15)$$

The inductance is in microhenrys (μH). The 'A_L' factor is a function of the properties of the core material; the units are microhenrys per one-hundred turns (μH/100 turns).

Although slug-tuned inductors are sometimes considered a bit beyond the hobbyist or ham, that is not actually true. The *Amidon Associates, Inc.* L-series coil forms are easily used to make almost any inductor that you are likely to need.

Using ferrite and powdered iron cores

Powdered iron and ferrite cores are used as forms to make a variety of inductors, transformers, baluns, EMI chokes, RF chokes and a host of other coil products. The nature of these forms makes it rather easy to construct workable and accurate components using a bit of wire and a touch of imagination. In the rest of this chapter we will take a look at how to use toroid cores, binocular cores, ferrite rods, choke bobbins and ferrite beads.

Materials used in cores

Understanding the nature and properties of powdered iron and ferrite cores is necessary to get the most out of your projects. Before looking at the cores themselves, let's first look at the materials that are used in their manufacture.

Powdered iron

Powdered iron cores are made of ferrous materials that are powdered and then formed into a shape with some sort of binding material. Two main types of material are used: *carbonyl irons* and *hydrogen reduced irons*. The carbonyl form offer superior temperature stability, and have permeability (μ) values from 1 μ to 35 μ. These components are often used in broadband transformers and broadband coils, with high *Q* values, up to frequencies of 200 MHz or 300 MHz. Carbonyl cores are often used in high power balun transformers. The hydrogen reduced types have lower *Q* values than carbonyl cores, but offer values of permeability up to about 90. They are used in low frequency chokes and inductors, and as electromagnetic interference (EMI) filters.

The various materials used for powdered iron cores are designated by both a number system and a colour-code system. Figure 10.7 shows the material type numbers, colour

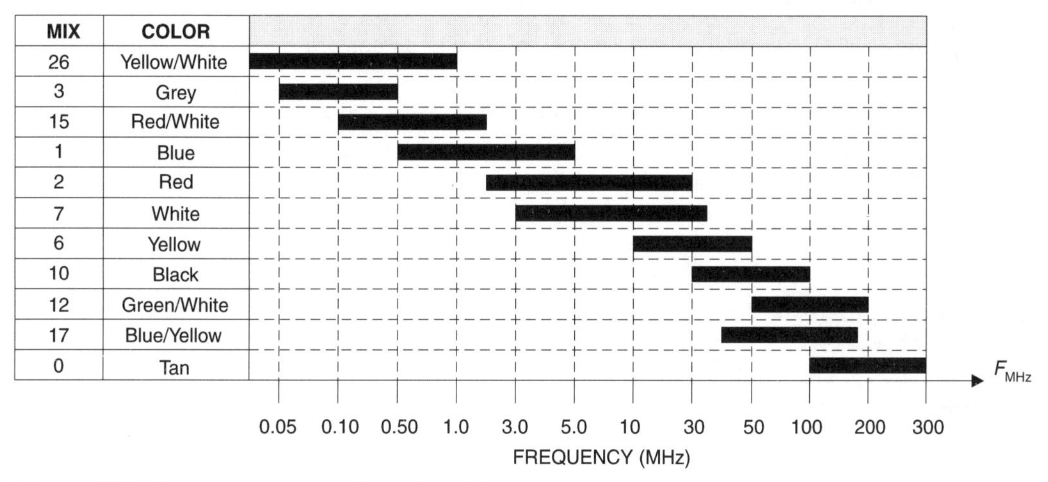

Figure 10.7 Frequency ranges of powdered iron cores.

Table 10.1

Type	μ	Colour code	Material
26	75	Yellow/White	Hydrogen reduced
3	35	Grey	Carbonyl HP
15	25	Red/White	Carbonyl GS6
1	20	Blue	Carbonyl C
2	10	Red	Carbonyl E
7	9	White	Carbonyl TH
6	8	Yellow	Carbonyl SF
10	6	Black	Powdered iron SF
12	3	Green/White	Synthetic oxide
17	3	Blue/Yellow	Carbonyl
0	1	Tan	Phenolic

Table 10.2

Type	Uses
26	High permeability, used in EMI filters, DC chokes and switched DC power supplies.
3	High-*Q* coils and transformers between 50 kHz and 500 kHz.
15	Good *Q*, high stability. Commonly used in AM BCB and 160 metre amateur applications.
1	High volume resistivity. Used for lower frequency applications.
2	High volume resistivity. Commonly used for inductors and transformers in the 3–30 MHz HF bands.
7	Used for HF and low end VHF inductors and transformers.
6	Offers higher *Q* between 30 and 50 MHz, but is used for HF and low VHF band inductors and transformers.
10	Good *Q* and high stability for use in inductors and transformers between 40 and 100 MHz.
12	Good *Q* but only moderate stability for inductors and transformers between 50 MHz and 100 MHz.
17	Similar to Type 12, but has better temperature stability and lower *Q*.
0	High-*Q* applications above 200 MHz. The actual inductance is more sensitive to winding technique than other types.

codes and approximate frequency ranges of the most common forms of powdered iron core. The characteristics of each type are given in Table 10.1, while some common uses of each type are given in Table 10.2.

Ferrite materials

Although the name 'ferrite' implies ferrous (iron-based) materials, these cores are actually made of some more exotic compounds of nickel–zinc and manganese–zinc. The nickel–zinc cores have a high volume resistivity, fairly decent stability, and relatively high *Q* factors. They typically have permeability (μ) values of 125 μ to 850 μ. The manganese–zinc cores have lower volume resistivity, with high *Q* values between 1 kHz and 1000 kHz. These materials are used in power transformers, switched power supplies, and EMI filters. They offer high attenuation to frequencies in the 20 MHz and up range.

Making the calculations

There are two basic issues when making coils and transformers:

1 How to calculate how many turns are needed to achieve a required inductance.
2 How do you find the inductance of a coil once the number of turns are known (short of measuring it on an inductance bridge, of course).

To find the number of turns required to achieve a required inductance requires knowledge of a parameter of the material called the A_L factor. The basic equations are:

1 *Ferrite materials*:

$$N = 1000\sqrt{\frac{L(\text{mH})}{A_L(\text{mH}/1000t)}} \tag{10.16}$$

2 *Powdered iron materials*:

$$N = 100\sqrt{\frac{L(\mu\text{H})}{A_L(\mu\text{H}/100t)}} \tag{10.17}$$

Where:
N is the number of turns
$L(\mu\text{H})$ is the inductance in microhenrys (μH)
$L(\text{mH})$ is the inductance in millihenrys (mH)
A_L is an attribute of the core material

To find the inductance of an existing coil, where the number of turns is known, solve Eqs (10.16) and (10.17) for the value of inductance. There are also some VHF/UHF materials in which the inductance is calculated in terms of nanohenrys per turn (nH/t).

Table 10.3 shows the properties and uses of several different popular ferrite cores.

Table 10.3

Material	μ	Uses
43	850	Inductors and wideband transformers up to 50 MHz.
61	125	Wideband transformers to 200 MHz, with high *Q* between 200 kHz and 15 MHz.
63	40	High-*Q* applications over 15 Mhz and 25 MHz.
67	40	High-*Q* applications between 10 MHz and 80 Mhz, and wideband applications to 200 MHz.
68	20	High-*Q* resonant LC tank circuits between 80 MHz and 180 MHz. It is also used in wideband amplifiers and transformers.
72	2000	High-*Q* applications to 500 kHz. It is also used for EMI reduction of frequencies between 500 kHz and 50 MHz.
75	5000	Low-loss operation between 1 kHz and 1000 kHz. Applications include pulse transformers, wideband transformers, and EMI filters to attenuate 500 kHz to 20 MHz.
77	2000	Wideband transformers between 1 kHz and 1000 kHz. EMI attenuation filters between 500 kHz and 50 MHz.
F	3000	Similar to type 77.
J	5000	Low-loss operation between 1 kHz and 1000 kHz. Applications include pulse transformers, wideband transformers, and EMI filters to attenuate 500 kHz to 20 MHz.

Now that we've laid the foundation by considering the materials, let's look at the different types of core that are available. Keep in mind that not all types of core are available in all types of powdered iron or ferrite material.

Toroid cores

The *toroid core* gets its name from the fact that its basic shape is the *torus*. The word torus is a fancy way of saying the core is doughnut shaped (at least, American doughnuts . . . the hole isn't punched out of the centre in some countries). Figure 10.8 shows the basic torus shape if you are not familiar with doughnuts.

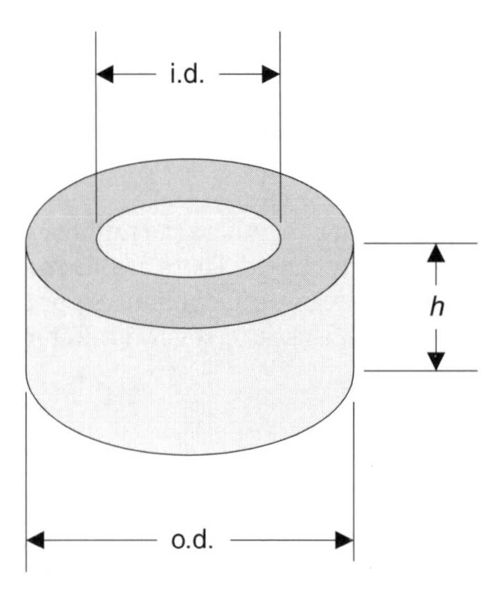

Figure 10.8 Toroidal core.

The main feature that makes toroid cores so attractive in radio construction is that they are inherently self-shielding, because magnetic flux stays in the core. As a result, it is possible to mount toroidal inductors in close proximity to each other without fear of coupling between them. The toroidal inductor can also be mounted closer to other components than other forms of inductor. Although in some cases you will want to mount adjacent coils at right angles to each other (as is done with normal slug-tuned or solenoid-wound air-core coils), it is often the case that the toroid coils can be mounted in the same plane.

Three dimensions and the A_L value are the critical attributes of a toroid core. The dimensions are *outside diameter* (o.d.), *inside diameter* (i.d.) and *height* (h). The A_L value is a function of the size of the core and the material.

Table 10.4 shows the standard forms of powdered iron toroid core and their critical parameters. The 'T-xx' number refers to the size of the core (the 'T' denotes a powdered iron core). The values in the main body of Table 10.4 are the A_L values used to calculate the number of turns.

Table 10.4

Mix	26	3	15	1	2	7	6	10	12	17	0
Colour	Yellow/ White	Grey	Red/ White	Blue	Red	White	Yellow	Black	Green/ White	Blue/ Yellow	Tan
Material	H reduced	Carb HP	Carb GS6	Carb C	Carb E	Carb TH	Carb SF	Powdered iron SF	Syn oxide	Carb	Phenolic
Frequency (MHz)	DC-1	0.05–0.50	0.10–2	0.5–5	2–30	3–35	10–50	30–100	50–200	40–180	100–300
μ	75	35	25	20	10	9	8	6	4	4	1
Temp. coef (PPM/C)	825	370	190	280	95	30	35	150	170	50	0
Core size						**A_L values**					
T-12	N/A	60	50	48	20	18	17	12	7.5	7.5	3
T-16	145	61	55	44	22	N/A	19	13	8	8	3
T-20	180	76	65	52	27	24	22	16	10	10	3.5
T-25	235	100	85	70	34	29	27	19	12	12	4.5
T-30	325	140	93	85	43	37	36	25	16	16	6
T-37	275	120	90	80	40	32	30	25	15	15	4.9
T-44	360	180	160	105	52	46	42	33	18.5	18.5	6.5
T-50	320	175	135	100	49	43	40	31	18	18	6.4
T-68	420	195	180	115	57	52	47	32	21	21	7.5
T-80	450	180	170	115	55	50	45	32	22	22	8.5
T-94	590	248	200	160	84	N/A	70	58	32	N/A	10.6
T-106	900	450	345	325	135	133	116	N/A	N/A	N/A	19
T-130	785	350	250	200	110	103	96	N/A	N/A	N/A	15
T-157	870	420	360	320	140	N/A	115	N/A	N/A	N/A	N/A
T-184	1640	720	N/A	500	240	N/A	195	N/A	N/A	N/A	N/A
T-200	895	425	N/A	250	120	105	100	N/A	N/A	N/A	N/A

H = Hydrogen
Carb = Carbonyl
Syn = Synthetic

Table 10.5

Core size	O.D. (in)	O.D. (mm)	I.D. (in)	I.D. (mm)	*H* (in)	*H* (mm)
T-12	0.125	3.175	0.062	1.575	0.05	1.270
T-16	0.160	4.064	0.078	1.981	0.06	1.524
T-20	0.200	5.080	0.088	2.235	0.07	1.778
T-25	0.250	6.350	0.12	3.048	0.096	2.438
T-30	0.307	7.798	0.151	3.835	0.128	3.251
T-37	0.375	9.525	0.205	5.207	0.128	3.251
T-44	0.440	11.176	0.229	5.817	0.159	4.039
T-50	0.500	12.700	0.300	7.620	0.190	4.826
T-68	0.690	17.526	0.370	9.398	0.190	4.826
T-80	0.795	20.193	0.495	12.573	0.250	6.350
T-94	0.942	23.927	0.560	14.224	0.312	7.925
T-106	1.060	26.924	0.570	14.478	0.437	11.100
T-130	1.300	33.020	0.780	19.812	0.437	11.100
T-157	1.570	39.878	0.950	24.130	0.570	14.478
T-184	1.840	46.736	0.950	24.130	0.710	18.034
T-200	2.000	50.800	1.250	31.750	0.550	13.970
T-200A	2.000	50.800	1.250	31.750	1.000	25.400
T-225	2.250	57.150	1.400	35.560	0.550	13.970
T-225A	2.250	57.150	1.400	35.560	1.000	25.400
T-300	3.000	76.200	1.920	48.768	0.500	12.700
T-300A	3.000	76.200	1.920	48.768	1.000	25.400
T-400	4.000	101.600	2.250	57.150	0.650	16.510
T-400A	4.000	101.600	2.250	57.150	1.000	25.400
T-500	5.200	132.080	3.080	78.232	0.800	20.320

The type number of any given toroid core is made up of the T-number (which gives size) and the material type. For example, a T-50-2 core is made of type 2 material, and will operate on the 2 to 30 MHz band. It has an A_L value of 49 (see Table 10.4). The dimensions (inches and millimetres) of the standard cores are given in Table 10.5.

Example

Calculate the number of turns required for a 3.3 μH inductor for an AM BCB interference filter for a receiver. Use the T-50-15 core. Table 10.4 shows that the T-50-15 core has an A_L value of 135.

$$N = 100\sqrt{\frac{L(\mu H)}{A_L}}$$

$$N = 100\sqrt{\frac{3.3\,\mu H}{135}}$$

$$N = 100\sqrt{0.024} = (100)(0.155) = 15.5 \text{ turns}$$

Table 10.6

Wire size (AWG)	12	14	16	18	20	22	24	26	28	30	32	34	36	38	40
Core size															
T-12	0	0	1	1	1	2	4	5	8	11	15	21	29	37	47
T-16	0	1	1	1	3	3	5	8	11	16	21	29	38	49	63
T-20	1	1	1	3	4	5	6	9	14	18	25	33	43	56	72
T-25	1	1	3	4	5	7	11	15	21	28	37	48	62	79	101
T-30	1	3	4	5	7	11	15	21	28	37	48	62	78	101	129
T-37	3	5	7	9	12	17	23	31	41	53	67	87	110	140	177
T-44	5	6	7	10	15	20	27	35	46	60	76	97	124	157	199
T-50	6	8	11	16	21	28	37	49	63	81	103	131	166	210	265
T-68	9	12	15	21	28	36	47	61	79	101	127	162	205	257	325
T-80	12	17	23	30	39	51	66	84	108	137	172	219	276	347	438
T-94	14	20	27	35	45	58	75	96	123	156	195	248	313	393	496
T-106	14	20	27	35	45	58	75	96	123	156	195	248	313	393	496
T-130	23	30	40	51	66	83	107	137	173	220	275	348	439	550	693
T-157	29	38	50	64	82	104	132	168	213	270	336	426	536	672	846
T-184	29	38	50	64	82	104	132	168	213	270	336	426	536	672	846
T-200	41	53	68	86	109	139	176	223	282	357	445	562	707	886	1115
T-225	46	60	77	98	123	156	198	250	317	400	499	631	793	993	1250
T-300	66	85	108	137	172	217	274	347	438	553	688	870	1093	1368	1721
T-400	79	100	127	161	202	255	322	407	513	648	806	1018	1278	1543	2013
T-520	110	149	160	223	279	349	443	559	706	889	1105	1396	1753	2192	2758

Table 10.6 shows the approximate number of turns that can be accommodated by each size toroid core as a function of wire size. The American Wire Gage (AWG) is used here, so convert to Standard Wire Gage (SWG) for an approximation of the number of turns that can be used with UK wire sizes. It is customary to use enamelled or similar insulation types of wire to wind the coils (although I've used PVC insulated hook-up wire, even though neither customary nor desirable).

The A_L values and sizes for ferrite toroid cores are shown in Tables 10.7 and 10.8, respectively. Keep in mind that the A_L values for ferrite are specified in millihenrys per thousand turns (mH/1000t), rather than the microhenrys per hundred turns (μH/100t) used for powdered iron cores.

Inductors and transformers

Figure 10.9 shows several different forms of inductor and transformers made with toroidal and other forms of core. The coil in Fig. 10.9A is a single-wound inductor. It is used as tuning inductors, lump inductance and RF chokes. These coils are characterized by a single winding of wire on the core.

A two-coil version is shown in Fig. 10.9B. This form of winding is used for transformers. The dots at the top of the winding are used to indicate the same-phase ends of the coils.

Table 10.7

Material	**43**	**61**	**63**	**67**	**68**	**72**	**75**	**77**	**F**	**J**
μ	**850**	**125**	**250**	**40**	**20**	**2M**	**5M**	**2M**	**3M**	**5M**
Core size										
FT-23	188	24.8	7.9	7.8	4	396	990	356	N/A	N/A
FT-37	420	55.3	17.7	17.7	8.8	884	2210	796	N/A	N/A
FT-50	523	68	22	22	11	1100	2750	990	N/A	N/A
FT-50A	570	75	24	24	12	1200	2990	1080	N/A	N/A
FT-50B	1140	150	48	48	12	2400	N/A	2160	N/A	N/A
FT-82	557	73.3	22.4	22.4	11.7	1170	3020	1060	N/A	3020
FT-87A	N/A	N/A	N/A	N/A	N/A	N/A	N/A	N/A	3700	6040
FT-114	603	79.3	25.4	25.4	N/A	1270	3170	1140	1902	3170
FT-114A	N/A	146	N/A	N/A	N/A	2340	N/A	N/A	N/A	N/A
FT-140	952	140	45	45	N/A	2250	6736	2340	N/A	6736
FT-150	N/A	N/A	N/A	N/A	N/A	N/A	N/A	N/A	2640	4400
FT-150A	N/A	N/A	N/A	N/A	N/A	N/A	N/A	N/A	5020	8370
FT-193A	N/A	N/A	N/A	N/A	N/A	N/A	N/A	N/A	4460	7435
FT-240	1240	173	53	53	N/A	3130	6845	3130	N/A	6845

M = 1 000 000

These dots are important when cross-connecting windings for transformers. Although the turns ratio is shown apparently 1:1, that is not universally the case. One winding could easily have a different number of turns than the other winding, depending on the application. For example, if the transformer is used as the RF input tuning coil of a receiver, the primary may have only 1 to 5 turns, while the secondary will have as many turns as are necessary to achieve the resonating inductance.

Table 10.8

Core size	**O.D. (in)**	**O.D. (mm)**	**I.D. (in)**	**I.D. (mm)**	***H* (in)**	***H* (mm)**
FT-23	0.230	5.842	0.120	3.048	0.060	1.524
FT-37	0.375	9.525	0.187	4.750	0.125	3.175
FT-50	0.500	12.700	0.281	7.137	0.188	4.775
FT-50A	0.500	12.700	0.312	7.925	0.250	6.350
FT-50B	0.500	12.700	0.312	7.925	0.500	12.700
FT-82	0.825	20.955	0.520	13.208	0.250	6.350
FT-87A	0.870	22.098	0.540	13.716	0.500	12.700
FT-114	1.142	29.007	0.750	19.050	0.295	7.493
FT-114A	1.142	29.007	0.750	19.050	0.545	13.843
FT-140	1.400	35.560	0.900	22.860	0.500	12.700
FT-150	1.500	38.100	0.750	19.050	0.250	6.350
FT-150A	1.500	38.100	0.750	19.050	0.500	12.700
FT-193A	1.932	49.073	1.250	31.750	0.750	19.050
FT-240	2.400	60.960	1.400	35.560	0.500	12.700

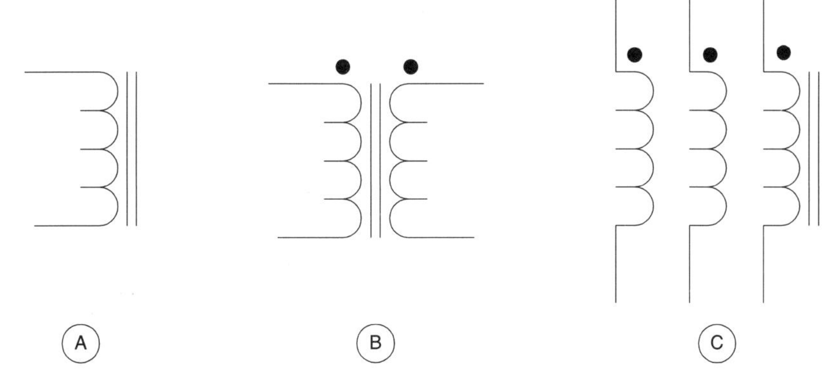

Figure 10.9 (A) Single coil; (B) bifilar wound transformer; (C) trifilar wound transformer.

The three-coil winding is shown in Fig. 10.9C. This sort of winding is used on multiple secondary transformers, balun (*bal*anced–*un*balanced) transformers, and phase reversing (or push-pull) transformers.

Broadband RF transformers

Perhaps the widest use for toroidal transformers is as a balun (or related) form of transformer. A true balun transformer provides the current paths to convert an unbalanced source (such as coaxial cable) to a balanced load (such as a dipole antenna), or vice versa. Some broadband transformers provide impedance transformation as well as load transformation. It is customary to install a 1:1 balun transformer at the feedpoint of a dipole or other balanced antenna in order to clean up the pattern. Otherwise, currents flowing in the shield of the coaxial cable are made part of the antenna and radiate (distorting the radiation pattern).

Figure 10.10 shows the basic forms of broadband transformers. In each case 'R' is not a physical resistor, but rather represents the load or source impedances. The version in Fig. 10.10A is a 1:1 balun transformer. It provides load transformation but not impedance transformation. It uses a three-coil winding. A 4:1 balun transformer is shown in Fig. 10.10B. This two-winding balun provides a 4:1 reduction or increase in impedance, depending upon which direction it is connected.

The transformers in Figs 10.10C and 10.10D are not baluns because they have both load and source unbalanced. These transformers are sometimes called by the ridiculous but descriptive term 'un–un'. The version in Fig. 10.10C uses a pair of two-winding transformers connected so as to provide a 9:1 impedance transformation ratio. The version in Fig. 10.10D is similarly constructed, but is connected to provide a 16:1 transformation. Both of these transformers can be used in solid-state amplifiers where the input impedance is low compared to the 50 ohm system impedance. They can also be used on antennas such as verticals that can have a low feedpoint impedance (2 to 37 ohms in the case of the vertical) that must be matched to 50 ohms.

The transformer in Fig. 10.11 is designed to provide two outputs that are 180 degrees apart. The transformer is wound with three identical coils, with one designated as the primary and the others as the two secondaries. These transformers are used in

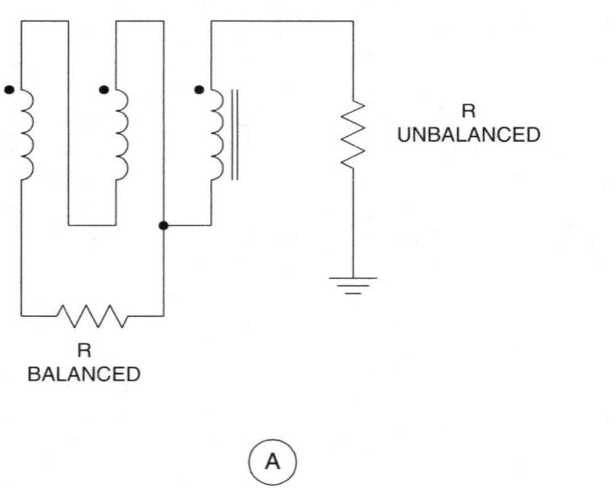

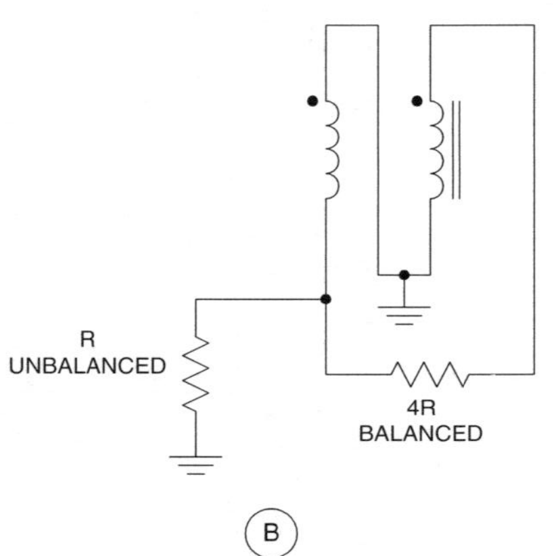

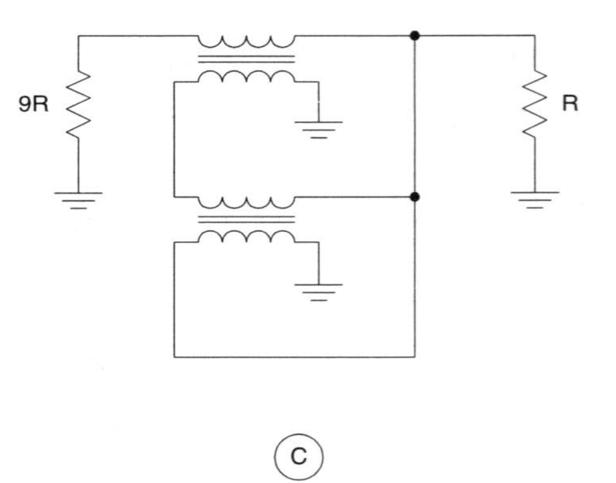

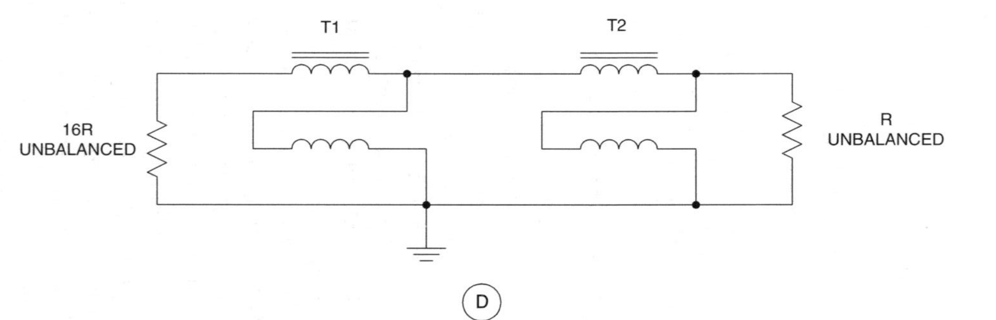

Figure 10.10 (A) 1:1 balun; (B) 4:1 balun; (C) 9:1 un–un; (D) 16:1 un–un.

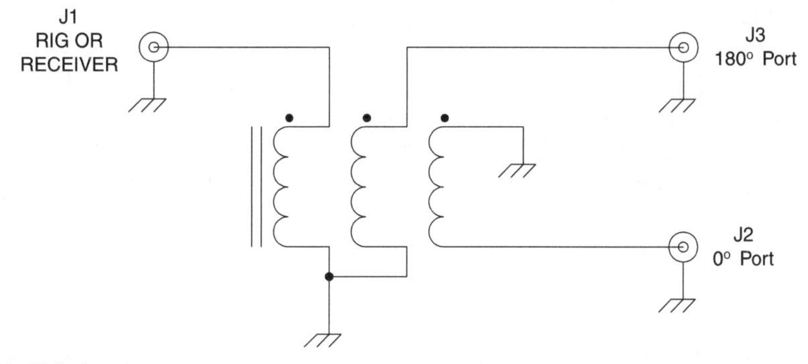

Figure 10.11 180 degree un–un.

instrumentation purposes, and for driving a pair of vertical radiators. In the latter application the directivity can be reversed by switching from parallel feed (no transformer) and phase-reversed feed with the transformer.

Winding toroid cores

Counting turns

Winding toroid cores seems to be a mystery to some. The number of turns is easily calculated, but the definition of 'turn' in the practical case seems at odds. The correct designation of 'turn' is a *pass through the toroid core*. Consider Fig. 10.12. The winding shown in Fig. 10.12A represents a one-turn winding, while that in Fig. 10.12B is a two-turn winding. Oddly, many people would see the coil in Fig. 10.12B and assume that a 'turn' is one complete round about the core. Unfortunately, that's wrong. In some coils, where the A_L value of the core is high, there is a high ratio of change of inductance per change in turns. If your toroids are consistently off in the final inductance, then consider how you are counting turns.

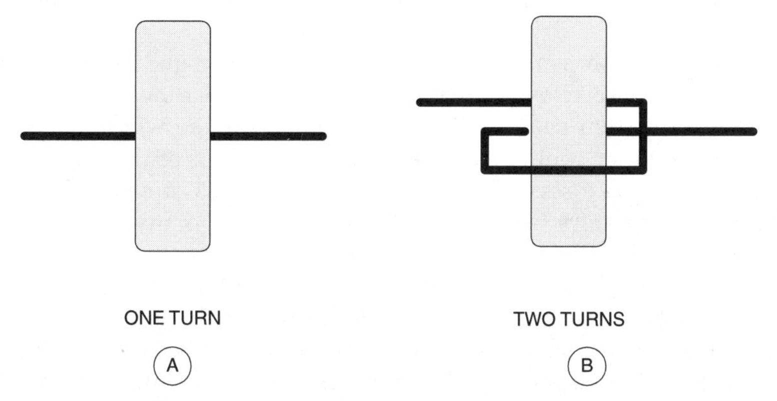

Figure 10.12 Turns counting on toroidal winding: (A) one turn; (B) two turns.

Winding styles

Now let's consider the various winding styles for toroidal cores. In the radio literature you will see transformers and coils described as *single wound, bifilar wound,* and *trifilar wound.* Figure 10.13 shows these styles. The coil in Fig. 10.13A is single wound. There is only one winding. The turns of the coil are spaced out up to 330 degrees of the core, leaving an arc distance of at least 30 degrees between the ends. If the ends are closer together, then there might be some capacitive coupling between the ends.

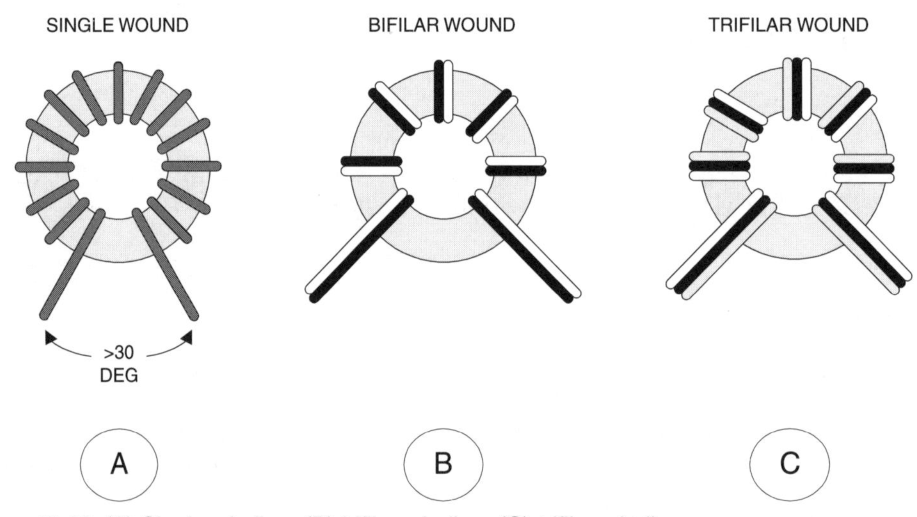

Figure 10.13 (A) Single winding; (B) bifilar winding; (C) trifilar winding.

Bifilar winding is shown in Fig. 10.13B. This form of winding uses two windings, but they are not simply scramble wound on the core, or wound separately. Rather, they are wound such that the two wires remain parallel and closely adjacent to each other throughout the entire winding. Again, observe the requirement of at least 30 degrees between the ends, and approximately equal spacing around the circumference of the core.

The trifilar winding is shown in Fig. 10.13C. This form of winding is like the bifilar in construction, but uses three closely adjacent and parallel windings.

It is wise to use different colours of wire insulation for the windings. Although you could use an ohmmeter or continuity tester to identify which of the four or six ends are paired, why bother? I keep several rolls of the various sizes of enamelled wire in Joe's Basement Therapy Laboratory (where I go to let the wind out of my head), each of slightly different colour enamel.

Two systems of designating the windings of a bifilar or trifilar wound toroid transformer in circuit diagrams are shown in Fig. 10.14. One method is to use phase dots. Another method is to use a letter–number system in which the winding is identified by the letter, and the ends are designated by numbers. The ends with the same numeral are of the same phase. Note that both ends designated '2' in Fig. 10.14 are dotted.

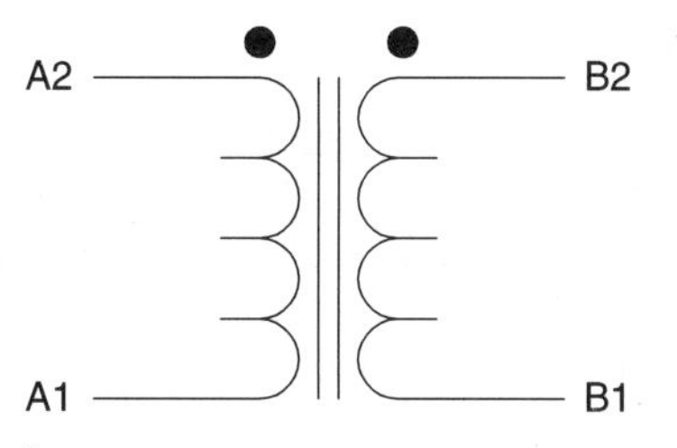

Figure 10.14 Bifilar wound transformer.

Figure 10.15 shows the method for winding transformers with a different number of turns in the two windings (Fig. 10.15A). There are basically three methods of winding these coils. First, you could wind the two windings in the bifilar manner until the smaller winding is out of wire. The other method is to intersperse the two windings in the manner of Fig. 10.15B. Finally, the smaller winding can be concentrated in one area of the core (Fig. 10.15C).

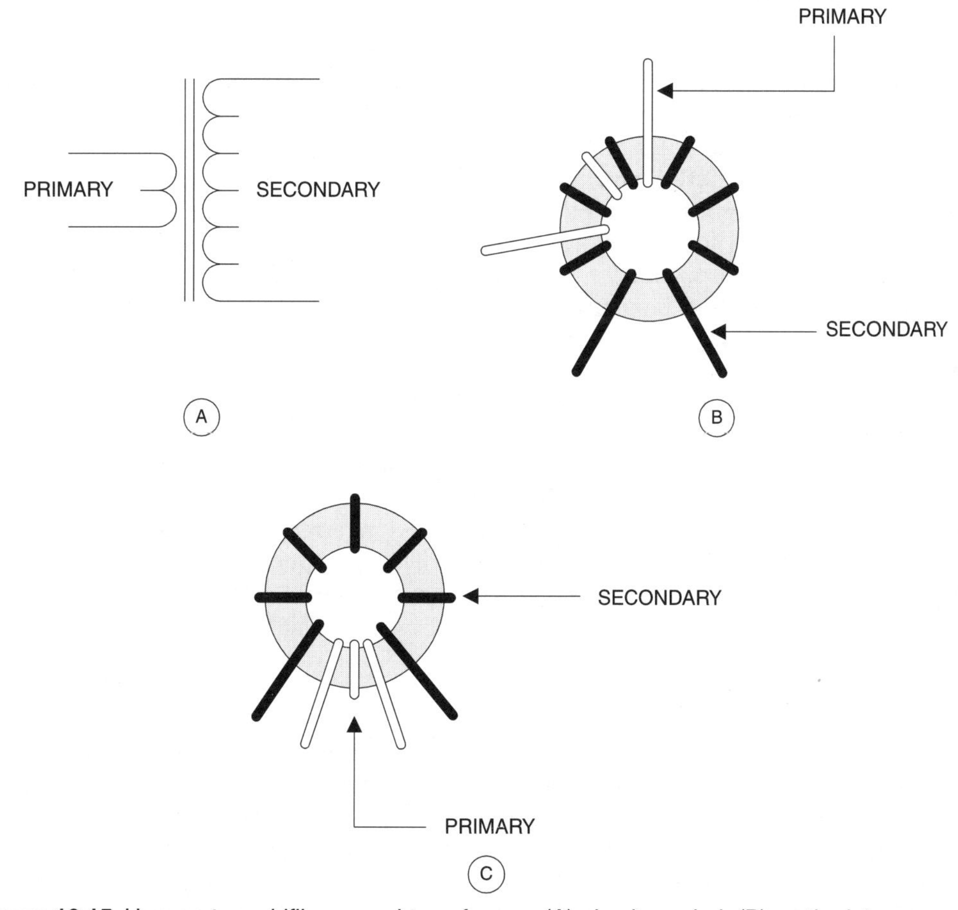

Figure 10.15 Uneven turns bifilar wound transformer: (A) circuit symbol; (B) method 1; (C) method 2.

Stabilizing the windings

If you build a large-core toroidal coil or transformer, then the heavy gauge of the wire can usually be counted on to keep the winding physically stable. But in the case of smaller cores, the windings sometimes have a habit of unravelling a bit. This can be handled either by applying a layer of *Q-dope* over the entire winding, or by placing a small spot of glue (Fig. 10.16) at the ends of the windings.

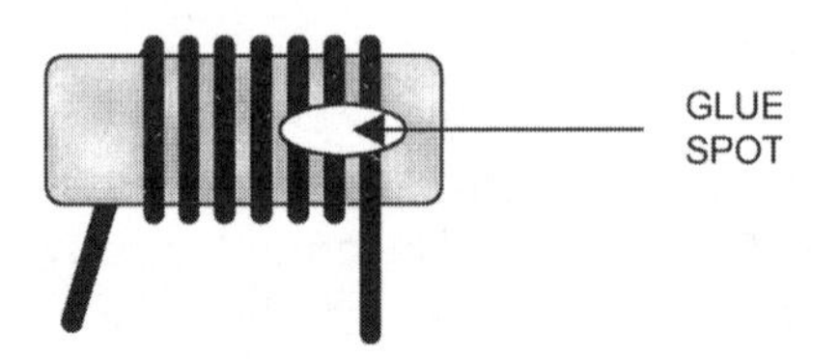

Figure 10.16 Stabilizing the windings on a toroid.

Mounting toroids

Now let's turn our attention to the matter of mounting toroids, both individually and in close proximity to other toroids. Figure 10.17A shows the method for mounting small toroids on a printed circuit board or perf board. The toroid is laid flat on the PCB, and the wires passed through the holes and soldered to the foil side.

The method of Fig. 10.17A works in cases where there is not a lot of vibration. In some more rugged applications a bit of silicone seal or caulk can be used to stabilize the toroid. However, in severe cases, or where moderately high powered large cores are used, it might be better to opt for the method of Fig. 10.17B. The toroid is held fast using a machine screw and hex nut. Although I have successfully used brass hardware, I normally use nylon screws and nuts. Similarly, the washers that sandwich the toroid, and the washer that is under the hex nut, are made of nylon or some other insulating material. Be careful not to overtighten the nut, or damage to the toroid core can occur (they are fragile).

Figure 10.17C shows the method for mounting a toroid on end. In this case, the edge of the form is placed against the PCB, and the wires brought through the holes as shown. In general, this method of mounting is less preferred than horizontal mounting because it is inherently less stable. However, when component density on the PCB is tight, or extraneous magnetic fields could cause a problem, this method becomes more reasonable.

Mounting multiple coils

Many circuits use two or more toroid coils or transformers in close proximity to each other. Examples include the input and output transformers of broadband amplifiers, and the inductors in RF filters. Figure 10.18 shows several different mounting schemes.

One of the glories of the toroidal core is that the magnetic field of a coil wound on it is self-contained. That means there is little or no interaction between adjacent components and the toroidal coil. Figures 10.18A through 10.18C show methods of mounting toroidal cores in close proximity on a PCB.

The toroid is said to have a self-contained magnetic field, and that is taken to mean that there can be no coupling to adjacent coils. That claim is true only when the winding is perfect and there are no manufacturing anomalies in the core itself. The typical core available to amateur builders is quite high quality. It is, however, often prudent to mount two or three adjacent coils orthogonal (i.e. at right angles) to each other. Figures 10.18D

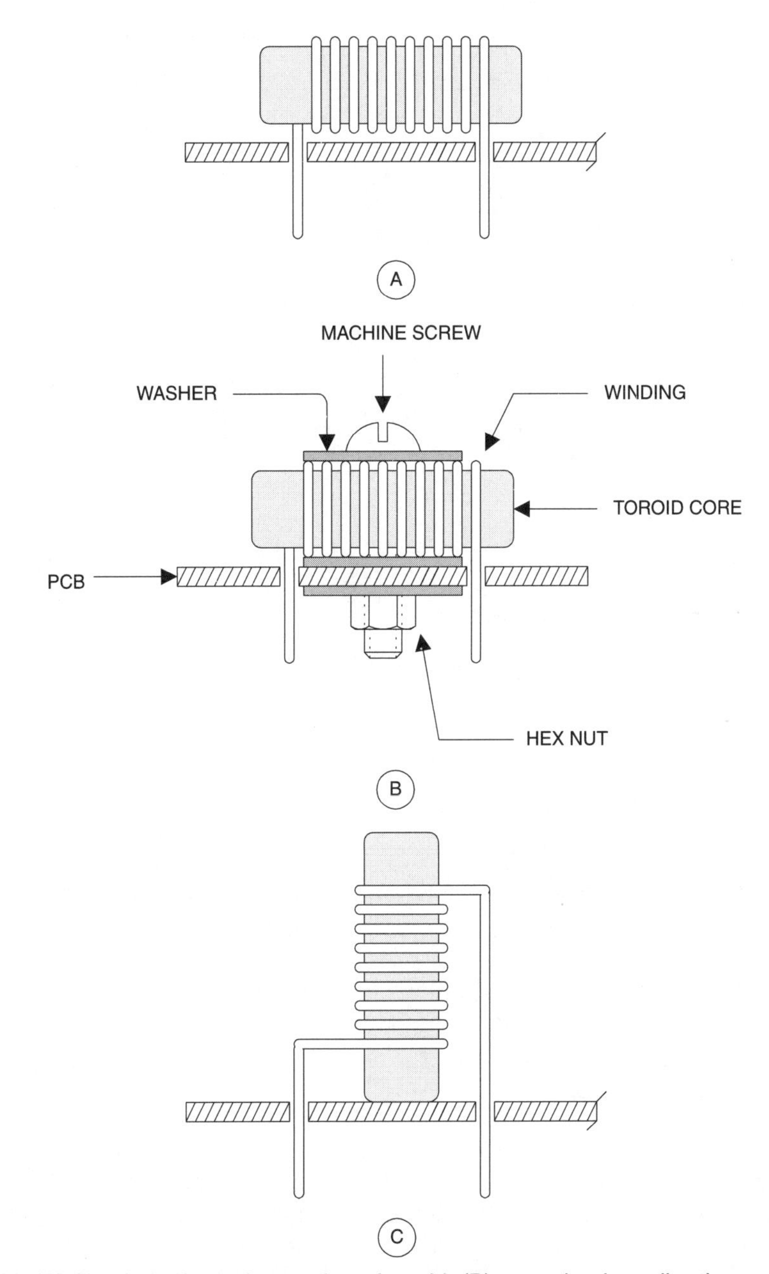

Figure 10.17 (A) Simple horizontal mounting of toroid; (B) mounting in a vibration environment; (C) simple vertical mounting of a toroid.

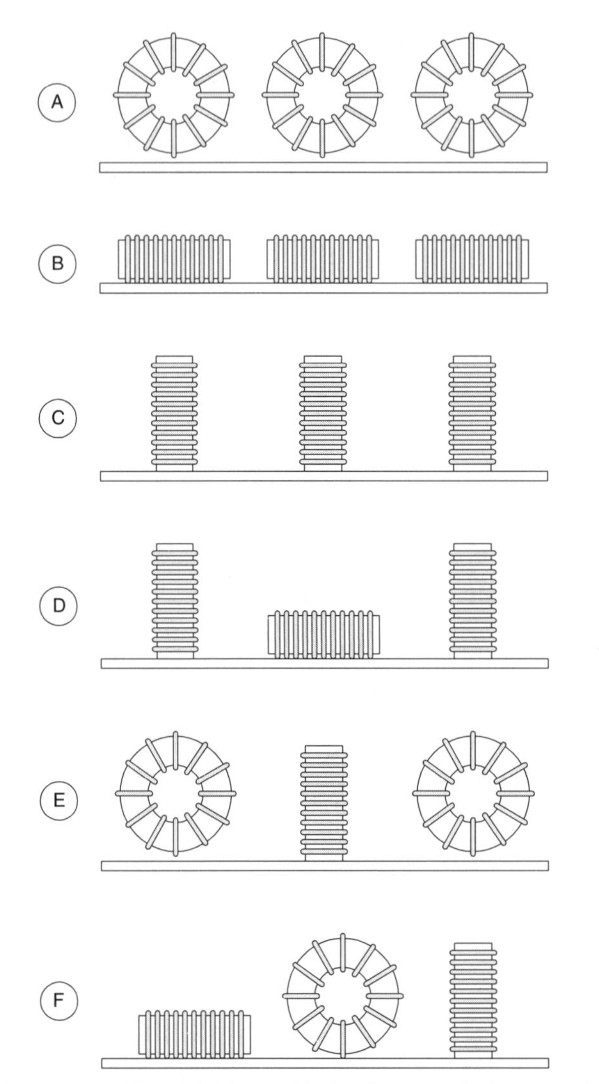

Figure 10.18 Methods for mounting multiple toroids in close proximity to each other.

through 10.18F show methods of mounting coils adjacent to each other, but in different planes. These methods of mounting minimize any stray coupling that might occur.

Special mounting methods

A lot of amateur applications, especially RF sensors for RF power meters and VSWR meters, use a toroidal current transformer with a single-turn primary winding and multi-turn secondary winding. Figure 10.19A shows this system schematically. The idea is to put the single-turn primary right in the centre of the hole, and therein lies the problem.

Figure 10.19B shows one solution to the problem. The single-turn primary is made of brass tubing or brazing rod. Select a size that is a slip fit for the hole in a smaller size

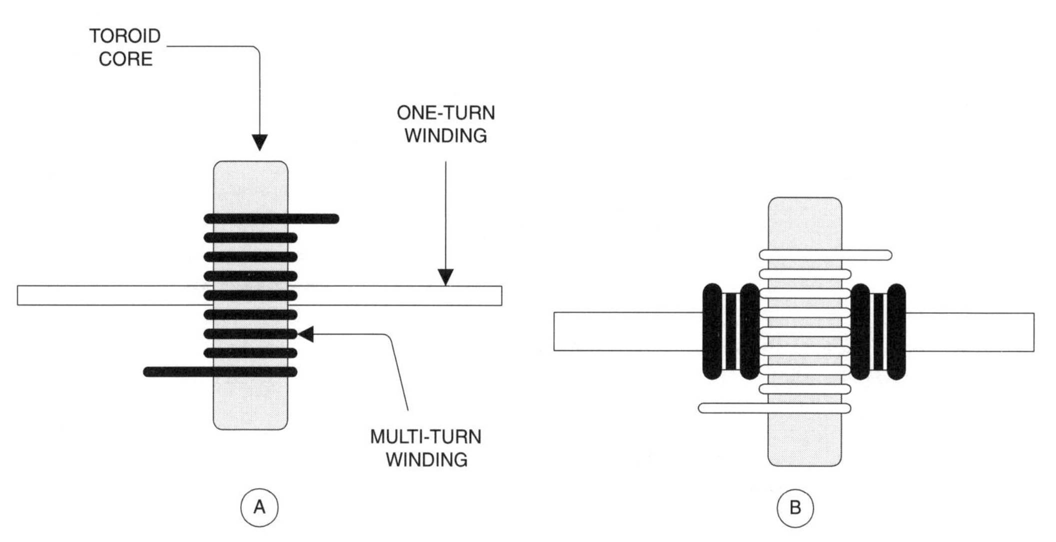

Figure 10.19 (A) Transmission line transformer; (B) physical implementation.

rubber grommet. The grommets are placed on either side of the toroid core, and then cemented into place. In another case a larger grommet is placed inside the through-hole of the toroid core, and then cemented into place.

High-power transformers

The volume and cross-sectional area of the cores are a determining factor in their power handling capacity. In order to boost the power capacity, two or more large size cores are often stacked together as one. Each toroid is wound with a single layer of fibreglass tape to insulate it from the other. The bifilar or trifilar windings are then placed over the two cores together.

I've seen (and used) nylon filament packing (or 'strapping') tape in place of the fibreglass, but only on low to moderate power. I have not seen anyone test this tape at the highest power levels authorized for amateur radio operators.

Mounting of a stacked high power toroid transformer is shown in Fig. 10.20. The cores are each wrapped in fibreglass tape, and then stacked on top of each other. Additional runs of tape are then used to secure the assembly together. The core assembly is then sandwiched between bakelite or plastic supports (washers can be used if available in those large sizes). The entire assembly is then mounted to a printed circuit board or metal chassis using a bolt and hex nut. Again, although brass bolts and nuts are sometimes seen, the use of nylon hardware is highly recommended.

Binocular cores

The *binocular core* (Fig. 10.21) is sometimes (erroneously) called a 'balun core'. Perhaps it gets this 'street name' from the fact that it was once used extensively in making wideband

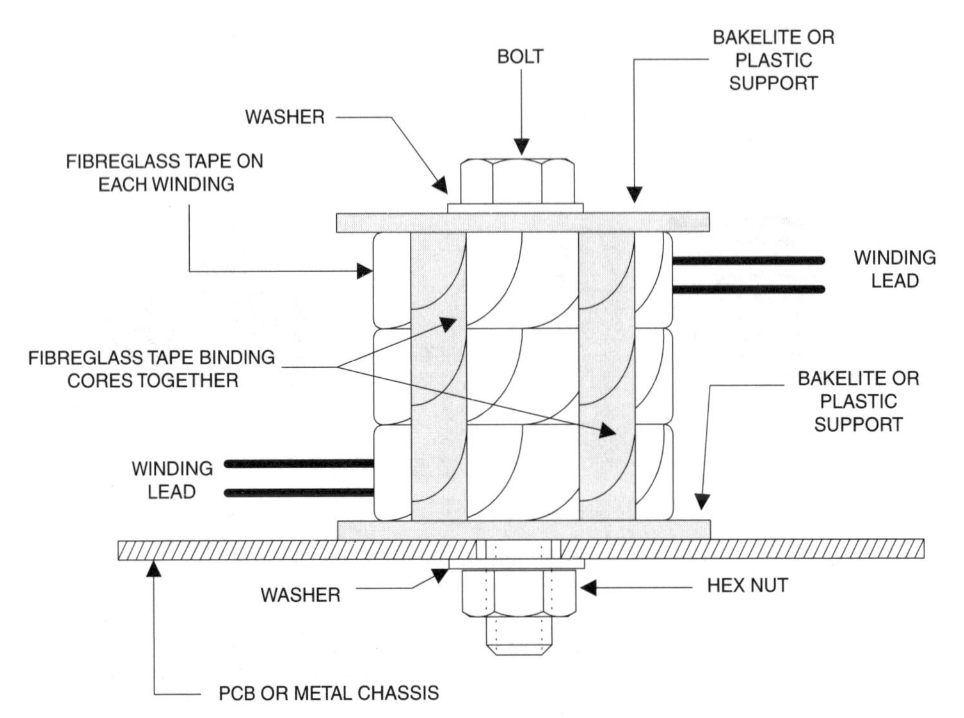

Figure 10.20 Mounting a power toroid.

balun input transformers for television and FM broadcast receivers. The 'two-hole balun core' designation used by Amidon Associates in the USA is a little nearer the case. The top and bottom view of the binocular core shows it to be square or rectangular, while a view of the ends shows a pair of through-holes. The binocular core is usually made of ferrite materials, although I suppose powdered iron versions can be found as well.

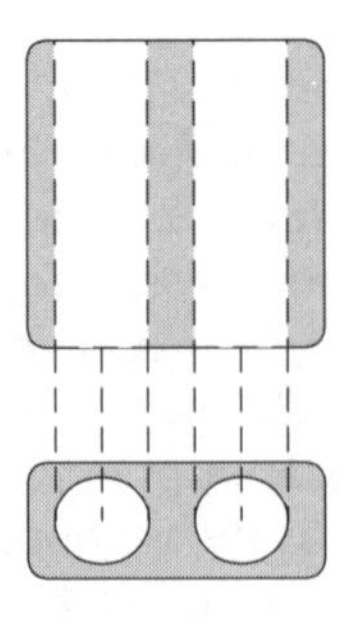

Figure 10.21 Binocular 'balun' core.

Table 10.9 shows the dimensions (inches and millimetres) and A_L values of some of the commonly available binocular cores. In the ferrite versions with very high A_L values, tremendous inductance values can be achieved (with correct selection of core material) using relatively few turns. The A_L values are in terms of millihenrys per 1000 turns, so the required number of turns is found from:

$$N = 1000\sqrt{\frac{L(\text{mH})}{A_L}} \quad (10.18)$$

Where:

N is the number of turns (t)

L(mH) is the inductance in millihenrys (mH)

A_L is a property of the ferrite (mH/1000t)

Table 10.9

Type No.	O.D. (in)	O.D. (mm)	I.D. (in)	I.D. (mm)	H (in)	H (mm)	T (in)	T (mm)	Type	A_L value
BN-43-202	0.525	13.335	0.150	3.810	0.550	13.970	0.295	7.493	1	2890
BN-43-2302	0.136	3.454	0.035	0.889	0.093	2.362	0.080	2.032	1	680
BN-43-2402	0.280	7.112	0.070	1.778	0.240	6.096	0.160	4.064	1	1277
BN-43-3312	0.765	19.431	0.187	4.750	1.000	25.400	0.375	9.525	1	5400
BN-43-7051	1.130	28.702	0.250	6.350	1.130	28.702	0.560	14.224	1	6000
BN-61-202	0.525	13.335	0.150	3.810	0.550	13.970	0.295	7.493	1	425
BN-61-2302	0.136	3.454	0.035	0.889	0.093	2.362	0.080	2.032	1	100
BN-61-2402	0.280	7.112	0.070	1.778	0.240	6.096	0.160	4.064	1	280
BN-61-1702	0.250	6.350	0.050	1.270	0.470	11.938	xxx	xxx	2	420
BN-61-1802	0.250	6.350	0.050	1.270	0.240	6.096	xxx	xxx	2	310
BN-73-202	0.525	13.335	0.150	3.810	0.550	13.970	0.295	7.493	1	8500
BN-73-2402	0.275	6.985	0.070	1.778	0.240	6.096	0.160	4.064	1	3750

Turns counting on binocular cores

A 'turn' on a binocular core is one pass through each hole. In Fig. 10.22A we see a core with a one-turn winding, while in Fig. 10.22B the two-turn case is shown. Additional turns are counted similarly.

Winding styles on binocular cores

Figure 10.23 shows three different methods for winding a binocular core. The preferred method (some would say 'correct' method) is shown in Fig. 10.23A. The winding is made through the two holes, and is not outside the core. If two more windings are used, then it is common practice to lay down the primary winding first, and then lay the secondary winding on top of it. Bifilar and trifilar winding methods can also be used, but these are a little more difficult to achieve than on toroid cores.

Alternate methods of winding are shown in Figs. 10.23B and 10.23C. In Fig. 10.23B, one winding is placed in the centre of the core in the manner of Fig. 10.23A. Two additional windings are placed on the outsides of the two holes. In Fig. 10.23C the two coils are wound separately through the two holes, but are orthogonal to each other. Both Figs 10.23B and 10.23C are used, especially when there is a wire size or fit problem, but are not highly recommended compared with Fig. 10.23A.

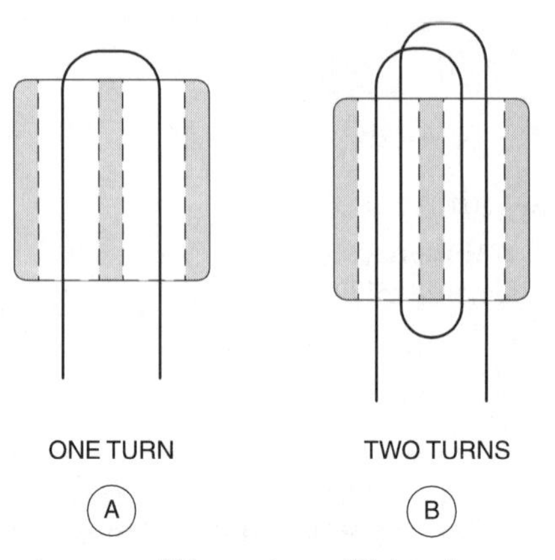

Figure 10.22 Winding a binocular core: (A) one turn; (B) two turns.

Winding a binocular core

In the 1950s there was a comedy act on American TV in which a chap – who obviously could be described as about a half-bubble off dead level – attempted to assemble a child's toy on Christmas Eve. To say that he was 'all thumbs' is to be charitable . . . it was more like 'all toes'. My first attempt to wind a binocular core was all too reminiscent of that early TV comedy. Later, I was shown a simple procedure that made it easy. Figure 10.24 shows this method.

The base of the assembly fixture is a piece of cardboard. I use the sort of cardboard insert that comes with men's shirts when they are returned 'boxed' from the cleaners. When cut into two halves the inserts are just about the right size and stiffness. The binocular core is fastened to the cardboard with a bit of masking tape (preferred over other types of tape because it is easy to remove). The first (inner) winding is started by taping one end of the wire to the cardboard a few centimetres from the core. The wire is snaked through both holes of the core as many times as needed for the coil being constructed. The end of the first winding is then taped to the cardboard with another bit of masking tape. The second winding is overlaid on the first using a similar

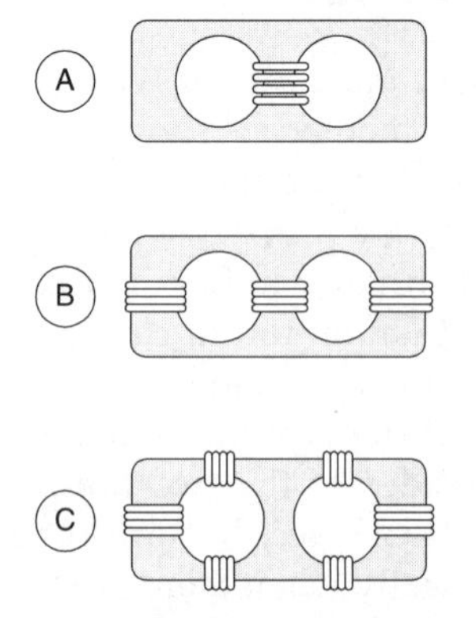

Figure 10.23 Three different winding styles for binocular balun.

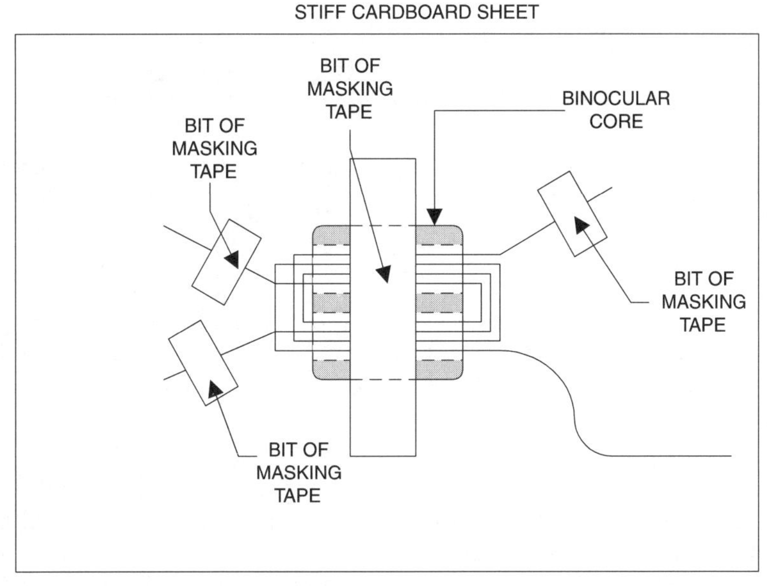

Figure 10.24 Method for winding binocular balun transformers.

procedure. After the windings are completed, a bit of *Q*-dope or cement can be used to secure the wires to the core.

Ferrite rods

Ferrite rods are used for high powered RF chokes, filament chokes in grounded grid linear amplifiers, and as directional antennas for receivers. Figure 10.25 shows the basic ferrite rod. The critical dimensions are its length (L), outside diameter (o.d.) and A_L value (mH/1000t). Table 10.10 shows the dimensions, permeability, A_L values and other data for commonly available 10 cm to 19 cm ferrite rods. The A_L values are only approximate because they are affected by other factors. Note that, although the usual formulas to find number of turns can be used, the actual inductance achieved depends somewhat on the position of the winding on the length of the rod.

The rods made of Type 61 material are used as radio antennas at frequencies from below the medium wave AM BCB to about 10 MHz. These antennas are highly directional, with two major lobes at right angles to the rod. As a result, they can be used for radio direction finding (RDF). The nulls off the ends of the antenna are used to discriminate station direction. Antennas for the VLF and LF frequency ranges are made using rods made of Type 33 material. When winding an RF choke, use Type 33 material for the 40 metre and 80 metre bands, and type 61 for 10 metres through 40 metres.

Figure 10.25 Ferrite rod.

Table 10.10

Rod type No.	Material	Permeability	O.D. (in)	O.D. (cm)	Length (in)	Length (cm)	A_L value	Ampere-turns
R61-025-400	61	125	0.25	0.64	4.00	10.16	26.00	110
R61-033-400	61	125	0.50	1.27	4.00	10.16	32.00	185
R61-050-400	61	125	0.50	1.27	4.00	10.16	43.00	575
R61-050-750	61	125	0.50	1.27	7.50	19.05	49.00	260
R33-037-400	33	800	0.37	0.94	4.00	10.16	62.00	290
R33-050-200	33	800	0.50	1.27	2.00	5.08	51.00	465
R33-050-400	33	800	0.50	1.27	4.00	10.16	59.00	300
R33-050-750	33	800	0.50	1.27	7.50	19.05	70.00	200

The inductance achieved for any given number of turns, the permeability of the rod, actual A_L value, the Q of the rod, and the ampere-turns rating are affected by both the length/diameter ratio of the rod and the position of the coil on the rod. Also affecting performance of the coil are the spacing between the turns of the coil and the spacing between the coil and the rod. Best A_L value and overall performance is achieved when the coil is centred on the rod's length, while highest Q is achieved when the coil is spread out over the length of the rod. Because of the inherent variation in rods, the inductance (if it is critical) should be checked before the coil is put into service.

Figures 10.26 and 10.27 show winding schemes for ferrite rod transformers. The trifilar method is shown in Fig. 10.26. All three wires are held parallel and adjacent to each other as they are wound along the length of the rod. A 4:1 bifilar wound balun transformer is shown in Fig. 10.27. The filament choke used in grounded grid amplifiers is wound in this same manner, although the connection scheme is different.

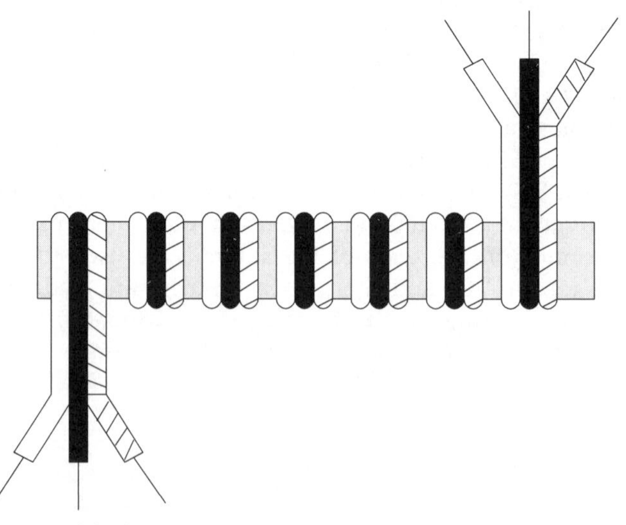

Figure 10.26 Trifilar wound ferrite rod.

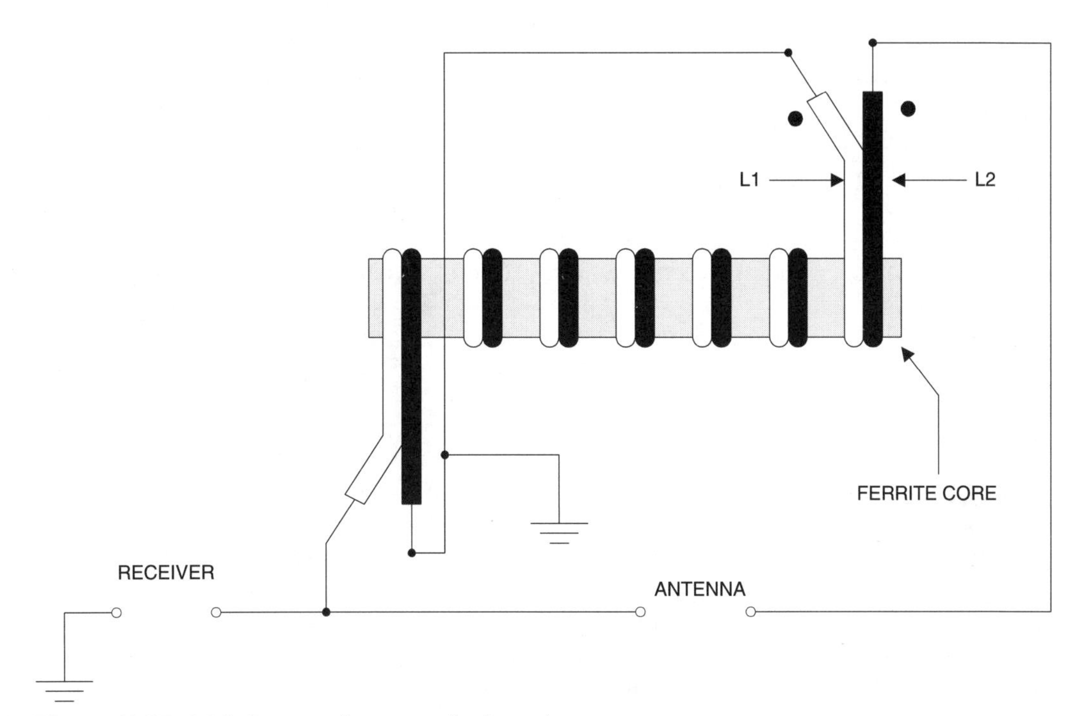

Figure 10.27 4:1 balun transformer on ferrite rod.

Bobbing along with a bobbin

Ferrite bobbins (Fig. 10.28) offer a convenient way to make small RF chokes and other coils. The bobbin is a former that permits the easy winding of such coils. They have a winding area, with end blocks to keep the wires in the correct space. They also have leads coming from each end to make connection to the circuits easier. The A_L values given in Table 10.11 are in mH/1000t.

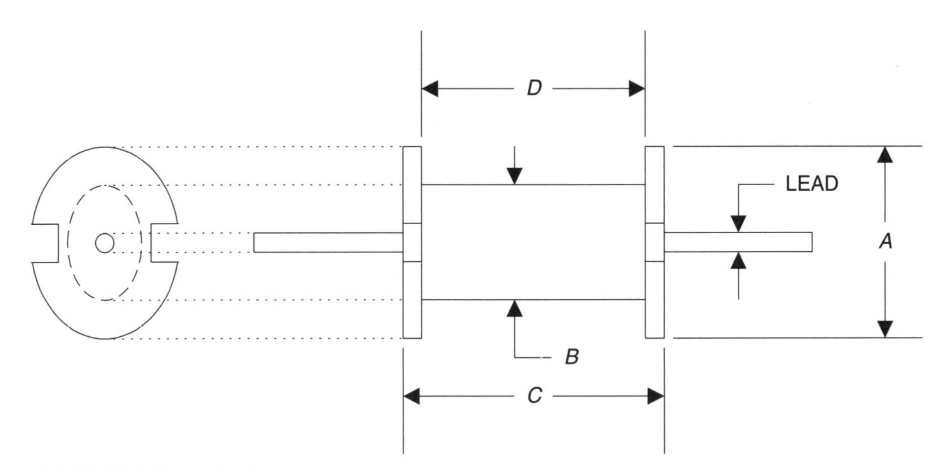

Figure 10.28 Bobbin style former.

Table 10.11

Bobbin No.	A_L	NI	20	22	24	26	28	30	32	34	36
B-72-1011	39	130	24	39	60	93	148	230	425	535	1050
B-72-1111	17	60	9	14	23	35	56	88	154	205	400

Bobbin No.	*A* (in)	*A* (mm)	*B* (in)	*B* (mm)	*C* (in)	*C* (mm)	*D* (in)	*D* (mm)	Lead
B-72-1011	0.372	9.449	0.187	4.750	0.75	19.05	0.50	12.70	20 AWG
B-72-1111	0.196	4.978	0.107	2.718	0.75	19.05	0.50	12.70	22 AWG

Ferrite beads

Ferrite beads (Fig. 10.29) are used to create small value RF chokes and EMI shielding between adjacent conductors. They are especially useful in higher frequency applications where even a small run of wire is a significant portion of a wavelength. Small ferrite beads, with a high o.d./i.d. ratio, produce a lossy inductance when inserted into a circuit. Such cores are low Q and lossy at higher frequencies, hence their use as an RF choke or EMI filter.

The three principal types of ferrite bead are shown in Fig. 10.29, while their characteristics are shown in Table 10.12. The A_L values are in terms of mH/1000t. Type 1 has a single through-hole, Type 2 has two through-holes in the manner of the binocular core, and Type 3 has six through-holes. The inductance of a coil based on these cores is dependent on the number of turns and the A_L value. Unlike other cores, however, the small size and high o.d./i.d. ratio makes it difficult to put more than one turn (or very few if smaller wire is used) in the core. For Type 3 cores, which have six holes, the specific inductance achieved for any given number of turns is dependent on the winding pattern through the holes.

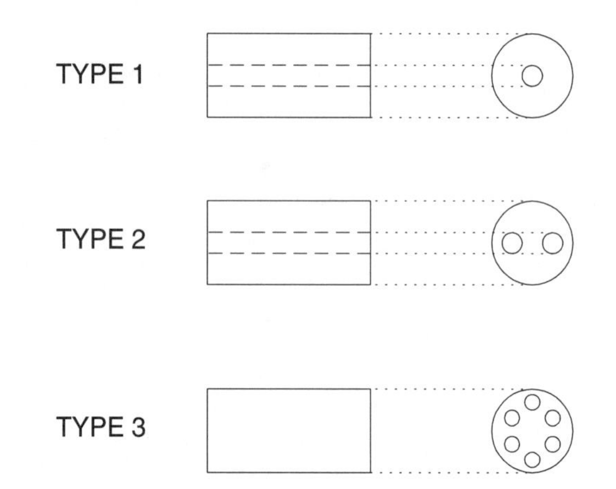

Figure 10.29 Three types of ferrite beads.

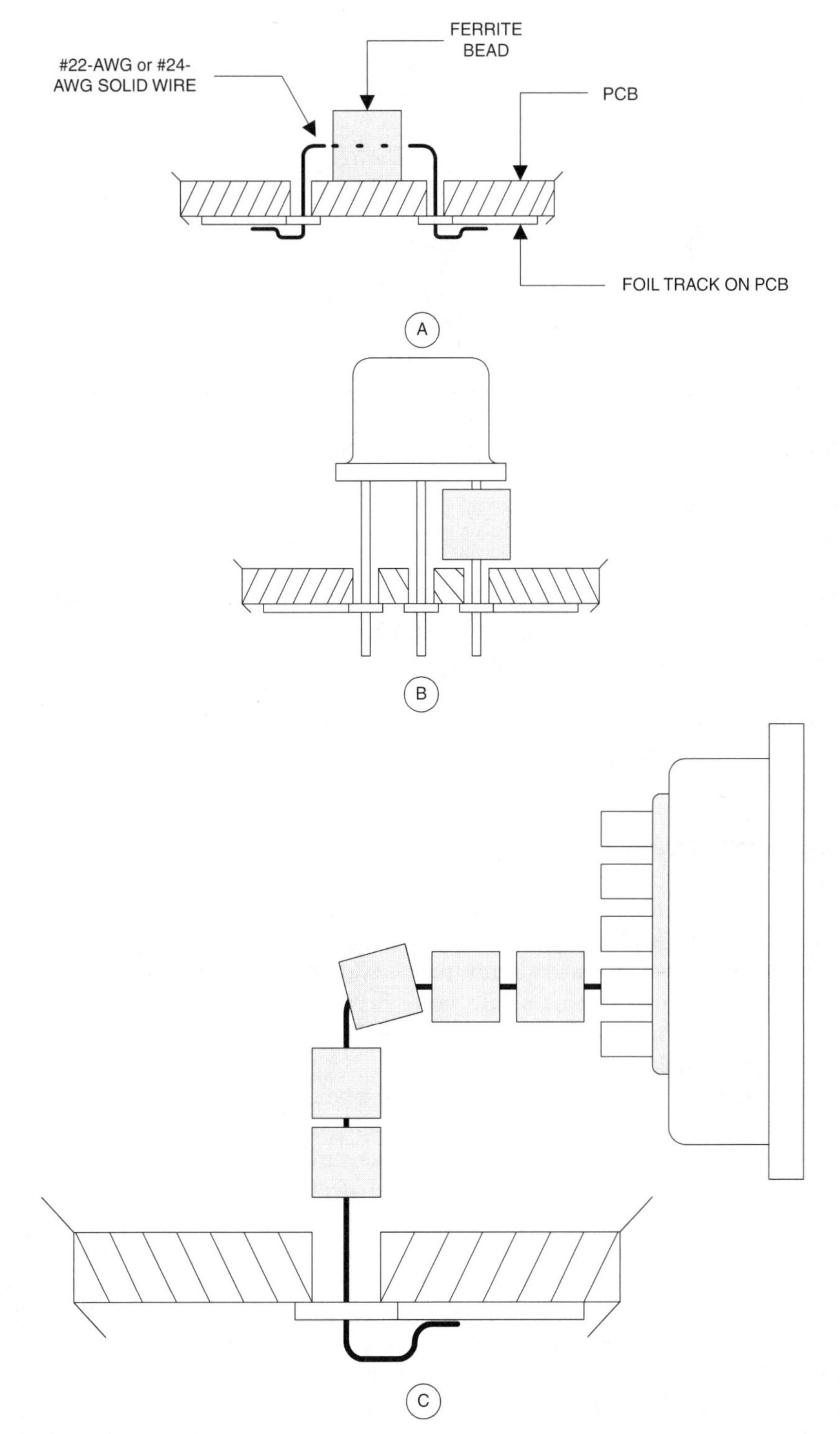

Figure 10.30 Mounting the ferrite beads: (A) on printed wiring board; (B) on transistor lead; (C) on connector.

Table 10.12

Part number	Type	43	64	73	75	77	*Z*-factor
FB-xx-101	1	510	150	1500	3000	N/A	1.00
FB-xx-201	1	360	110	1100	N/A	N/A	0.70
FB-xx-301	1	1020	300	3000	N/A	N/A	2.00
FB-xx-801	1	1300	390	3900	N/A	N/A	2.60
FB-xx-901	2	N/A	1130	N/A	N/A	N/A	7.50
FB-xx-1801	1	2000	590	5900	N/A	N/A	3.90
FB-xx-2401	1	520	N/A	1530	N/A	N/A	1.02
FB-xx-5111	3	3540	1010	N/A	N/A	N/A	6.70
FB-xx-5621	1	3800	N/A	N/A	N/A	9600	6.40
FB-xx-6301	1	1100	N/A	N/A	N/A	2600	1.70
FB-xx-1020	1	3200	N/A	N/A	N/A	N/A	6.20
FB-xx-1024	1	N/A	N/A	N/A	N/A	5600	3.70

xx = Material type number

Another advantage of the ferrite bead choke is that it provides low value of inductance and capacitance, so there is little chance of spurious self-resonance at frequencies within the pass band of the amplifier or circuit in question. Amidon's literature recommends type #73 and #77 material for RFI resulting from HF amateur radio operation. Type #43 material is used for 30 to 400 MHz operation, and type #64 above 400 MHz. Type #75 material is recommended for VLF through 20 MHz operation.

Mounting ferrite beads

Ferrite beads are mounted in a variety of ways. In wideband MMIC amplifiers such as the MAR-x series of devices, a ferrite bead is often used as an RF 'peaking coil' to smooth the frequency response over the entire band (there tends to be a roll-off at high frequencies). Figure 10.30A shows how a ferrite bead 'choke' is mounted to a printed circuit board. The single-turn is a piece of solid 'hook-up' wire (#22-AWG, #24-AWG, or UK equivalents) passed through the hole in the bead.

If the top of the PCB has a copper foil ground plane, then some ferrite materials must be mounted in a special manner. Certain ferrite materials are semiconductive, so can cause a short-circuit to a metal conductor. In those cases, either space the bead a short distance off the foil, or place an insulating layer below it. The latter, however, can cause an unwanted capacitance. If you control the design of the printed circuit board, then etch out a small area of metal around the bead to reveal the underlying insulating material.

The mounting in Fig. 10.30B shows how an RF choke can be placed in series with the collector (or other element) of a transistor. Slip the ferrite bead over the transistor lead, and then insert the lead in the hole through the printed circuit board.

Figure 10.30C shows the use of several ferrite beads slipped over a lead to the pin of a connector to or from the outside world. This method is used to either keep RF at home, or prevent it from entering the shielded enclosure. Keep the lead as short as possible, or the whole exercise will be for nothing.

11 Tuning and matching

In this chapter we are going to look at how inductor–capacitor (L–C) circuits can be used for tuning frequencies and matching impedances. But first here is a useful technique for visualizing how these circuits work.

Vectors for RF circuits

A *vector* (Fig. 11.1A) is a graphical device that is used to define the *magnitude* and *direction* (both are needed) of a quantity or physical phenomena. The *length* of the arrow defines the magnitude of the quantity, while the direction in which it is pointing defines the direction of action of the quantity being represented.

Vectors can be used in combination with each other. For example, in Fig. 11.1B we see a pair of displacement vectors that define a starting position ($P1$) and a final position ($P2$) for a person who travelled from point $P1$ 12 miles north and then 8 miles east to arrive at point $P2$. The *displacement* in this system is the hypotenuse of the right triangle formed by the 'north' vector and the 'east' vector. This concept was once illustrated pungently by a university bumper sticker's directions to get to a rival school: *'North 'til you smell it, east 'til you step in it.'*

Figure 11.1C shows a calculation trick with vectors that is used a lot in engineering, science and especially electronics. We can *translate* a vector parallel to its original direction, and still treat it as valid. The 'east' vector (E) has been translated parallel to its original position so that its tail is at the same point as the tail of the 'north' vector (N). This allows us to use the *Pythagorean theorem* to define the vector. The magnitude of the displacement vector to $P2$ is given by:

$$P2 = \sqrt{N^2 + E^2} \tag{11.1}$$

But recall that the magnitude only describes part of the vector's attributes. The other part is the *direction* of the vector. In the case of Fig. 11.1C the direction can be defined as the angle between the 'east' vector and the displacement vector. This angle (θ) is given by:

$$\theta = \arccos\left(\frac{E1}{P}\right) \tag{11.2}$$

In generic vector notation there is no 'natural' or 'standard' frame of reference, so the vector can be drawn in any direction so long as the users understand what it means. In the system above, we have adopted – by convention – a method that is basically the same as the old-fashioned *Cartesian coordinate system* X–Y graph. In the example of Fig. 11.1B the X axis is the 'east' vector, while the Y axis is the 'north' vector.

In electronics, the vectors are used to describe voltages and currents in AC circuits are standardized (Fig. 11.2) on this same kind of Cartesian system in which the inductive reactance (X_L), i.e. the opposition to AC exhibited by inductors, is graphed in the 'north'

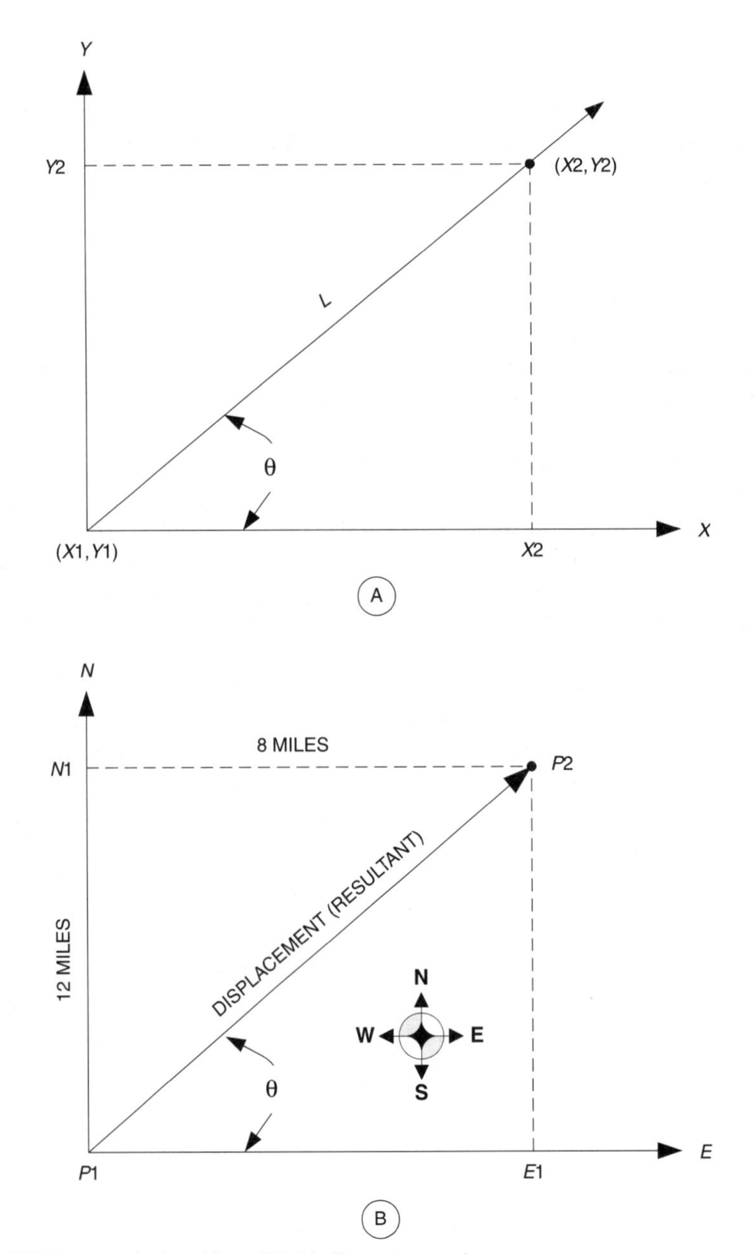

Figure 11.1 (A) Vector relationships; (B) N–E vector analogy;

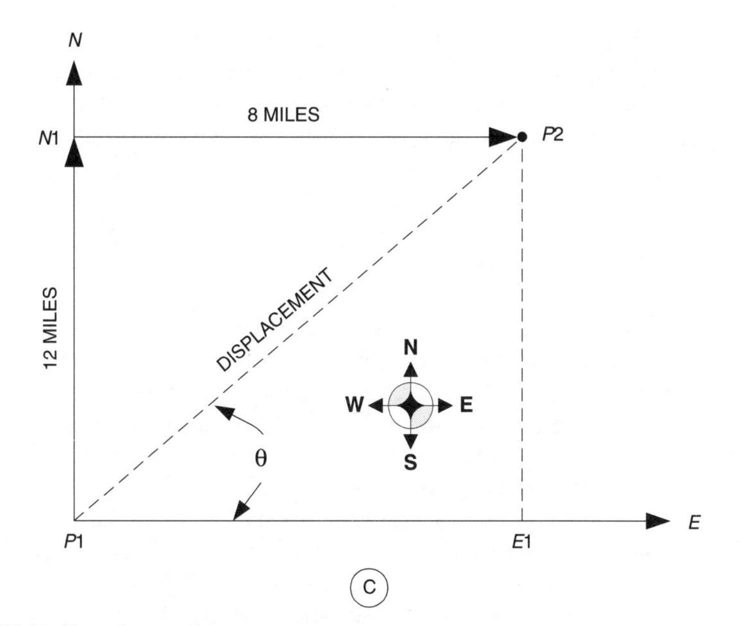

Figure 11.1 (C) N–E vector analogy.

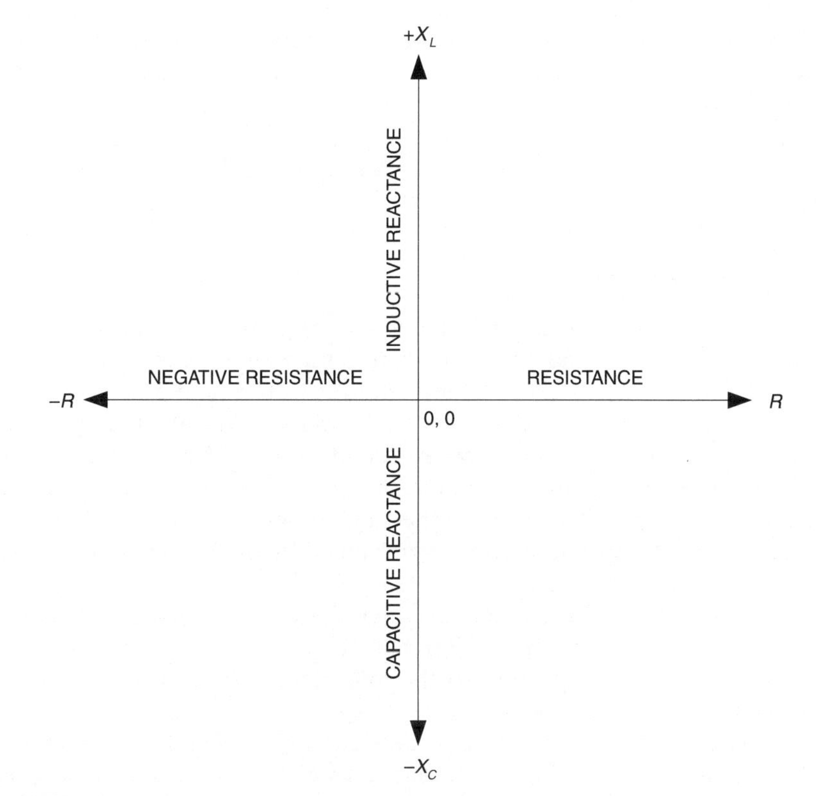

Figure 11.2 Electrical vectors.

direction, the capacitive reactance (X_C) is graphed in the 'south' direction and the resistance (R) is graphed in the 'east' direction. Negative resistance ('west' direction) is sometimes seen in electronics. It is a phenomenon in which the current *decreases* when the voltage increases. RF examples of negative resistance include tunnel diodes and Gunn diodes.

L–C resonant tank circuits

When you use an inductor (L) and a capacitor (C) together in the same circuit, the combination forms an *L–C resonant circuit*, also sometimes called a *tank circuit* or *resonant tank circuit*. These circuits are used to select one frequency, while rejecting all others (as in to tune a radio receiver). There are two basic forms of L–C resonant tank circuit: *series* (Fig. 11.3A) and *parallel* (Fig. 11.3B). These circuits have much in common, and much that makes them fundamentally different from each other.

The condition of *resonance* occurs when the capacitive reactance (X_C) and inductive reactance (X_L) are *equal in magnitude* ($|+X_L| = |-X_C|$). As a result, the resonant tank circuit shows up as purely resistive at the resonant frequency (Fig. 11.3C), and as a complex impedance at other frequencies. The L–C resonant tank circuit operates by an oscillatory exchange of energy between the magnetic field of the inductor, and the electrostatic field of the capacitor, with a current between them carrying the charge.

Because the two reactances are both frequency dependent, and because they are inverse to each other, the resonance occurs at only one frequency (f_r). We can calculate the standard resonance frequency by setting the two reactances equal to each other and solving for f. The result is:

$$f = \frac{1}{2\pi\sqrt{LC}} \tag{11.3}$$

Series resonant circuits

The series resonant circuit (Fig. 11.3A), like other series circuits, is arranged so that the terminal current (I) from the source (V) flows in both components equally.

In a circuit that contains a resistance, inductive reactance and a capacitive reactance, there are three vectors to consider (Fig. 11.4), plus a resultant vector. Using the parallelogram method, we first construct a resultant for the R and X_C, which is shown as vector '*A*'. Next, we construct the same kind of vector ('*B*') for R and X_C. The resultant ('*C*') is made using the parallelogram method on '*A*' and '*B*'. Vector '*C*' represents the impedance of the circuit; the magnitude is represented by the length, and the phase angle by the angle between '*C*' and R.

In Fig. 11.4, the inductive reactance is larger than the capacitive reactance, so the excitation frequency is greater than f_r. Note that the voltage drop across the inductor is greater than that across the capacitor, so the total circuit looks like it contains a small inductive reactance.

Figure 11.5A shows a series resonant L–C tank circuit, and Fig. 11.5B shows the current and impedance as a function of frequency. The *series resonant circuit has a low impedance at its resonant frequency, and a high impedance at all other frequencies*. As a result, the line current

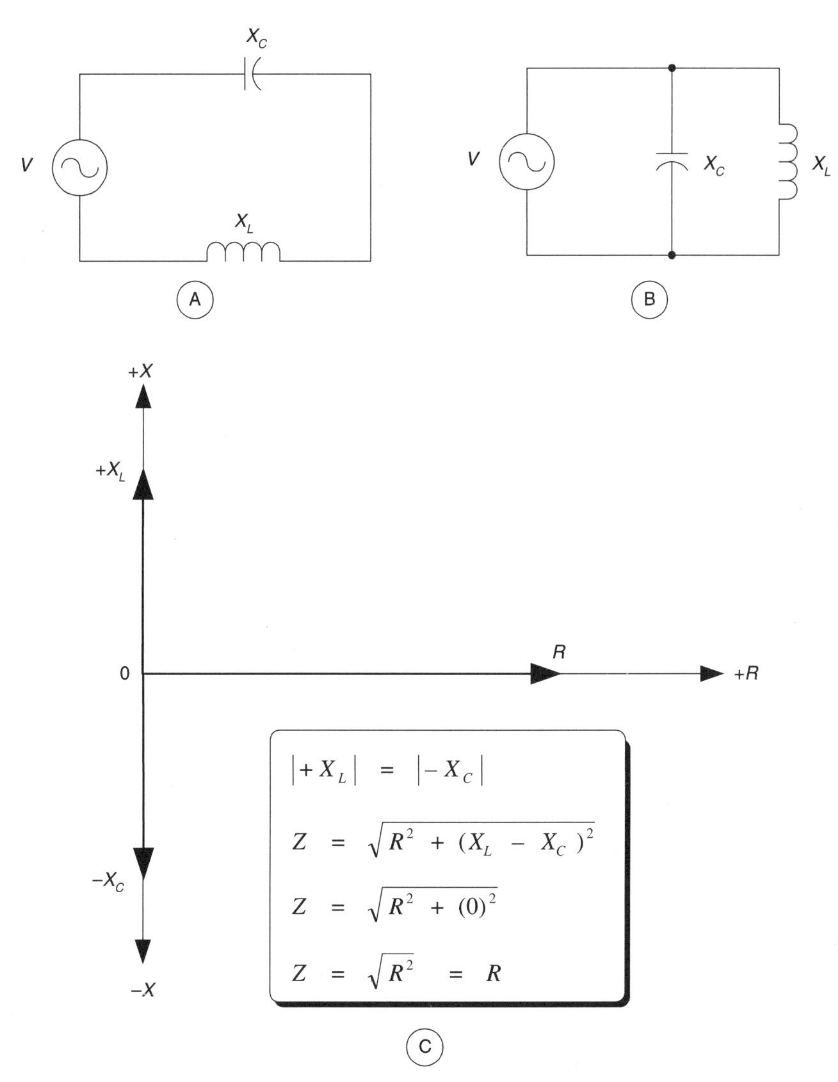

Figure 11.3 (A) Series resonant circuit; (B) parallel resonant circuit; (C) vector relationship.

(I) from the source is maximum at the resonant frequency and the voltage across the source is minimum.

Parallel resonant circuits

The parallel resonant tank circuit (Fig. 11.6A) is the inverse of the series resonant circuit. The line current (I) from the source splits and flows in inductor and capacitor separately. The *parallel resonant circuit has its highest impedance at the resonant frequency, and a low impedance at all other frequencies* (Fig. 11.6B). Thus, the line current from the source is minimum at the resonant frequency, and the voltage across the L–C tank circuit is maximum. This fact is important in radio tuning circuits, as you will see in due course.

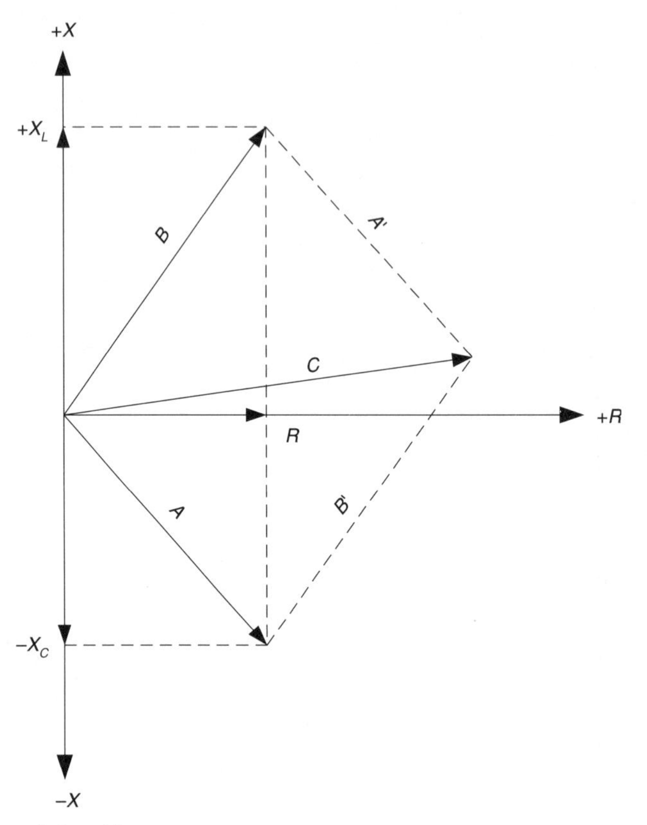

Figure 11.4 Vector relationship.

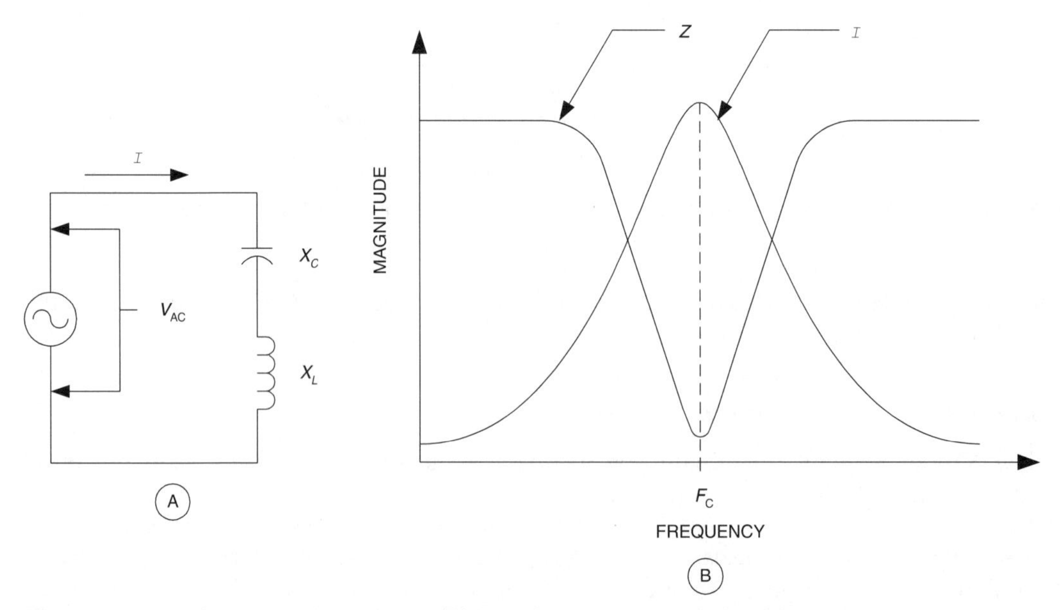

Figure 11.5 (A) Series resonant circuit; (B) impedance-current relationship.

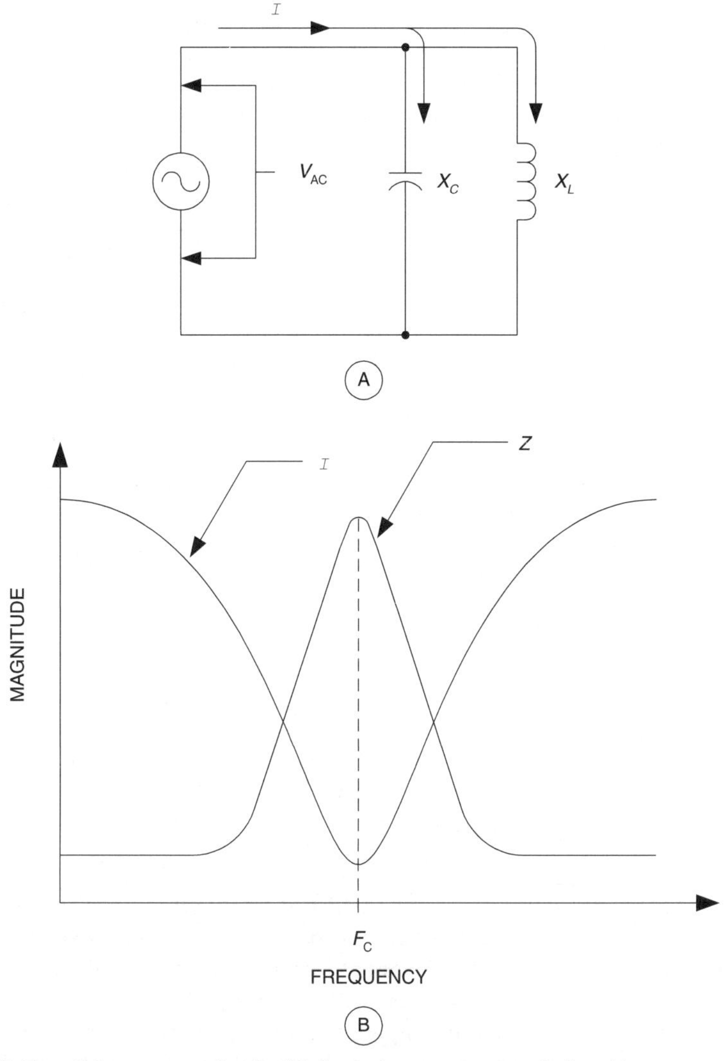

Figure 11.6 (A) Parallel resonant circuit; (B) impedance-current relationship.

Tuned RF/IF transformers

Many of the resonant circuits used in RF circuits, and especially radio receivers, are actually transformers that couple signals from one stage to another. Figure 11.7 shows several popular forms of tuned RF/IF tank circuits. In Fig. 11.7A, one winding is tuned while the other is untuned. In the configurations shown, the untuned winding is the secondary of the transformer. This type of circuit is often used in transistor and other solid-state circuits, or when the transformer has to drive either a crystal or mechanical bandpass filter circuit. In the reverse configuration (L1 = output, L2 = input), the same circuit is used for the antenna coupling network, or as the interstage transformer between RF amplifiers in TRF radios.

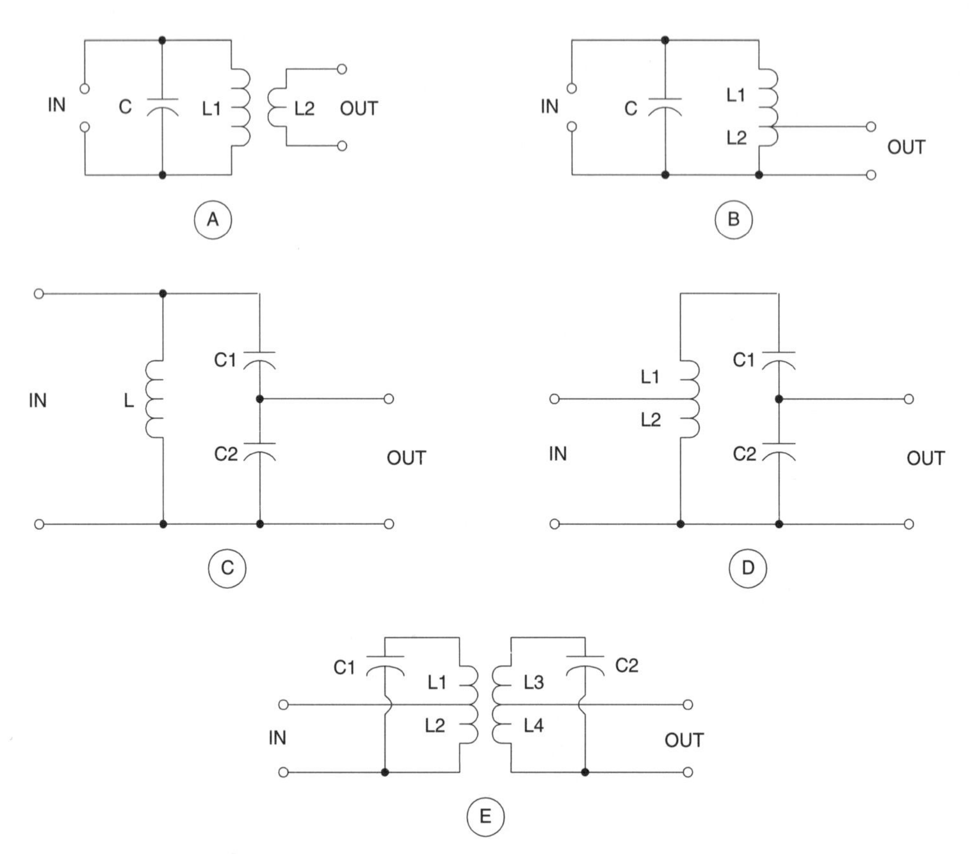

Figure 11.7 Various L–C resonant circuits.

The circuit in Fig. 11.7B is a parallel resonant L–C tank circuit that is equipped with a low impedance tap on the inductor. This type of circuit is often used to drive a crystal detector or other low impedance load. Another circuit for driving a low impedance load is shown in Fig. 11.7C. This circuit splits the capacitance that resonates the coil into two series capacitors. As a result, we have a capacitive voltage divider. The circuit in Fig. 11.7D uses a tapped inductor for matching low impedance sources (e.g. antenna circuits), and a tapped capacitive voltage divider for low impedance loads.

Finally, the circuit in Fig. 11.7E uses a tapped primary and tapped secondary winding in order to match two low impedance loads; this is an example of a *double-tuned* circuit. As we will soon see, this gives an improved bandpass characteristic.

Construction of RF/IF transformers

The tuned RF/IF transformers built for radio receivers are typically wound on a common cylindrical form, and surrounded by a metal shield can that prevents interaction of the fields of coils that are in close proximity to each other.

Figure 11.8A shows the schematic for a typical RF/IF transformer, while the sectioned view (Fig. 11.8B) shows one form of construction. This method of building the

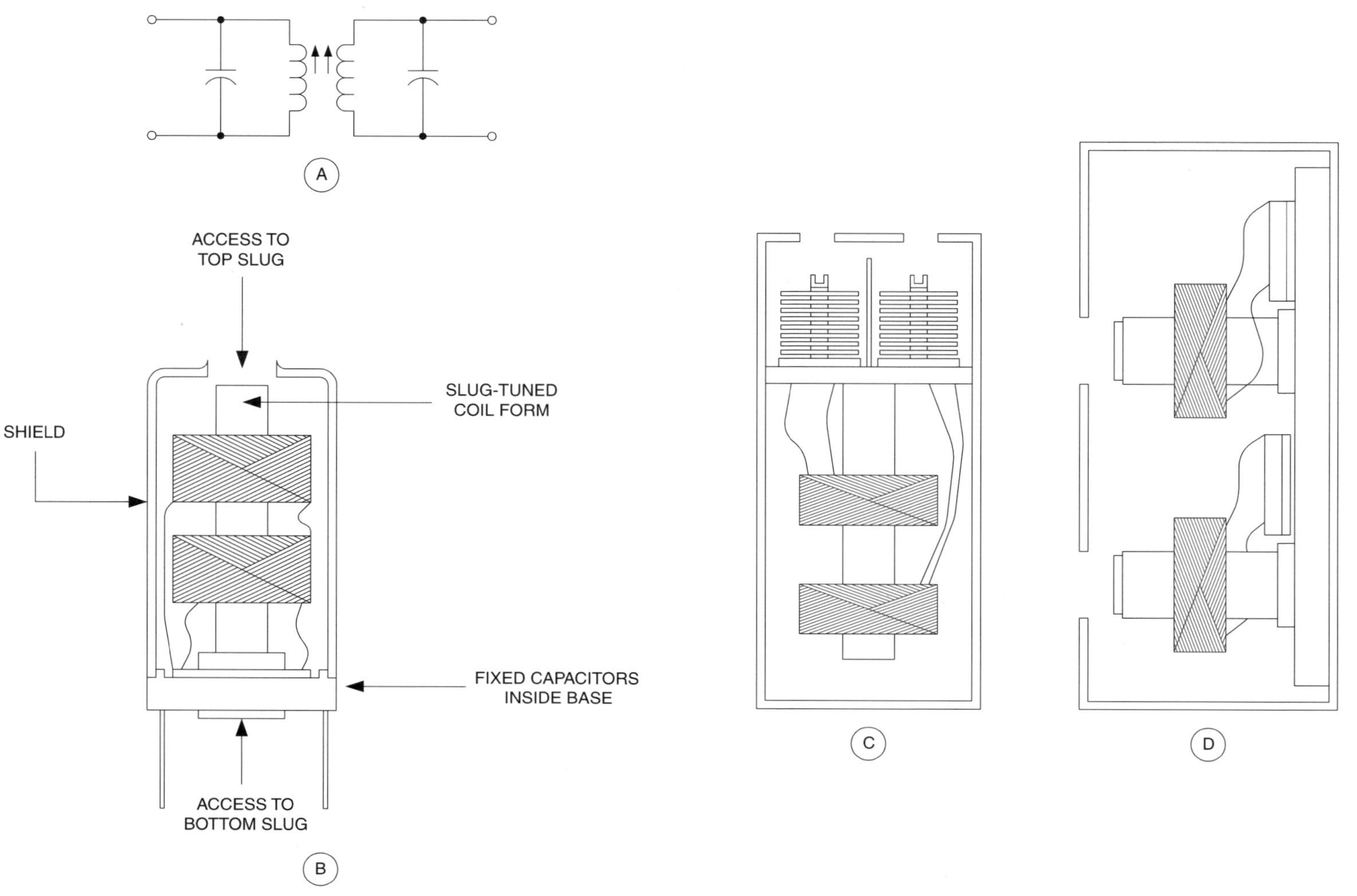

Figure 11.8 (A) L–C resonant transformer; (B) one mechanical implementation; (C) & (D) older forms of mechanical implementation.

transformers was common at the beginning of World War II, and continued into the early transistor era. The methods of construction shown in Figs 11.8C and 11.8D were popular prior to World War II. The capacitors in Fig. 11.8B were built into the base of the transformer, while the tuning slugs were accessed from holes in the top and bottom of the assembly. In general, expect to find the secondary at the bottom hole, and the primary at the top hole.

The term *universal wound* refers to a cross-winding system that minimizes the interwinding capacitance of the inductor, and therefore raises the self-resonant frequency of the inductor (a good thing).

Bandwidth of RF/IF transformers

Figure 11.9A shows a parallel resonant RF/IF transformer, while Fig. 11.9B shows the usual construction in which the two coils (L1 and L2) are wound at distance *d* apart on a common cylindrical form.

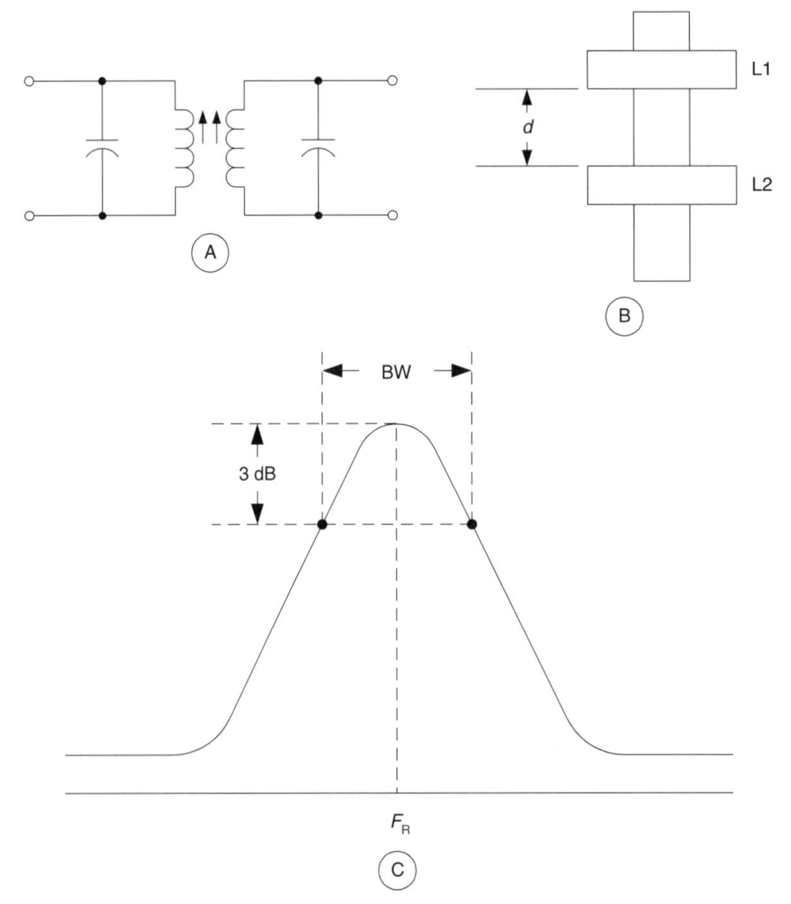

Figure 11.9 (A) Circuit symbol; (B) physical implementation; (C) bandwidth of critically coupled circuit.

The *bandwidth* of the RF/IF transformer is the frequency difference between the frequencies where the signal voltage across the output winding falls off –3 dB (i.e. half power, or roughly 71 per cent voltage) from the value at the resonant frequency (f_r), as shown in Fig. 11.9C. If *F*1 and *F*2 are the –3 dB frequencies, then the bandwidth (BW) is *F*2 – *F*1. The shape of the frequency response curve in Fig. 11.9C is said to represent *critical coupling*. It has a flatter top and steeper sides than a similar single-tuned circuit would have.

An example of a *subcritical* or *undercoupled* RF/IF transformer is shown in Fig. 11.10. As shown in Fig. 11.10B, the windings are farther apart than in the critically coupled case, which makes the bandwidth (Fig. 11.10C) much narrower than in the critically coupled case. The subcritically coupled RF/IF transformer is often used in shortwave or communications receivers in order to allow the narrower bandwidth to discriminate against adjacent channel stations.

The *overcritically coupled* RF/IF transformer is shown in Fig. 11.11. Here we note in Fig. 11.11B that the windings are closer together, which makes the bandwidth (Fig. 11.11C)

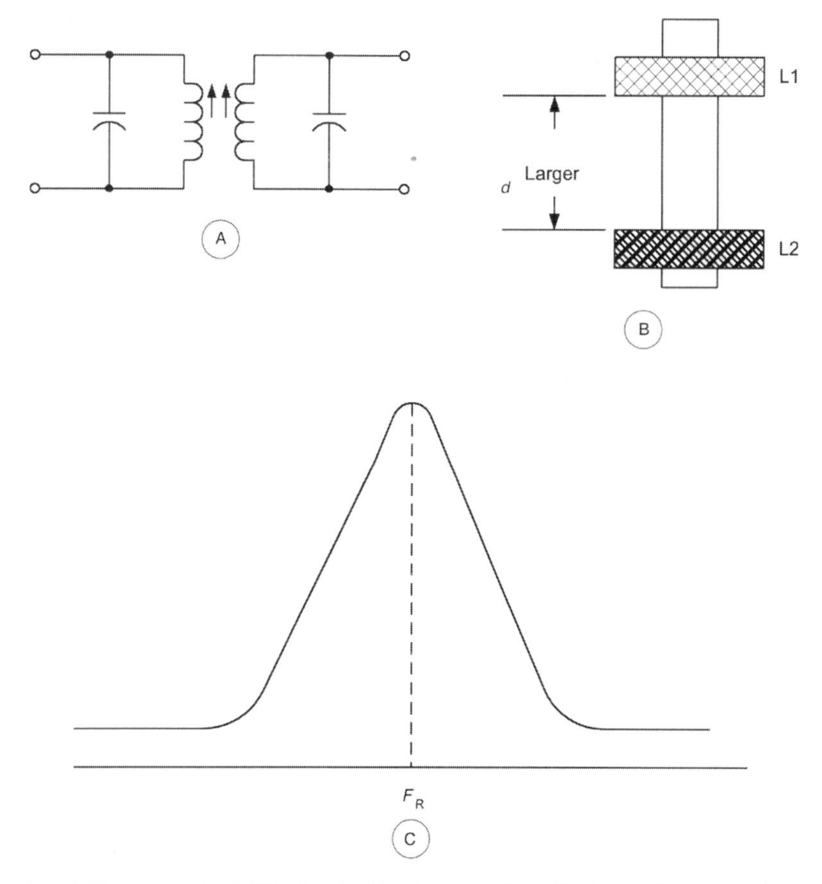

Figure 11.10 (A) Circuit symbol; (B) physical implementation; (C) bandwidth of undercoupled circuit.

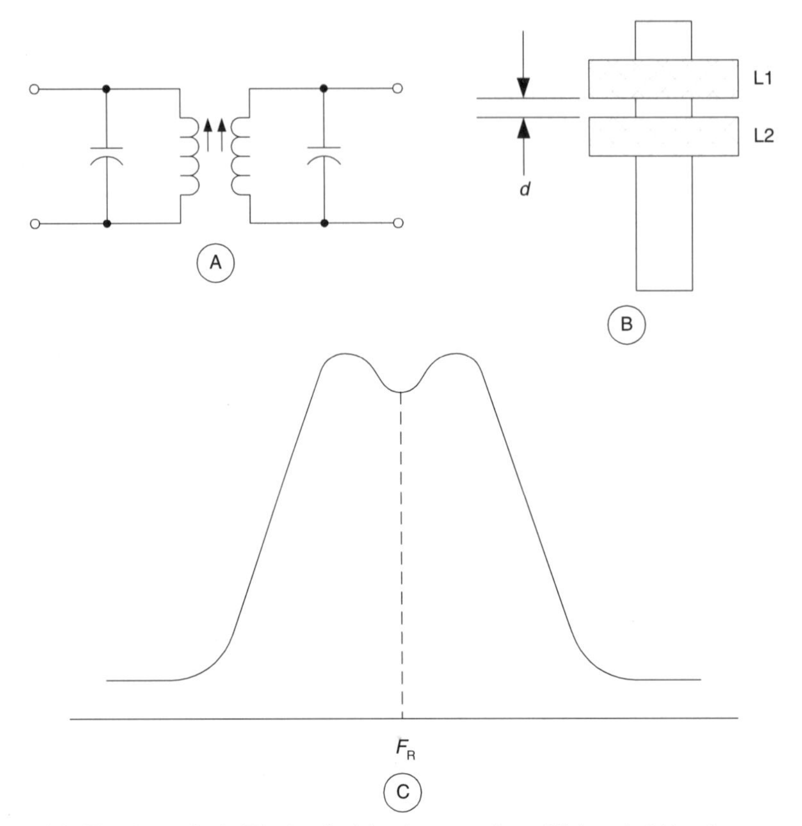

Figure 11.11 (A) Circuit symbol; (B) physical implementation; (C) bandwidth of overcoupled circuit.

much broader but with a dip in the centre. In some radio schematics and service manuals (not to mention early textbooks), this form of coupling was sometimes called 'high fidelity' coupling because it allowed more of the sidebands of the signal (which carry the audio modulation) to pass with less distortion of frequency response.

The bandwidth of a single-tuned tank circuit can be summarized in a *figure of merit* called Q. The Q of the circuit is the ratio of the bandwidth to the resonant frequency: $Q = BW/f_r$.

$$Q = \frac{BW}{F_r} \tag{11.4}$$

A critically coupled pair of tank circuits has a bandwidth which is √2 greater than a single-tuned circuit with the same value of Q.

A resistance in the L–C tank circuit will cause it to broaden, that is to lower its Q. The resistor is sometimes called a 'de-Qing resistor'. The 'loaded Q' (i.e. Q when a resistance is present, as in Fig. 11.12A) is always less than the unloaded Q. In some radios, a switched resistor (Fig. 11.12B) is used to allow the user to broaden or narrow the bandwidth. This switch might be labelled 'fidelity' or 'tone' or something similar.

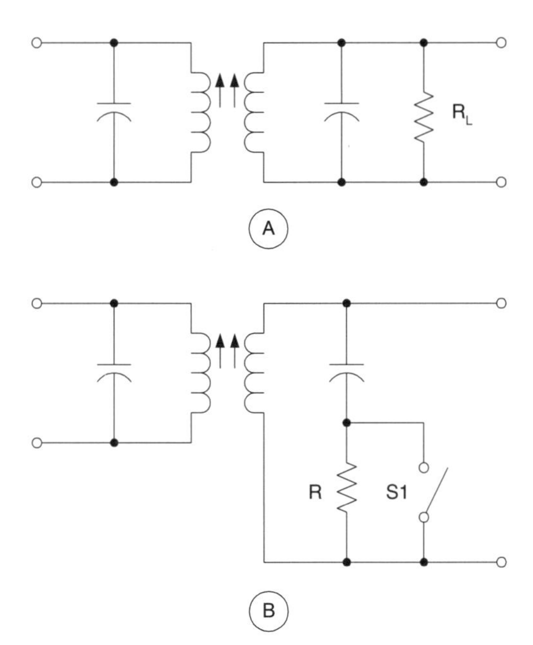

Figure 11.12 Bandwidth in L–C resonant circuit is altered by the resistor.

Choosing component values for L–C resonant tank circuits

Resonant L–C tank circuits are used to tune radio receivers; it is these circuits that select the station to be received, while rejecting others. A superheterodyne radio receiver (the most common type) is shown in simplified form in Fig. 11.13. According to the superhet principle, the radio frequency being received (F_{RF}) is converted to another frequency, called the *intermediate frequency* (F_{IF}), by being mixed with a *local oscillator* signal (F_{LO}) in a non-linear mixer stage. The output spectrum will consist mainly of F_{RF}, F_{LO}, $F_{RF} - F_{LO}$ (difference frequency), and $F_{RF} + F_{LO}$ (sum frequency). In older radios, for practical reasons the difference frequency was selected for F_{IF}; today either sum or difference frequencies can be selected depending on the design of the radio.

There are several L–C tank circuits present in this notional superhet radio. The antenna tank circuit (C1/L1) is found at the input of the RF amplifier stage, or if no RF amplifier is used it is at the input to the mixer stage. A second tank circuit (L2/C2), tuning the same range as L1/C1, is found at the output of the RF amplifier, or the input of the mixer. Another L–C tank circuit (L3/C3) is used to tune the local oscillator; it is this tank circuit that sets the frequency that the radio will receive.

Additional tank circuits (only two shown) may be found in the IF amplifier section of the radio. These tank circuits will be fixed tuned to the IF frequency, which in common AM broadcast band (BCB) radio receivers is typically 450 kHz, 455 kHz, 460 kHz, or 470 kHz depending on the designer's choices (and sometimes country of origin) other IF frequencies are also seen, but these are most common. FM broadcast receivers typically use a 10.7 MHz IF, while shortwave receivers might use a 1.65 MHz, 8.83 MHz, 9 MHz or an IF frequency above 30 MHz.

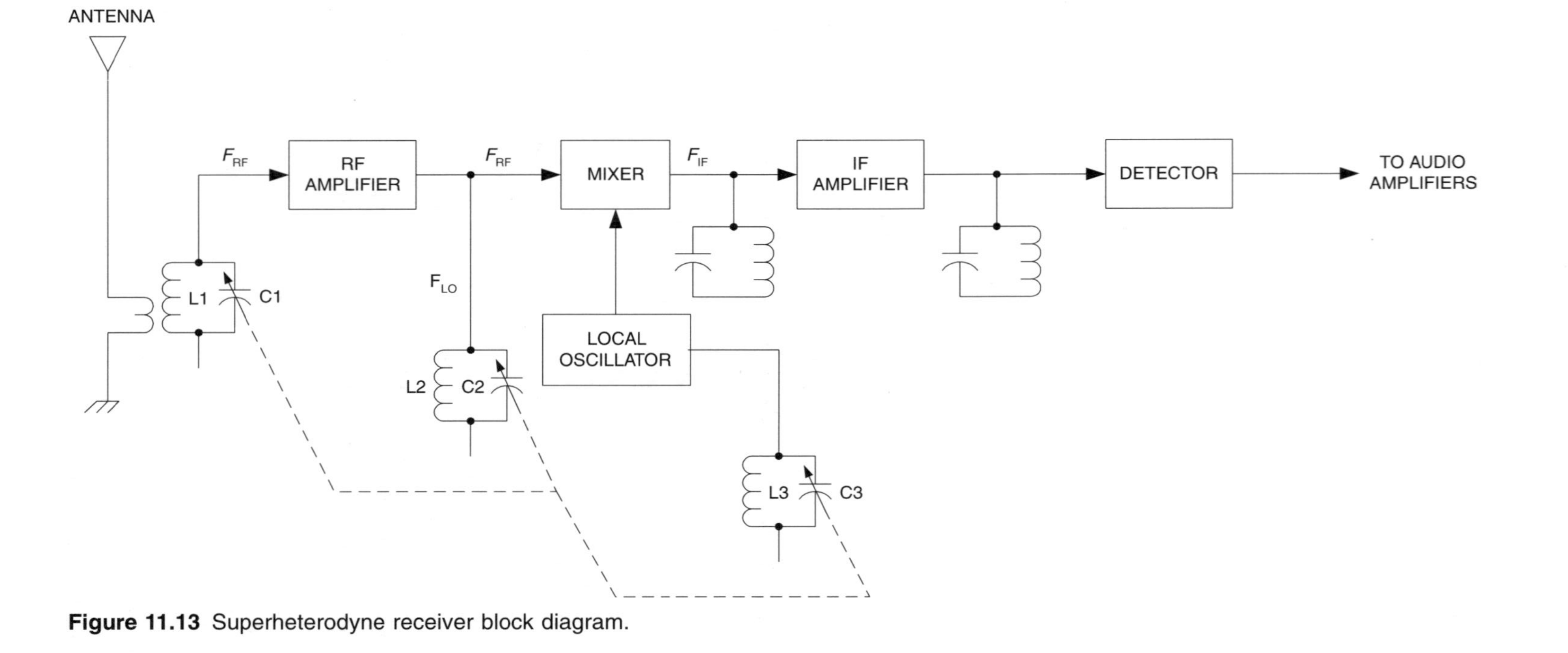

Figure 11.13 Superheterodyne receiver block diagram.

The tracking problem

On a radio that tunes the front-end with a single knob, which is almost all receivers today, the three capacitors (C1–C3 in Fig. 11.13) are typically *ganged*, i.e. the capacitors are mounted on a single rotor shaft. These three tank circuits must *track* each other; i.e. when the RF amplifier is tuned to a certain radio signal frequency, the LO must produce a signal that is different from the RF frequency by the amount of the IF frequency. Perfect tracking is probably impossible, but the fact that your single knob tuned radio works is testimony to the fact that the tracking isn't too terrible.

The issue of tracking LC tank circuits for the AM BCB receiver has not been a major problem for many years: the band limits are fixed over most of the world, and component manufacturers offer standard adjustable inductors and variable capacitors to tune the RF and LO frequencies. Indeed, some even offer three sets of coils: antenna, mixer input/RF amp output and LO. The reason why the antenna and mixer/RF coils are not the same, despite tuning to the same frequency range, is that these locations see different distributed or 'stray' capacitances. In the USA, it is standard practice to use a 10 to 365 pF capacitor and a 220 μH inductor for the 540 to 1600 kHz AM BCB. In some other countries, slightly different combinations are sometimes used: 320 pF, 380 pF, 440 pF, 500 pF and others are seen in catalogues.

Recently, however, two events coincided that caused me to examine the method of selecting capacitance and inductance values. First, I embarked on a design project to produce an AM DXers receiver that had outstanding performance characteristics. Second, the AM broadcast band was recently extended so that the upper limit is now 1700 kHz, rather than 1600 kHz. The new 540 to 1700 kHz band is not accommodated by the now-obsolete 'standard' values of inductance and capacitance. So I calculated new candidate values. Shortly, we will see the result of this effort.

The RF amplifier/antenna tuner problem

In a typical RF tank circuit, the inductance is kept fixed (except for a small adjustment range that is used for overcoming tolerance deviations) and the capacitance is varied across the range. Figure 11.14 shows a typical tank circuit main tuning capacitor (C1), trimmer capacitor (C2) and a fixed capacitor (C3) that is not always needed. The stray capacitances (C_s) include the interwiring capacitance, the wiring to chassis capacitance, and the amplifier or oscillator device input capacitance. The frequency changes as the square root of the capacitance changes. If $F1$ is the minimum frequency in the range, and $F2$ is the maximum frequency, then the relationship is:

$$\frac{F2}{F1} = \sqrt{\frac{C_{max}}{C_{min}}} \tag{11.5}$$

or, in a rearranged form that some find more congenial:

$$\left(\frac{F2}{F1}\right)^2 = \frac{C_{max}}{C_{min}} \tag{11.6}$$

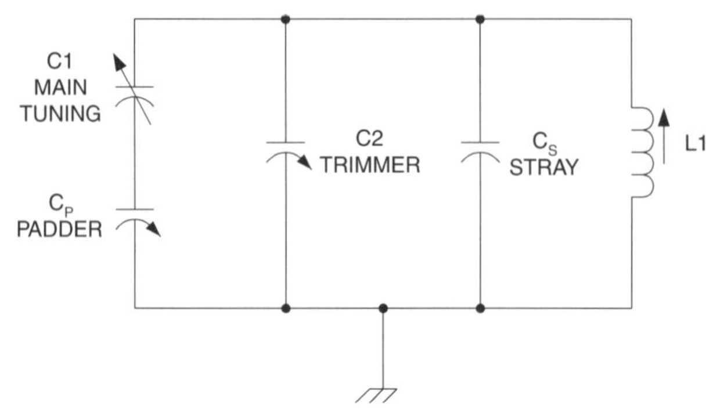

Figure 11.14 Tuning arrangement.

In the case of the new AM receiver, I wanted an overlap of about 15 kHz at the bottom end of the band, and 10 kHz at the upper end, so needed a resonant tank circuit that would tune from 525 kHz to 1710 kHz. In addition, because variable capacitors are widely available in certain values based on the old standards, I wanted to use a 'standard' AM BCB variable capacitor. A 10 to 380 pF unit from a vendor was selected.

The minimum required capacitance, C_{min}, can be calculated from:

$$\left(\frac{F2}{F1}\right)^2 C_{min} = C_{min} + \Delta C \tag{11.7}$$

Where:

$F1$ is the minimum frequency tuned
$F2$ is the maximum frequency tuned
C_{min} is the minimum required capacitance at $F2$
ΔC is the difference between C_{max} and C_{min}

Example

Find the minimum capacitance needed to tune 1710 kHz when a 10 to 380 pF capacitor ($\Delta C = 380 - 10$ pF = 370 pF) is used, and the minimum frequency is 525 kHz.

Solution:

$$\left(\frac{F2}{F1}\right)^2 C_{min} = C_{min} + \Delta C$$

$$\left(\frac{1710\text{ kHz}}{525\text{ kHz}}\right)^2 C_{min} = C_{min} + 370\text{ pF}$$

$$10.609 C_{MIN} = C_{MIN} + 370\text{ pF} = 38.51\text{ pF}$$

The maximum capacitance must be $C_{min} + \Delta C$, or 38.51 + 370 pF = 408.51 pF. Because the tuning capacitor (C1 in Fig. 11.14) does not have exactly this range, external capacitors must be used, and because the required value is higher than the normal value additional capacitors are added to the circuit in parallel to C1. Indeed, because somewhat unpredictable 'stray' capacitances also exist in the circuit, the tuning capacitor values should be a little less than the required values in order to accommodate strays plus tolerances in the actual – versus published – values of the capacitors. In Fig. 11.14, the main tuning capacitor is C1 (10 to 380 pF), C2 is a small value trimmer capacitor used to compensate for discrepancies, C3 (not shown) is an optional capacitor that may be needed to increase the total capacitance, and C_s is the stray capacitance in the circuit.

The value of the stray capacitance can be quite high, especially if there are other capacitors in the circuit that are not directly used to select the frequency (e.g. in Colpitts and Clapp oscillators the feedback capacitors affect the L–C tank circuit). In the circuit that I was using, however, the L–C tank circuit is not affected by other capacitors. Only the wiring strays and the input capacitance of the RF amplifier or mixer stage need be accounted. From experience I apportioned 7 pF to C_s as a *trial* value.

The minimum capacitance calculated above was 38.51, there is a nominal 7 pF of stray capacitance, and the minimum available capacitance from C1 is 10 pF. Therefore, the combined values of C2 and C3 must be 38.51 pF – 10 pF – 7 pF, or 21.5 pF. Because there is considerable reasonable doubt about the actual value of C_s, and because of tolerances in the manufacture of the main tuning variable capacitor (C1), a wide range of capacitance for C2 + C3 is preferred. It is noted from several catalogues that 21.5 pF is near the centre of the range of 45 pF and 50 pF trimmer capacitors. For example, one model lists its range as 6.8 pF to 50 pF, its centre point is only slightly removed from the actual desired capacitance. Thus, a 6.8 to 50 pF trimmer was selected, and C3 is not used.

Selecting the inductance value for L1 is a matter of picking the frequency and associated required capacitance at one end of the range, and calculating from the standard resonance equation solved for L:

$$L_{\mu H} = \frac{10^6}{4\pi^2 f_{low}^2 C_{max}}$$

$$L_{\mu H} = \frac{10^6}{(4)(\pi^2)(525\,000)^2(4.085 \times 10^{-10})} = 224.97 \approx 225\,\mu H$$

The RF amplifier input L–C tank circuit and the RF amplifier output L–C tank circuit are slightly different cases because the stray capacitances are somewhat different. In the example, I am assuming a JFET transistor RF amplifier, and it has an input capacitance of only a few picofarads. The output capacitance is not a critical issue in this specific case because I intend to use a 1 mH RF choke in order to prevent JFET oscillation. In the final receiver, the RF amplifier may be deleted altogether, and the L–C tank circuit described above will drive a mixer input through a link coupling circuit.

The local oscillator (LO) problem

The local oscillator circuit must track the RF amplifier, and must also tune a frequency range that is different from the RF range by the amount of the IF frequency (455 kHz). In keeping with common practice I selected to place the LO frequency 455 kHz *above* the RF frequency. Thus, the LO must tune the range 980 kHz to 2165 kHz.

There are three methods for making the local oscillator track with the RF amplifier frequency when single shaft tuning is desired: the *trimmer capacitor* method, the *padder capacitor* method, and the *different-value cut-plate capacitor* method.

Trimmer capacitor method

The trimmer capacitor method is shown in Fig. 11.15, and is the same as the RF L–C tank circuit. Using exactly the same method as before, but with a frequency ratio of (2165/980) to yield a capacitance ratio of $(2165/980)^2 = 4.88{:}1$, solves this problem. The results were a minimum capacitance of 95.36 pF, and a maximum capacitance of 465.36 pF. An inductance of 56.7 μH is needed to resonate these capacitances to the LO range.

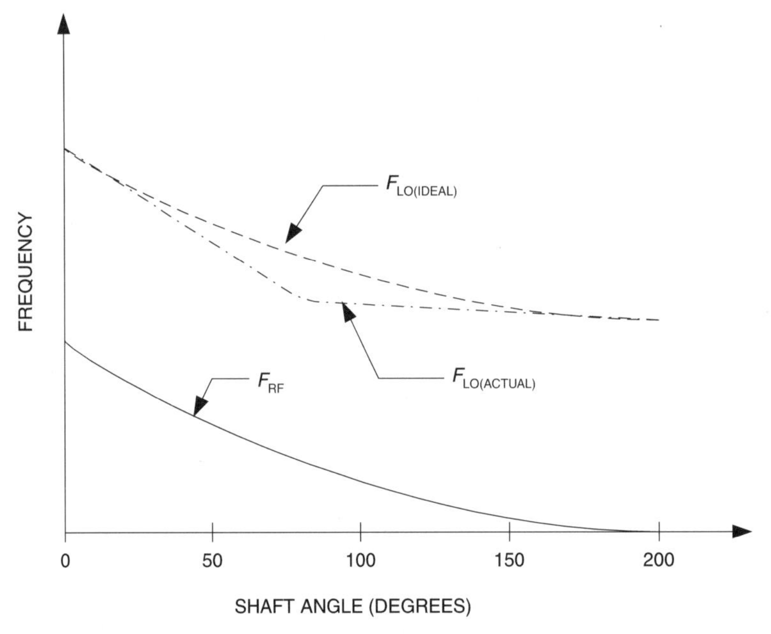

Figure 11.15 Tracking diagram.

There is always a problem associated with using the same identical capacitor for both RF and LO. It seems that there is just enough difference that tracking between them is always a bit off. Figure 11.15 shows the ideal LO frequency and the calculated LO frequency. The difference between these two curves is the degree of mistracking. The curves overlap at the ends, but are awful in the middle. There are two cures for this problem. First, use a *padder capacitor* in series with the main tuning capacitor (Fig. 11.16). Second, use a *different-value cut-plate capacitor.*

Figure 11.16 shows the use of a padder capacitor (C_p) to change the range of the LO section of the variable capacitor. This method is used when both sections of the variable

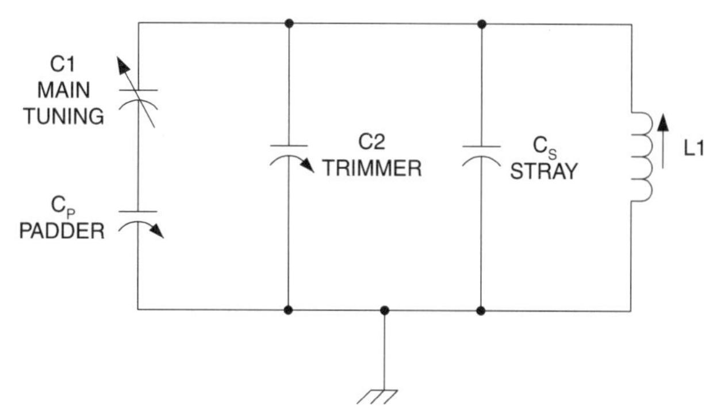

Figure 11.16 Tuning arrangement.

capacitor are identical. Once the reduced capacitance values of the $C1/C_p$ combination are determined the procedure is identical to the above. But first, we have to calculate the value of the padder capacitor and the resultant range of the $C1/C_p$ combination. The padder value is found from:

$$\frac{C1_{max}C_p}{C1_{max} + C_p} = \left(\frac{F2}{F1}\right)^2 \left(\frac{C1_{min}C_p}{C1_{min} + C_p}\right) \tag{11.8}$$

And solving for C_p. For the values of the selected main tuning capacitor and LO:

$$\frac{(380\,\text{pF})(C_p)}{(380 + C_p)\,\text{pF}} = (4.88)\left(\frac{(10\,\text{pF})(C_p)}{(10 + C_p\,\text{pF})}\right) \tag{11.9}$$

Solving for C_p by the least common denominator method (crude, but it works) yields a padder capacitance of 44.52 pF. The series combination of 44.52 pF and a 10 to 380 pF variable yields a range of 8.2 pF to 39.85 pF. An inductance of 661.85 μH is needed for this capacitance to resonate over 980 kHz to 2165 kHz.

A practical solution to the tracking problem that comes close to the ideal is to use a *cut-plate* capacitor. These variable capacitors have at least two sections, one each for RF and LO tuning. The shape of the capacitor plates are especially cut to a shape that permits a constant change of *frequency* for every degree of shaft rotation. With these capacitors it is possible to produce three-point tracking, or better.

Impedance matching in RF circuits

Impedance matching is necessary in RF circuits to guarantee the maximum transfer of power between a source and a load. If you have a source it will have a source resistance. That source resistance must be matched to the load impedance for maximum power transfer to occur. That does not mean that maximum voltage transfer occurs (which requires the load to be very large with respect to the source impedance), but rather the power transfer is maximized.

There are several methods used for impedance matching in RF circuits. The simple transformer is one such method. The broadband transformer is another. Added to that are certain resonant circuits.

Before we get to the different types of impedance transformation, let's talk a little about notation. R1 is the source resistance, and R2 is the load resistance. Any inductors will be labelled C1, L1 and so forth. The primary impedance of a transformer is Z_P, while the secondary is Z_S. In a transformer, the number of primary turns is N_P and the number of secondary turns is N_S.

Transformer matching

Transformer impedance matching (Fig. 11.17) is simple and straightforward. In an ideal transformer (one without losses or leakage reactance), we know that:

$$\frac{Z_P}{Z_S} = \left[\frac{N_P}{N_S}\right]^2 \tag{11.10}$$

This relationship tells us that the impedance ratio is equal to the square of the turns ratio of the transformer. If the load impedance is purely resistive, then the impedance reflected across the primary windings will be a pure resistance. Similarly, if the load impedance is reactive (capacitive or inductive) then the reflected impedance seen across the primary will be reactive. The phase angle of the signal will be the same.

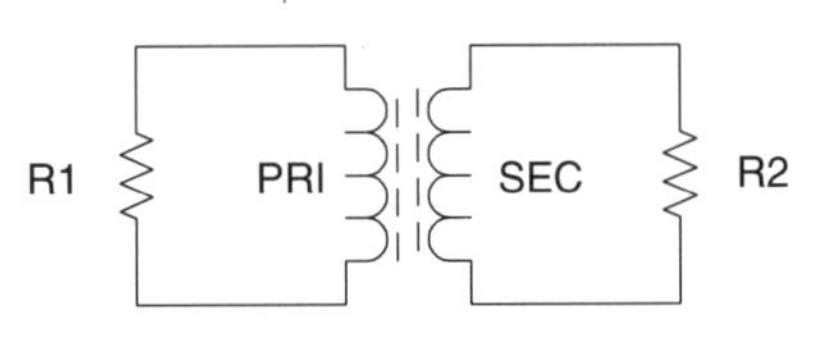

Figure 11.17 Simple transformer.

From the equation above we can derive a different equation that relates the turns ratio required to the impedance ratio:

$$\frac{N_P}{N_S} = \sqrt{\frac{Z_P}{Z_S}} \tag{11.11}$$

For example, suppose an RF transistor has a collector impedance of 150 ohms, and you want to match it to a 10 ohm next stage impedance. This requires a turns ratio of

$$\frac{N_P}{N_S} = \sqrt{\frac{150}{10}} = 3.9:1 \tag{11.12}$$

The primary of the transformer must have 3.9 times as many turns in the primary as in the secondary.

Resonant transformers

The transformer is made up of at least one inductor (usually two or more), so it can be made resonant. Figures 11.18 and 11.19 show two transformers that are resonant. The transformer in Fig. 11.18 has tuning in both primary and secondary windings. It also has taps to accommodate lower impedance situations than can be accommodated across the entire transformer. The use of a high impedance for the tuned portion raises the loaded Q of the circuit, while the tap allows lower impedance transistor and integrated circuits to be used.

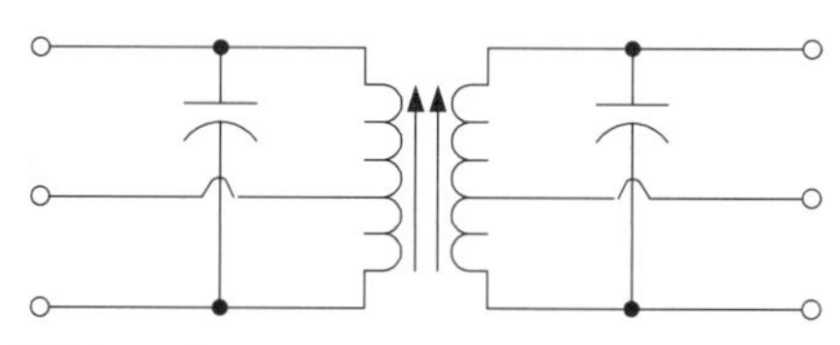

Figure 11.18 Parallel-tuned RF transformer.

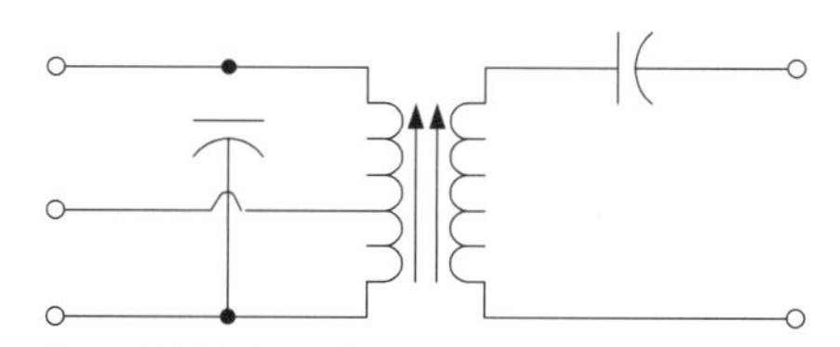

Figure 11.19 Parallel–series-tuned RF transformer.

In the transformer shown in Fig. 11.19 the primary winding is the same as in the previous case. The difference comes in the secondary winding, which is series tuned. This connection blocks any DC the transformer is connected to, while retaining the RF capabilities of the transformer.

Figure 11.20 shows the application of transformers such as Fig. 11.18. This is an IF or RF amplifier using a single transistor for the active element. Transformer T1 has a tapped secondary to accommodate the lower impedance of the transistor base circuit. Similarly with the primary of transformer T2, it is tapped to a lower impedance to accommodate the low impedance of the transistor collector circuit. In both cases, both primary and secondary windings are tuned.

Figure 11.21 shows a circuit in which the primary winding is resonant, but the secondary circuit is not. Point X can be a source of DC, for example the DC power supply to accommodate a transistor or IC. The primary winding is tapped to accommodate the low impedance of a transistor.

The primary winding of Fig. 11.21 appears to be series tuned, but in fact it is parallel tuned. There will be a low impedance path to AC around point X, such as another capacitor, and that puts the value of C1 in parallel with the inductance of the transformer primary.

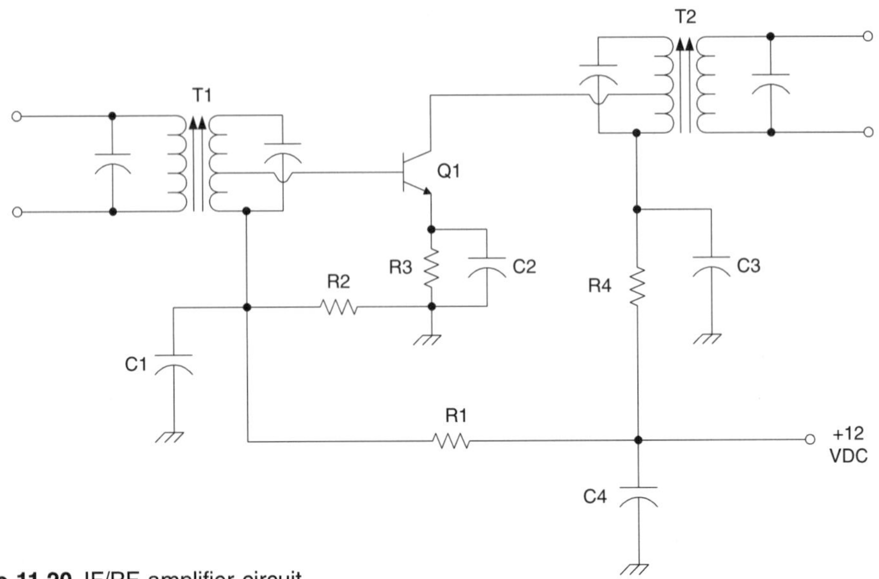

Figure 11.20 IF/RF amplifier circuit.

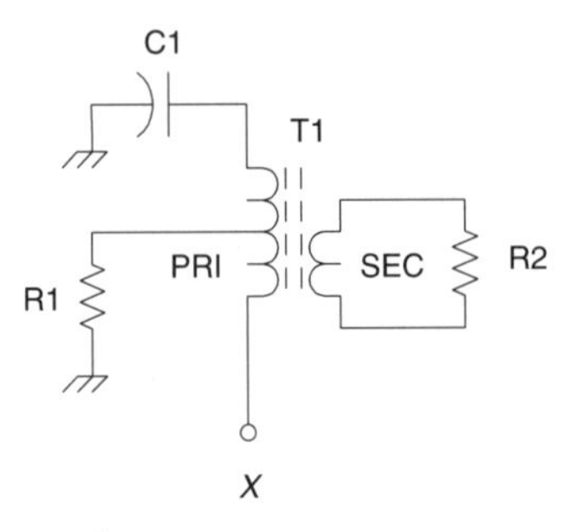

Figure 11.21 Tuned primary RF transformer.

Resonant networks

There are several different forms of network that can be used for impedance matching. For example, the *reverse-L section* circuit (Fig. 11.22) consists of a capacitor and an inductor in a circuit that has the capacitor in the input circuit. This circuit has a requirement that $R1 > R2$. It is also possible to connect the capacitor across the output, but the resistance ratio reverses ($R1 < R2$).

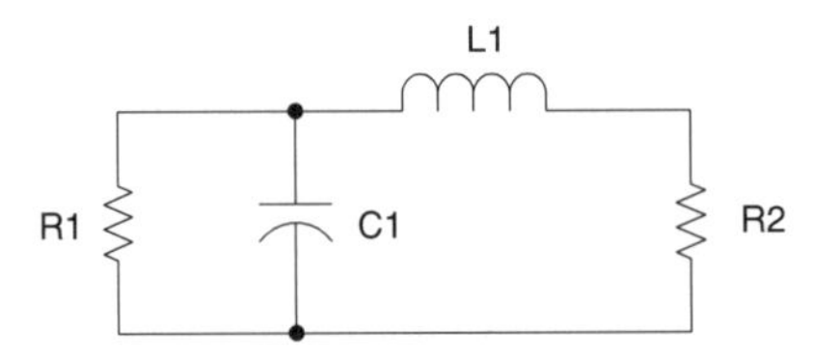

Figure 11.22 L-section coupler.

The value of the inductive reactance is given by:

$$X_L = \sqrt{R1R2 - R2^2} \qquad (11.13)$$

And the reactance of the capacitor is given by:

$$X_C = \frac{-R1R2}{X_L} \qquad (11.14)$$

Inverse-L network

A somewhat better solution for the $R1 < R2$ case is given by the network in Fig. 11.23. This network has the capacitor in series with the signal line, and the inductor in parallel with the output. In this case, the inductive reactance required is:

$$X_L = R2\sqrt{\frac{R1}{R2 - R1}} \qquad (11.15)$$

And the capacitive reactance is:

$$X_C = \frac{-R1R2}{X_L} \qquad (11.16)$$

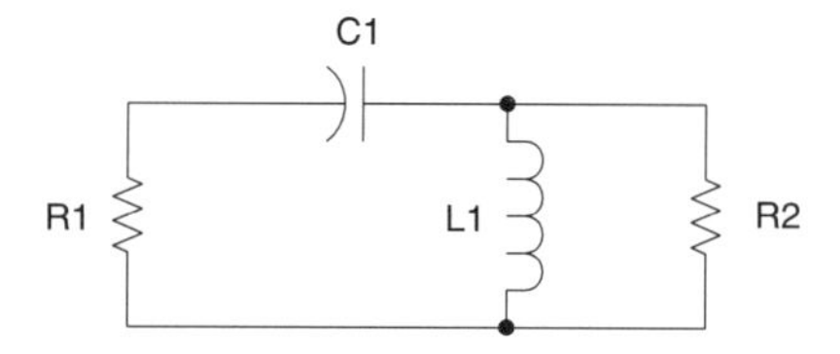

Figure 11.23 Different L-section coupler.

π-network

The π-network (Fig. 11.24) gets its name from the fact that the network looks like the Greek letter 'π'. It consists of a series inductor, flanked by two capacitors shunted across the signal line. The π-network is useful where the impedance $R1$ is greater than $R2$, and

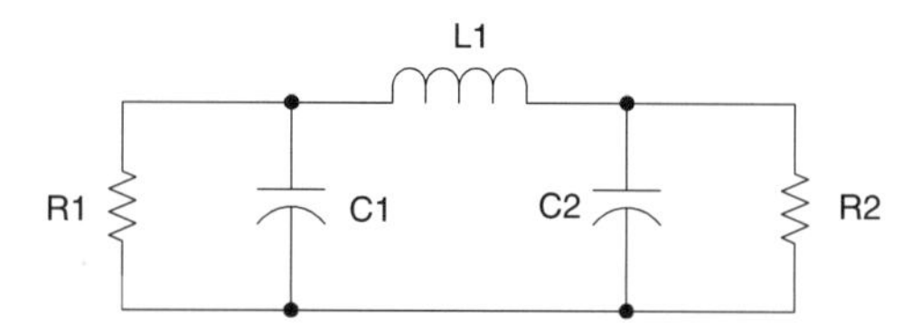

Figure 11.24 Pi-section coupler.

works best when it is considerably larger than $R2$. For example, in a vacuum tube anode coupling network where the impedance is on the order of 4000 ohms, and the amplifier is coupled to a 50 ohm load. It can be used backwards if $R2$ is greater than $R1$.

It is necessary to set the Q of the network (usually between 5 and 20) to greater than:

$$Q > \sqrt{\frac{R1}{R2} - 1} \tag{11.17}$$

After we meet the Q requirement, we can calculate the values of the inductive reactance and the two capacitive reactances:

$$X_{C1} = \frac{R1}{Q} \tag{11.18}$$

$$X_{C2} = R2 \times \sqrt{\frac{R1/R2}{Q^2 + 1 - (R1/R2)}} \tag{11.19}$$

$$X_{L1} = R1 \times \left[\frac{Q + (R2/X_{C2})}{Q^2 + 1}\right] \tag{11.20}$$

Split-capacitor network

The split-capacitor network (Fig. 11.25) may be used whenever $R1 < R2$. For example, when a 50 ohm impedance must be matched to a 1000 or 1500 ohm input impedance to an integrated circuit. The value of Q is greater than:

$$Q > \sqrt{\frac{R2}{R1} - 1} \tag{11.21}$$

In this case, the values of the capacitive reactances and the inductive reactance is given by:

$$X_L = \frac{R2}{Q} \tag{11.22}$$

$$X_{C2} = \frac{R1}{\sqrt{\frac{R1(Q^2 + 1)}{R2} - 1}} \tag{11.23}$$

$$X_{C1} = \left[\frac{R2Q}{Q^2 + 1}\right]\left[1 - \frac{R1}{QX_{C2}}\right] \tag{11.24}$$

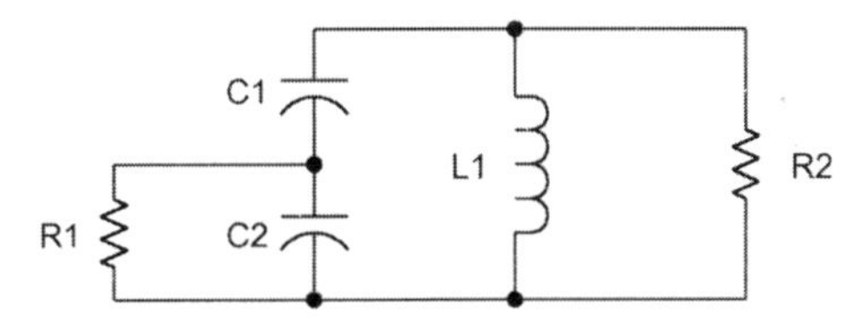

Figure 11.25 Split-capacitor coupler.

Transistor-to-transistor impedance matching

The three networks shown in Figs 11.26 through 11.28 are used to convert the collector impedance of a transistor (R1) to the base impedance of a following transistor (R2).

The circuit in Fig. 11.26 is used where $R1 < R2$. The first thing to do is select a value of Q between 2 and 20. We can also accommodate the case where the output contains not just resistance ($R1$) but capacitance as well (R_{Cs}). In that case,

$$X_{L1} = QR1 + X_{C_s} \tag{11.25}$$

$$X_{C2} = Q_L R2 \tag{11.26}$$

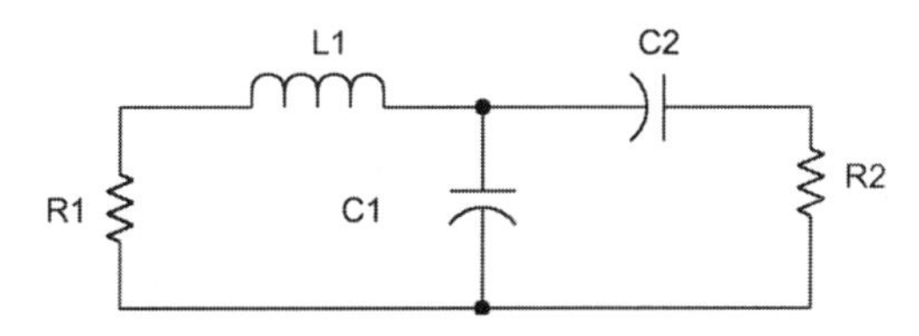

Figure 11.26 Coupler circuit.

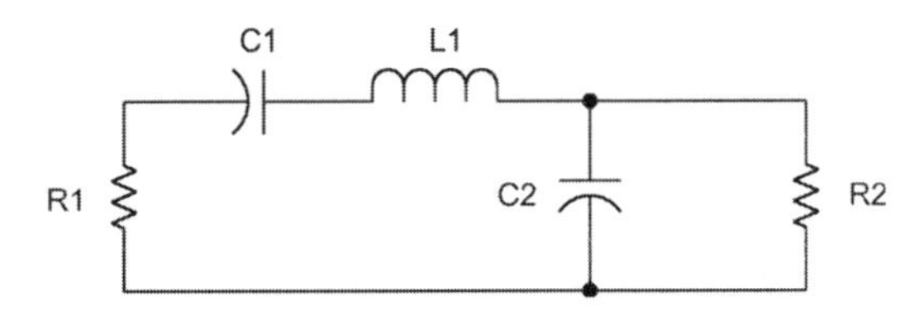

Figure 11.27 Coupler circuit.

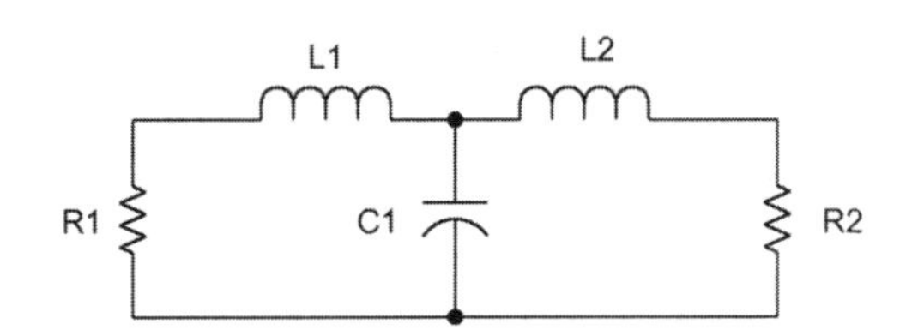

Figure 11.28 Coupler circuit.

$$X_{C1} = \frac{R_V}{Q - Q_L} \tag{11.27}$$

Where:

$$R_V = R1(1 + Q^2) \tag{11.28}$$

and,

$$Q_L = \sqrt{\frac{R_V}{R2} - 1} \tag{11.29}$$

The network in Fig. 11.27 is also used when $R1 < R2$, and Q is between 2 and 20, and the output impedance has a capacitive reactance (X_{Cs}). If this network is selected, then the values of the capacitors are found by:

$$X_{C1} = QR1 \tag{11.30}$$

$$X_{C2} = R2\sqrt{\frac{R1}{R2 - R1}} \tag{11.31}$$

$$X_{L1} = X_{C1} + \left[\frac{R1R2}{X_{C2}}\right] + X_{Cs} \tag{11.32}$$

Finally, the network in Fig. 11.28 has two inductors in series with the signal line and a capacitor in parallel with the signal line. The network is used whenever $R1 < R2$ and the source impedance has a capacitive reactance (X_{Cs}). In that case,

$$X_{L1} = (R1Q) + X_{Cs} \tag{11.33}$$

$$X_{L2} = R2Q_1 \tag{11.34}$$

$$X_{C1} = \frac{R_V}{Q + Q_L} \tag{11.35}$$

Where Q_L and R_V are as defined above in Eqns 11.29 and 11.28 respectively.

12 Splitters and hybrids

RF power combiners and splitters

The principal difference between power combiners and power splitters is in the application. Otherwise, they are the same circuits.

Definition: A *combiner* is a passive electronic device that will linearly mix two or more signal sources into a common port. The combiner is not a mixer because it is linear, and thus does not produce additional frequency products.

Definition: A *splitter* performs exactly the opposite function of a combiner. It directs RF power from a single input source to two or more loads.

Characteristics of splitter/combiner circuits

Splitter/combiners are typically passive electronic networks that provide one common port and two or more independent ports. When power is applied to the common port, and delivered to the independent ports, then the circuit operates as a splitter. When power is applied to the independent ports the combination of individual signals is added linearly at the common port.

These devices typically provide a 0 degree phase shift between ports. When used as a splitter equal amplitude signals are delivered to the respective independent ports. Also in the splitter mode, except for the purely resistive network, there is a high degree of port-to-port isolation between the independent ports.

The minimum theoretical splitter mode insertion loss (Table 12.1) occurs because the power is split into N different channels, and is calculated from:

$$\text{Insertion loss (dB)} = 10\log_{10}(N) \tag{12.1}$$

where N is the number of independent ports.

The splitter mode is used for a number of different purposes in RF circuits or test set-ups. It can be used to provide a number of identical output signals from one input signal applied to the common port. In the combiner mode it can be used for vector addition or subtraction of signals.

A simple 'Tee' or 'Y' connection can be used for splitting or combining, but there are problems with that approach. The impedance, for example, will be the parallel combination of all source impedance connected to the Tee/Y junction. The impedance will

Table 12.1 Insertion Loss

Ports (N)	Loss (dB)
2	3.0
3	4.8
4	6.0
5	7.0
6	7.8
7	8.5
8	9.0
10	10.0
12	10.8
15	11.8
20	13.0
30	14.8

be Z/N when all impedances are equal. For example, if a 50 ohm impedance is used for all devices, then the impedance at the junction is $50/N$. For the two-way Tee/Y, then, a pair of 50 ohm loads will be seen as a single 12 ohm load at the common port.

The simplest, and least desirable, proper splitter/combiner is the resistive network discussed in the next section.

Resistive splitter/combiner

The simplest usable splitter/combiner circuit is the resistive network in Fig. 12.1. This circuit uses three resistors in a Y-network to provide three ports. It can also be extended to higher numbers of ports. The resistor value is given by:

$$R = R_0\left(\frac{N-2}{N}\right) \tag{12.2}$$

and the insertion loss is:

$$loss = -20\log(N-1) \tag{12.3}$$

where R_0 is the system impedance and N is the number of ports. The loss is higher than an ideal splitter, and gets worse as more ports are added.

For example, if the system impedance is 75 ohms, then for three ports, $R = 25$ ohms and the loss is -6 dB.

The resistors used in this circuit must be non-inductive. This limits selection to carbon composition or metal film resistors (and not all metal film types!). If a higher power than 2 watts is needed, then each arm of the Y-network must be made from multiple resistors in series or parallel. The value calculated for R might not be a standard value, except perhaps in certain lines of 1 per cent or less tolerance precision resistors. R can, however, be approximated using standard values.

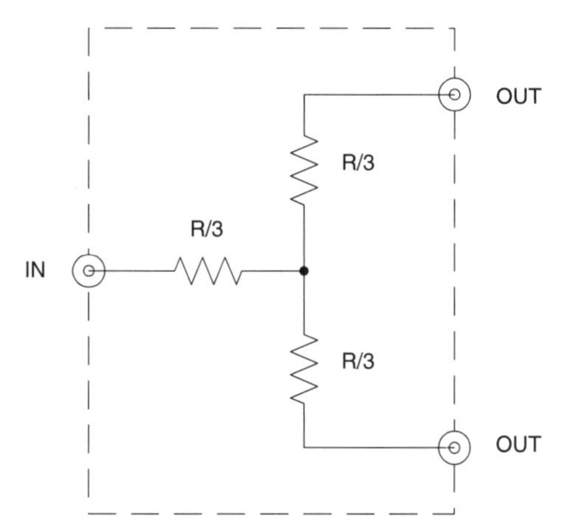

Figure 12.1 Simple resistive coupler for signal generators.

It is also possible to approximate the values by using series or parallel combinations of standard value resistors. For example, a pair of 51 ohm standard value resistors in parallel will make a good match for 25 ohms.

The advantage of the resistive splitter/combiner is its broadband operation. The bandwidth can extend into the UHF region with discrete resistors, and into the gigahertz region if implemented with surface mount resistors and appropriate printed circuit technology. The upper frequency limit in either case is set by the stray inductance and capacitance.

The disadvantages include a relatively high insertion loss (–6 dB for three ports), and only about 6 dB isolation between output ports. If these limitations are not important in a given application, then this form of splitter/combiner is ideally suited.

Transformer splitter/combiner

Figure 12.2A shows a somewhat better form of splitter/combiner circuit. This circuit can be used from 500 kHz to over 1000 MHz if the proper transformers and capacitor are provided. In this circuit, the common port goes to L1, while OUT-1 and OUT-2 are the independent ports.

The main network consists of a centre-tapped inductor, L2, and a resistor equal to 2R. Inductor L1 provides impedance matching at the common port, and C1 compensates for stray reactances and flattens the broadband response. L2 is centre tapped, with the input signal applied to the tap and the outputs taken from the ends. This transformer can be wound on either T-50-2 or T-50-6 toroidal cores for the HF bands, or a T-50-15 core for the AM BCB and medium wave bands. Use 18 turns of #26 AWG wire for the HF bands, and 22 turns for MW bands. The resistor across the ends of L2 should be twice the system impedance. That means 100 ohms for 50 ohm systems, and 150 ohms for 75 ohm systems (both are standard values).

Some impedance transformation is needed if the system impedance is to be maintained, so L1 must be provided. This transformer is tapped, but not at the centre. The inset detail

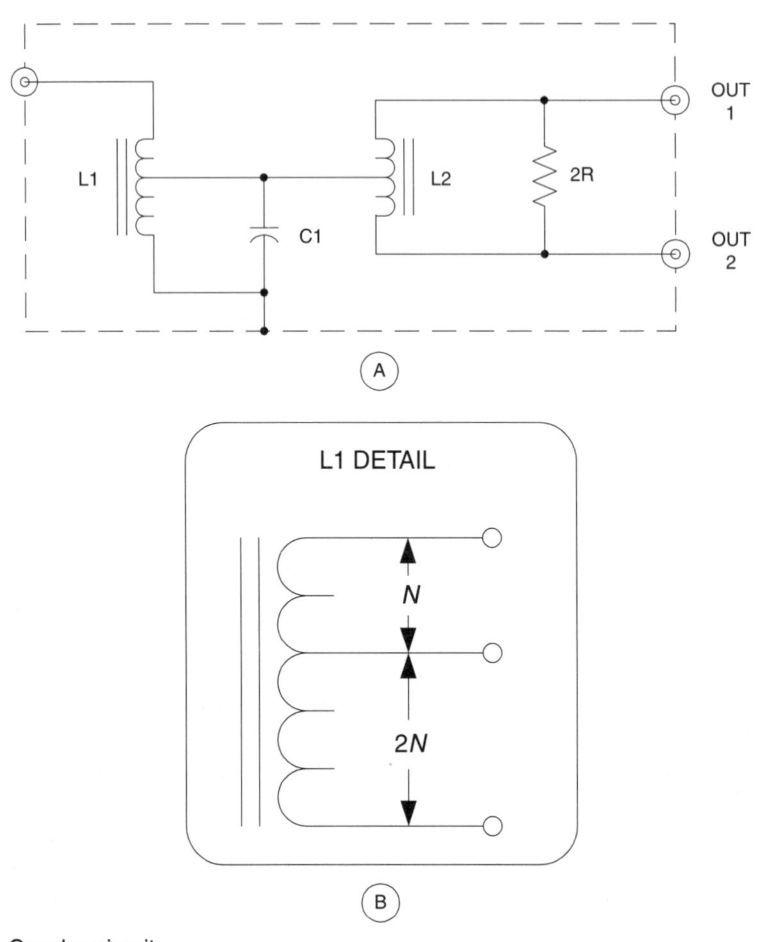

Figure 12.2 Coupler circuit.

in Fig. 12.2B shows the relationship of the tap to the winding: it is located at the one-third point on the winding. If the bottom of the coil is grounded, then the tap is at the two-thirds point (2*N* turns), and the input is at the top (*N* + 2*N* turns).

It is important to use toroid core inductors for the combiner. The most useful core types are listed above, although for other applications other cores could also be used. Figure 12.3A shows one way the cores can be wound. This is the linear winding approach, i.e. uses a single coil of wire. The turns are wound until the point where the tap occurs. At that point one of two approaches is taken. One, you could end the first half of the winding and cut the wire. Adjacent to the tap start the second half of the winding. Scrape the insulation off the ends at the tap, and then twist the two ends together to form the tap. Alternatively, you can loop the wire (see detail inset to Fig. 12.3B), and then continue the winding. The loop then becomes the tap. Scrape the insulation off the wire, and solder it. Although the tap here is a centre tap (which means L2), it also serves for L1 if you offset the tap a bit to the left or right.

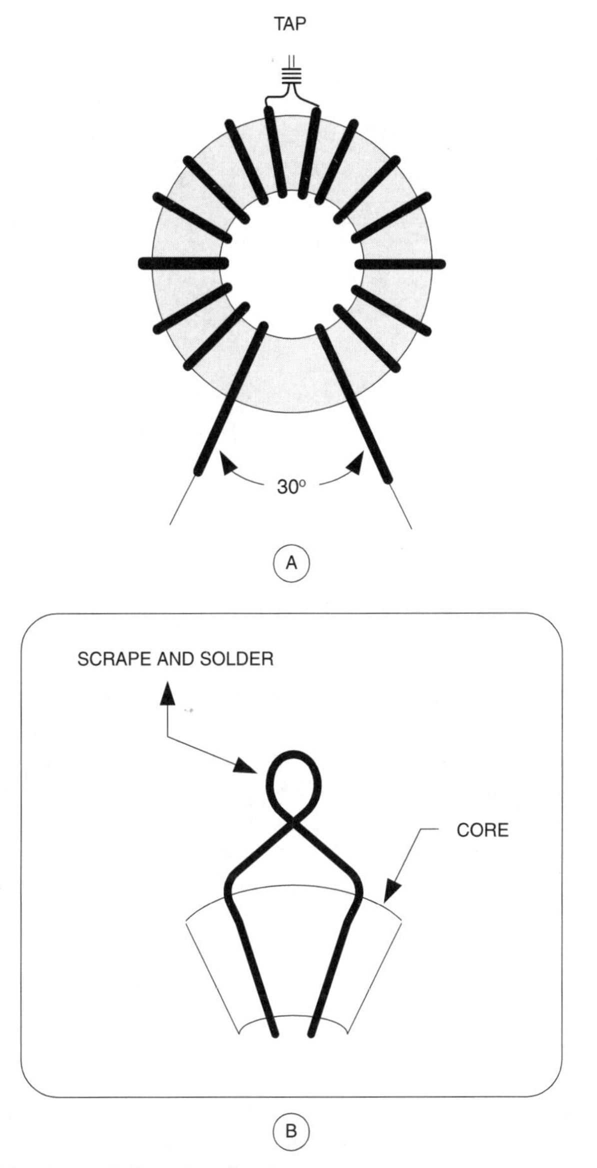

Figure 12.3 Physical implementation showing tap.

An alternate method for L2 is shown in Fig. 12.4. This is superior to the other form for L2, but it is a little more difficult. Either wind the two wires together side by side (Fig. 12.4A), or twist them together before winding (Fig. 12.4B). Make a loop at the centre-tap, and scrape it for soldering.

The capacitor usually has a value of 10 pF, although people with either a sweep generator, or a CW RF signal generator and a lot more patience than I've got, can optimize performance by replacing it with a 15 pF trimmer capacitor. Adjust the trimmer for flattest response across the entire band.

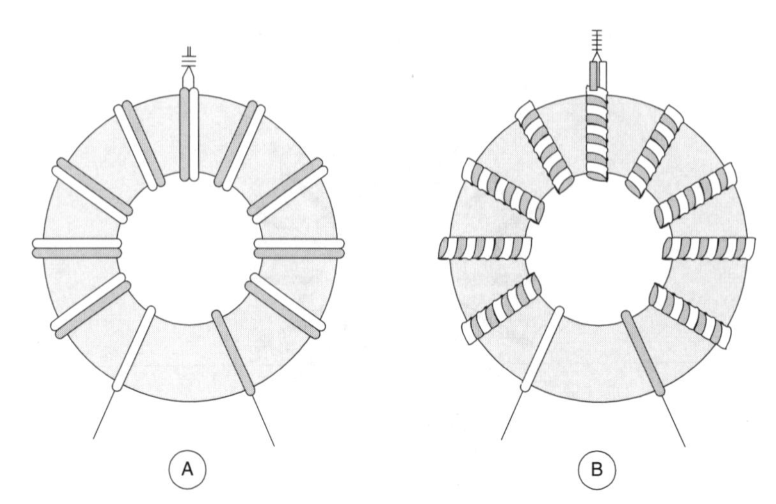

Figure 12.4 Bifilar winding styles.

How it works

A signal entering at the common port goes equally to the two output ports. In this case L2 and 2R don't really do very much. The insertion loss is about −3 dB.

The clever part comes if a signal enters at, say, port OUT-1. Then provided the common port sees the right impedance, the signals to port OUT-2 via the resistor 2R and the inductor L2 cancel each other out. So the signal just goes to the common port, again with a −3 dB loss. There is a high degree of isolation between OUT-1 and OUT-2. Although theoretically it is possible to attain infinite cancellation 20 to 60 dB is more reasonable in practical circuits.

Mismatch losses

It is important in any RF system to ensure impedance matching at all ports. However, in practical terms it often happens that some degree of mismatch will occur. Let's examine a scenario of what happens when there is a mismatch loss at port S.

In general, with circuits such as Fig. 12.2A it is very important to ensure an impedance match at the common port, although some degree of mismatch can be tolerated at both OUT-1 and OUT-2.

There is an imbalance of behaviour between ports when a serious mismatch occurs at either OUT-1 or OUT-2. Assume that the common port and OUT-2 are properly and perfectly matched by 50 ohm loads, but OUT-1 is short-circuited (0 ohms). Because of the isolation, OUT-2 is not affected by the short-circuit, and continues to see a 50 ohm impedance. But the impedance seen by the common port is reduced 3:1. If it is normally 50 ohms, then the effect of one shorted output port is a reduction to 50/3 = 16.67 ohms.

Modified VSWR bridge splitter/combiner

Figures 12.5 and 12.6 show 6 dB splitter/combiner circuits based on the popular bridge used to measure voltage standing wave ratio (VSWR). Each of these circuits uses a bridge made of three resistors and one winding of a transformer. In both cases, the transformers have a 1:1 turns ratio. Also, in both cases $R1 = R2 = R3 = R_0$. In Fig. 12.5 the transformer is not tapped. It is a straight 1:1 turns ratio toroid transformer. The circuit in Fig. 12.6, however, uses a centre tap on the primary of T1. Note that there is a difference in the location of the summation output between the two circuits. These circuits have been

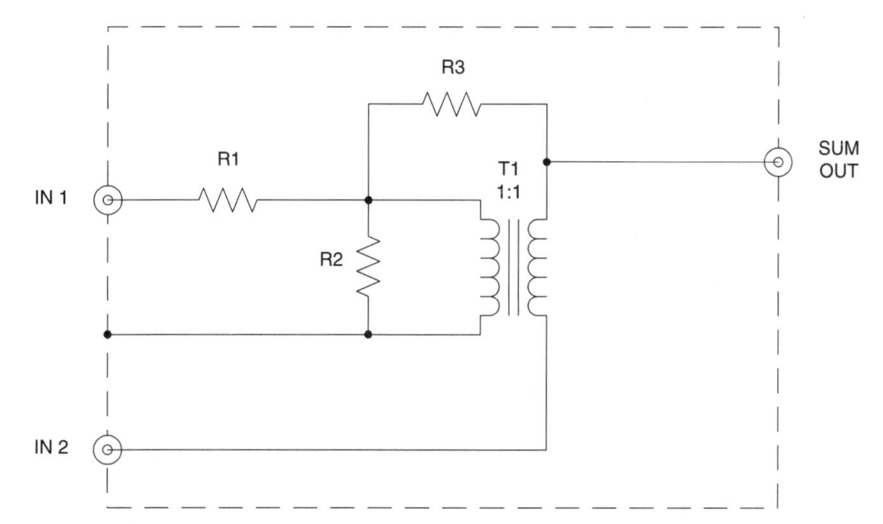

Figure 12.5 VSWR type coupler circuit.

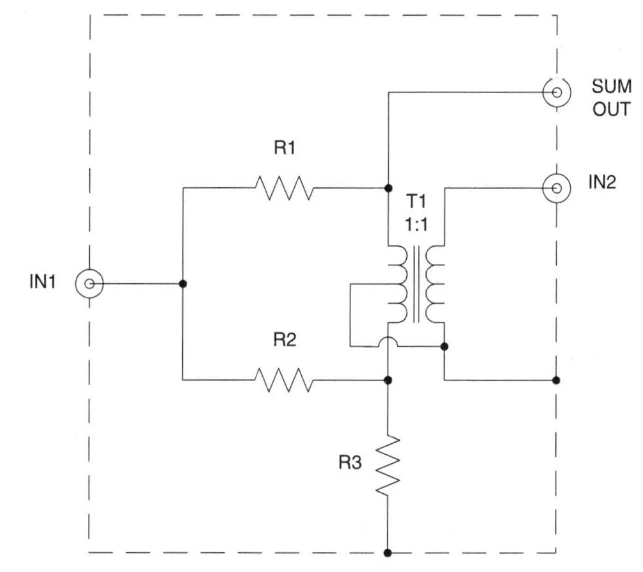

Figure 12.6 Coupler circuit.

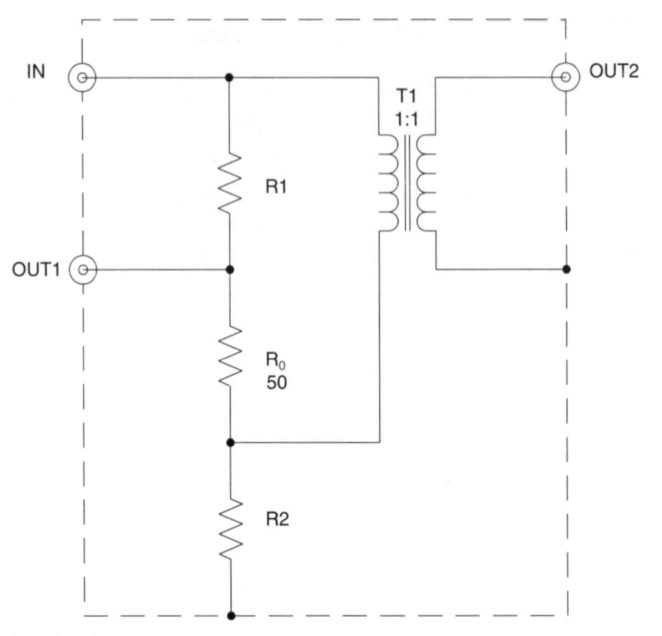

Figure 12.7 Coupler circuit.

popular for combining a two-signal generator, e.g. a sweep generator and a marker generator.

A variation on the theme is shown in Fig. 12.7. This circuit is sometimes also used as a directional coupler. RF power applied to the input port appears at OUT-2 with only an insertion loss attenuation. A sample of the input signal appears at OUT-1. Alternatively, if RF power is applied to OUT-2, it will appear at the input, but does not appear at the OUT-1 port due to cancellation.

For the case where the device has a –3.3 dB output at OUT-2 and a –10 dB output at OUT-1, and a 50 ohm system impedance (R_0), the value of $R1$ = 108 ohms, and $R2$ = 23 ohms. The equations for this device are:

$$R_0 = \sqrt{R1R2} \tag{12.4}$$

$$CF = 20\log\left(\frac{R_0}{R1 + R_0}\right) \tag{12.5}$$

$$L_i = -20\log\left(\frac{R_0}{R2 + R_0}\right) \tag{12.6}$$

Where:
R_0 is the system impedance (e.g. 50 Ω)
CF is the coupling factor
L_i is the insertion loss from IN to OUT-2
$R1$ and $R2$ are the resistances of R1 and R2

Taking the negative of Eq. (12.4) gives the insertion loss from IN to OUT-1.

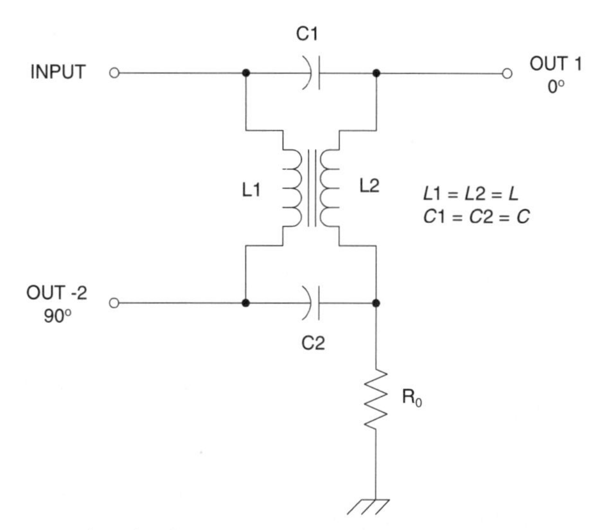

Figure 12.8 90 degree coupler circuit.

90 degree splitter/combiner

Figure 12.8 shows a 3 dB splitter/combiner made of lumped L and C elements, and which produces a 0 degree output at OUT-1, and a 90 degree output at OUT-2. A closely coupled 1:1 transformer is used to supply two inductances, *L*1 and *L*2. This transformer is wound in the bifilar manner to ensure tight coupling. The values of inductance and capacitance, assuming that $L1 = L2 = L$, and $C1 = C2 = C$, are given by:

$$L = \frac{R_0}{2.828\pi f_{3\,\text{dB}}} \tag{12.7}$$

$$C = \frac{1}{2.828\pi f_{3\,\text{dB}} R_0} \tag{12.8}$$

Where:
L is the inductance of L1 and L2
C is the capacitance of C1 and C2
R_0 is the system impedance (e.g. 50 ohms)
$f_{3\,\text{dB}}$ is the 3 dB coupling frequency

The bandwidth of this circuit is approximately 20 per cent for 1 dB amplitude balance.

Transmission line splitter/combiners

The Wilkinson power splitter/combiner is shown in Fig. 12.9. This network can achieve 20 dB isolation between the two output ports over a bandwidth that is approximately ±20 per cent of the design frequency. It consists of two transmission lines, TL1 and TL2, and a bridging resistor (R), which has a value of $R = 2R_0 = 2 \times 50\,\Omega = 100\,\Omega$.

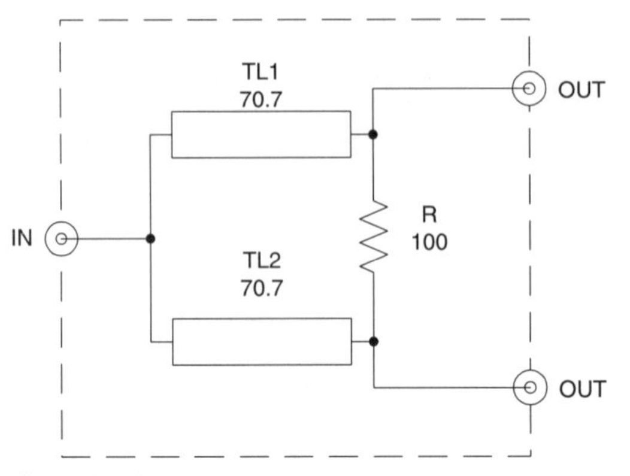

Figure 12.9 Wilkinson splitter circuit.

Transmission lines TL1 and TL2 are each quarter wavelength, and have a characteristic impedance equal to 1.414 times the system impedance (e.g. 70.7 Ω for a 50 Ω system).

The Wilkinson network can be implemented using coaxial cable at VHF and below, although at higher frequencies printed circuit transmission line segments are required. If coaxial cable is used, then the physical length of TL1 and TL2 are shortened by the *velocity factor* (VF) of the cable. The values of VF will be 0.66 for polyethylene dielectric coax, 0.80 for polyfoam dielectric, and 0.70 for *Teflon™ dielectric cable. The physical length is:*

$$Length = \frac{75YF}{F_{\text{MHz}}} \text{ metres} \tag{12.9}$$

An *N*-way version of the same idea is shown in Fig. 12.10. In this network a transmission line, TL1–TL(n), and resistor are used in each branch. The resistor values are

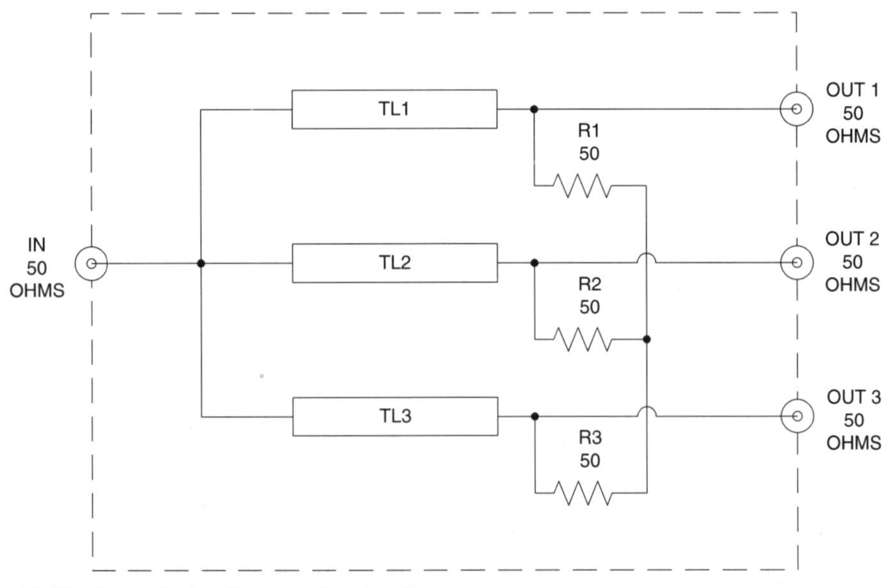

Figure 12.10 Transmission line coupler circuit.

the value of R_0. In the case shown, the values of resistors are 50 ohms because it is designed for standard 50 ohm systems. The characteristic impedance of the transmission lines used in the network is:

$$Z_0 = R_0\sqrt{N} \tag{12.10}$$

Where:
Z_0 is the characteristic impedance of the transmission lines
R_0 is the system standard impedance
N is the number of branches

In the case of a 50 ohm system with three branches, the characteristic impedance of the lines is $(50\,\Omega)(\sqrt{3}) = (50\,\Omega)(1.73) = 86.5\,\Omega$.

90 degree transmission line splitter/combiner

Figure 12.11 shows the network for producing 0 degree to 90 degree outputs, with −3 dB loss, using transmission line elements. The terminating resistor at one node of the bridge is the system impedance, R_0 (e.g. 50 ohms).

Each transmission line segment is a quarter wavelength ($\lambda/4$), so have physical lengths calculated from Eq. (12.9) above. The characteristic impedance of TL1 and TL2 is the system impedance, R_0, while the impedances of TL3 and TL4 are $0.707R_0$. In the case of 50 ohm systems, the impedance of TL3 and TL4 is 35 ohms.

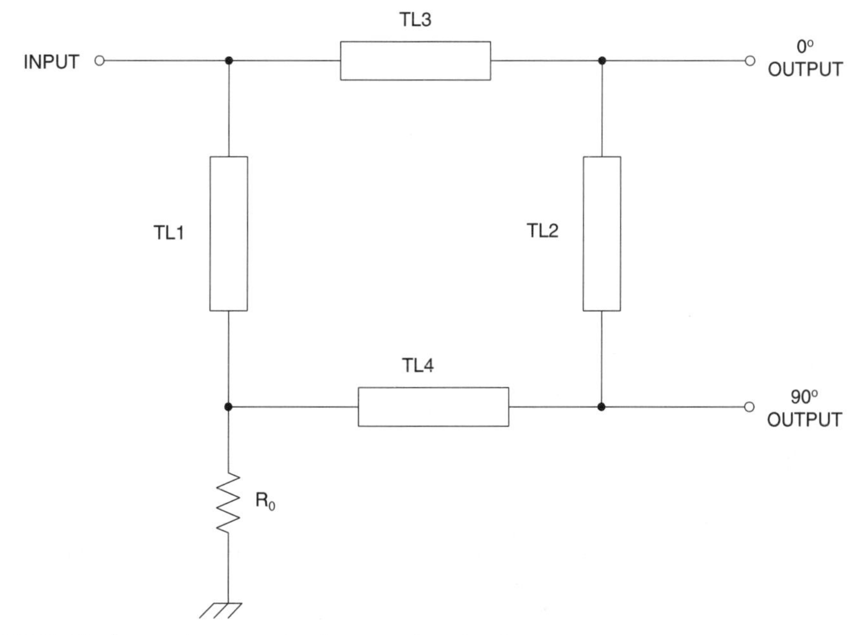

Figure 12.11 90 degree transmission line coupler circuit.

Hybrid ring 'rat-race' network

The 'rat race' network of Fig. 12.12 has a number of applications in communications. It consists of five transmission line segments, TL1 through TL5. At VHF, UHF and microwave frequencies this form is often implemented in printed circuit board transmission lines.

Four of these transmission line segments (TL1–TL4) are quarter wavelength, while TL5 is half wavelength. The characteristic impedance of all lines is $1.414R_0$. Each quarter wavelength segment creates a 90 degree phase shift, while the half wavelength produces a 180 degree phase shift. TL4 and TL5 together produce a 270 degree phase shift.

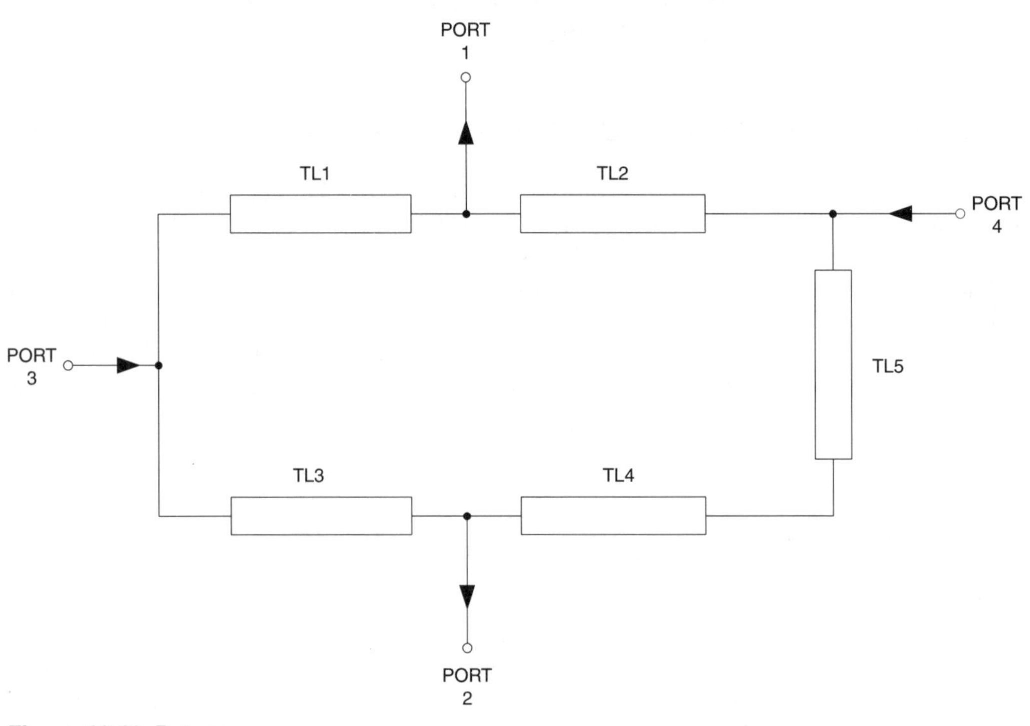

Figure 12.12 Rat race.

It is necessary to terminate all ports of the rat race network in the system characteristic impedance, R_0, whether they are used or not. The bandwidth of this network is approximately 20 per cent.

Different applications use different ports for input and output. Table 12.2 shows some of the relationships found in this network.

A coaxial cable version is shown in Fig. 12.13. This network is implemented using coaxial cable sections and tee connectors. In this case, there are three-quarter wavelength sections (90 degree) and one three-quarter wavelength (270 degree) section. Applications of this network include those where a high degree of isolation is required between ports.

Table 12.2

Input	Use
Port 3	0° splitter, −3 dB at ports 1 and 2.
Port 4	180° splitter, −3 dB, −90° at port 1, and −270° at port 2.

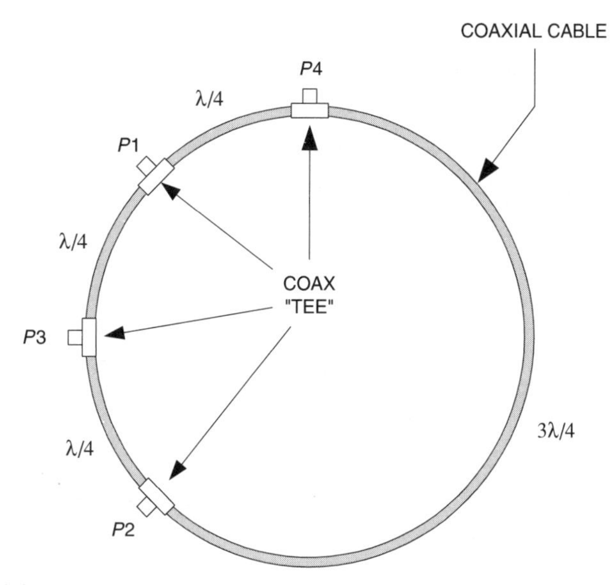

Figure 12.13 Coaxial rat race.

RF hybrid couplers

The hybrid coupler is an RF device that will either: (a) split a signal source into two directions, or (b) combine two signal sources into a common path. The circuit symbols shown in Fig. 12.14A are essentially signal path schematics. Consider the situation where an RF signal is applied to port 1. This signal is divided equally, flowing to both port 2 and port 3. Because the power is divided equally the hybrid is called a 3 dB divider, i.e. the power level at each adjacent port is one-half (−3 dB) of the power applied to the input port.

The RF hybrid is a transformer circuit (Fig. 12.14B) in which the primary has N turns and the two secondary windings have $N/\sqrt{2}$ turns. If the input is port 1 (shown with one side grounded), then the outputs at ports 2 and 3 exhibit 0 degree and 180 degree phase shifts respectively, while port 4 shows cancellation.

If the ports are properly terminated in the system impedance, then all power is absorbed in the loads connected to the ports adjacent to the injection port. None travels to the opposite port. The termination of the opposite port is required, but it does not dissipate power because the power level is zero.

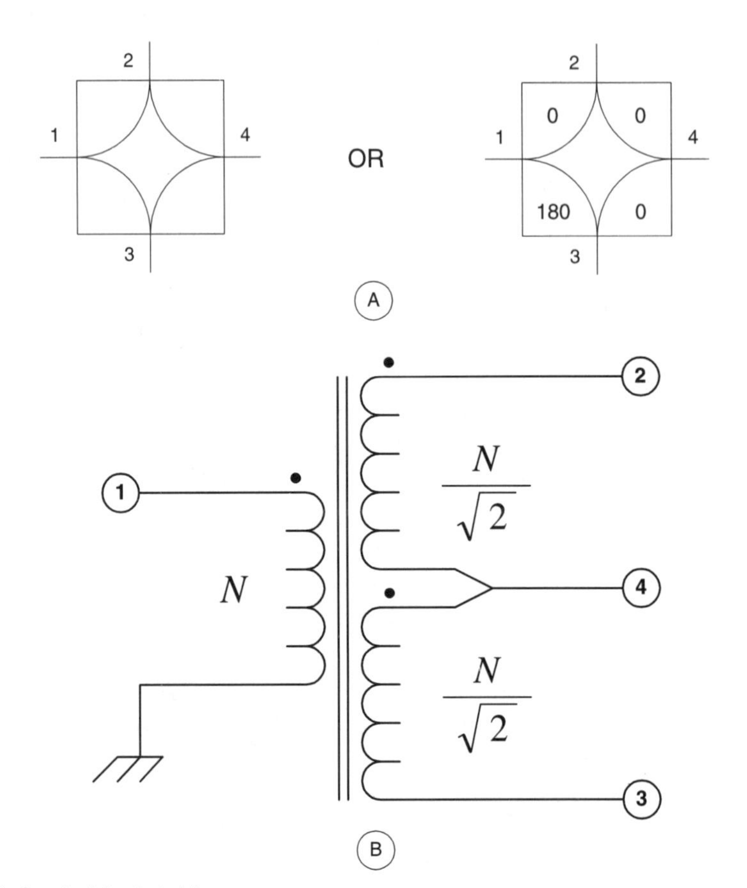

Figure 12.14 Symbol for hybrid.

The one general rule to remember about hybrids is that *opposite ports cancel*. That is, *power applied to one port in a properly terminated hybrid will not appear at the opposite port*. In the case cited above the power was applied to port 1, so no power appeared at port 4. One of the incredibly useful features of the hybrid is that it accomplishes this task while allowing all devices connected to it to see the system impedance, R_0.

Applications of hybrids

The hybrid can be used for a variety of applications where either combining or splitting signals are required.

Combining signal sources

In Fig. 12.15 there are two signal generators connected to opposite ports of a hybrid (port 2 and port 3). Power at port 2 from signal generator no. 1 is therefore cancelled at port 3,

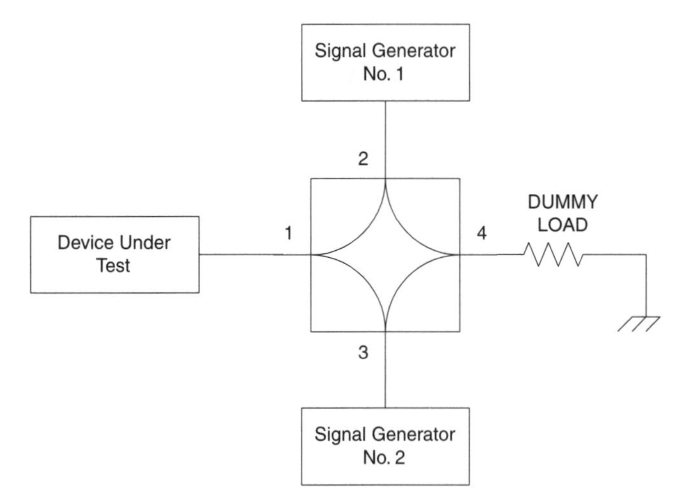

Figure 12.15 Combining two signal sources.

and power from signal generator no. 2 (port 3) is cancelled at port 2. Therefore, the signals from the two signal generators will not interfere with each other.

In both cases, the power splits two ways. For example, the power from signal generator no. 1 flows into port 2 and splits two ways. Half of it (3 dB) goes to port 1, while the other half goes to the dummy load on port 4. The output from signal generator no. 2 splits in the same way.

Bi-directional amplifiers

A number of different applications exist for bi-directional amplifiers, i.e. amplifiers that can handle signals from two opposing directions on a single line. The telecommunications industry, for example, uses such systems to send full duplex signals over the same lines.

Similarly, cable tv systems that use two-way (e.g. cable modem) require two-way amplifiers. Figure 12.16 shows how the hybrid coupler can be used to make such an amplifier. In some telecommunications textbooks the two directions are called *east* and *west*, so this amplifier is occasionally called an *east–west amplifier*. At other times this circuit is called a *repeater*.

In the bi-directional E–W amplifier of Fig. 12.16 amplifier A1 amplifies the signals travelling west to east, while A2 amplifies signals travelling east to west. In each case, the amplifiers are connected to hybrids HB1 and HB2 via opposite ports, so will not interfere with each other. Otherwise, connecting two amplifiers input-to-output-to-input-to-output is a recipe for disaster . . . if only a large amount of destructive feedback.

Transmitter/receiver isolation

One of the problems that exists when using a transmitter and receiver together on the same antenna is isolating the receiver input from the transmitter input. Even a weak transmitter will burn out the receiver input if its power were allowed to reach the receiver input circuits. One solution is to use some form of transmit/receive (T/R) switch.

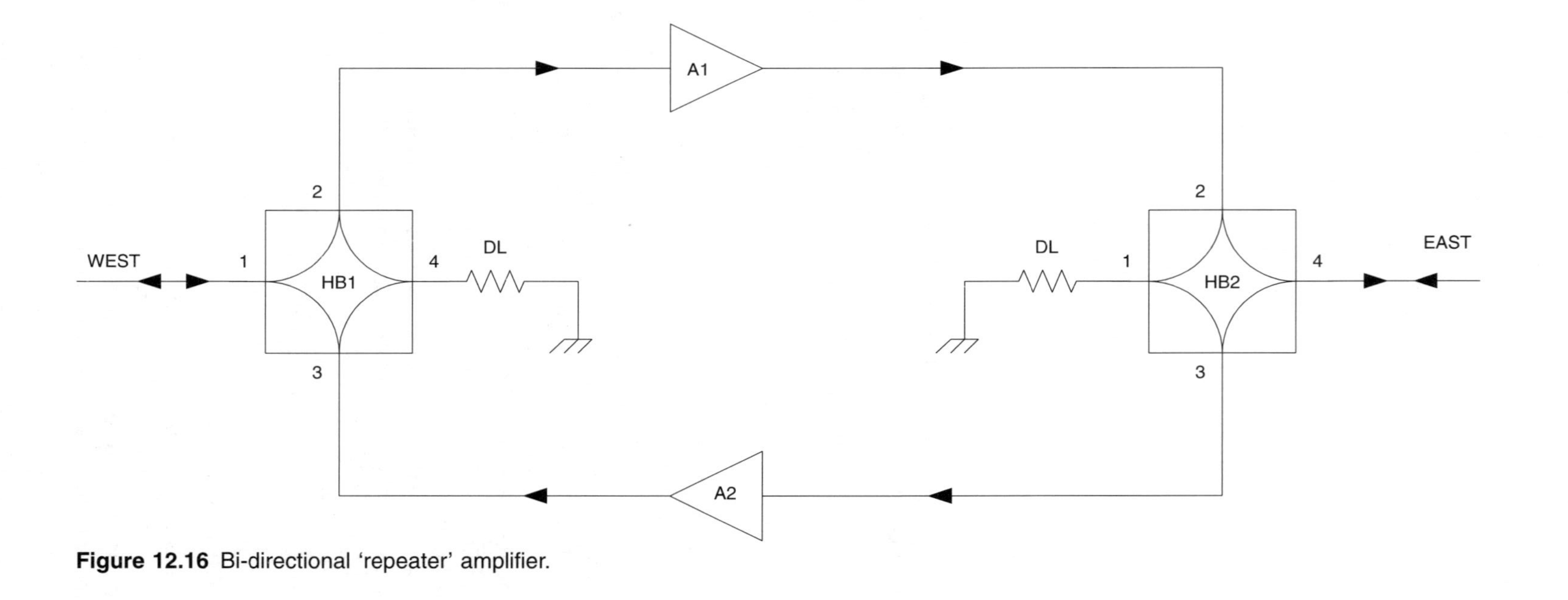

Figure 12.16 Bi-directional 'repeater' amplifier.

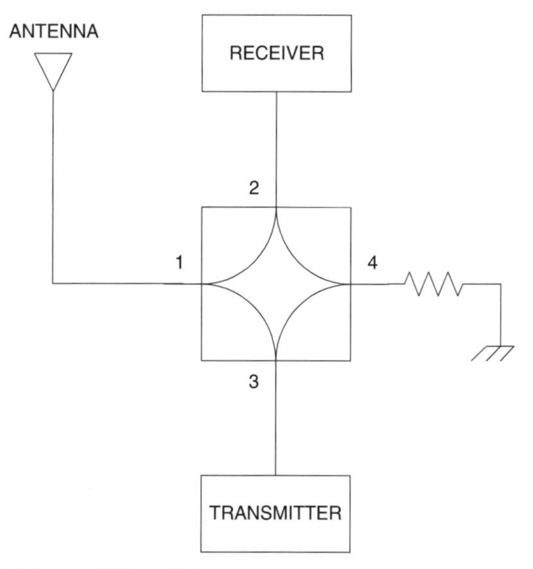

Figure 12.17 Use of hybrid as a T/R switch.

Another option is to use a hybrid as shown in Fig. 12.17. In this circuit the transmitter output and receiver input are connected to opposite ports of a hybrid device. Thus, the transmitter power does not reach the receiver input.

The antenna is connected to the adjacent port between the transmitter port and the receiver port. Signal from the antenna will flow over the port 1 to port 2 path to reach the receiver input. Transmitter power, on the other hand, will enter at port 3, and is split into two equal portions. Half the power flows to the antenna over the port-3 to port-1 path, while half the power flows to a dummy load through the port-3 to port-4 path.

There is a problem with this configuration. Because half the power is routed to a dummy load, there is a 3 dB reduction in the power available to the antenna. A solution is shown in Fig. 12.18. In this configuration a second antenna is connected in place of the dummy load. Depending on the spacing (S), and the phasing, various directivity patterns can be created using two identical antennas.

If the hybrid produces no phase shift of its own, then the relative phase shift of the signals exciting the antennas is determined by the difference in length of the transmission line between the hybrid and the antennas. This is the basis for the phased array antenna. Consult an antenna book for details.

Quadrature hybrids

The hybrids discussed thus far split the power half to each adjacent port, but the signals at those ports are 180 degrees out of phase with each other. That is, there is a zero degree phase shift over the paths from the input to two output ports, and a 180 degree phase shift to the third port. There are, however, two forms of phase shifted hybrids. The form shown in Fig. 12.19A is the standard 0 degree to 180 degree hybrid. The signal over the port-1 to

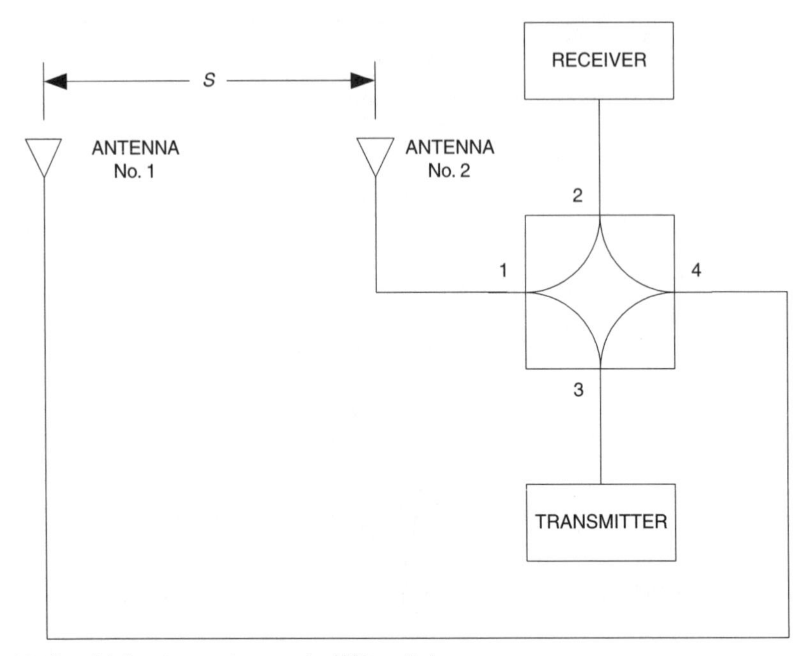

Figure 12.18 Combining two antennas in T/R switch.

port-2 path is not phase shifted (0 degrees), while that between port 1 and port 3 is phase shifted 180 degrees. Most transformer-based hybrids are inherently 0 degree to 180 degree hybrids.

A 0 degree to 90 degree hybrid is shown in Fig. 12.19B. This hybrid shows a 90 degree phase shift over the port-1 to port-2 path, and a 0 degree phase shift over the port-1 to port-3 path. This type of hybrid is also called a *quadrature hybrid*.

One application for the quadrature hybrid is the balanced amplifier shown in Fig. 12.20. Two amplifiers, A1 and A2 are used to process the same input signal arriving via hybrid HB1. The signal splits in HB1, so goes to both A1 and A2. If the input impedances of the amplifiers are not matched to the system impedance, then signal will be reflected from them back towards HB1. The reflected signal from A2 arrives back at the input in-phase (0 degrees), but that reflected from A1 has to pass through the 90 degree phase shift arm

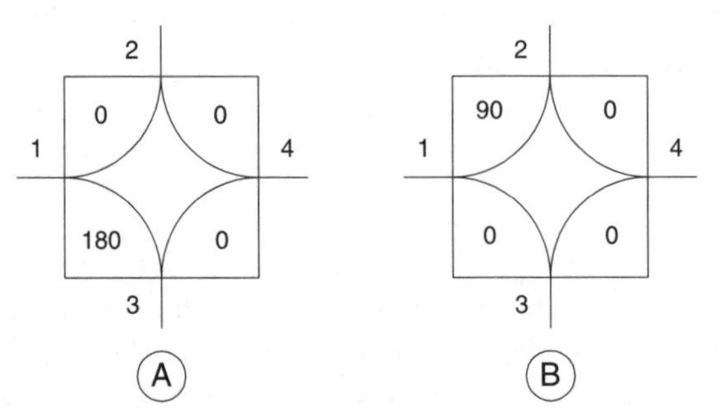

Figure 12.19 (A) 180 degree hybrid; (B) 90 degree hybrid.

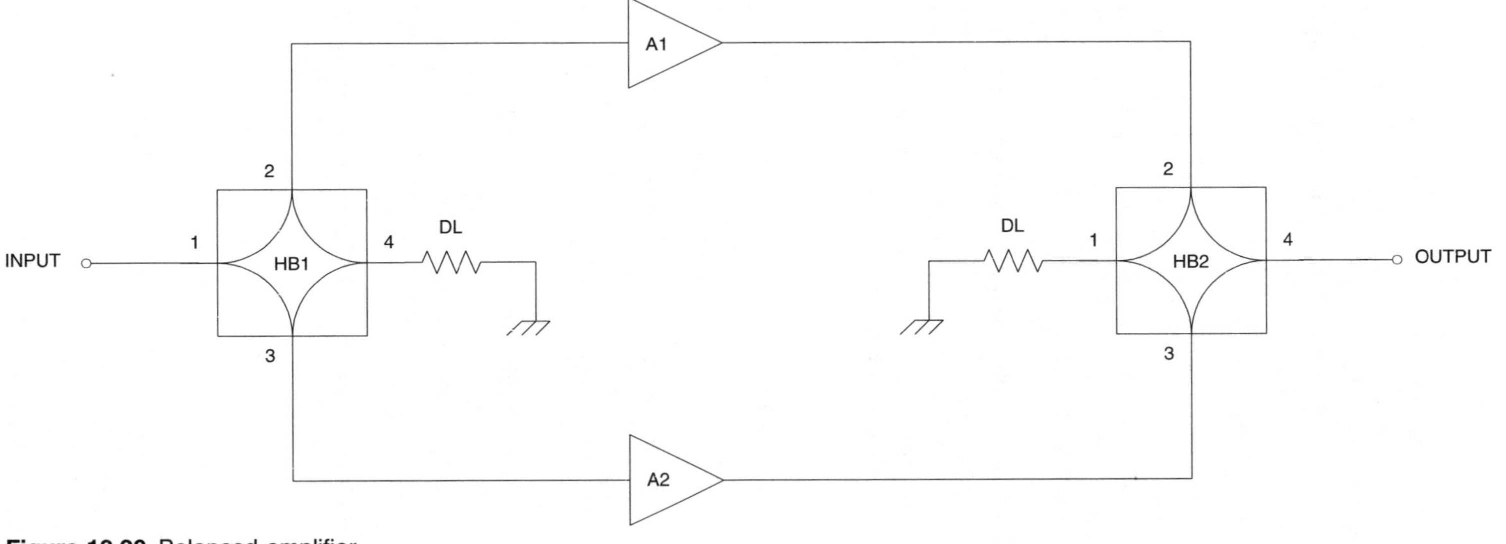

Figure 12.20 Balanced amplifier.

twice, so has a total phase shift of 180 degrees. Thus, the reflections caused by mismatching the amplifier inputs are cancelled out.

The output signals of A1 and A2 are combined in hybrid HB2. The phase balance is restored by the fact that the output of A1 passes through the 0 degree leg of HB2, while the output of A2 passes through the 90 degree leg. Thus, both signals have undergone a 90 degree phase shift, so are now restored to the in-phase condition.

RF directional couplers

Directional couplers are devices that will pass signal across one path, while passing a much smaller signal along another path. One of the most common uses of the directional coupler is to sample a RF power signal either for controlling transmitter output power level or for measurement. An example of the latter use is to connect a digital frequency counter to the low level port, and the transmitter and antenna to the straight through (high power) ports.

The circuit symbol for a directional coupler is shown in Fig. 12.21. Note that there are three outputs and one input. The IN–OUT path is low loss, and is the principal path between the signal source and the load. The coupled output is a sample of the forward path, while the isolated shows very low signal. If the IN and OUT are reversed, then the roles of the coupled and isolated ports also reverse.

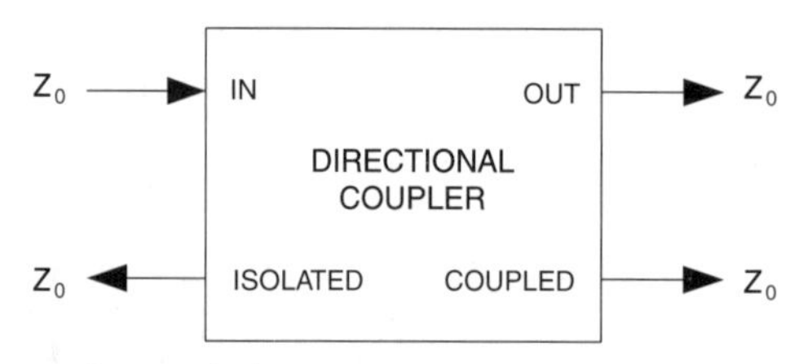

Figure 12.21 Directional coupler symbol.

An implementation of this circuit using transmission line segments is shown in Fig. 12.22. Each transmission line segment (TL1 and TL2) has a characteristic impedance, Z_0, and is one quarter wavelength long. The path from port 1 to port 2 is the low-loss signal direction. If power flows in this direction, then port 3 is the coupled port and port 4 is isolated.

For a coupling ratio (port 3/port 4) $\leq -15\,\text{dB}$ the value of coupling capacitance must be:

$$C_c < \frac{0.18}{\omega Z_0} \text{ farads} \qquad (12.11)$$

The coupling ratio is

$$CR = 20\log(\omega C Z_0)\ \text{dB} \qquad (12.12)$$

$$(\omega = 2\pi f)$$

The bandwidth is about 12 per cent.

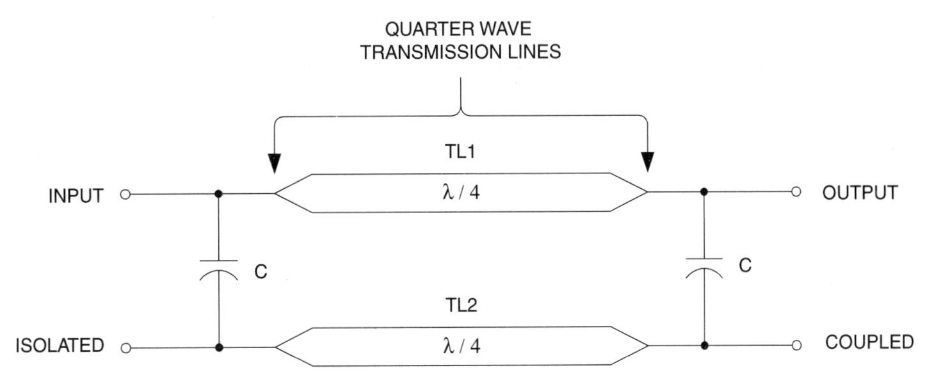

Figure 12.22 Transmission line directional coupler.

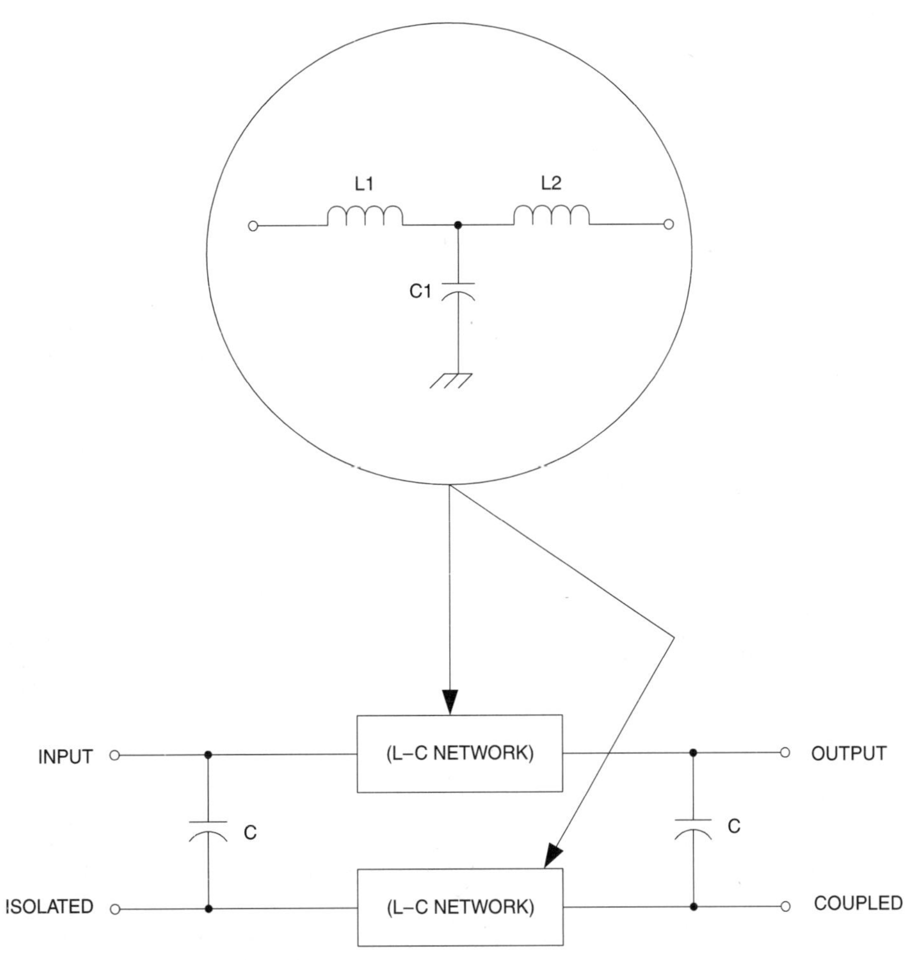

Figure 12.23 L–C network directional coupler.

The circuit shown in Fig. 12.23 is an L–C lumped constant version of the transmission lines. This network can be used to replace TL1 and TL2 in Fig. 12.22. The values of the components are:

$$L1 = \frac{Z_0}{\omega_0} \tag{12.13}$$

$$C1 = \frac{1}{\omega_0 Z_0} \tag{12.14}$$

Figure 12.24 shows a directional coupler used in a lot of RF power meters and VSWR meters. The transmission lines are implemented as printed circuit board tracks. It consists of a main transmission line (TL1) between port 1 and port 2 (the low-loss path), and a coupled line (TL2) to form the coupled and isolated ports. The coupling capacitance (pF) is approximated by 9.399X (X is in metres) when implemented on G-10 epoxy fibreglass printed circuit board.

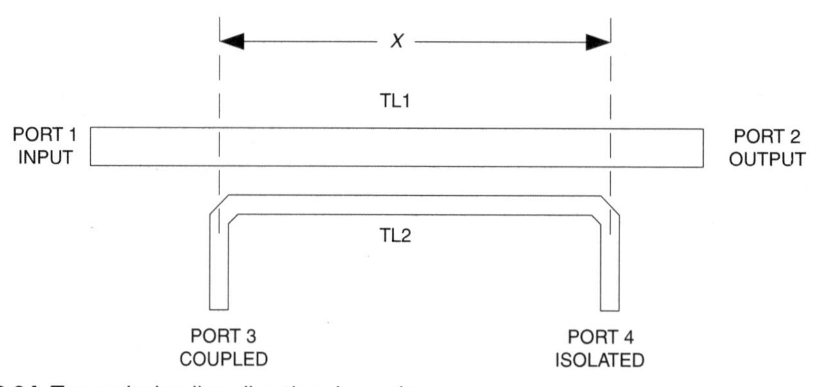

Figure 12.24 Transmission line directional coupler.

A reflectometer directional coupler is shown in Fig. 12.25A. This type of directional coupler is at the heart of many commercial VSWR meters and RF power meters used in the HF through low VHF regions of the spectrum. This circuit is conceptually similar to the previous transmission line, but is designed around a toroid transmission line transformer. It consists of a transformer in which the low-loss path is a single-turn primary winding, and a secondary wound of enamelled wire.

Details of the pick-up sensor are shown in Fig. 12.25B. The secondary is wound around the rim of the toroid in the normal manner, occupying not more than 330 degrees of circumference. A rubber or plastic grommet is fitted into the centre hole of the toroid core. The single-turn primary is formed by a single conductor passed once through the hole in the centre of the grommet. It turns out the 3/16 inch o.d. brass tubing (the kind sold in hobby shops that cater for model builders) will fit through several standard grommet sizes nicely, and will slip-fit over the centre conductor of SO-239 coaxial connectors.

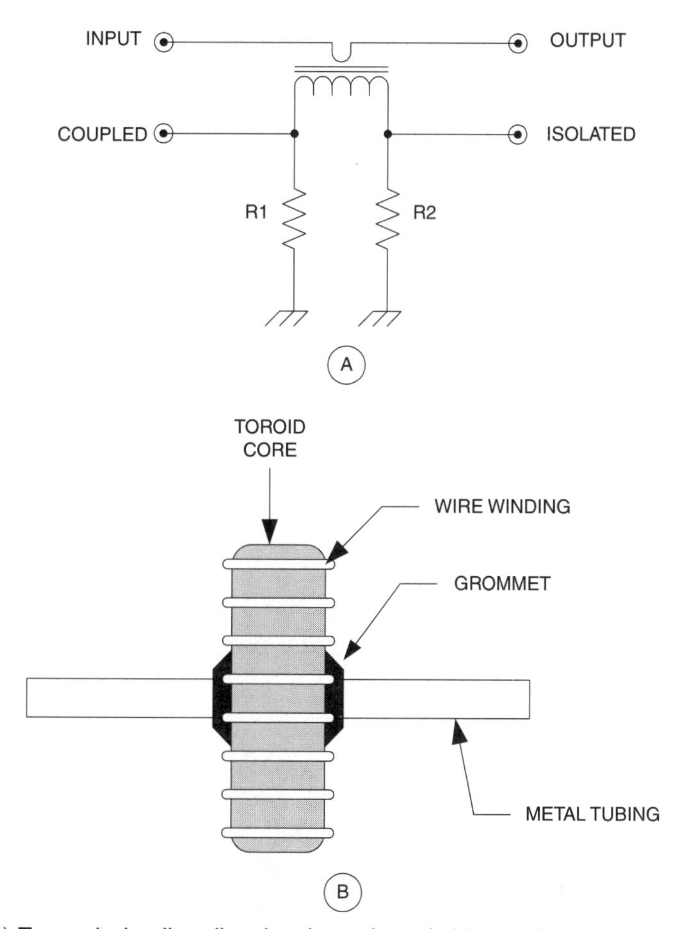

Figure 12.25 (A) Transmission line directional coupler suitable for HF; (B) physical implementation.

Another transmission line directional coupler is shown in Fig. 12.26. Two short lengths of RG-58/U transmission line (≈6 in) are passed through a pair of toroid coils. Each coil is wound with eight to 12 turns of wire. Note that the shields of the two transmission line segments are grounded only at one end.

Each combination of transmission line and toroid core form a transformer similar to the previous case. These two transformers are cross-coupled to form the network shown. The XMTR-ANTENNA path is the low-loss path, while (with the signal flow direction shown) the other two coupled ports are for forward and reflected power samples. These samples can be rectified and used to indicate the relative power levels flowing in the forward and reverse directions. Taken together these indications allow us to calculate VSWR.

Directional couplers are used for RF power sampling in measurement and transmitter control. They can also be used in receivers between the mixer or RF amplifier and the antenna input circuit. This arrangement can prevent the flow of LO signal and mixer products back towards the antenna, where they could be radiated and cause electromagnetic interference (EMI) to other devices.

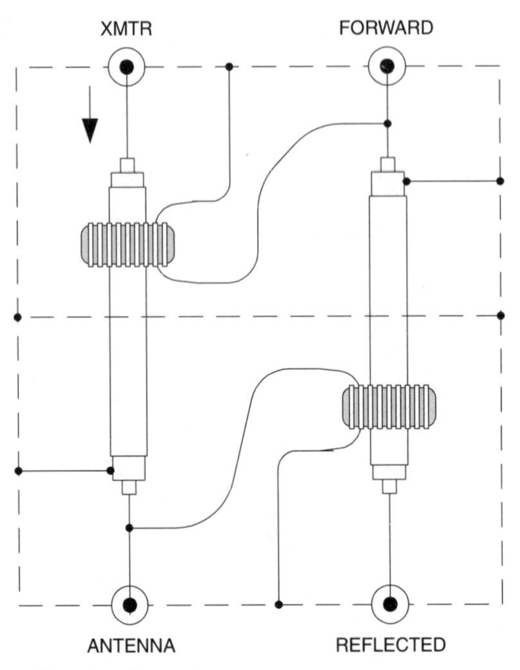

Figure 12.26 Dual sensor directional coupler.

Conclusion

RF combiners, splitters and hybrids solve a number of problems, especially in the instrumentation of the RF laboratory. They will also solve a number of problems in antenna systems, receiver/transmitter systems and other situations.

13 Monolithic microwave integrated circuits

Monolithic microwave integrated circuits (MMICs) and similar devices are used in a wide variety of applications. These devices may be very wideband, or relatively narrow band. Very wideband amplifiers have a bandpass (frequency response) of several hundred megahertz, or more, typically ranging from sub-VLF to the low end of the microwave spectrum. An example might be a range of 100 kHz to 1000 MHz (i.e. 1 GHz). These circuits have a variety of practical uses: receiver preamplifiers, signal generator output amplifiers, buffer amplifiers in RF instrument circuits, cable television line amplifiers, and many others in communications and instrumentation.

Very wideband amplifiers were difficult to design and build until the advent of monolithic microwave integrated circuit (MMIC) devices. Several factors contribute to the difficulty of designing and building very wideband amplifiers. For example, there are too many stray capacitances and inductances in a typical circuit layout, and these form resonances and filters that distort and restrict the frequency response characteristic.

If you have ever tried to build a very wideband amplifier, then it was likely to be a very frustrating experience. Because of new, low cost devices called *silicon monolithic microwave integrated circuits* (SMMICs), it is possible to design and build amplifiers that cover the spectrum from near-DC to about 2000 MHz, and that use seven or fewer components. These devices offer gains of 13 to 30 dB of gain (see Table 13.1), and produce output power levels up to 40 mW (+16 dBm). Noise figures range from 3.5 to 7 dB. In this chapter we will use the MAR-x series of MMICs by Mini-Circuits Laboratories to be representative.

Figure 13.1 shows the circuit symbol for the MAR-x devices. Note that it is a very simple device. The only connections are *RF input*, *RF output* and two *ground* connections. The use

Table 13.1

Type number	Colour dot	Gain @ 500 MHz (dB)	Max. freq.
MAR-1	Brown	17.5	1000 MHz
MAR-2	Red	12.8	2000 MHz
MAR-3	Orange	12.8	2000 MHz
MAR-4	Yellow	8.2	1000 MHz
MAR-6	White	19.0	2000 MHz
MAR-7	Violet	13.1	2000 MHz
MAR-8	Blue	28.0	1000 MHz

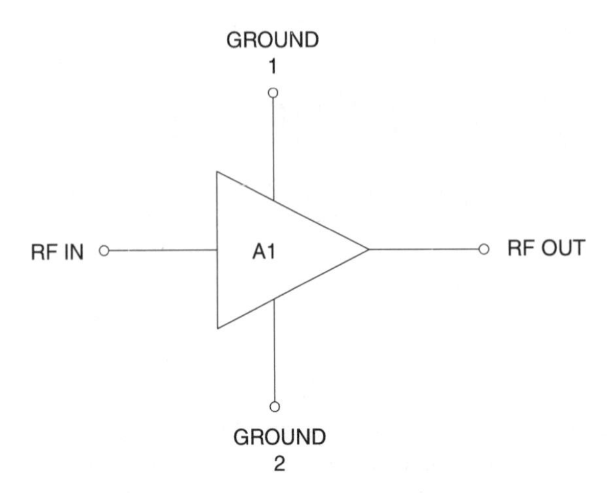

Figure 13.1 MMIC device circuit diagram.

of dual grounds reduces overall inductance and thereby improves the stability and frequency response. Direct current (DC) power is applied to the output terminal through an external network.

The package for the MAR-x device is shown in Fig. 13.2. Although it is an IC, the device looks very much like a small UHF/microwave transistor package. The body is made of plastic, and the leads are wide metal strips (rather than wire) in order to reduce the stray inductance that narrower wire leads would exhibit. These devices are small enough that handling can be difficult; I found that hand forceps ('tweezers') were necessary to position

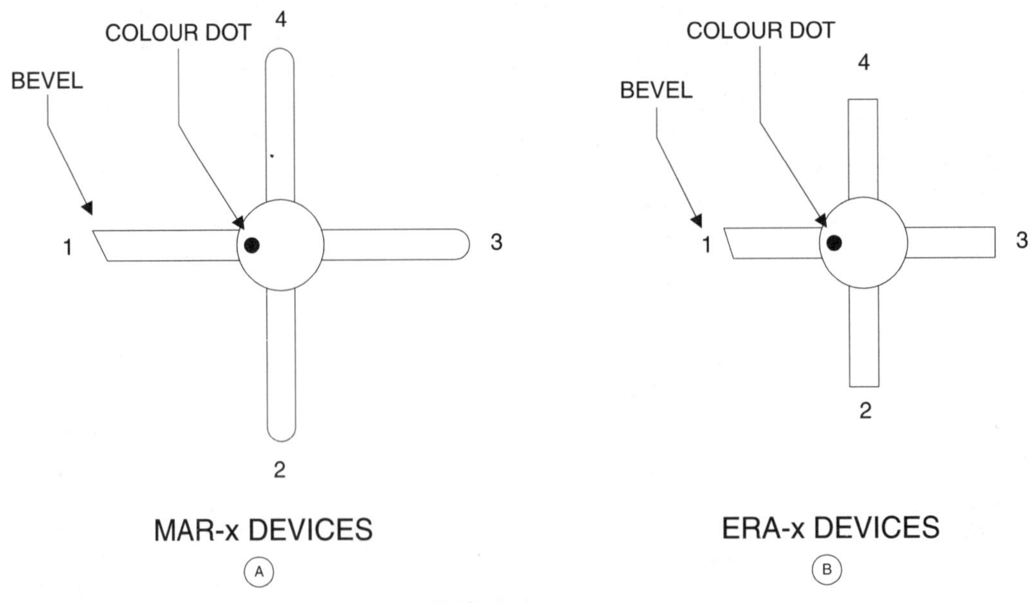

Figure 13.2 Mini-Circuits Laboratories MMIC devices.

the device on a prototype printed circuit board. A magnifying glass or jeweller's eye loupe are not out of order for those with poor close-in eyesight. A colour dot, and a bevelled tip on one lead, are the keys that identify pin no. 1 (which is the RF input connection). When viewed from above, pin numbering (1, 2, 3, 4) proceeds counterclockwise from the keyed pin.

Internal circuitry

The MAR-x series of devices inherently match 50 ohm input and output impedances without external impedance transformation circuitry, making them an excellent choice for general RF applications. Figure 13.3 shows the internal circuitry for the MAR-x devices. These devices are silicon bipolar monolithic ICs in a two transistor Darlington amplifier configuration. The MAR-x devices act like transistors with very high gain. Because the transistors are biased internal to the MAR-x package, the overall gains are typically 13 to 33 dB, depending on the device selected and operating frequency. No external bias resistors are needed, although a collector load resistor to $V+$ is used.

The good match to 50 ohms for both input and output impedances (R) is due to the circuit configuration, and is approximately:

$$R = \sqrt{R_F R_E} \tag{13.1}$$

If R_F is about 500 ohms, and R_E is about 5 ohms, then the square root of their product is the desired 50 ohms.

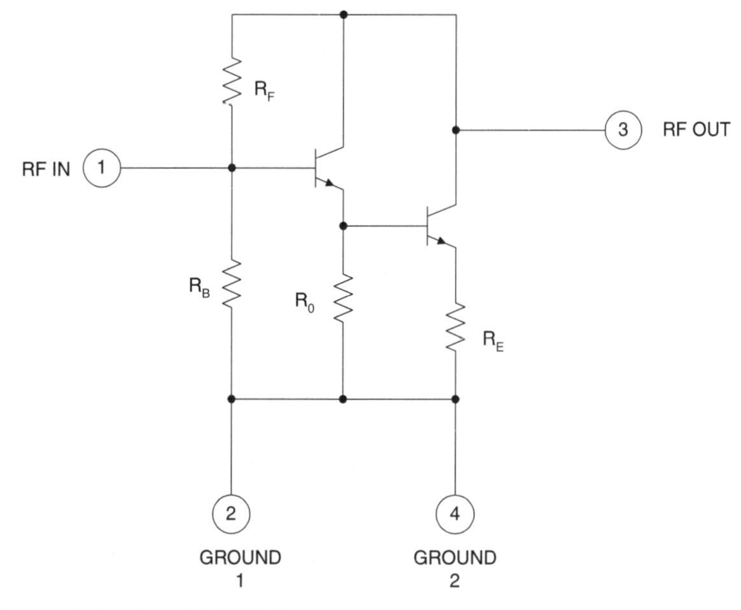

Figure 13.3 Internal circuitry of MMIC device.

Basic amplifier circuit

The basic circuit for a wideband amplifier project based on the MAR-x device is shown in Fig. 13.4. The RF IN and RF OUT terminals are protected by DC blocking capacitors C1 and C2. For VLF and MW applications, use 0.01 μF disk ceramic capacitors, and for HF through the lower VHF (≤100 MHz) use 0.001 μF disk ceramic capacitors. If the project must work well into the high VHF through low microwave region (>100 MHz to 1000 MHz or so), then opt for 1000 pF 'chip' capacitors. If there is no requirement for lower frequencies, then chip capacitors in the 33 to 100 pF range can be used.

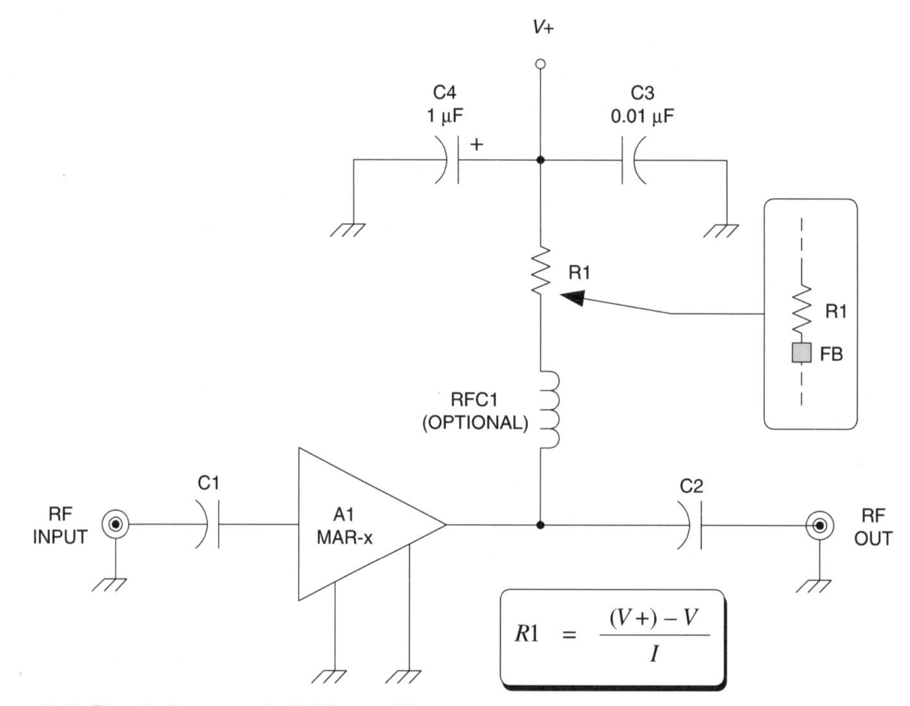

Figure 13.4 Circuit diagram of MMIC amplifier.

The capacitors used for C1 and C2 should be chip capacitors in all but the lower frequency (<100 MHz) circuits. Chip capacitors can be a bit bothersome to use, but their use pays ever greater dividends as operating frequency increases.

Capacitor C3 is used for two purposes. It will prevent signals from being coupled to the DC power supply, and from there to other circuits. It will also prevent higher frequency signals and noise spikes from outside sources from affecting the amplifier circuit. In some cases, a 0.001 μF chip capacitor is used at C3, but for the most part a 0.01 μF disk ceramic will suffice.

The other capacitor at the DC power supply is a 1 μF tantalum electrolytic. It serves to decouple low frequency signals, and smooth out short duration fluctuations in the DC supply voltage. Higher values than 1 μF may be required if the amplifier is used in particularly noisy environments.

Direct current is fed to the amplifier through a current limiting resistor (R1), via the RF OUT terminal on the MAR-x (lead no. 3). The maximum allowable DC potential is +7.5 VDC for MAR-8, +5 VDC for MAR-1 through MAR-4, +4 VDC for MAR-7, and 3.5 VDC for MAR-6. If a minimum voltage $V+$ power supply is used, e.g. +5 VDC for MAR-1, then make R1 a 47 ohm to 100 ohm resistor. Use only $\frac{1}{4}$ watt or $\frac{1}{2}$ watt non-inductive resistors, such as the carbon composition or metal film types. If the use of higher $V+$ potentials (e.g. +9 or +12 VDC) is necessary, then use a higher value resistor for R1. To determine the value of $R1$, decide on a current level (I), and do an Ohm's law calculation:

$$R1(\text{ohms}) = \frac{V + -V_{\text{MMIC}}}{I} \tag{13.2}$$

In most cases, a good operating current level for the popular MAR-1 is about 15 mA (or 0.015 A).

An optional inductor, RFC1, is shown in the circuit of Fig. 13.4. This inductor serves two purposes. First, it improves the decoupling isolation of the MAR-x output from the DC power supply by blocking RF signals. Second, it acts as a 'peaking coil' to improve gain on the high frequency end of the frequency response curve. It does this latter job by adding its inductive reactance (X_L) to the resistance of R1 to form a load impedance that increases with frequency because $X_L = 2pFL$. Suitable values of inductance range from less than 0.5 μH to about 100 μH, depending on the application and frequency range. Sometimes, however, the coil forms the total load impedance. In those cases, a decoupling capacitor is used at the junction of RFC1 and R1.

Inductor coils are not without problems in very wideband amplifiers because the stray capacitances between the coil windings form unintended self-resonances with the coil inductance. These resonances can distort the frequency response curve and may cause oscillations. A popular solution to this problem is to use a small ferrite bead ('FB' in the inset to Fig. 13.4). The bead acts as a small value RF choke. These beads have a small hole in them that fits nicely over the radial lead of a quarter watt resistor.

Alternative DC power schemes are shown in Fig. 13.5. The circuit of Fig. 13.5A splits the load resistance into two components, R1 and R2. The value of R1 will represent most of the required resistance, with R2 typically being 33 to 100 ohms. This circuit, like the basic circuit, works well to $V+$ of 7 to 9 VDC, but is not recommended for higher supply voltages.

Power feed schemes that work well at $V+$ voltages greater than 9 VDC are shown in Figs 13.5B and 13.5C. Both use voltage regulation to stabilize the supply voltage to the MAR-x device.

Other MAR-x circuits

The simple circuit of Fig. 13.4 will work well in most cases, especially where the input and output impedance are reasonably stable. But if the source or load impedances vary, then the amplifier may suffer a degradation of performance, or show some instability. One solution to the problem is to use resistive attenuator pads in the input and output signal lines. *Attenuators* in an amplifier circuit? Yes, that's right. A 1 dB or 2 dB attenuator in the

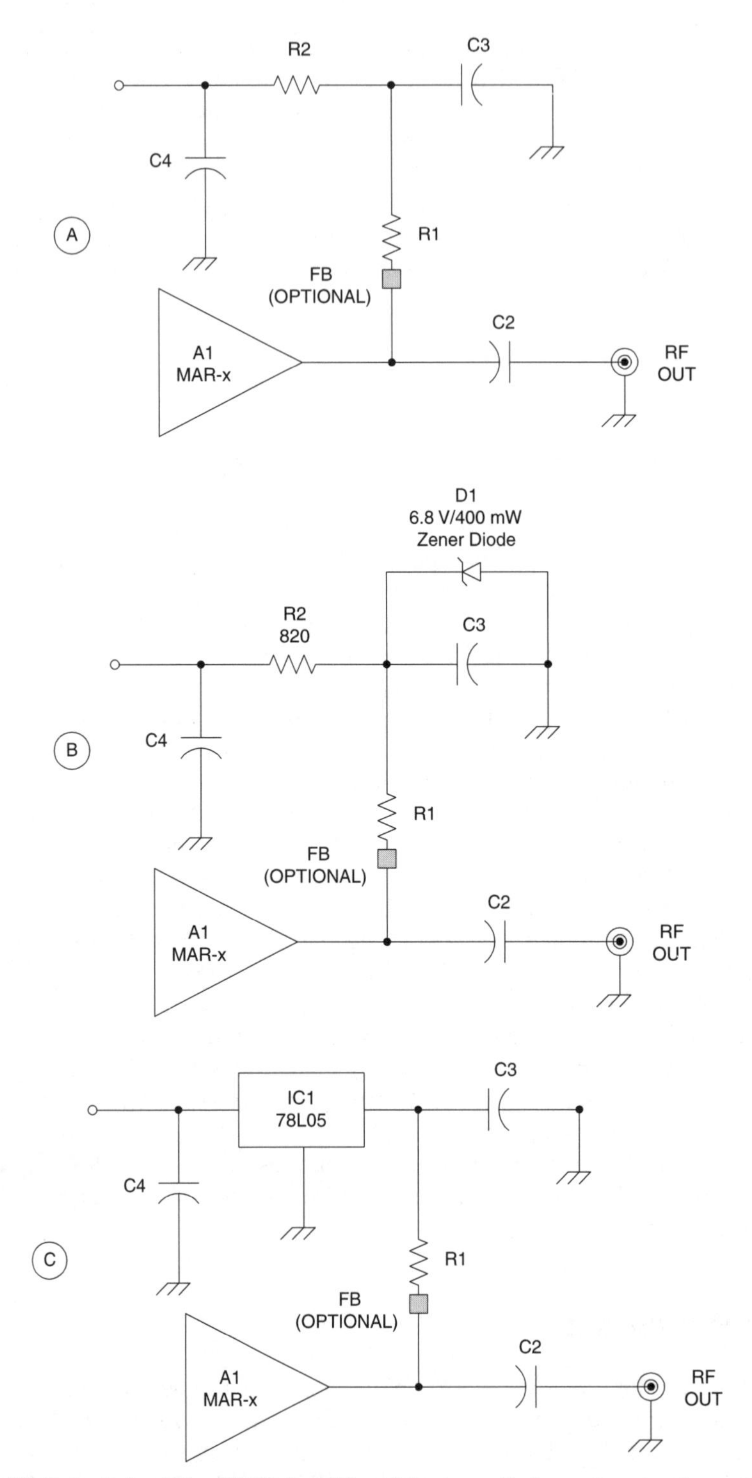

Figure 13.5 (A) Output circuit for MMIC amplifier; (B) zener diode power supply regulation; (C) three-terminal IC voltage regulator circuit.

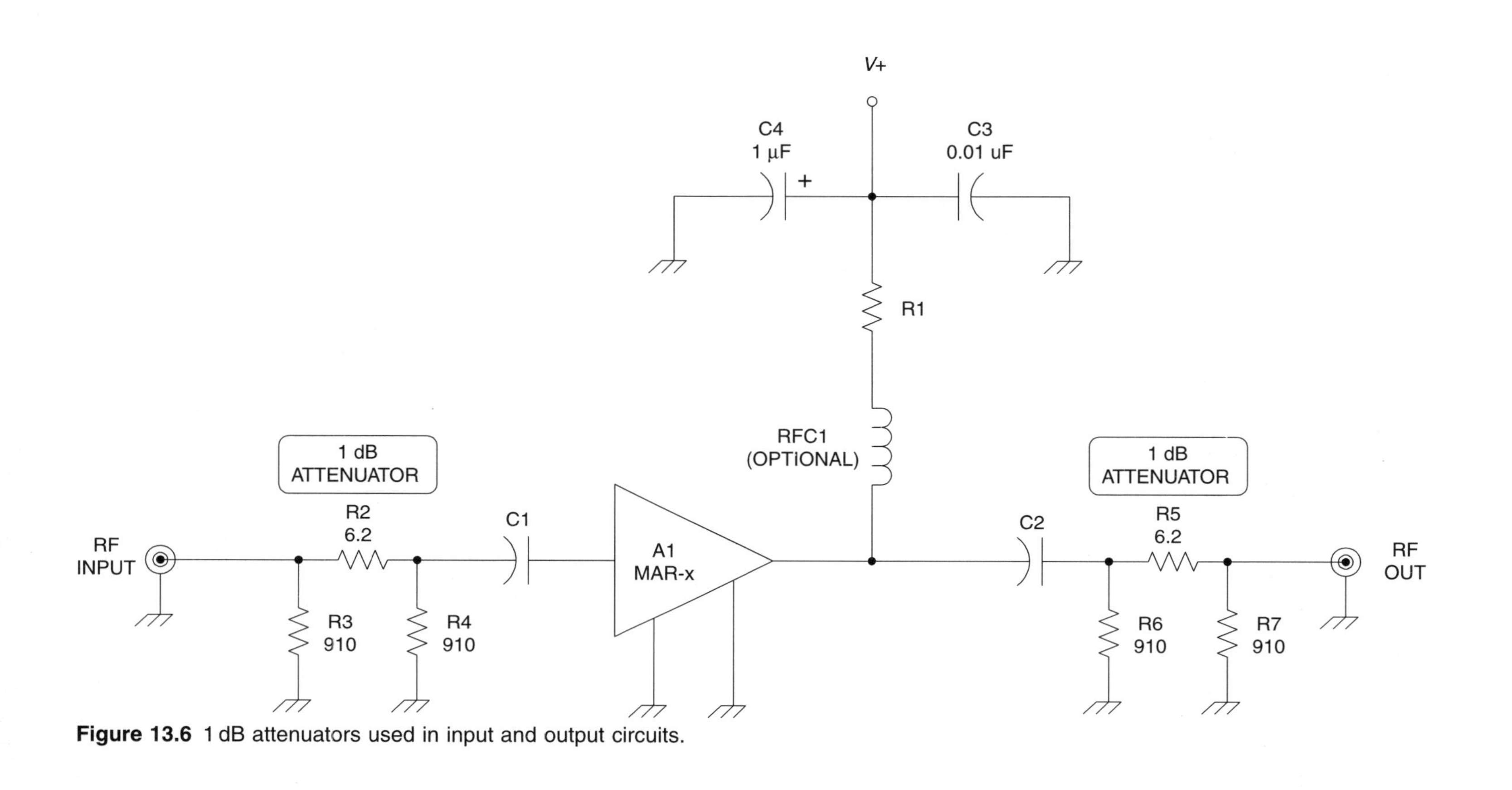

Figure 13.6 1 dB attenuators used in input and output circuits.

input and output signal lines will stabilize the impedances seen by the amplifier, but only marginally affect the overall gain of the circuit. In vacuum tube days, we called this type of technique 'swamping'.

Figure 13.6 shows the circuit of Fig. 13.4 revised to reflect the use of simple resistive attenuator pads in the input and output lines. If 1 dB attenuators are used, then the overall gain is the natural gain of the MAR-x device less 2 dB. The resistors used for these attenuator pads must be non-inductive types, such as carbon composition or metal film units. If the amplifier is to be used at the higher end of its range, then chip resistors are preferable to ordinary axial lead resistors.

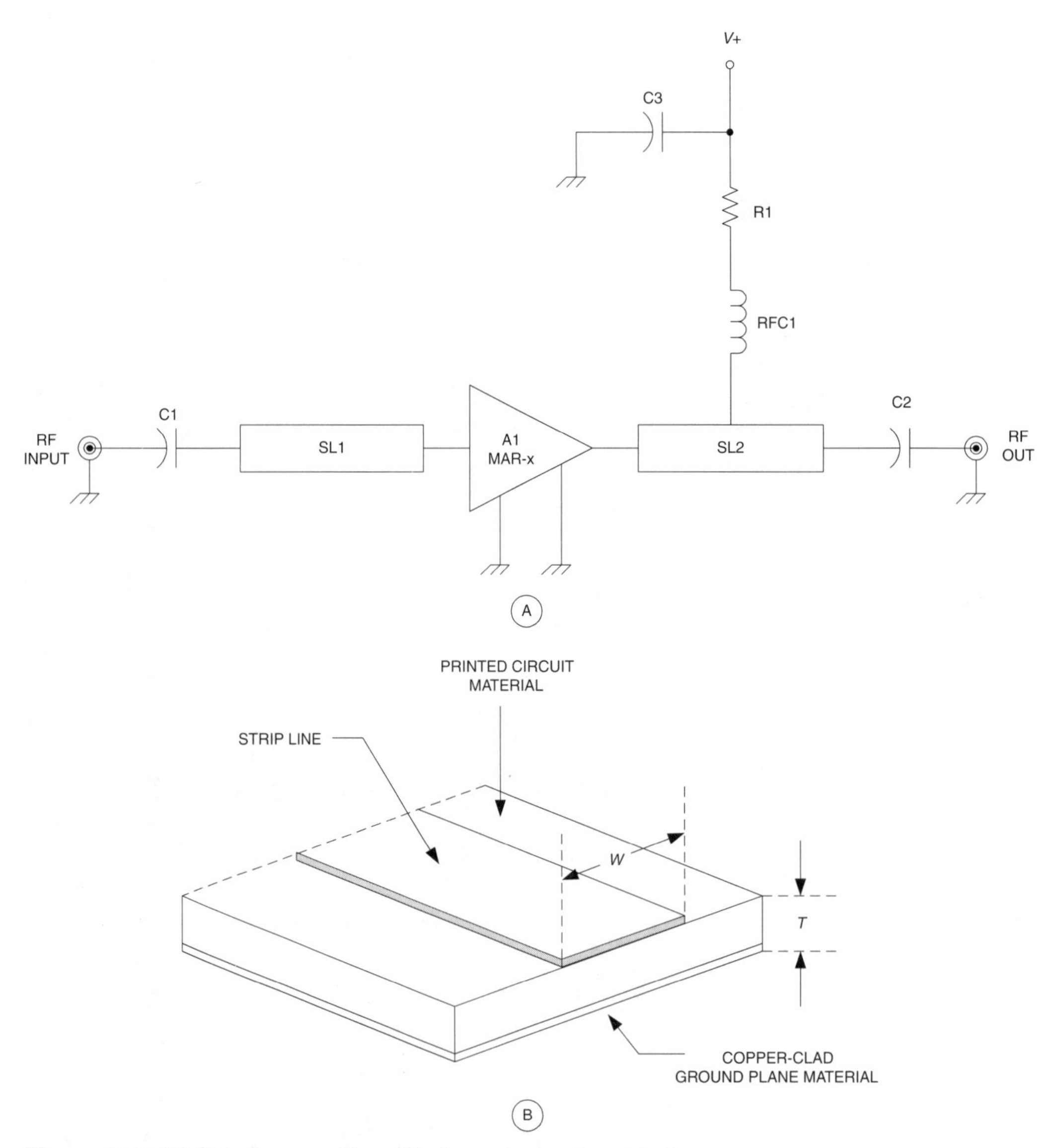

Figure 13.7 (A) Strip-line amplifier; (B) dimensioning the strip line.

An alternate approach is to use manufactured shielded RF 50 ohm attenuator pads. Another of *Mini-Circuits* products are the AT-1 and MAT-1 1-dB attenuators; they are suitable for the purpose, and match the frequency range of most of the MAR-x products. These low cost devices are similar except for size, and are intended for mounting on printed circuit boards.

Keep in mind that the use of attenuators is not for free (TANSTAFL principle: There Ain't No Such Thing As a Free Lunch). The resistive attenuators reduce the gain (as mentioned before), but also increase the noise factor by an amount set by the loss factor of the input attenuator pad.

For VHF, UHF and low end microwave amplifiers it may be preferable to use a printed circuit *strip line transmission line* for the input and output circuits. Figure 13.7A shows such a circuit with input (SL1) and output (SL2) strip lines, while Fig. 13.7B shows detail of how these lines are made. The characteristic impedance (Z_0) of the line is a function of the relative dielectric constant of the printed circuit material (e), the thickness of the material (T), and the width (W) of the strip line conductor. Common Epoxy G-10 printed circuit boards ($e \approx 4.8$) are usable to 1000 MHz and work well to about 300 MHz. Above 300 MHz the losses increase significantly. PTFE woven glass fibre printed circuit board ($e \approx 2.55$) operates to well over 2000 MHz, which is higher than the upper limit of the MAR-x devices. Widths required for 50 ohm strip lines for various printed circuit board materials are shown in Table 13.2.

Figure 13.8A shows the circuit layout of a typical printed circuit board for a MAR-x wideband amplifier. The printed circuit board should be double clad, i.e. copper on both top and bottom. The strip lines at the input and output are etched from the component side of the printed circuit material, not the bottom side as is common practice in lower frequency projects. The reason for this approach is to reduce the inductance of the leads to the MAR-x device.

Strip lines should not contain abrupt discontinuities, or else parasitic losses will increase. It is common practice to taper the line over a short distance from the strip line to the width of the MAR-x leads right at the body of the device.

Another tactic to keep stray lead inductances to a minimum is to drill a small hole in the printed circuit to hold the body of the MAR-x (Fig. 13.8B). The diameter of the MAR-x

Table 13.2 Values for 50 Ω

Material	ε	T	W
G-10 epoxy fibreglass	4.8	0.062 in (1.58 mm)	0.108 in (2.74 mm)
PTFE woven glass fibre	2.55	0.010 in (0.254 mm)	0.025 in (0.635 mm)
		0.031 in (0.787 mm)	0.079 in (0.20 mm)
		0.062 in (1.58 mm)	0.158 in (4 mm)

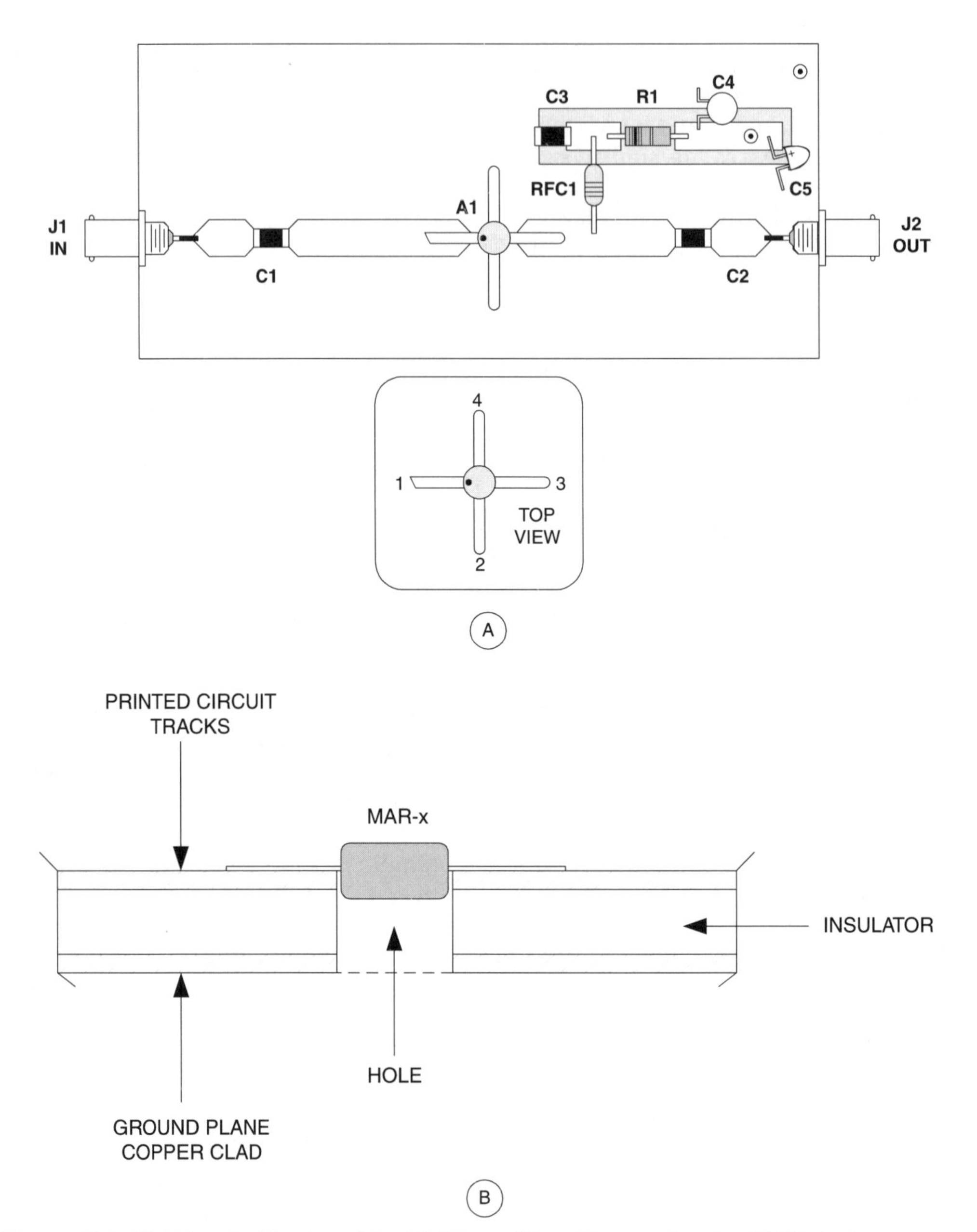

Figure 13.8 (A) Printed wiring board for MMIC amplifier; (B) mounting the MMIC device.

package is 0.085 in (2.15 mm), and the hole should be only slightly larger than this value.

The capacitors in the input and output circuit, as well as the decoupling capacitor at the junction of RFC1 and R1, are chip capacitors. The break in the strip line to accommodate these capacitors should be just wide enough to separate the ohmic contacts at either end of the capacitor body. For the 1000 pF (0.001 μF) chip capacitors that I used in making a model in preparation for this chapter, the insulated centre section between contacts on the capacitors averaged 0.09 inch (2.3 mm) as measured on a vernier caliper set.

It is essential to keep ground returns as short as possible, especially when the amplifier operates in the higher end of its range. If you opt to use the ground plane cladding for the DC and signal return, then plated-through holes are required between the two sides of the board. These plated-through holes must be placed directly below the ground leads of the MAR-x package.

Multiple device circuits

The MAR-x devices can be connected in cascade, parallel or push-pull. The cascade connection increases the overall gain of the amplifier, while the parallel and push-pull configurations increase the output power available.

The simplest cascade scheme is to connect two stages such as Fig. 13.4 in series such that the output capacitor of the first stage becomes the input capacitor of the second stage. Figure 13.9 shows a somewhat better approach. This circuit uses strip line matching sections at the inputs and outputs, and between stages. Table 13.3 gives the dimensions of these lines for two different cases: case A is for a 100 to 500 MHz amplifier, and case B is for a 500 to 2000 MHz amplifier. In both cases the MAR-8 device is used.

The parallel case is shown in Fig. 13.10. The MAR-x devices can be connected directly in parallel to increase the output power capacity of the amplifier. In the case of Fig. 13.10 there are four MAR-x devices connected in parallel; this will give four times the power available from a single device. Other combinations are also possible. I built a two-up version for a signal generator output stage.

Table 13.3

Component	Case A	Case B
R1	124 Ω	69.1 Ω
R2	69.8 Ω	69.1 Ω
C1, C4	470 pF	68 pF
C2	1.5 pF	2 pF
C3	7.5 pf	2 pF
Capacitors are chip type. Resistors are 1% chip type		
	W × *L*	
SL1	0.10 × 0.10 in	0.04 × 0.10 in
	2.54 × 2.54 mm	1.02 × 2.54 mm
SL2	0.10 × 0.05 in	0.04 × 0.10 in
	2.54 × 1.27 mm	1.02 × 2.54 mm
SL3	0.10 × 0.20 in	0.04 × 0.10 in
	2.54 × 5.08 mm	1.02 × 2.54 mm
SL4	0.10 × 0.10 in	0.04 × 0.10 in
	2.54 × 2.54 mm	1.02 × 2.54 mm
SL5	0.05 × 0.20 in	0.05 × 0.20 in
	1.27 × 5.08 mm	1.27 × 5.08 mm

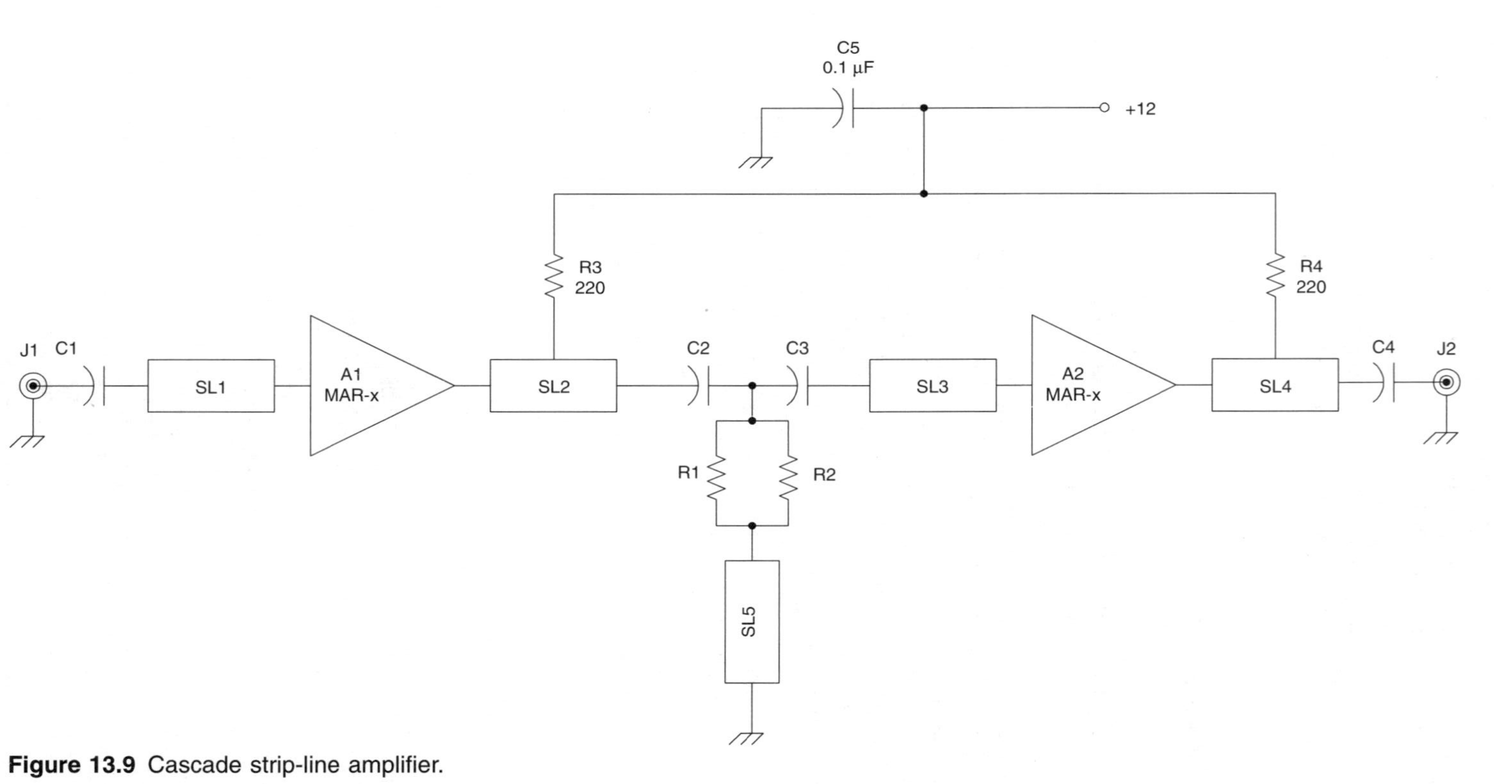

Figure 13.9 Cascade strip-line amplifier.

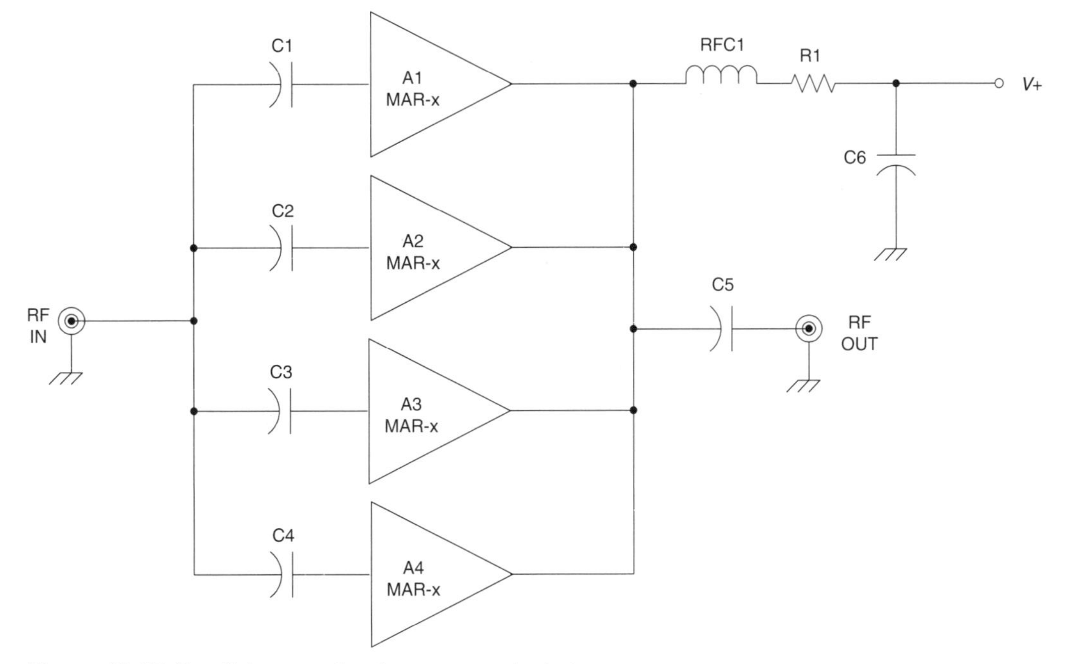

Figure 13.10 Parallel connecting increases output power.

Unfortunately, in the parallel amplifier the input and output impedances are no longer 50 ohms, but rather it is $50/N$ where N is the number of devices connected in parallel. In Fig. 13.10, the input and output impedances are 50/4 or 12.5 ohms. An impedance matching device must be used to transform the impedances to the 50 ohm standard for RF systems. Most impedance transformation devices do not have the same wide bandwidth as the MAR-x devices, so there will be a degradation of the bandwidth of the overall circuit.

The push-pull configuration is shown in Fig. 13.11. In this circuit there are two banks of two MAR-x devices each. The two banks are connected in push-pull, so this circuit is correctly called a push-pull parallel amplifier. It retains the gain and the increase in power level of the parallel connection, but improves the second harmonic distortion that some parallel configurations exhibit. Push-pull amplifiers inherently reduce even-order harmonic distortion.

The input and output transformers (T1 and T2) for the circuit of Fig. 13.11 are balun types, and provide a 180 degree phase shift of the signals for the two halves of the amplifier. The balun transformers are typically wound on ferrite toroidal coil forms with #26 AWG or finer wire. Because the balun transformers are limited in frequency response, this circuit is typically used in medium and shortwave applications. A common specification for these transformers is to wind six or seven bifilar turns on a toroidal form, the turns made of #28 enamelled wire twisted together at about five twists to the inch (≈2 twists per cm).

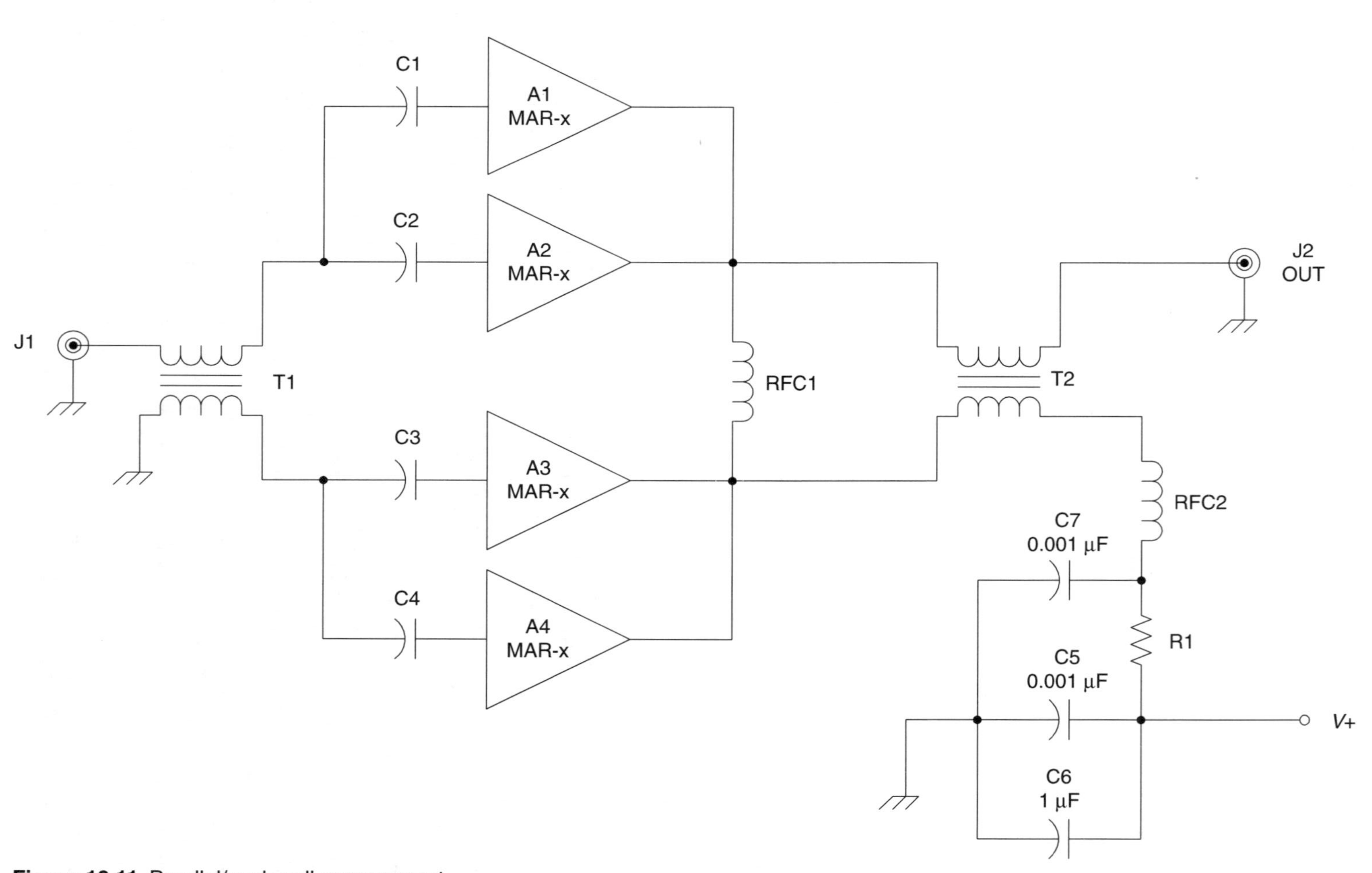

Figure 13.11 Parallel/push-pull arrangement.

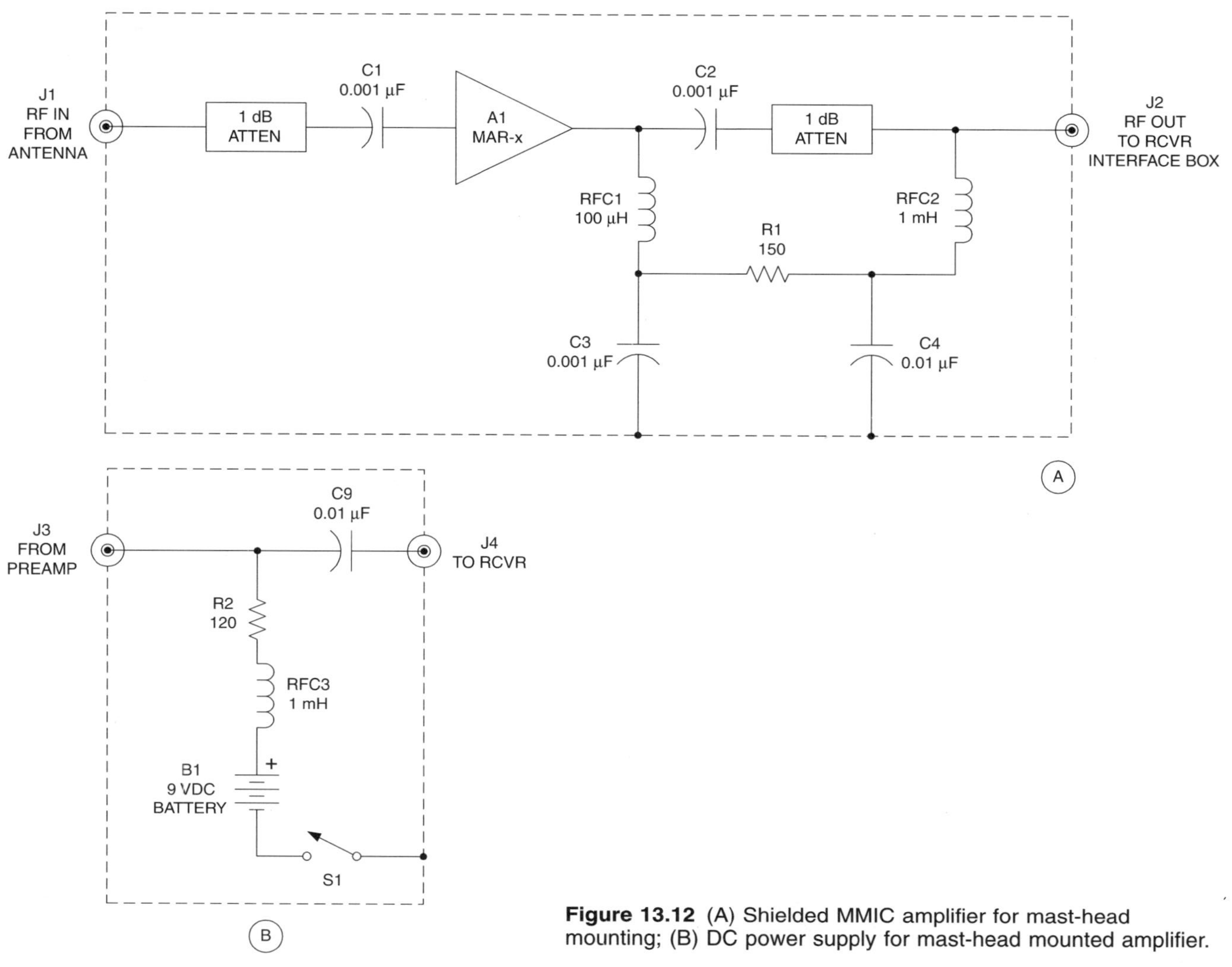

Figure 13.12 (A) Shielded MMIC amplifier for mast-head mounting; (B) DC power supply for mast-head mounted amplifier.

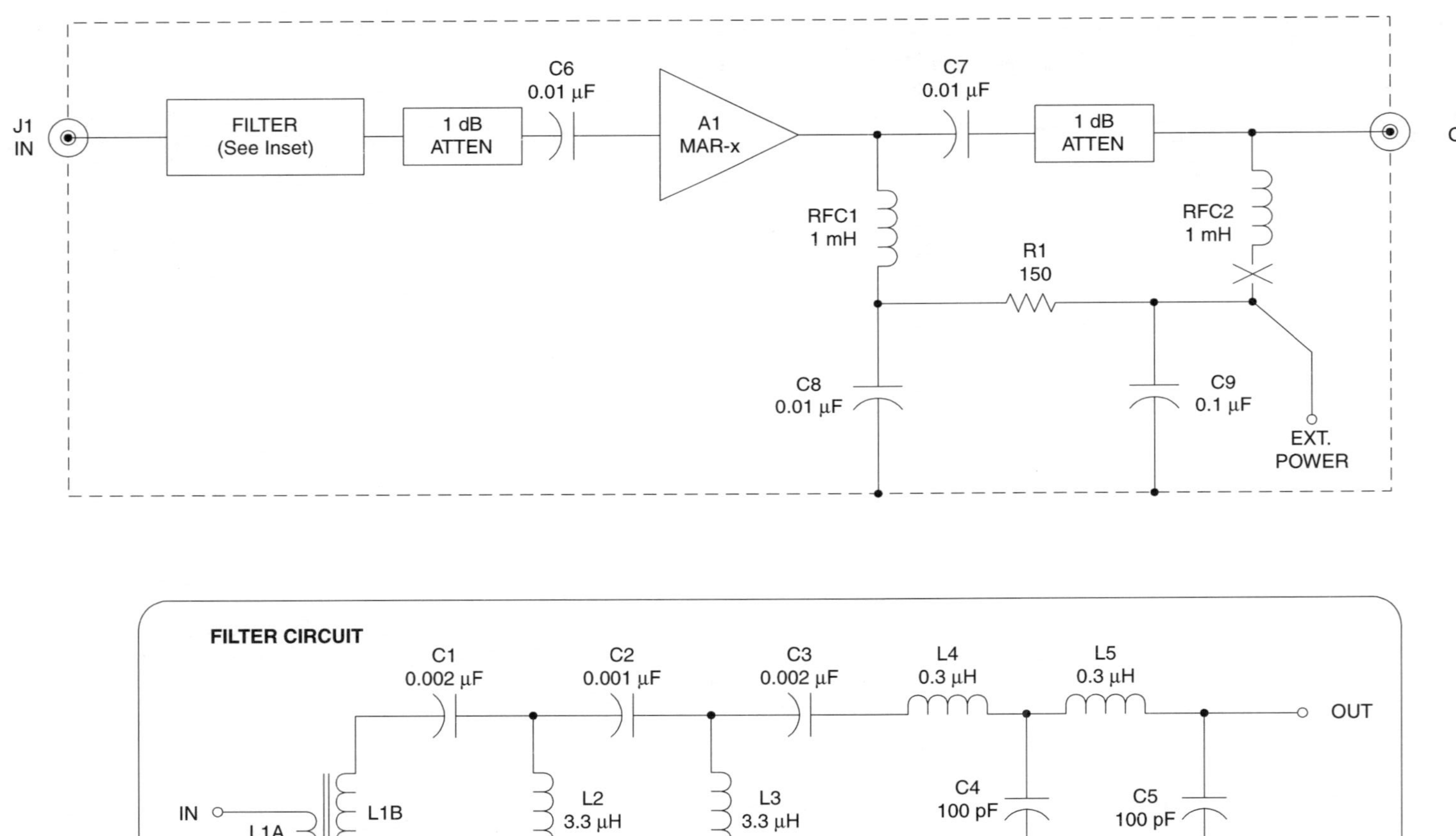

Figure 13.13 Filter inserted in the circuit.

Mast-mounted wideband preamplifier

This project describes a preamplifier, based on the MAR-1 device, that will work at remote locations and is powered from its own coaxial cable feedline.

Figure 13.12A shows the circuit of the remote portion. It consists of a standard MAR-1 device, but with optional –1 dB attenuators in series with input and output lines. The attenuators are used to enhance stability, but not all builders will want to use them. If you are willing to risk oscillation, and need the extra 2 dB of gain stolen by the attenuators, then delete them.

Note that the power supply end of resistor R1 is connected not to the power supply, but rather to the output coaxial connector (J2). A 1 mH RF choke (RFC2) isolates the DC circuit path from the RF flowing in the coaxial cable. Similarly, a DC blocking capacitor (C5) isolates the MAR-1 and the attenuator from the DC voltage on the coaxial line.

The entire remote circuit must be built inside a shielded weatherproof box. If no suitable box is available, then use an ordinary aluminium box, and coat the seams and joints with silicone seal, or something similar.

At the receiver end of the coaxial line a DC/RF combiner box is needed; a design is shown in Fig. 13.12B. The primary DC power in this case is a 9 VDC battery. Again, an RF choke isolates the DC power supply from the RF in the circuit, and a DC blocking capacitor (C9) is used to keep the DC voltage from the receiver.

Broadband HF amplifier

Figure 13.13 shows the MAR-1 device used in a broadband preamplifier for the 3 to 30 MHz high frequency (HF) bands. Like the previous circuit, it is intended for mast mounting, but if the power circuit is broken at 'X' can be used other than remotely. It can be powered by the circuit shown in Fig. 13.12B.

The key feature that differentiates this circuit from the previous circuit is the bandpass filter in the input circuit. It consists of a 1600 kHz high-pass filter followed by a 32 MHz low-pass filter. The circuit keeps strong, out-of-band signals from interfering with the operation of the preamplifier. Because the MAR-1 is a very wideband device, it will easily respond to AM and FM broadcast band signals.

Part 4

Measurement and techniques

14 Measuring inductors and capacitors

The measurement of the values of inductors (L) and capacitors (C) at radio frequencies differs somewhat from the same measurements at low frequencies. Although similarities exist, the RF measurement is a bit more complicated. One of the reasons for this situation is that stray or 'distributed' inductance and capacitance values of the test set-up will affect the results. Another reason is that capacitors and inductors are not ideal components, but rather all capacitors have some inductance, and all inductors have capacitance. In this chapter we will take a look at several methods for making such measurements.

VSWR method

When a load impedance ($R + jX$) is connected across an RF source the maximum power transfer occurs when the load impedance (Z_L) and source (Z_S) impedances are equal ($Z_L = Z_S$). If these impedances are not equal, then the *voltage standing wave ratio* (VSWR) will indicate the degree of mismatch. We can use this phenomenon to measure values of inductance and capacitance using the scheme shown in Fig. 14.1A. The instrumentation required includes a signal generator or other signal source, and a VSWR meter or VSWR analyser.

Some VSWR instruments require a transmitter for excitation, but others will accept the lower signal levels that can be produced by a signal generator. An alternative device is the SWR analyser type of instrument. It contains the signal generator and VSWR meter, along with a frequency counter to be sure of the actual test frequency. Whatever signal source is used, however, it must have a variable output frequency. Further, the frequency readout must be accurate.

The load impedance inside the shielded enclosure consists of a non-inductive resistor (R1) that has a resistance equal to the desired system impedance resistive component (50 ohms in most RF applications, and 75 ohms in television and video). An inductive reactance (X_L) and a capacitive reactance (X_C) are connected in series with the load. The circuit containing a resistor, capacitor and inductor simulates an antenna feedpoint impedance. The overall impedance is:

$$Z_L = \sqrt{R^2 + (X_L - X_c)^2} \tag{14.1}$$

Note the reactive portion of Eq. (14.1). When the condition $|X_L| = |X_C|$ exists, the series network is at resonance, and VSWR is minimum (Fig. 14.1B). This gives us a means for

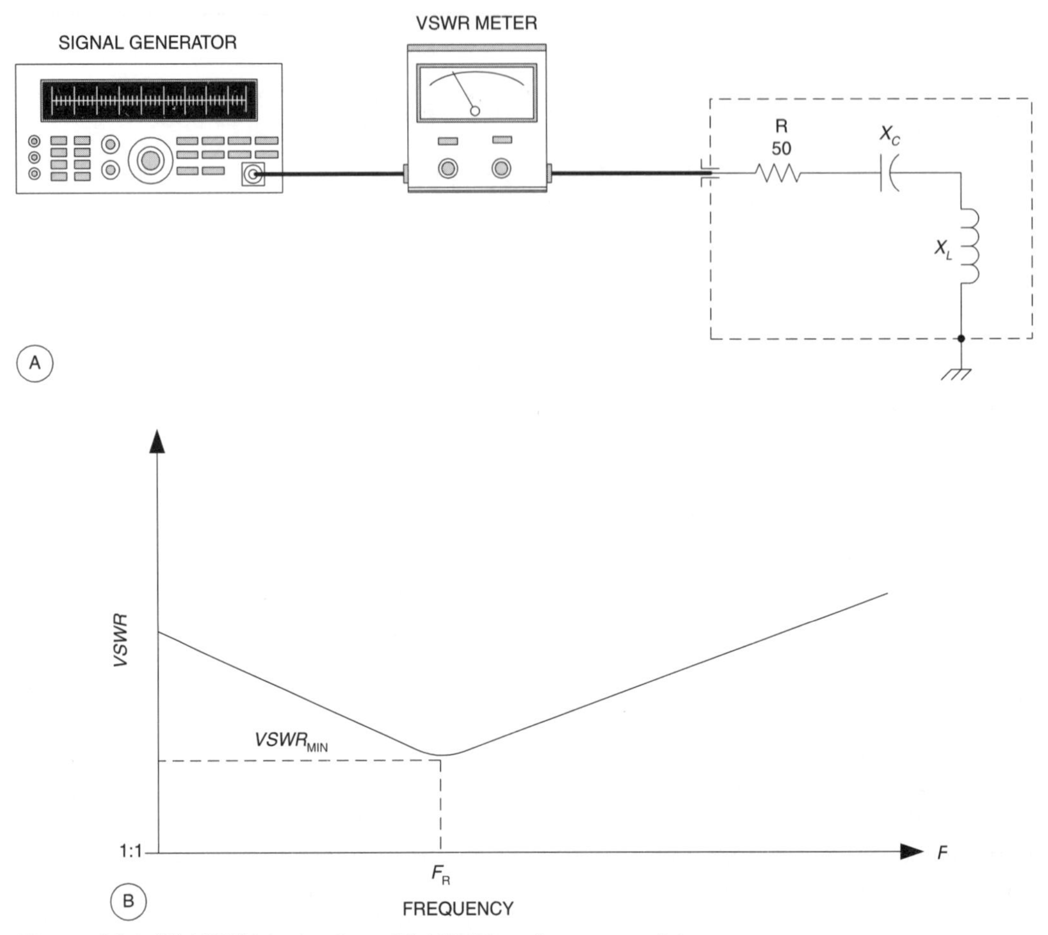

Figure 14.1 (A) VSWR test set-up; (B) VSWR-vs-frequency plot.

measuring the values of the capacitor or inductor, provided that the other is known. For example, if you want to measure a capacitance, then use an inductor of known value.

Using the test set-up in Fig. 14.1A, adjust the frequency of the signal source to produce minimum VSWR.

1 For finding an inductance from a known capacitance:

$$L_{\mu H} = \frac{10^6}{4\pi^2 f^2 C_{pF}} \tag{14.2}$$

Where:
$L_{\mu H}$ = inductance in microhenrys (μH)
C_{pF} is the capacitance in picofarads (pF)
f is the frequency in MHz

2 For finding a capacitance from a known inductance:

$$C_{pF} = \frac{10^6}{4\pi^2 f^2 L_{\mu H}} \qquad (14.3)$$

The accuracy of this approach depends on how accurately the frequency and the reactance are known, and how accurately the minimum VSWR frequency can be found.

Voltage divider method

A resistive voltage divider is shown in Fig. 14.2A. The voltage drops across R1 and R2 are $V1$ and $V2$, respectively. We know that either voltage drop is found from:

$$V_x = \frac{VR_x}{R1 + R2} \qquad (14.4)$$

Where:
V_X is $V1$ and R_X is $R1$ or, V_X is $V2$ and R_X is $R2$, depending on which voltage drop is being measured

We can use the voltage divider concept to find either inductance or capacitance by replacing R2 with the unknown reactance. In Fig. 14.2B resistor R2 has been replaced by an inductor (L). The resistor R_S is the inductor series resistance. If we measure the voltage drop across R1 (i.e. 'E' in Fig. 14.2B), then we can calculate the inductance from:

$$L = \frac{R}{2\pi f} \times \sqrt{\left(\frac{V}{E}\right)^2 - \left(1 + \frac{R_S}{R1}\right)^2} \qquad (14.5)$$

As can be noted in Eq. (14.5) if $R1 >> R_S$, then the quotient $R_S/R1$ becomes negligible.

In capacitors the series resistance is typically too small to be of consequence. We can replace L in the model of Fig. 14.2B with a capacitor, and again measure voltage E. The value of the capacitor will be:

$$C = \frac{2\pi f}{R \times \sqrt{\left(\frac{V}{E}\right)^2 - 1}} \qquad (14.6)$$

The value of resistance selected for R1 should be similar to, or a bit smaller than, the expected reactance of the capacitor or inductor being measured. This will keep the voltage values manageable, and help maintain accuracy.

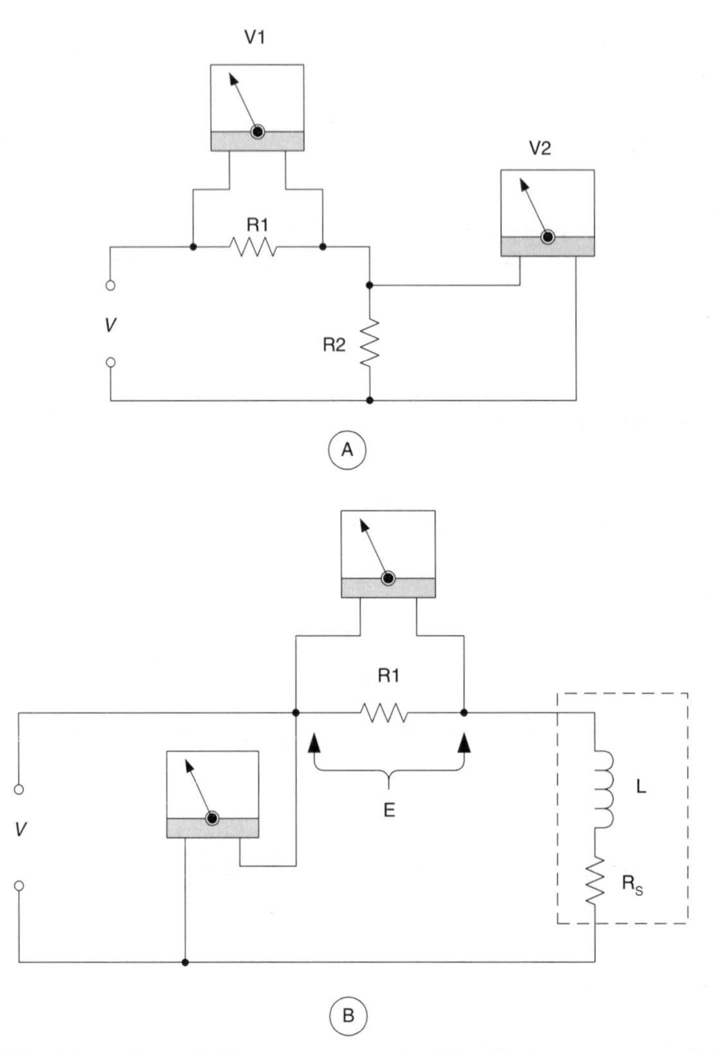

Figure 14.2 (A) Resistor voltage divider measurements; (B) voltage measurements in complex impedance situation.

Signal generator method

If the frequency of a signal generator is accurately known, then we can use a known inductance to find an unknown capacitance, or a known capacitance to find an unknown inductance. Figure 14.3 shows the test set-up for this option. The known and unknown components (L and C) are connected together inside a shielded enclosure. The parallel tuned circuit is lightly coupled to the signal source and the display through very low value capacitors (C1 and C2). The rule is that the reactance of C1 and C2 should be very high compared with the reactances of L and C at resonance.

The signal generator is equipped with a 6 dB resistive attenuator in order to keep its output impedance stable. The output indicator can be any instrument that will read the RF

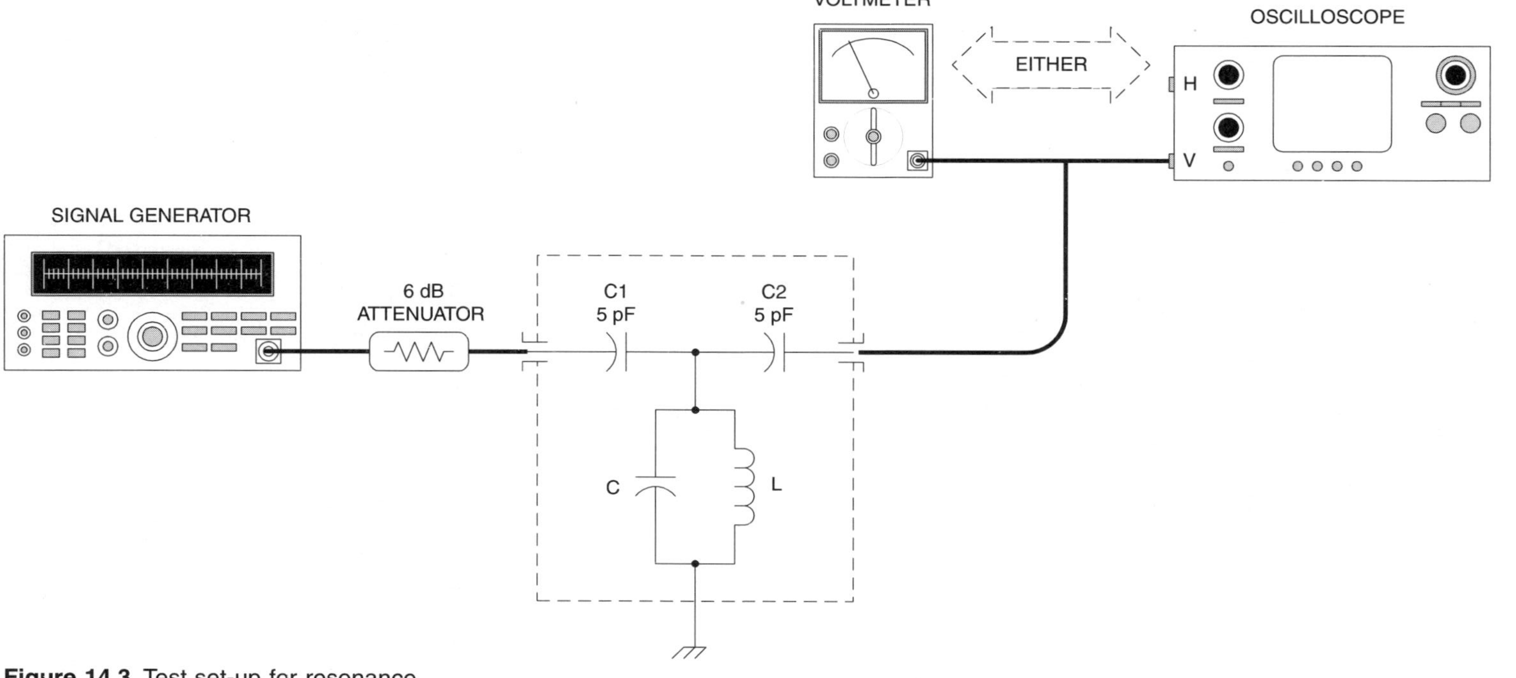

Figure 14.3 Test set-up for resonance.

voltage at the frequency of resonance. For example, you could use either an RF voltmeter or an oscilloscope.

The procedure requires tuning the frequency of the signal source to provide a peak output voltage reading on the voltmeter or 'scope. If the value of one of the components (L or C) is known, then the value of the other can be calculated using Eqns (14.2) or (14.3), as appropriate.

Alternate forms of coupling are shown in Fig. 14.4. In either case, the idea is to isolate the instruments from the L and C elements. In Fig. 14.4A the isolation is provided by a pair of high value (10 k to 1 meg) resistors, R1 and R2. In Fig. 14.4B the coupling and isolation are provided by a one- or two-turn link winding near the inductor. The links and the main inductor are lightly coupled to each other.

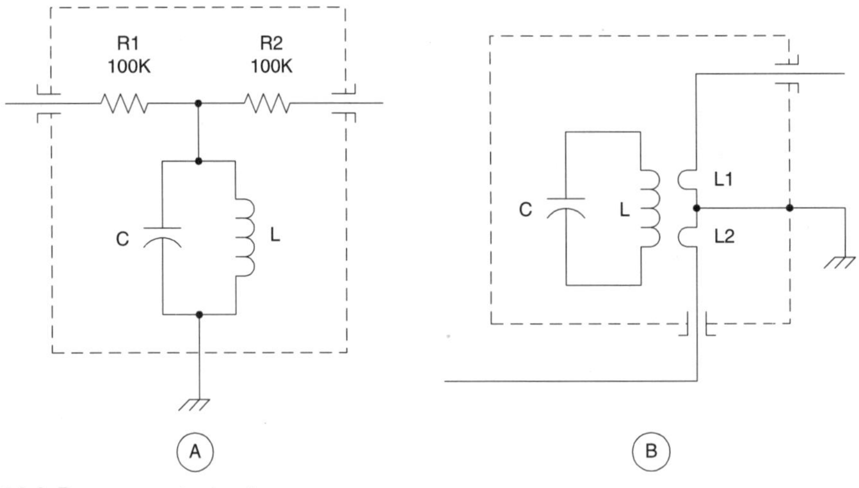

Figure 14.4 Resonance test set-up.

Frequency shifted oscillator method

The frequency of a variable frequency oscillator (VFO) is set by the combined action of an inductor and a capacitor. We know that a change in either capacitance or inductance produces a frequency change equal to the square root of the component ratio. For example, for an inductance change:

$$L2 = L1 \times \left[\left(\frac{F1}{F2} \right)^2 - 1 \right] \tag{14.7}$$

Where:

$L1$ is the original inductance
$L2$ is the extra inductance
$F1$ is the original frequency
$F2$ is the new frequency

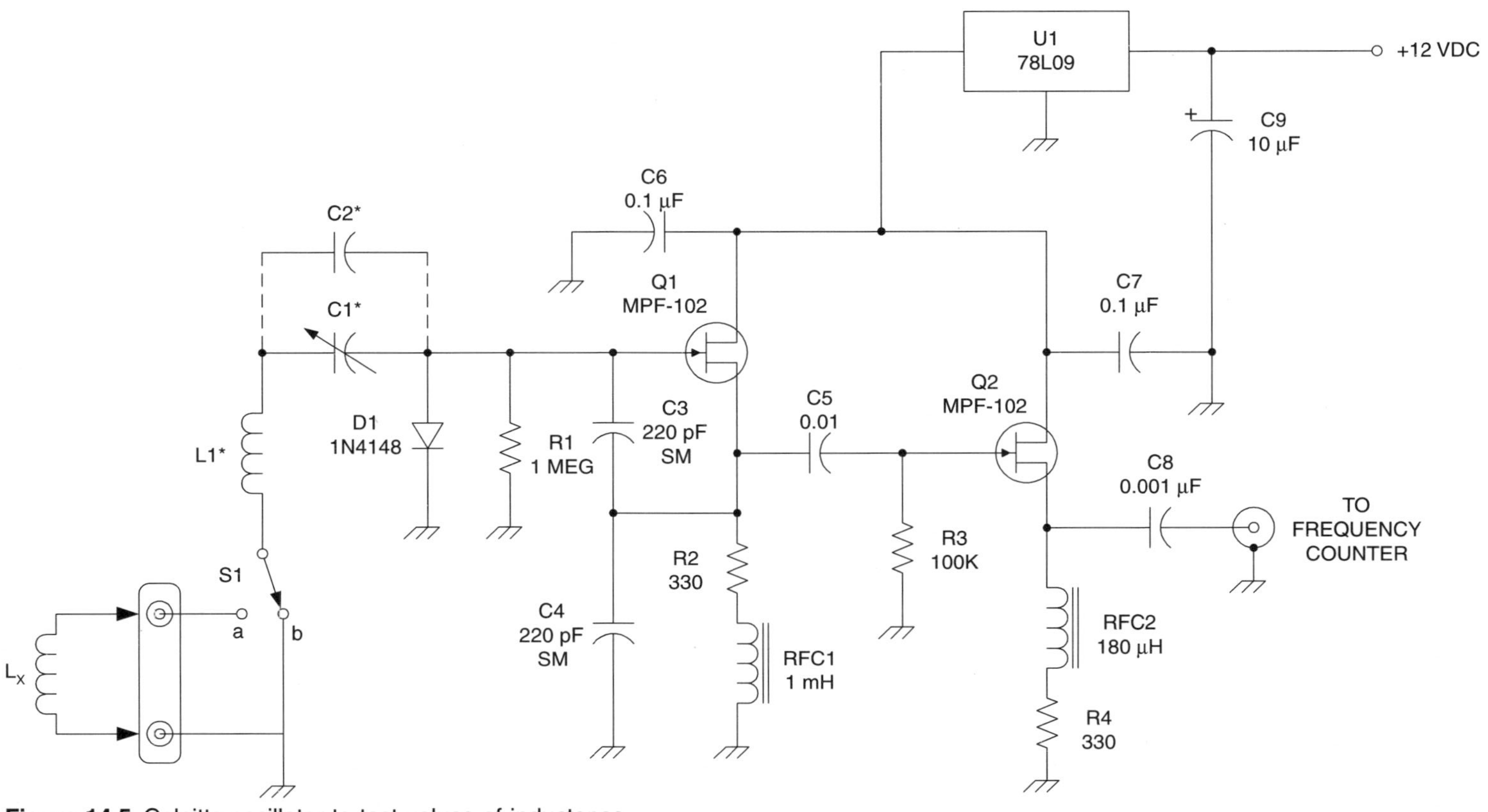

Figure 14.5 Colpitts oscillator to test values of inductance.

From this equation we can construct an inductance meter such as Fig. 14.5. This circuit is a Clapp oscillator designed to oscillate in the high frequency (HF) range up to about 12 MHz. The components L1, C1 and C2 are selected to resonate at some frequency. Inductor L1 should be of the same order of magnitude as L_X. The idea is to connect the unknown inductor across the test fixture terminals. Switch S1 is set to position 'b' and the frequency ($F1$) is measured on a digital frequency counter. The switch is then set to position 'a' in order to put the unknown inductance L_X in series with the known inductance $L1$. The oscillator output frequency will shift to $F2$. When we know $L1$, $F1$ and $F2$ we can apply Eq. (14.7) to calculate $L2$.

If we need to find a capacitance, then modify the circuit to permit a capacitance to be switched into the circuit across C1 instead of an inductance as shown in Fig. 14.5. Replace the 'L' terms in Eq. (14.7) with the corresponding 'C' terms.

Using RF bridges

Most RF bridges are based on the *Wheatstone bridge* circuit (Fig. 14.6). In use since 1843, the Wheatstone bridge has formed the basis for many different measurement instruments. The *null condition* of the Wheatstone bridge exists when the voltage drop of R1/R2 is equal to the voltage drop of R3/R4; then the voltmeter (M1) will read zero. The basic measurement scheme is to know the values of three of the resistors, and use them to measure the value of the fourth.

The Wheatstone bridge works well for finding unknown resistances from DC up to relatively low RF frequencies, but to measure L and C values at higher frequencies we need to modify the bridges.

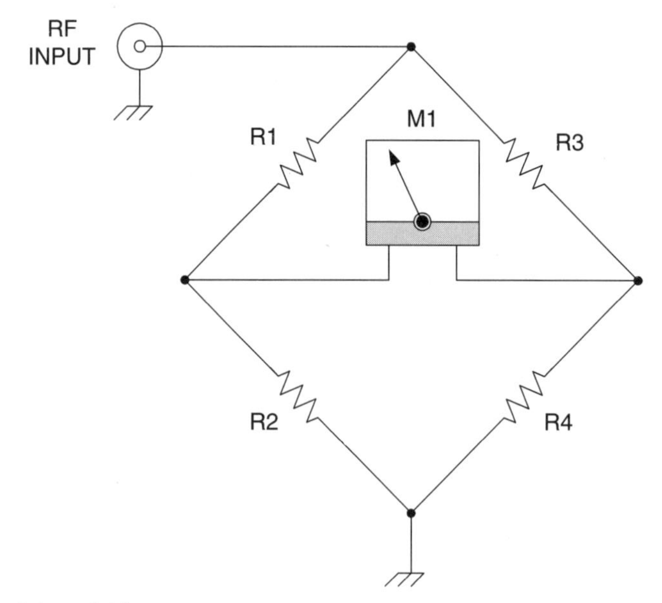

Figure 14.6 Wheatstone bridge.

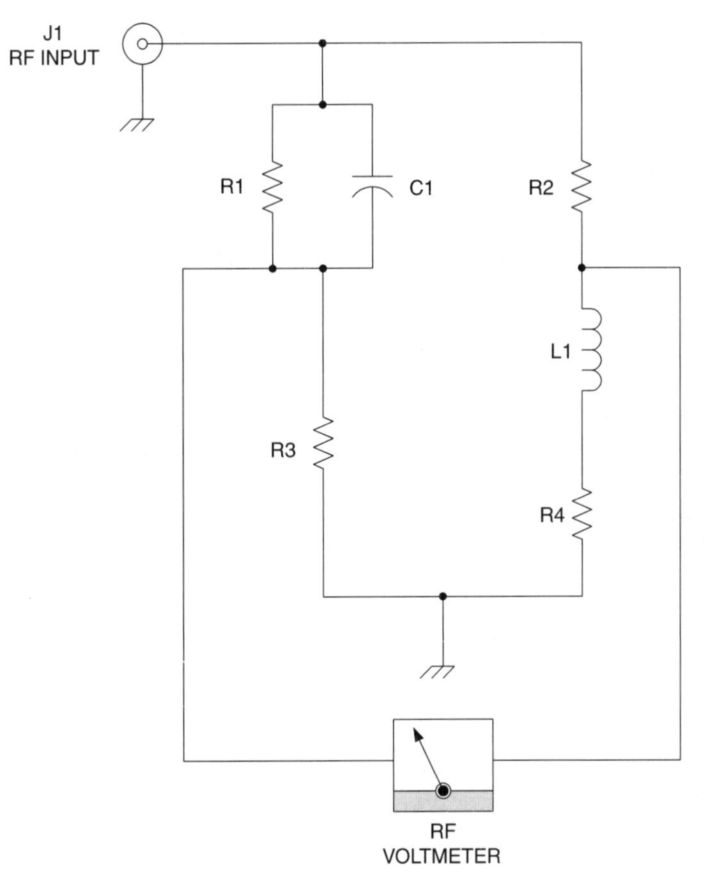

Figure 14.7 Maxwell bridge.

Maxwell bridge

The Maxwell bridge is shown in Fig. 14.7. The null condition for this bridge occurs when:

$$L1 = R2 \times R3 \times C1 \tag{14.8}$$

and,

$$R4 = \frac{R2 \times R3}{R1} \tag{14.9}$$

Note that the balance equations are totally independent of frequency. The bridge is also not too sensitive to resistive losses in the inductor (a failing of some other methods). Additionally, it is much easier to obtain calibrated standard capacitors for C1 than it is to obtain standard inductors for L1. As a result, one of the principal uses of this bridge is inductance measurements.

Maxwell bridge circuits are often used in measurement instruments called *Q-meters*, which measure the quality factor (Q) of inductors. The equation for Q is, however, frequency sensitive:

$$Q = 2\pi \times F \times R1 \times C1 \tag{14.10}$$

Where:
F is in hertz
$R1$ is in ohms
C1 is in farads

Hay bridge

The Hay bridge (Fig. 14.8) is similar to the Maxwell bridge, except that the R1/C1 combination is connected in series rather than parallel. The Hay bridge is, unlike the Maxwell bridge, frequency sensitive. The balance equations for the null condition are also a little more complex:

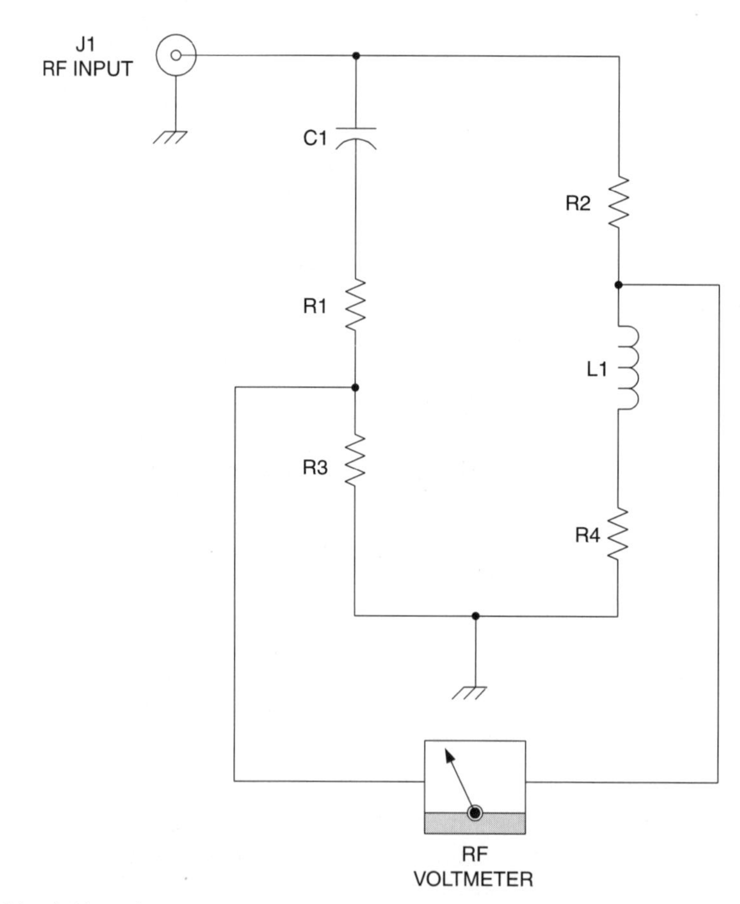

Figure 14.8 Hay bridge.

$$L1 = \frac{R2 \times R3 \times C1}{1 + \left[\frac{1}{Q}\right]^2} \tag{14.11}$$

$$R4 = \left[\frac{R2 \times R3}{R1}\right] \times \left[\frac{1}{Q^2 + 1}\right] \tag{14.12}$$

Where:

$$Q = \frac{1}{\omega \times R1 \times C1} \tag{14.13}$$

The Hay bridge is used for measuring inductances with high Q figures, while the Maxwell bridge is best with inductors that have a low Q value.

Note: A frequency independent version of Eq. (14.11) is possible when Q is large:

$$L1 = R2 \times R3 \times C1 \tag{14.14}$$

Schering bridge

The Schering bridge circuit is shown in Fig. 14.9. The balance equation for the null condition is:

$$C3 = \frac{C2 \times R1}{R2} \tag{14.15}$$

$$R3 = \frac{R2 \times C1}{C2} \tag{14.16}$$

The Schering bridge is used primarily for finding the capacitance and the power factor of capacitors. In the latter applications no actual R3 is connected into the circuit, making the series resistance of the capacitor being tested (e.g. C3) the only resistance in that arm of the bridge. The capacitor's Q factor is found from:

$$Q_{C3} = \frac{1}{\omega \times R1 \times C1} \tag{14.17}$$

Finding parasitic capacitances and inductances

Capacitors and inductors are not ideal components. A capacitor will have a certain amount of series inductance (called 'parasitic inductance'). This inductance is created by the conductors in the capacitor, especially the leads. In older forms of capacitor, such as the wax paper dielectric devices used prior to about 1960, the series inductance was very

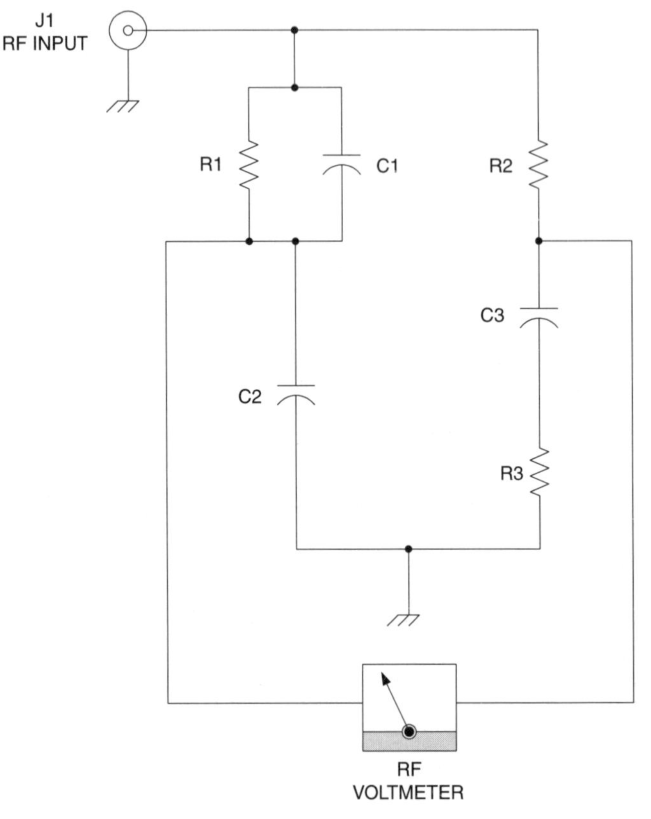

Figure 14.9 Schering bridge.

large. Because the inductance is in series with the capacitance of the capacitor, it forms a series resonant circuit.

Figure 14.10 shows a test set-up for finding the series resonant frequency. A *tracking generator* is a special form of sweep generator that is synchronized to the frequency sweep of a spectrum analyser. They are used with spectrum analysers in order to perform stimulus response measurements such as Fig. 14.10.

The nature of a series resonant circuit is to present a low impedance at the resonant frequency, and a high impedance at all frequencies removed from resonance. In this case, that impedance is across the signal line. The display on the spectrum analyser will show a pronounced, sharp dip at the frequency where the capacitance and the parasitic inductance are resonant. The value of the parasitic series inductance can be found from Eq. (14.2).

Inductors are also less than ideal. The adjacent turns of wire form small capacitors, which when summed up can result in a relatively large capacitance value. Figure 14.11 shows a method for measuring the parallel capacitance of an inductor.

Because the capacitance is in parallel with the inductance, it forms a parallel resonant circuit. These circuits will produce an impedance that is very high at the resonant

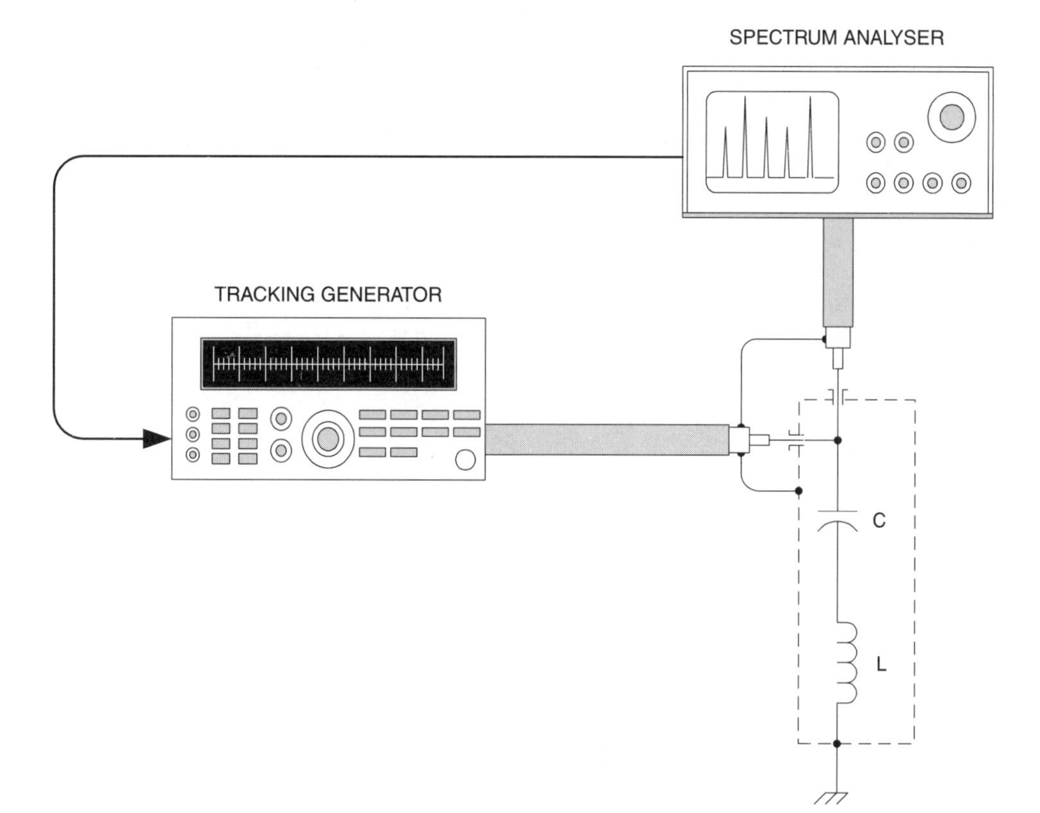

Figure 14.10 Measuring series parasitics.

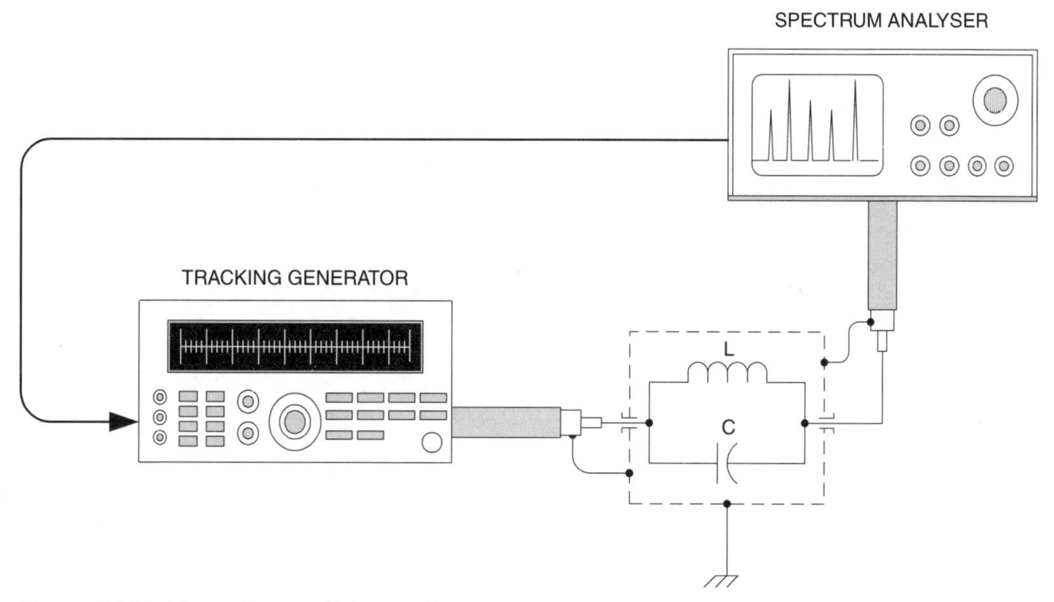

Figure 14.11 Measuring parallel parasitics.

frequency, and very low at frequencies removed from resonance. In Fig. 14.11 the inductor and its parasitic parallel capacitance are in series with the signal line, so will (like the other circuit) produce a pronounced dip in the signal at the resonant frequency. The value of the parasitic capacitance can be found from Eq. (14.3).

Conclusion

There are other forms of bridge, and other methods, for measuring L and C elements in RF circuits, but those discussed above are very practical, especially in the absence of specialist instrumentation.

15 RF power measurement

Power units

Electrical power is defined as *energy flow per unit of time*. The internationally accepted standard unit of power is the *watt* (W), which is defined as an energy flow of *one joule per second*. The unit watt is not always suitable for some applications. As a result, it is common to find sub-units and super-units in use (e.g. milliwatts, kilowatts and megawatts).

It is common practice to express power relationships in terms of *decibel notation* (dB), which allows gains and losses to be added and subtracted, rather than multiplied and divided, somewhat simplifying the arithmetic. In low power RF circuitry a useful power unit is dBm, which is the power level in dB relative to 1 mW.

Relative power levels:

$$dB = 10 \log \left[\frac{P1}{P2} \right] \tag{15.1}$$

Absolute power levels

$$dBm = 10 \log \left[\frac{P_W}{0.001} \right] \tag{15.2}$$

or,

$$dBm = 10 \log P_{mW} \tag{15.3}$$

Where:

dBm is power level relative to one milliwatt
$P1$ and $P2$ are two power levels (same units)
P_W is power in watts (W)
P_{mW} is power in milliwatts (mW)

Types of RF power measurement

Measuring RF power is essentially the same as measuring low frequency AC power, but certain additional problems present themselves. For a continuous wave (CW) signal, the issue is relatively straightforward because the signal is a constant amplitude sine wave. For on/off telegraphy, the problem gets somewhat more difficult because the waves are not constant amplitude, and the average RF power depends on the ratio of on time to off time. In the case of the sine wave, a peak voltage reading instrument such as a diode detector can be calibrated for root mean square (rms) power by the simple expedient of dividing the indication by the square root of two (1.414). If the meter measures power directly (e.g. a thermally based instrument), then no waveform correction factor is needed.

Methods for measuring RF power

RF power meters use a number of different approaches to making the measurement. Some instruments measure the current or the voltage at a resistive load, and depend on the equations $P = I^2R$ or $P = V^2/R$.

Other methods are based on the fact that power dissipated in a resistive load is converted to heat, so the *temperature change* before and after the RF power is applied can be used as the indicator of RF power. This approach has the advantage of directly measuring power, rather than deducing it.

Figure 15.1 shows the basic scheme. A load resistor, R_0, with a resistance equal to the system impedance, is enclosed in an isolated environment with some sort of temperature

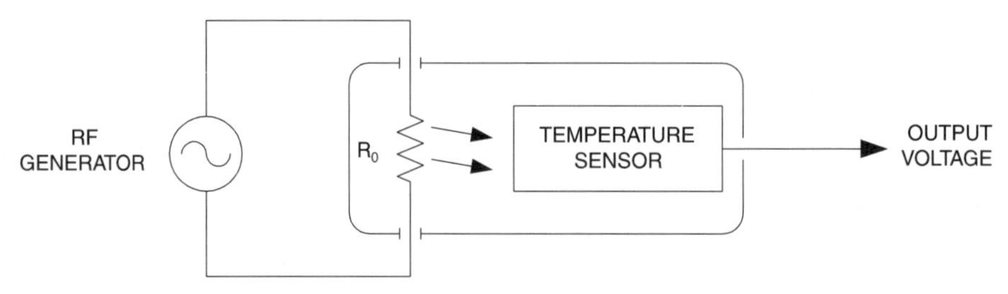

Figure 15.1 Thermal sensing circuit.

sensor. Theoretically, one could place a dummy load resistor in a workshop room, and then use a glass mercury thermometer and stopwatch to measure the rise in temperature and elapsed time to find the power. That's hardly practical, however. The basic idea is to find a sensor, such as a thermistor or thermocouple, that will convert the heat generated in the load resistor to a DC (or low frequency AC) signal that is easily measured with ordinary electronic instruments.

Thermistor RF power meters

A *thermistor* is a resistor that has a large change of its electrical resistance with changes in temperature. Thermistors are usually made of a metallic oxide compound. Figure 15.2A shows the resistance-vs-temperature curve for a typical thermistor device. A *negative temperature coefficient* (NTC) device will decrease resistance with increases in temperature.

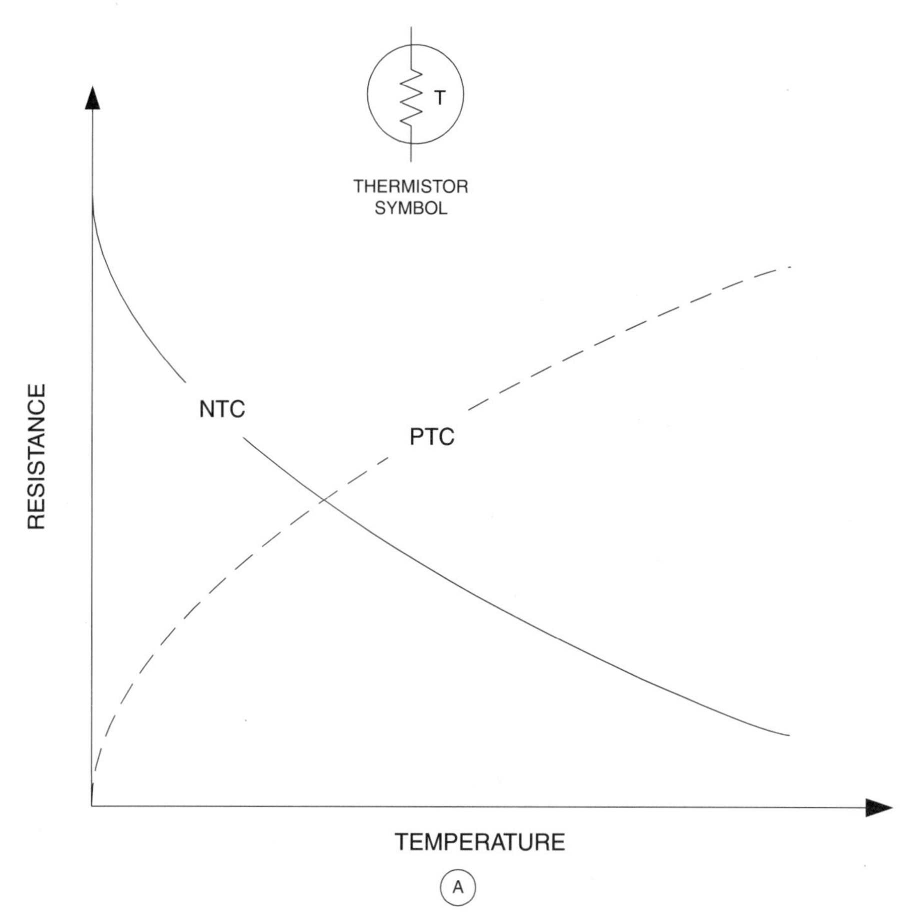

Figure 15.2 (A) Positive and negative temperature coefficients;

A *positive temperature coefficient* (PTC) device is the opposite: resistance rises with increases in temperature.

Bolometers

Figure 15.2B shows a family of resistance-vs-self-heating power curves for a single thermistor operated at different temperatures. The resistance is not only non-linear, which makes measurements difficult enough in its own right, the shape and placement of the curve varies with temperature. As a result, straight thermistor instruments can be misleading. *Bolometry* is a method that takes advantage of this problem to create a more accurate RF power measurement system.

Self-heating power is caused by a DC current flowing in the thermistor. Figure 15.2C shows how self-heating can be used in bolometry. The thermistor (RT1) is adjusted to a specified self-heating point when no RF power is applied to the dummy load R_0. The resistance of thermistor RT1 can be calculated from the voltmeter because the DC current is also known.

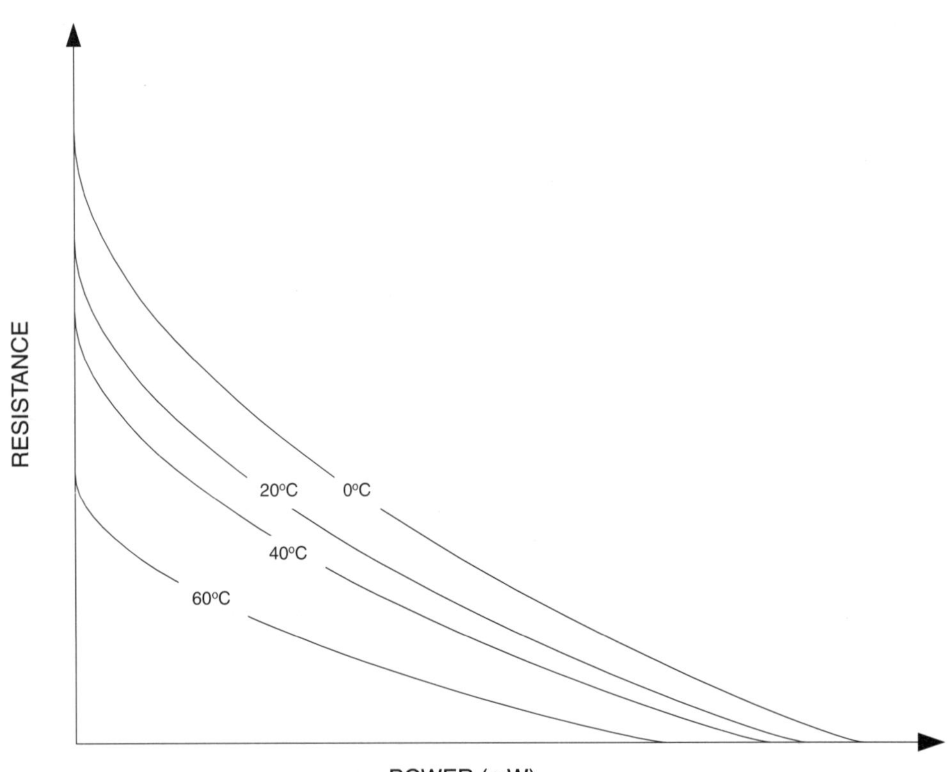

Figure 15.2 (B) resistance at various power and temperature levels;

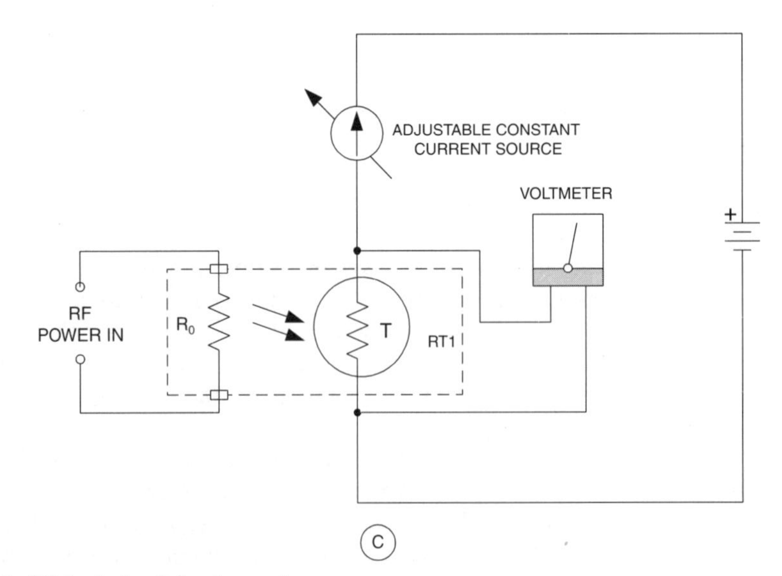

Figure 15.2 (C) test circuit for thermal sensor.

When RF power is applied to the dummy load, heat radiated from the load will cause the resistance of RT1 to decrease. The bolometer current source is then adjusted to decrease the bias until the resistance rises back to the value it had before power was applied. The change of bias power required to restore the thermistor to the same resistance is therefore equal to the power dissipated in the dummy load.

Self-balancing bridge instruments

The self-balancing (aka autobalancing or autonull) bridge shown in Fig. 15.3 uses a Wheatstone bridge thermistor to perform bolometry measurement of RF power. The thermistor mount sensor assembly contains a dummy load and a thermistor (R_T). The null condition is created when $R1/R_T = R2/R3$.

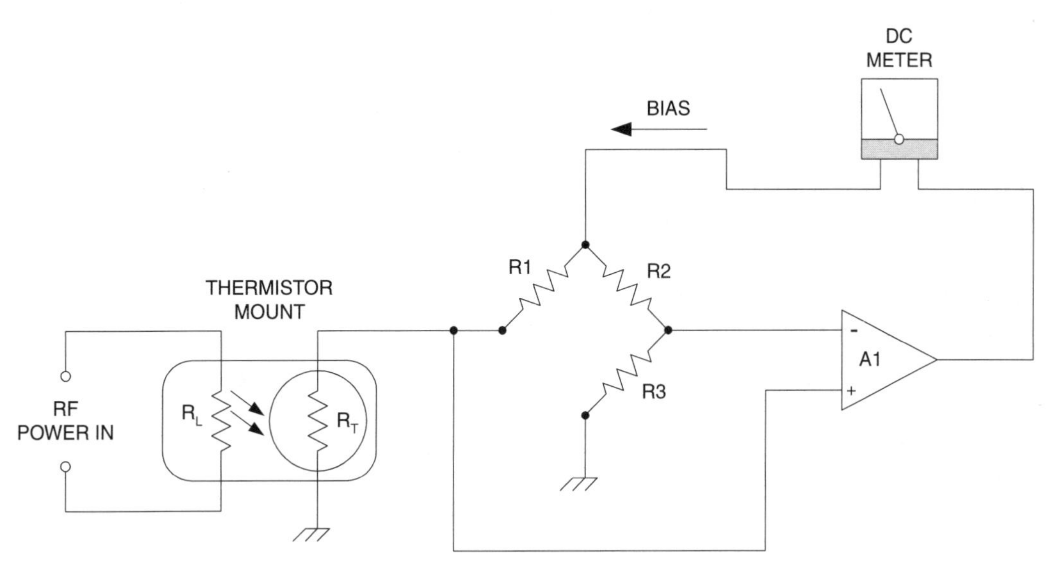

Figure 15.3 Wheatstone bridge circuit.

The self-balancing bridge uses a differential amplifier (A1) to perform the balancing. A differential amplifier produces an output voltage proportional to the difference in two input voltages. When the Wheatstone bridge is in balance, then the output of the differential amplifier is zero. The bias for the Wheatstone bridge, hence the thermistor in the bolometer sensor, is derived from the output of the amplifier. A change in the resistance of the thermistor unbalances the bridge, and this moves the amplifier's differential input voltage away from zero. The amplifier output voltage goes up, thereby changing the bias current by an amount and direction necessary to restore balance. Thus, by reading the bias current we can infer the power level that changed the thermistor resistance.

Because the thermistor will have a different characteristic curve at different ambient temperatures it is necessary to either control the ambient temperature, or correct for it. It is very difficult to control the ambient temperature. Although it is done, it is also not terribly practical in most cases. As a result, it is common to find RF power meters using

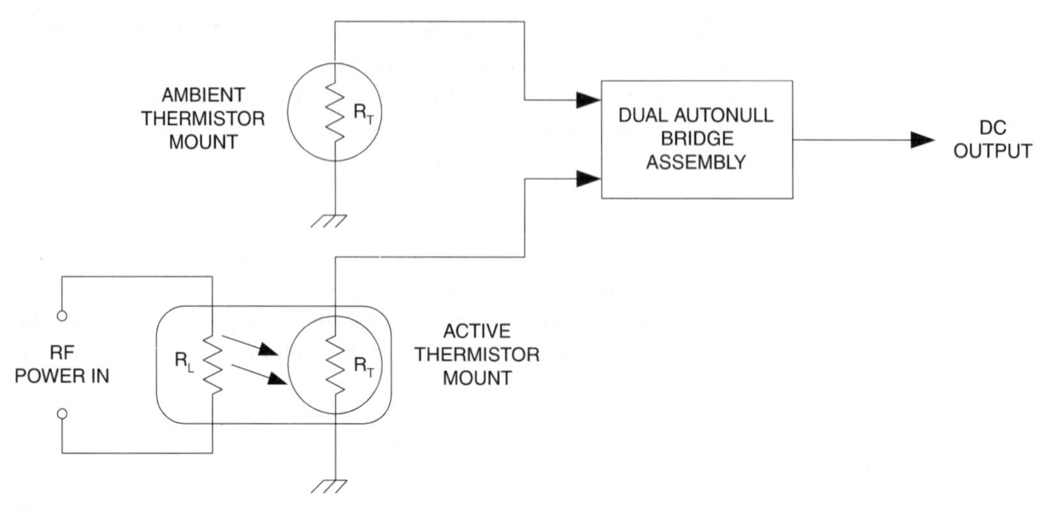

Figure 15.4 Dual auto-null bridge.

two thermistors in the measurement process (Fig. 15.4). One thermistor is mounted in the thermistor sensor mount used to measure RF power, while the other is used to measure the ambient temperature. The readings of the ambient thermistor are used to correct the readings of the sensor thermistor.

Thermocouple RF power meters

The thermocouple is one of the oldest forms of temperature sensor. When two *dissimilar metals* are connected together to form a junction, and the junction is heated, then the potential across the free ends (V_T) is proportional to the temperature of the hot junction. This phenomenon is called the *Seebeck effect*.

A thermocouple RF ammeter is constructed using thermocouples and a small value resistance heating element (Fig. 15.5). The meter will have a small wire resistance element in close proximity to a thermocouple element. The thermocouple is, in turn, connected to

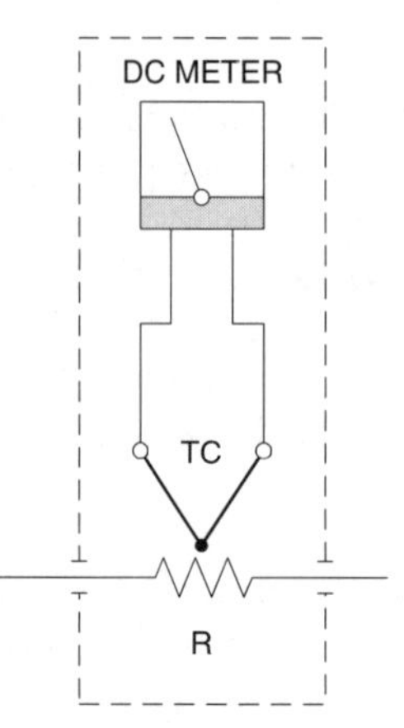

Figure 15.5 Thermocouple RF ammeter.

a DC meter. When current flows through the resistance heating element, the potential across the ends of the thermocouple changes proportional to the rms value of the current. Thus, the RF ammeter measures the rms value of the RF current.

If the RF ammeter is used to measure the current flowing from an RF source to a resistive load, then the product I^2R indicates the true RF power. These meters can measure RF current up to 50 or 60 MHz, depending on the instrument.

Thermocouples and thermistors share the ability to measure true RF power. Although thermocouple RF ammeters have been used since the 1930s, or earlier, the use of thermocouples in higher frequency and microwave power meters started in the 1970s. Thermocouples are more sensitive than thermistor sensors, and are inherently square law devices.

Figure 15.6 shows a solid-state thermocouple sensor that can be used well into the microwave region. Two semiconductor thermocouples are connected such that they are in

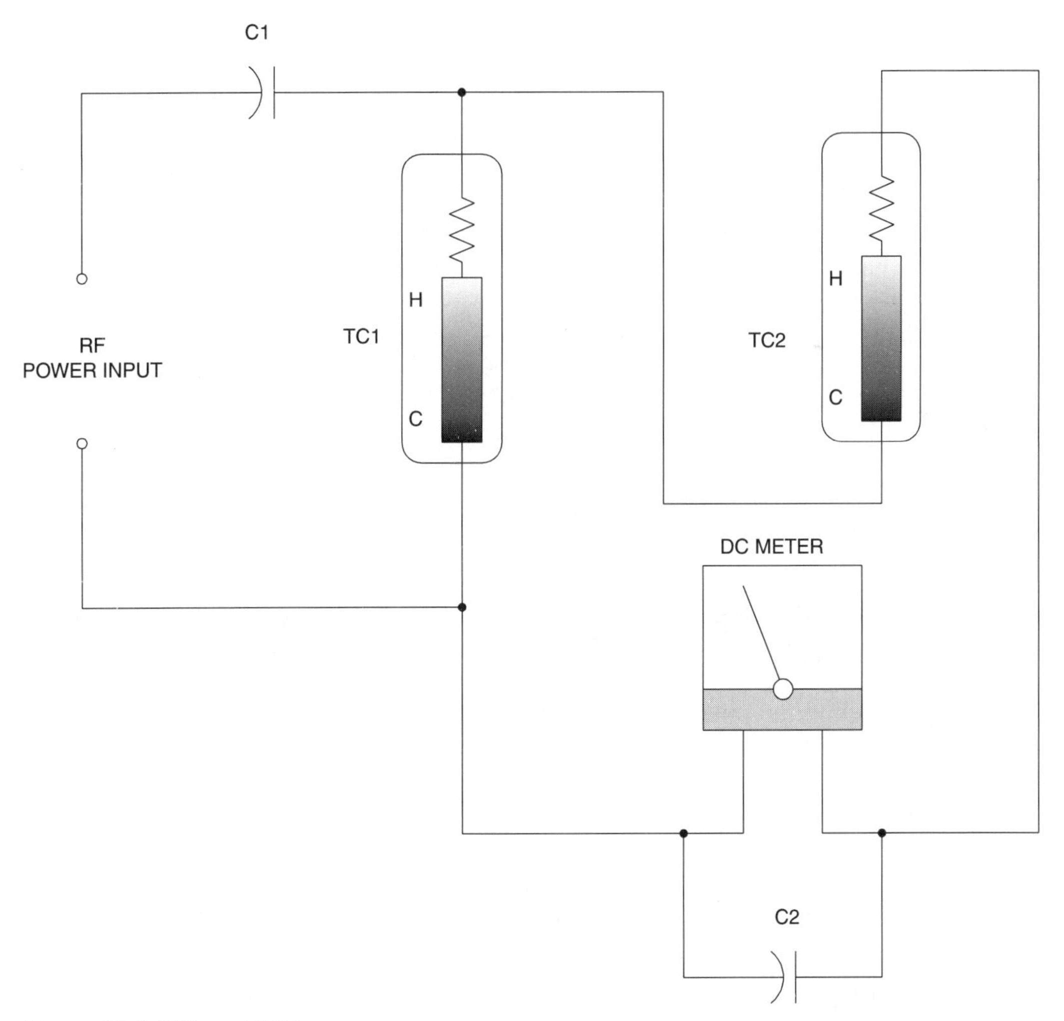

Figure 15.6 TC1 and TC2 power sensor.

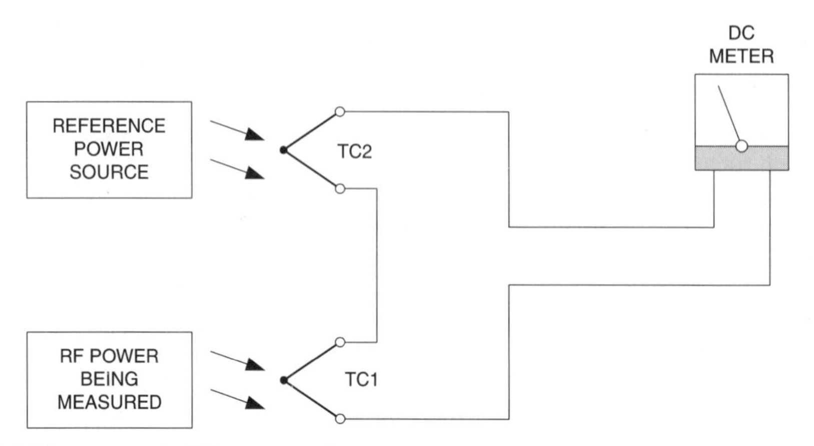

Figure 15.7 Thermocouple RF power meter.

series for DC, and parallel for RF frequencies. Thus, their combined output voltages are read on the DC voltmeter. Because of the capacitors, however, they are in parallel for RF frequencies, and if designed correctly will make a 50 ohm termination for a transmission line.

Thermocouples suffer the same reliance on knowing the ambient temperature as thermistors. Figure 15.7 shows a method for overcoming this problem. A pair of thermistors are used. One is used either in a bolometry circuit or as a terminating sensor to measure the unknown RF power. The other sensor is used to measure a highly controlled reference power source. Depending on the implementation, the reference power might be DC, low frequency AC or another RF oscillator with a highly controlled, accurately calibrated output power level.

Diode detector RF power meters

Rectifying diodes convert bi-directional alternating current to unidirectional pulsating DC. When filtered, the output side of a diode is a DC level that is proportional to the amplitude of the applied AC signal. Figure 15.8 shows the unidirectional action in the form of the *I-vs-V* curve. When the applied bias is positive ('forward bias') the current will begin to flow, but not proportionally. At some point (200 to 300 mV in germanium diodes and 600 to 700 mV in silicon diodes), marked $V\gamma$ in Fig. 15.8, the response enters a linear region.

When the applied voltage reverse biases the diode, the current flow ceases, except for a very small *leakage current* (I_L). One indicator of the quality of diodes is that I_L is minimized on higher quality units.

The non-linear region of the *I*-vs-*V* curve is called the *square law region*. In this region the rectified output voltage from the diode is proportional to the input power (Fig. 15.9). This behaviour is seen from power levels of –70 to –20 dBm.

In low cost RF power measuring instruments silicon and even germanium diodes are often used, but these are not highly regarded for professional measurements. Low barrier

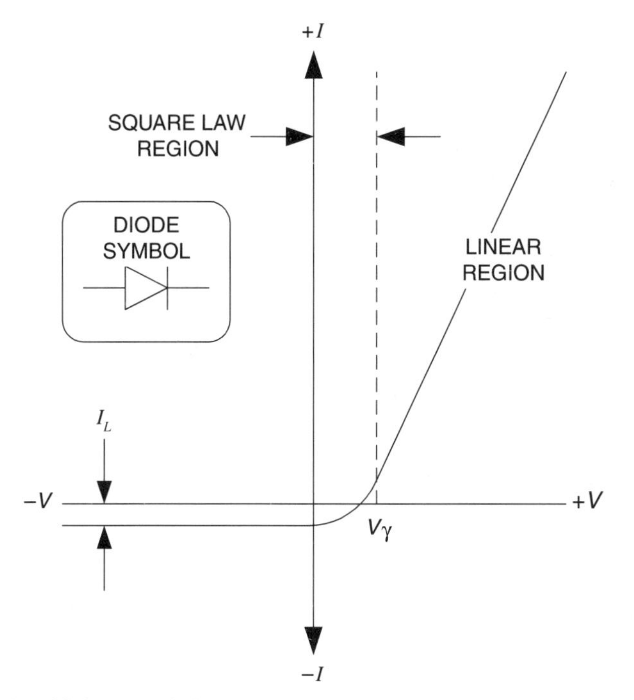

Figure 15.8 Diode *I*-vs-*V* characteristic.

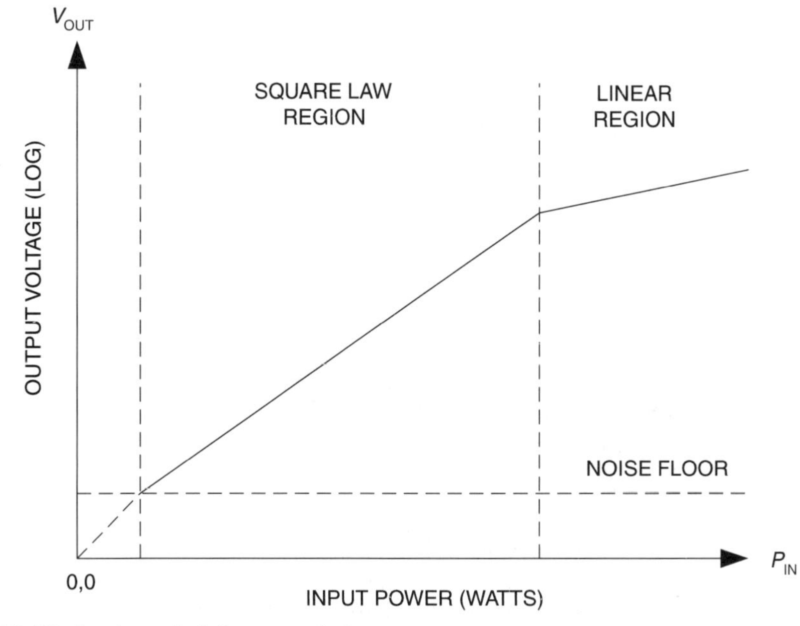

Figure 15.9 Diode characteristic expanded.

Schottky diodes are widely used up to well into the microwave region. For higher frequencies in the microwave region planar doped barrier (PDB) diodes are preferred. They work up to 18 GHz or better, and power levels of –70 dBm. It is claimed that PDB diodes are more than 3000 times more efficient than thermocouple detectors.

Circuits

Figures 15.10A and 15.10B show two similar circuits using a diode detector. Resistor R_L is a *dummy load* that has a resistance value equal to the characteristic impedance of the transmission line connecting the system (e.g. 50 ohms). Diode D1 is the rectifier diode, while capacitor C1 is used to filter the pulsations at the rectifier output into pure DC. A problem with this circuit is that it is limited to power levels consistent with the native characteristics of the diode.

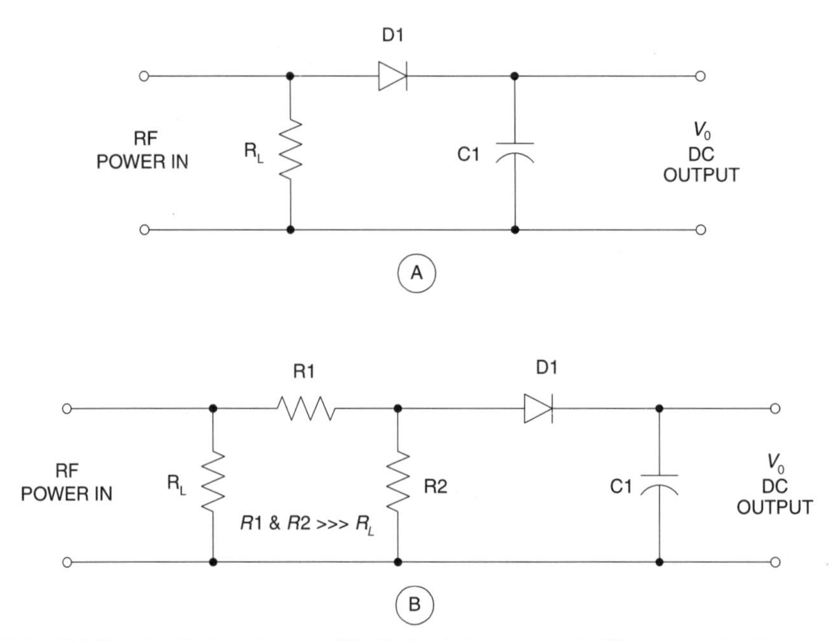

Figure 15.10 (A) Simple diode detector; (B) diode detector used with dummy load.

Figure 15.10B shows the same circuit with a resistor voltage divider (R1/R2) to reduce the voltage associated with higher power levels to the characteristic of the diode. The actual voltage applied to the diode will be $V_{RL} \times [R2/(R1 + R2)]$. This circuit is similar to the metering circuit built into a number of low cost amateur radio dummy loads.

Practical in-line bridge circuits

Low cost in-line RF power meters (Fig. 15.11) are available using a number of different forms of bridge circuit. These are superior to the classical Wheatstone bridge because they

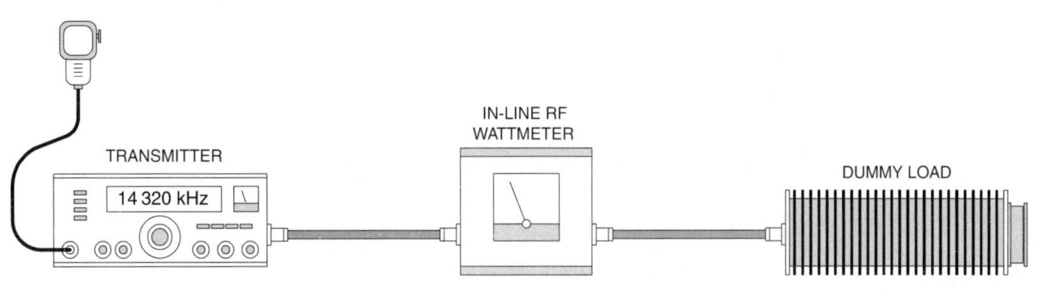

Figure 15.11 Testing a transmitter.

can be left in-line while transmitting. Although the illustration shows a dummy load, it could just as easily use a radiating antenna.

Micromatch

One form of in-line RF power meter is the *micromatch* circuit of Fig. 15.12. This device is similar to a Wheatstone bridge in which the antenna impedance represents one arm, and a pair of capacitive reactances (X_{C1} and X_{C2}) represent two other arms. The output voltage

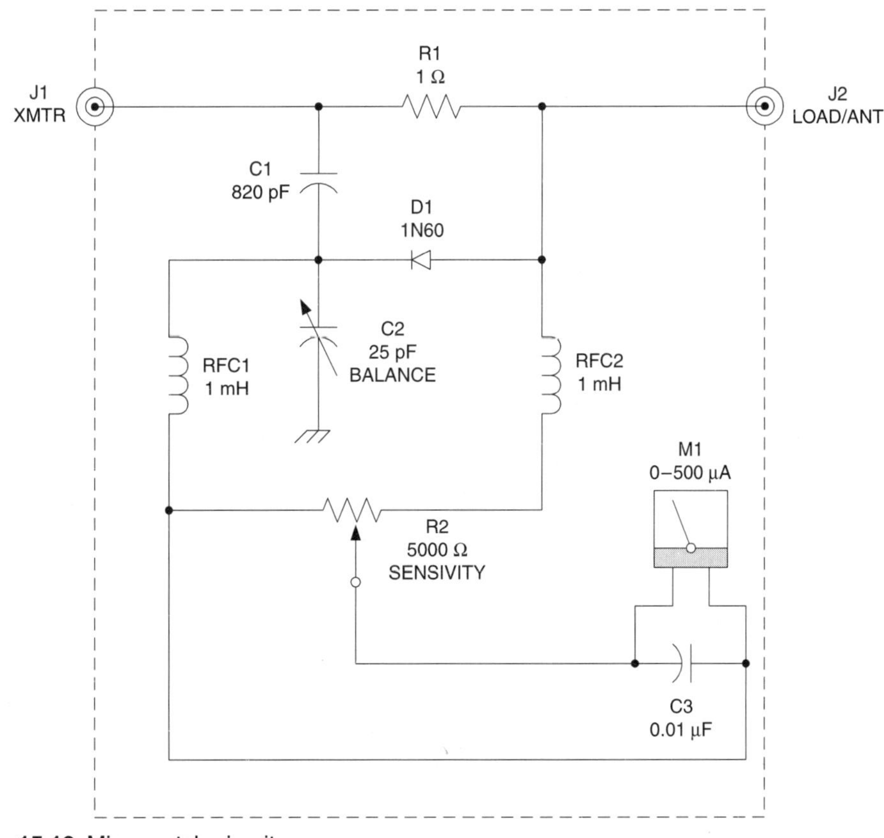

Figure 15.12 Micromatch circuit.

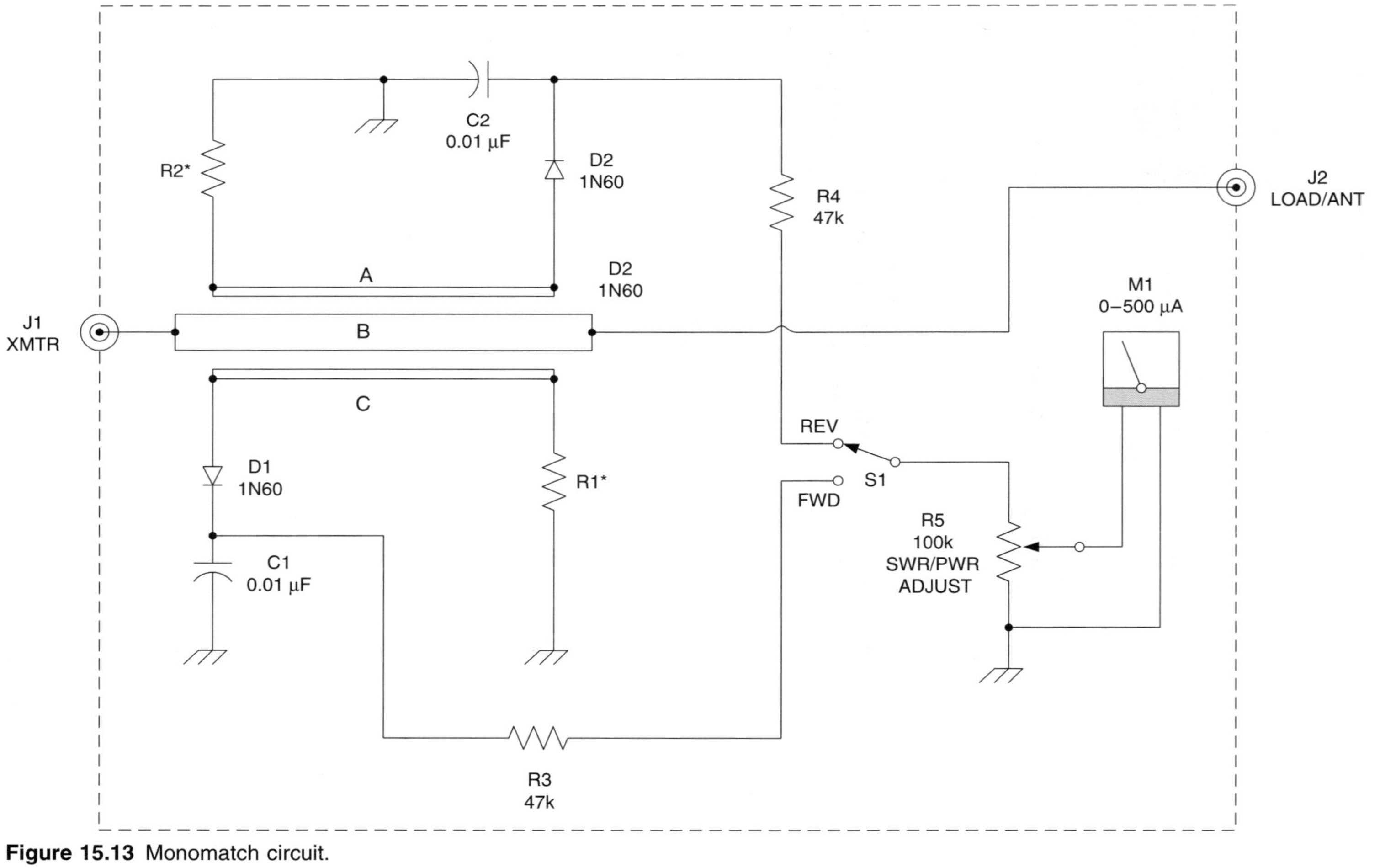

Figure 15.13 Monomatch circuit.

of the bridge is rectified by D1, and filtered by R2/C3, before being applied to a microammeter (note: any meter from 100 μA to 1 mA full scale).

The bridge consists of X_{C1}, X_{C2}, $R1$ and R_L (the antenna or load resistance). The null condition exists when $X_{C1}/X_{C2} = R1/R_L$. For 50 ohm antenna systems the ratio $R1/R_L$ is 1/50, so a value of C2 around 15 pF is needed to produce the correct C1/C2 ratio. For a 75 ohm system, about 10 pF is needed. A number of people prefer to make a compromise by assuming a 68 ohm load, so the capacitance needed in C2 for a 1/68 ratio is about 12 pF.

The series resistor (R1) is a one ohm unit. In commercial micromatch RF wattmeters this resistor is made using ten 2 watt, 10 ohm resistors connected in parallel.

The RF power level is calibrating by adjusting the sensitivity control, R2. In at least one commercial micromatch there are actually three switch selectable sensitivity controls. These are calibrated for 10 watt, 100 watt and 1000 watt ranges.

Monomatch

The classical transmission line monomatch RF wattmeter is shown in Fig. 15.13. It can be used in the high frequency (HF) through VHF ranges. It consists of three transmission line segments (A, B and C) connected as a directional coupler. In older instruments, the transmission line directional coupler was made using a length of RG-8/U coaxial cable with a pair of thin enamel insulated wires slipped between the shield and inner insulator. In more recent instruments the three transmission line segments are etched on a printed circuit board.

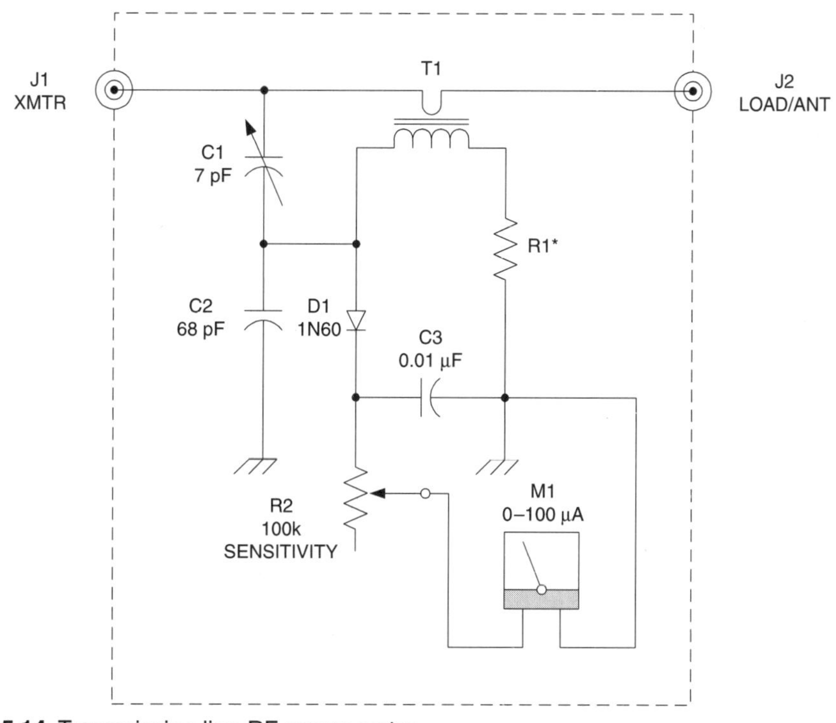

Figure 15.14 Transmission line RF power meter.

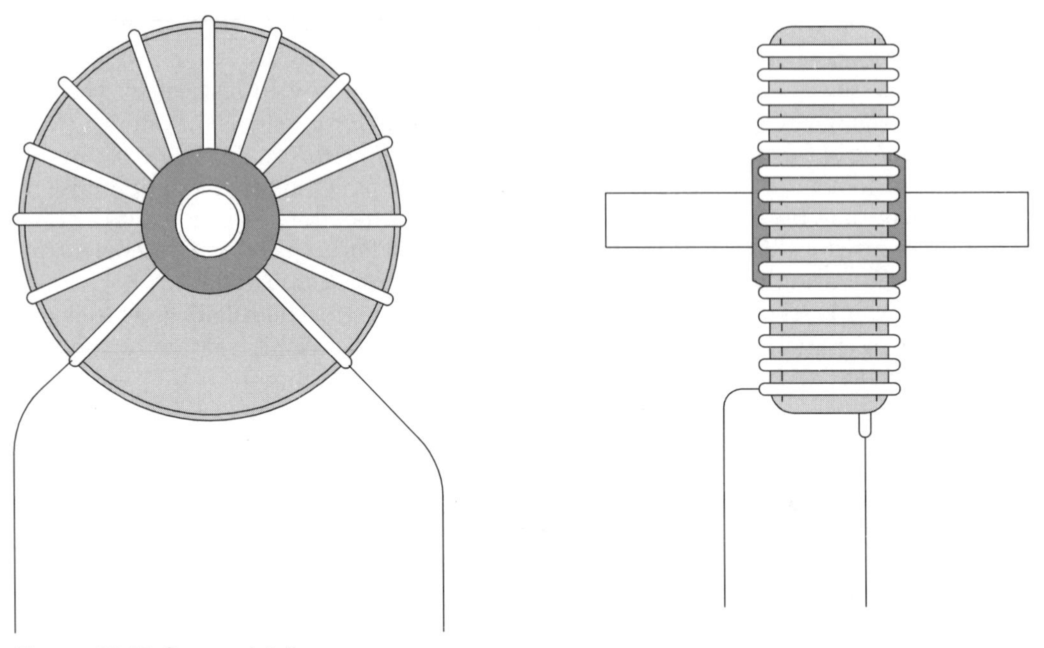

Figure 15.15 Sensor detail.

The termination of sampling lines A and C are terminated in either 50 ohm or 75 ohm non-inductive resistors (e.g. carbon composition or metal film). Again, a compromise value of 68 ohms is often seen, so that either 50 ohm or 75 ohm antennas can be measured with only a small error.

Figure 15.14 shows an alternate monomatch system that uses a broadband transmission line transformer (T1) made using a ferrite or powdered iron toroidal core. This circuit is usable throughout the HF range. Detail for the construction of the transformer is shown in Fig. 15.15. A 12 mm to 40 mm toroid core is wound with 10 to 30 turns of #22 through #30 enamelled wire, leaving a gap of at least 30 degrees between wire ends. A rubber grommet is inserted in the hole to receive the through transmission line. Small diameter copper or brass tubing can be used, provided that it is a snug fit to the grommet.

The Bird Thruline® sensor

The Bird Electronics (30303 Aurora Road, Cleveland, OH, 44139, USA) Thruline® sensor is shown in Fig. 15.16A. The sensor consists of a coaxial transmission line section, and a wire loop directional coupler that connects to a diode detector (D1).

Consider the equivalent circuit in Fig. 15.16B. The factor M is the mutual coupling between the loop and the centre conductor of the coaxial line section, as well as the voltage divider consisting of R1 and C1. Potential E is the voltage between the inner and outer conductors of the coaxial line, while E_R is the voltage drop across the resistor, e_M is the voltage across the inductor, while e is the output potential. We know that the R1–C1

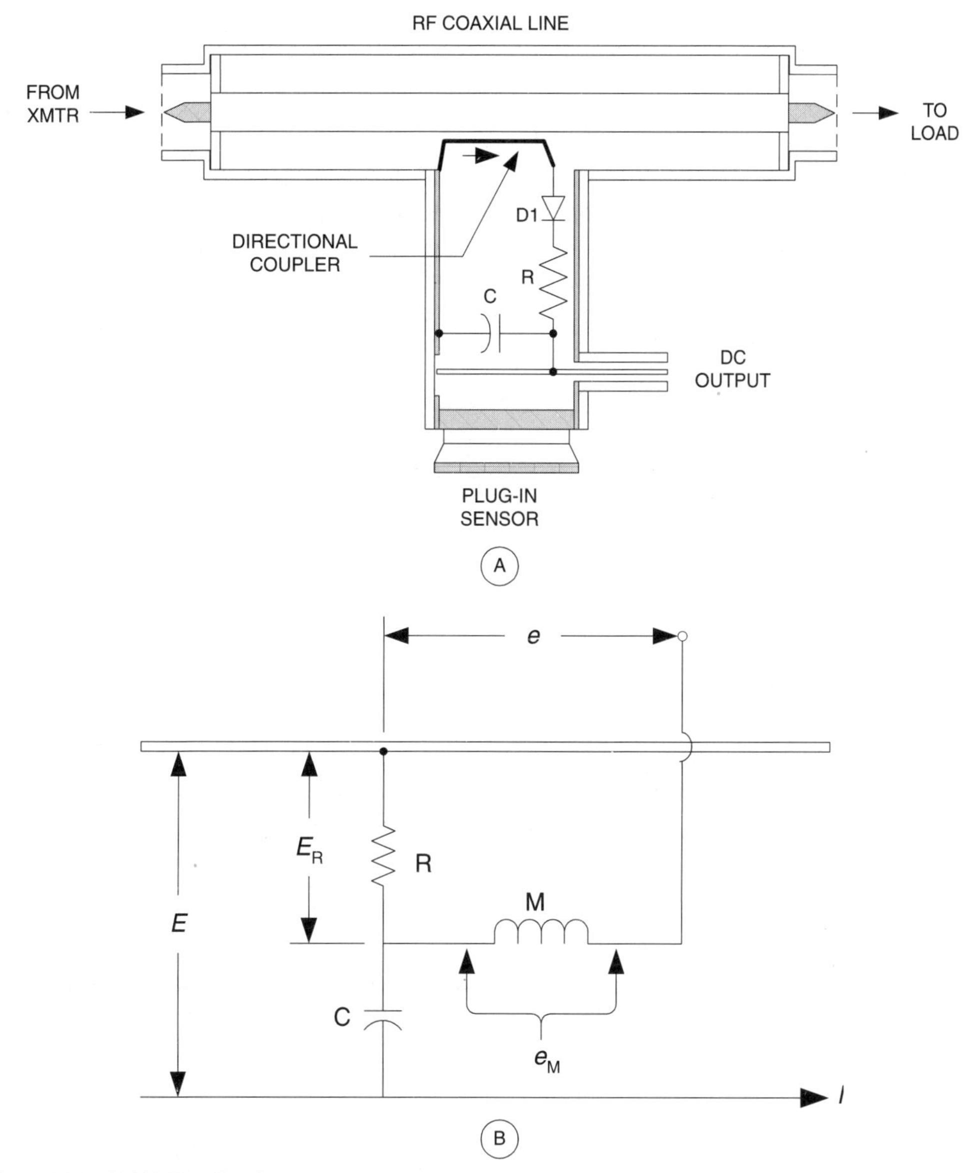

Figure 15.16 (A) Thruline® structure; (B) equivalent circuit.

voltage divider produces a potential given by Eq. (15.4) provided that $R << X_C$ and $e_m = Ij\omega \pm M$.

$$e_R = \frac{RE}{X_C} = REJ\omega C \qquad (15.4)$$

The output voltage is:

$$e = e_R + e_M = j\omega(CRE \pm MI) \qquad (15.5)$$

The values of the components are selected such that $R << X_C$ and $CR = M/Z_0$. We can now state that the DC output voltage is:

$$e = j\omega\left[\frac{EM}{Z_0} \pm MI\right] = j\omega M(E/Z_0 \pm I) \tag{15.6}$$

At any point along a transmission line the voltage appearing between the centre conductor and outer conductor (E) is a function of the forward voltage (E_F) and the reflected voltage (E_R). By combining equations, we see that when the directional coupler is pointed at the load, the output voltage of the sensor reads the forward voltage, and produces an output voltage of

$$e = \frac{j\omega ME_F}{Z_0} \tag{15.7}$$

And when pointed at the source:

$$e = \frac{J\omega ME_R}{Z_0} \tag{15.8}$$

Thus, this sensor produces a voltage that is a function of the direction of the RF signal flowing in the transmission line.

The Bird 4410A offers an insertion VSWR of 1.05:1 up to 1000 MHz. The sensor elements are plug-in. Each element has an arrow on it to indicate the direction of the measurement (pointed towards the load or the source, depending on whether you measure P_F or P_R). Once you know P_F and P_R, you can computer the VSWR from Eq. (15.9):

$$VSWR = \frac{1 + \sqrt{P_R/P_F}}{1 - \sqrt{P_R/P_F}} \tag{15.9}$$

One difference between this instrument and earlier instruments is that there is a calibration factor control on the meter to optimize performance for the specific sampling element inserted.

Calorimeters

Calorimeters are capable of making very accurate measurements of RF power, especially at high power levels where other methods tend to fall down. These instruments measure the heating capability of the RF waveform, so produce an output proportional to the average power level that is independent of the applied waveform. The First Law of Thermodynamics (numbering of the laws of thermodynamics differs in various parts of the world: in the USA we start with the '0th law') is the basis for the operation of calorimeters: *energy can neither be created nor destroyed, only changed in form.*

There are two basic forms of RF power calorimeter: dry and flow (or wet). The dry calorimeters are used at lower power levels, and are represented by the thermistor and thermocouple methods discussed earlier. Flow calorimeters are used at higher power levels.

Flow calorimeters come in two varieties: *substitution flow* and *absolute flow*. Power can be measured using the following relationship:

$$P = F_{Mass} \times (T_{OUT} - T_{IN}) \times C_P(T) \quad (15.10)$$

Where:
P is the power level
F_{Mass} is the mass flow rate of the fluid used in the calorimeter
T_{OUT} is the fluid temperature after being heated by the RF load resistor
T_{IN} is the fluid temperature before being heated by the RF load resistor
$C_P(T)$ is the fluid specific heat as a function of temperature (T)

Substitution flow calorimeters

This form of RF power meter (Fig. 15.17) uses two fluid loops. Each fluid loop is heated by a separate termination resistor. Termination 'A' is heated by a low frequency AC power source, and the power applied to this termination is measured by an AC power meter. The unknown RF power is applied to termination 'B'. The differential temperature ($T_{OUT} - T_{IN}$) is measured by a differential thermocouple.

When the temperatures of the two fluids are equal to each other, then the output of the thermocouple is zero. When the AC power is adjusted to balance the temperatures while RF is applied, producing a zero output voltage from the thermocouple, the RF power is equal to the more easily measured AC power. A temperature stabilizer and heat exchanger return the temperature of the fluid to base level after it is used to measure power.

This method will produce an error of 0.28 per cent or better, up to RF power levels of one kilowatt. Both water and oil are used as fluid coolants in various instruments.

Absolute flow calorimeters

Figure 15.18 shows the absolute flow calorimeter. This type of RF power meter measures the mass flow rate of the coolant, as well as the temperatures before (T_{IN}) and after (T_{OUT}) the RF load resistor. The mass flow rate is:

$$F_{Mass} = fW_S(T_{IN}) \quad (15.11)$$

Where:
W_S is the specific weight of the fluid at the input temperature
f is the volume flow rate (litres/min)
All other terms as previously defined

By combining equations we get the equation for measuring RF power by this means:

$$P = k \times f \times W_S(T_{IN}) \times C_P(T_{AVE}) \times (T_{OUT} - T_{IN}) \quad (15.12)$$

Where:
$T_{AVE} = (T_{OUT} - T_{IN})/2$
All other terms are as previously defined

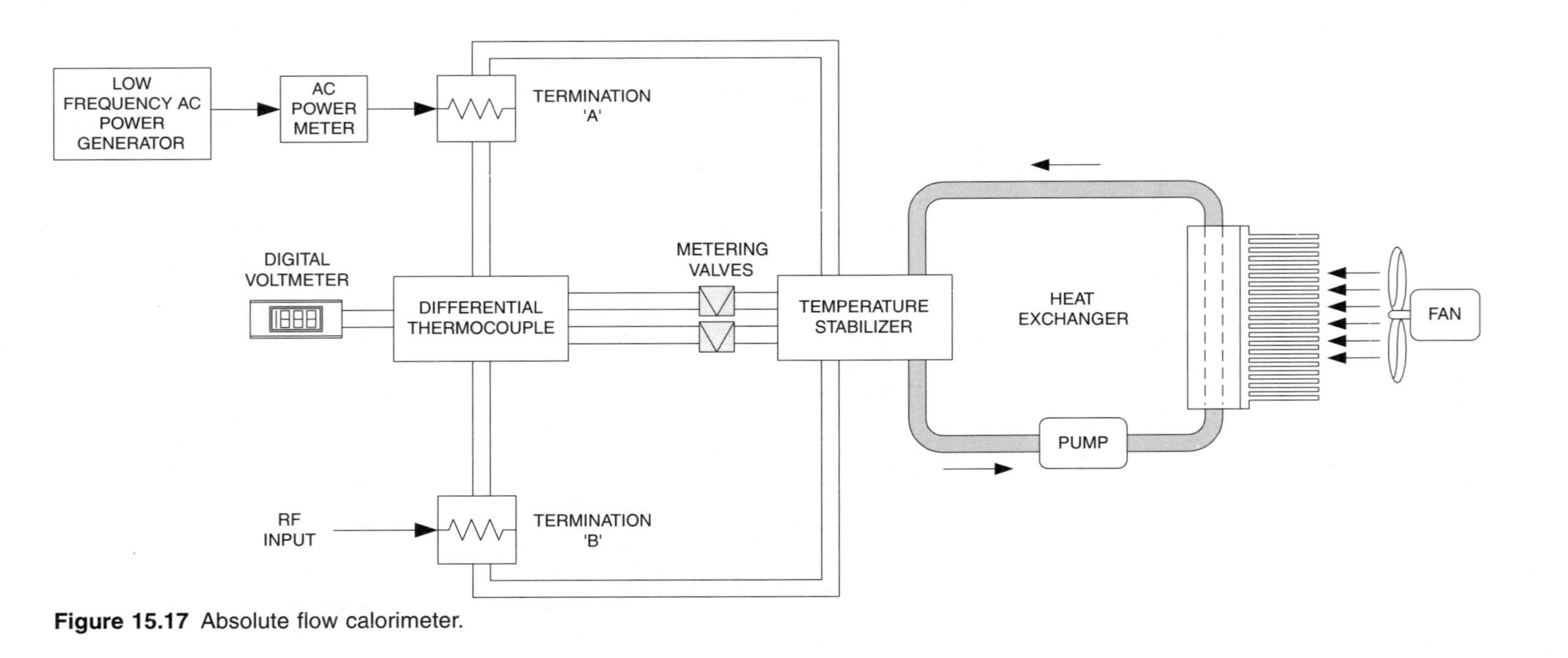

Figure 15.17 Absolute flow calorimeter.

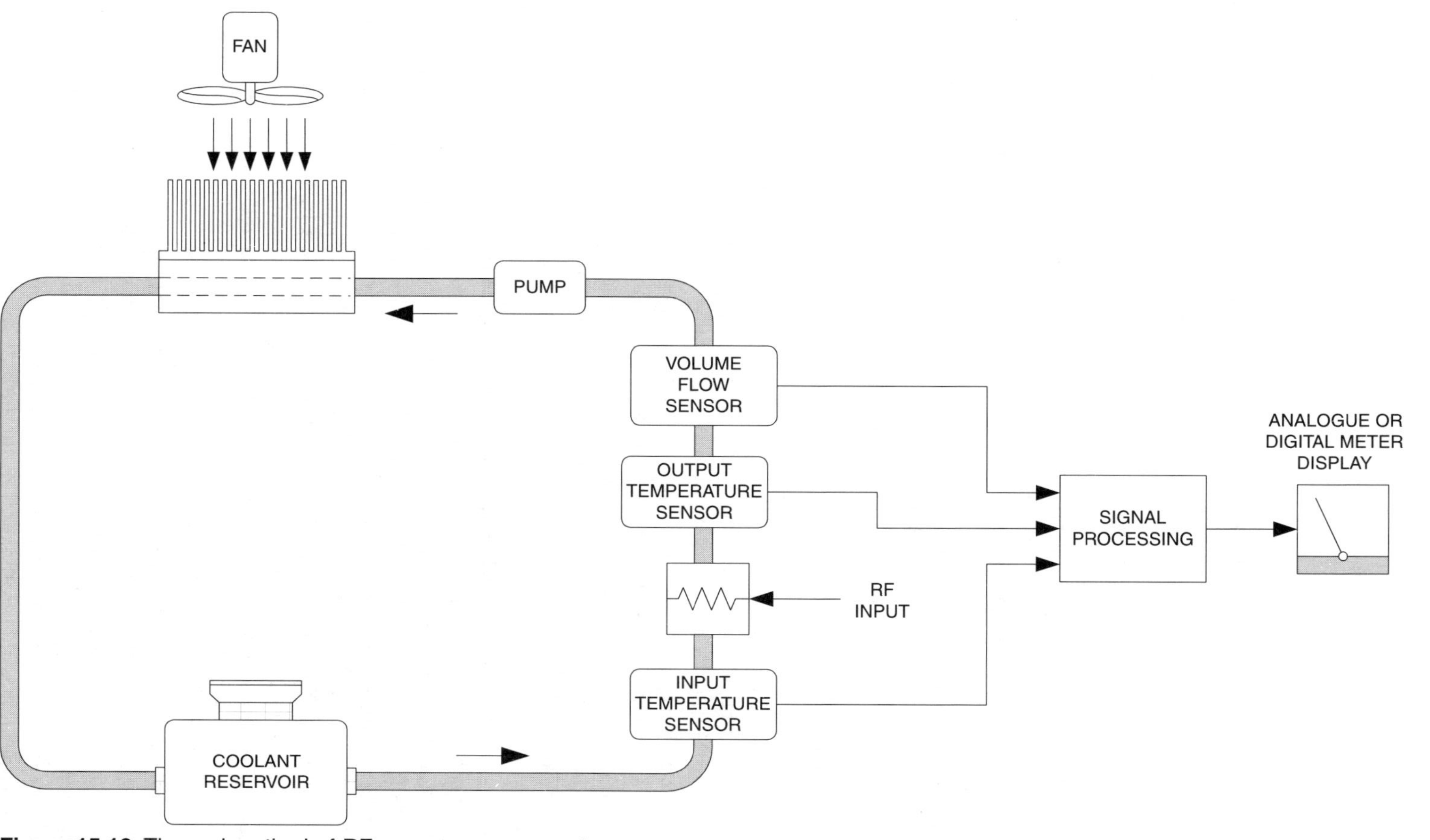

Figure 15.18 Thermal method of RF power measurement.

One of the advantages of the absolute flow approach is that is does not depend on nulling or calibration of a low frequency power source, yet it produces good accuracy at high power levels (up to 80 kW).

Micropower and low power measurements

At very low power levels the diode output voltage drops very low. For example, at a power level of –70 dBm the diode produces about 50 nanovolts (nV) output potential. This level is too close to the noise and the drift values of typical DC amplifiers to be of use. A solution is to use a *chopper circuit* as shown in Fig. 15.19.

A chopper is an electronic switch that turns the DC signal from the diode output on and off at a high rate (typically 100 to 10 000 times per second). The switching action is created by either a square wave generator or sine wave oscillator applied to the toggle input of the electronic switch.

The chopped signal is essentially an AC signal, so it can be amplified in an AC amplifier, which has a much smaller feedback controlled drift figure than DC amplifiers. Also, the AC signal can be bandpass filtered to remove noise. The bandpass filter is centred on the frequency of the carrier oscillator.

The chopped, amplified and filtered signal is applied to a synchronous detector that is controlled by the same carrier oscillator that performed the chopping action. A low-pass filter following the synchronous detector removes residual components of the switching action at the carrier frequency. Finally, a DC amplifier provides scaling to the correct DC level, or as level translation for an analogue-to-digital converter (ADC).

Micropower measurements pose special problems because they are made at levels below the range of most practical RF power sensors. In some cases, the chopper approach can be used with a diode detector. At lower levels, however, some other method is needed. Figure 15.20 shows a comparison method using a calibrated RF signal generator. The instrument selected must have a calibrated output attenuator that provides accurate outputs in either dBm or microvolts (μV). The signal generator and the unknown micropower source are connected to a receiver equipped with an S-meter through a hybrid coupler. The coupler must have either equal port-to-port losses for the two inputs, or at least accurately known different losses. Optional calibrated step attenuators are also sometimes used to balance the power levels. The receiver acts as a micropower wattmeter or voltmeter because it will produce an S-meter reading of even very weak signals.

Two methods can be used: either *equal deflection* or *double deflection*. In the equal deflection method, the unknown source is turned on, and the S-meter reading noted. The unknown source is then turned off, and the signal generator is turned on. The output of the signal generator is adjusted to produce the same S-meter deflection. The power level of the unknown source is therefore equal to the calibrated signal generator output level.

The double deflection method sets the signal generator output to zero, and then applies the unknown RF power to the receiver. The S-meter reading is noted (for practical reasons, adjust the attenuator to let the meter fall on a specific indicator marking). The signal generator output is then increased until the S-meter reading goes up one S-unit (which will be either 3 dB or 6 dB, depending on the design of the receiver). The output level of the signal generator is therefore equal to the unknown power source.

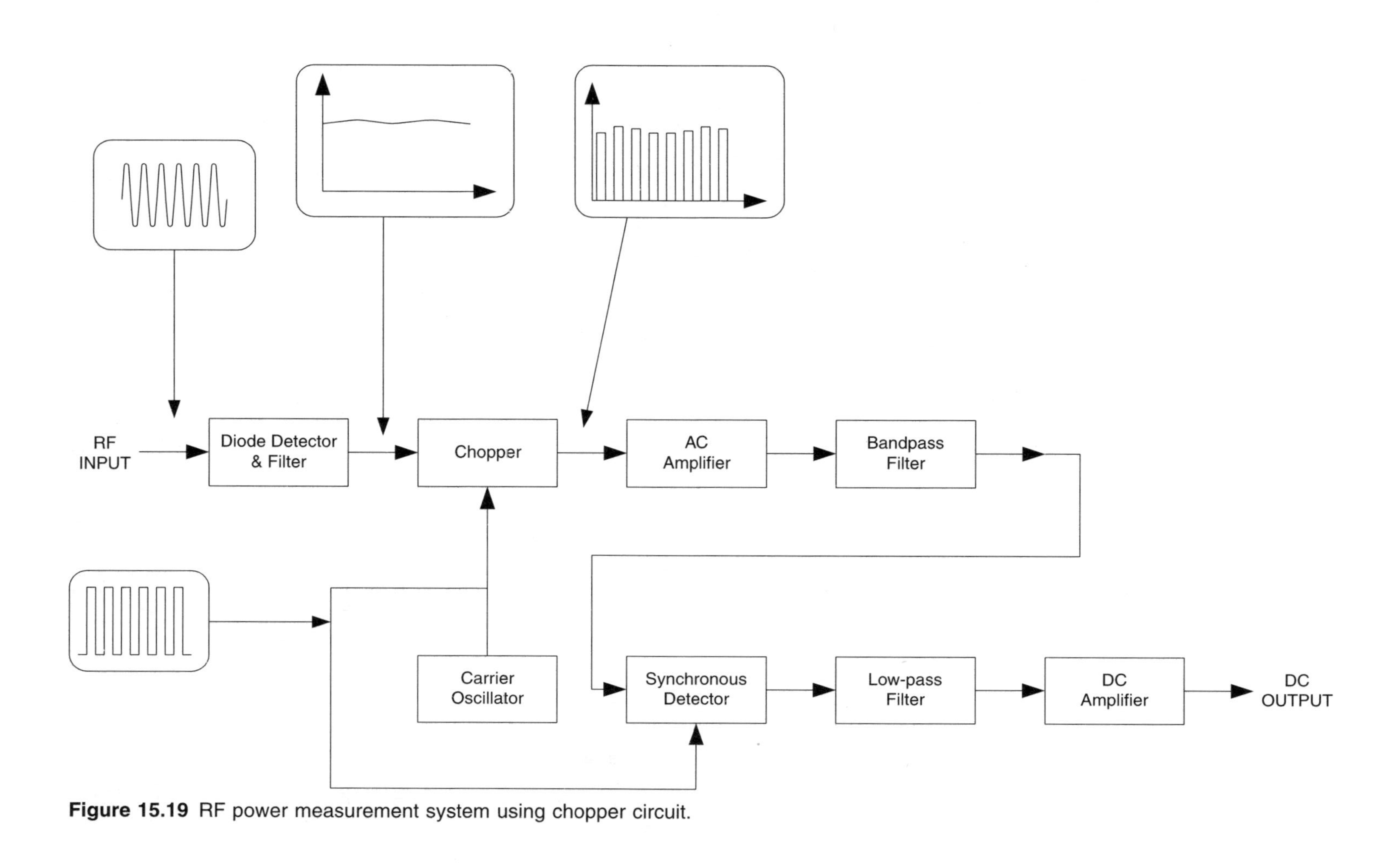

Figure 15.19 RF power measurement system using chopper circuit.

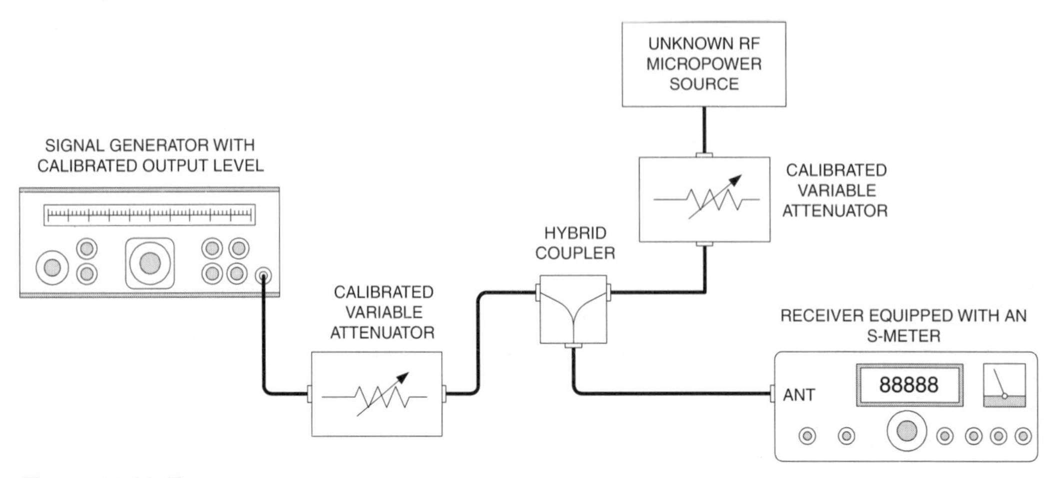

Figure 15.20 Test set-up.

Error and uncertainty sources

All measurements have some basic error, i.e. a difference between the actual value of a variable and the value read from a meter. The three dominant classes of error in RF power measurements are *mismatch uncertainty, sensor uncertainty* and *meter uncertainty.*

Meter uncertainty is error due to problems in the meter indicating device itself. It might be a measurement error, i.e. a difference between the actual output voltage and the displayed output voltage (which represents power). You might see zero set error, zero carryover, drift, noise and other sources of instrument error.

On analogue meters there are also additional error sources. For example, the width of the pointer covers a certain distance on the scale, so creates a bit of ambiguity. Also, there may be a parallax error if the meter is not read straight.

Digital meters exhibit quantization error and last digit bobble error. The quantization error comes from the fact that the digital representation of a value can only assume certain discrete values, and an actual value might be halfway between the two authorized levels. Last digit bobble (±1 count) error results from the fact that the least significant digit tends to bounce back and forth between two adjacent values.

Sensor error may come in a variety of guises, depending on the nature of the sensor. Thermistors and thermistors, for example, have different forms of error. Most sensors, though, exhibit a effective efficiency error due to losses in the sensor. It occurs when some of the applied RF energy is radiated as heat rather than being used to affect the output reading. This problem may be expressed as a calibration uncertainty or calibration factor by the manufacturer of the sensor.

Mismatch loss and mismatch uncertainty

The mismatch loss occurs when a standing wave ratio (SWR or VSWR) exists in the system. Maximum power transfer occurs when a source impedance and a load impedance

Table 15.1

VSWR	ρ
1	0.000
1.2	0.091
1.4	0.167
1.6	0.231
1.8	0.286
2	0.333
2.5	0.429
3	0.500

are matched. If these impedances are not matched, then a portion of the power sent from the source to the load is reflected. The *reflection coefficient* (ρ) is:

$$\rho = \frac{VSWR - 1}{VSWR + 1} \tag{15.13}$$

Table 15.1 shows the reflection coefficient for VSWR values from 1:1 to 3:1. The single-ended *mismatch loss* in decibels is

$$ML = 10 \log[1 \pm \rho^2] \text{ dB} \tag{15.14}$$

If the system is mismatched on both ends, then mismatch loss is:

$$ML = 20 \log [1 \pm (\rho_1 \times \rho_2)] \tag{15.15}$$

The mismatch uncertainty, expressed as a per cent:

$$MU = \pm 2 \times \rho_1 \times \rho_2 \times 100\% \tag{15.16}$$

Let's assume that there is a 1.75:1 VSWR at the source end ($\rho 1 = 0.27$) and a VSWR of 1.15:1 ($\rho 2 = 0.07$) at the sensor/load end. The mismatch uncertainty is:

$$MU = \pm 2 \times 0.27 \times 0.07 \times 100\% = 3.78\%$$

16 Filtering against EMI/RFI

Electronic circuits must do two different functions: (1) they must respond to desired signals, and (2) they must reject undesired signals. But the world is full of a variety of interfering signals, all of which are grouped under the headings *electromagnetic interference* (EMI) or *radio frequency interference* (RFI). These EMI/RFI sources can ruin the performance of, or destroy, otherwise well-functioning electronic circuits. Oddly, most circuits are designed for function (1): they do, in fact, respond properly to desired signals. But many devices fail miserably on function (2): they will, in fact, respond to undesired signals in an inappropriate manner. Let's take a look at some of the techniques that can be used to EMI/RFI-proof electronic devices and circuitry.

Shielding

The first step in providing protection against EMI/RFI is shielding the circuitry. The entire circuit is placed inside a metallic enclosure that both prevents external EMI/RFI fields from interacting with the internal circuits, and also prevents internal fields from doing the opposite. Let's assume for this discussion that the circuits are well shielded. So what's the problem? Unfortunately, the shield is not perfect because of power and signal leads entering and/or leaving the enclosure. For those, some sort of filtering is needed.

Filter circuits

Filter circuits can be active or passive, but for this present discussion let's consider the passive varieties only. Such filters are made of either R–C, L–C or R–L–C components in appropriate types of network.

Figure 16.1 shows simple resistor–capacitor (R–C) networks in both low-pass and high-pass filter configurations. Note that the two circuits are similar except that the positioning of the R and C components are reversed. The cut-off frequency of these circuits is found from:

$$F_C = \frac{1}{2\pi R1C1} \tag{16.1}$$

Where:
F_C is the cut-off frequency in hertz (Hz)
$R1$ is in ohms
$C1$ is in farads

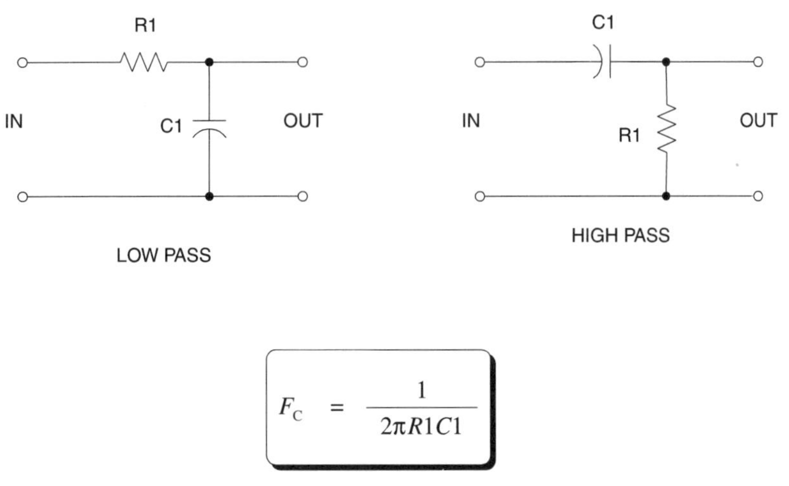

Figure 16.1 R–C filters: (A) low pass; (B) high pass.

These circuits provide a frequency roll-off beyond F_C of –6 dB/octave, although sharper roll-off can be obtained by cascading two or more sections of the same circuit.

Figure 16.2 shows four different Chebyshev filters (two LPF and two HPF). These filters are:

Fig. 16.2A Low-pass pi-configuration

Fig. 16.2B Low-pass Tee-configuration

Fig. 16.2C High-pass pi-configuration

Fig. 16.2D High-pass Tee-configuration

Each of these filter circuits is a 'five-element' circuit, i.e. they each have five L or C components. Lesser (e.g. three element) and greater (e.g. seven and nine element) are also used. Fewer elements will give a poorer frequency roll-off, while more elements give a sharper roll-off. The component values given in these circuits are normalized for a 1 MHz cut-off frequency. To find the required values for any other frequency, divide these values by the desired frequency in megahertz. For example, to make a high-pass Tee-configuration filter for, say, 4.5 MHz, take the values of Fig. 16.2D and divide by 4.5 MHz:

$L1 = 5.8\,\mu\text{H}/4.5\,\text{MHz} = 1.29\,\mu\text{H}$

$L2 = L1$

$C1 = 2776\,\text{pf}/4.5\,\text{MHz} = 617\,\text{pF}$

$C2 = 1612\,\text{pF}/4.5\,\text{MHz} = 358\,\text{pF}$

$C3 = C1$

If the desired frequency is less than 1 MHz, then it must still be expressed in MHz, i.e. 100 kHz = 0.1 MHz and 10 kHz = 0.01 MHz.

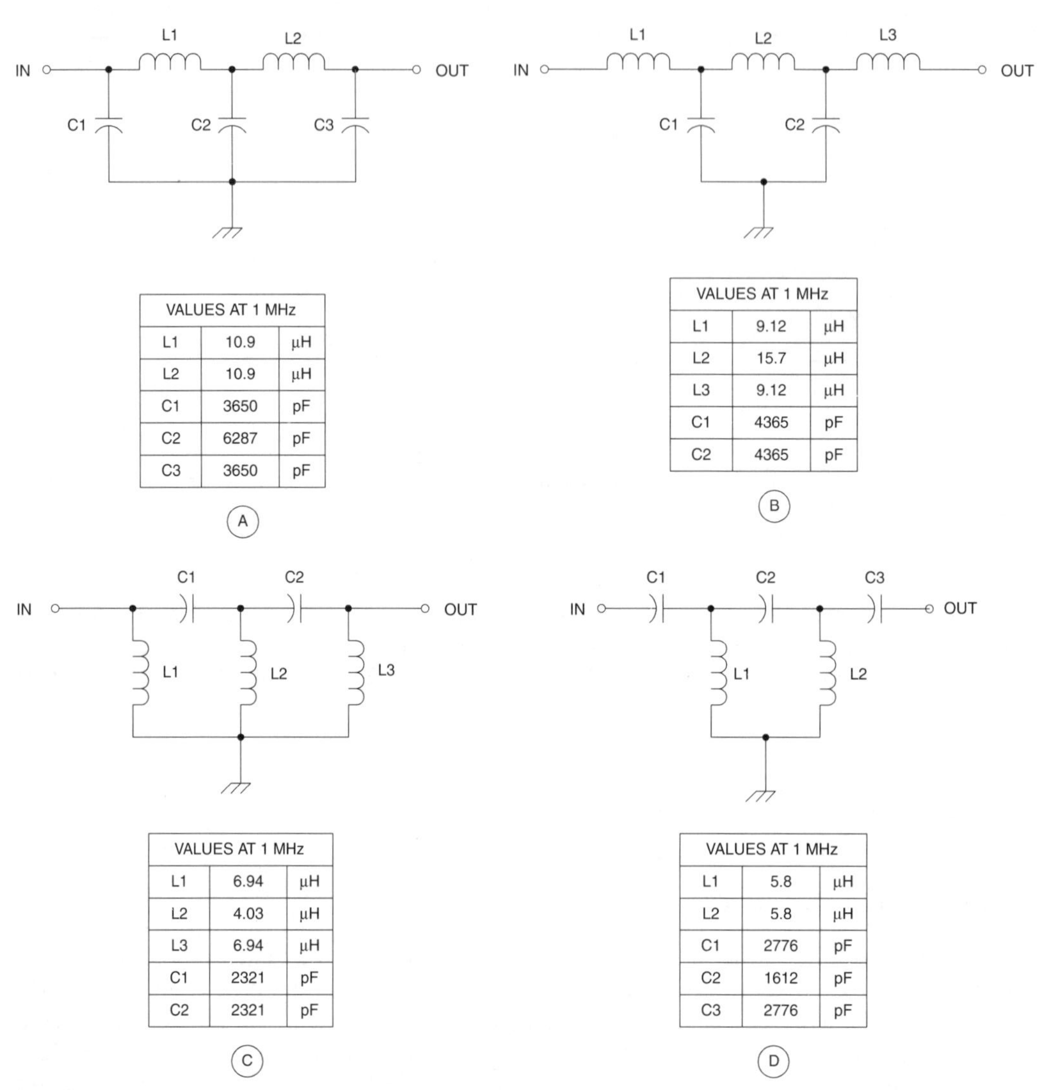

VALUES AT 1 MHz		
L1	10.9	μH
L2	10.9	μH
C1	3650	pF
C2	6287	pF
C3	3650	pF

(A)

VALUES AT 1 MHz		
L1	9.12	μH
L2	15.7	μH
L3	9.12	μH
C1	4365	pF
C2	4365	pF

(B)

VALUES AT 1 MHz		
L1	6.94	μH
L2	4.03	μH
L3	6.94	μH
C1	2321	pF
C2	2321	pF

(C)

VALUES AT 1 MHz		
L1	5.8	μH
L2	5.8	μH
C1	2776	pF
C2	1612	pF
C3	2776	pF

(D)

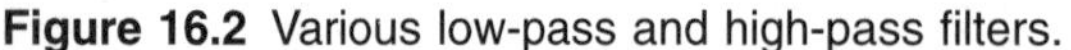

Figure 16.2 Various low-pass and high-pass filters.

R–C EMI/RFI protection

Some circuits, especially those that operate at low frequencies, may use R–C low-pass filtering for the EMI/RFI protection function. Consider the differential amplifier in Fig. 16.3. This circuit is representative of a number of scientific and medical instrument amplifier input networks. A medical electrocardiogram (ECG) amplifier, for example, picks up the human heart's electrical activity as seen from skin electrodes on the surface.

There are a number of problems that will afflict the recording, other than the obvious 50/60 Hz problem. The ECG must often be used in the presence of strong radio frequency

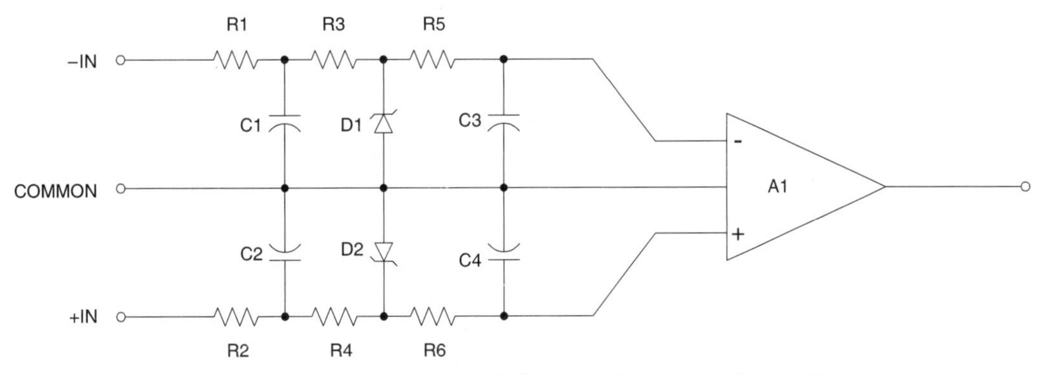

Figure 16.3 Defibrillator protection and RF EMI filtering of a bioelectric amplifier.

RF fields from *electrosurgery machines*. These 'electronic scalpels' are used by surgeons to cut and cauterize, and will produce very strong fields on frequencies of 500 kHz to 3 MHz. They must also survive high voltage DC jolts from a defibrillator machine.

Figure 16.3 contains both RF filtering and a means for limiting the defibrillator jolt. The resistors and capacitors form a three-stage cascade RC filter, one for each input of the differential amplifier. These components will filter the RF component. Typical values range from 100 K to 1 megohm for the resistors, and 100 pF to 0.01 μF for the capacitors. The high voltage protection is provided by a combination of the input resistors and a pair of zener diodes (D1 and D2) shunting the signal and common lines.

Feedthrough capacitors

One effective way to reduce the effects of EMI/RFI that pass into a shielded compartment via power and signal lines is the *feedthrough capacitor* (Fig. 16.4), sometimes called an *EMI filter*. These capacitors typically come in 500 pF, 1000 pF and 2000 pF values. Both solder-on and screw-on (as shown in Fig. 16.4) are available. In some catalogues these capacitors are referred to as 'EMI filters' rather than 'feedthrough capacitors'. I may be a cynic, but the 'EMI' designation seems to add considerably to the price without any apparent advantage

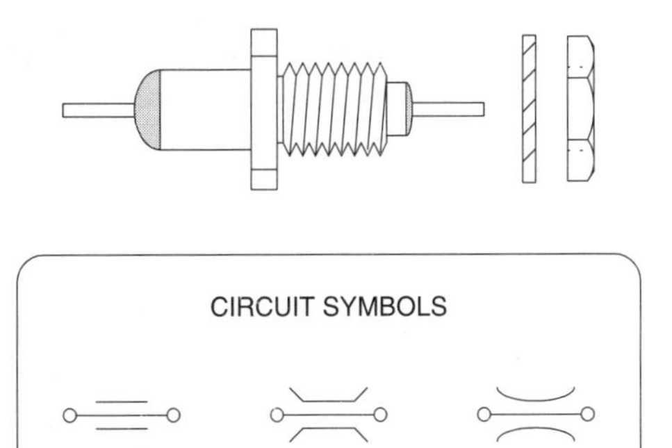

Figure 16.4 Feedthrough capacitor.

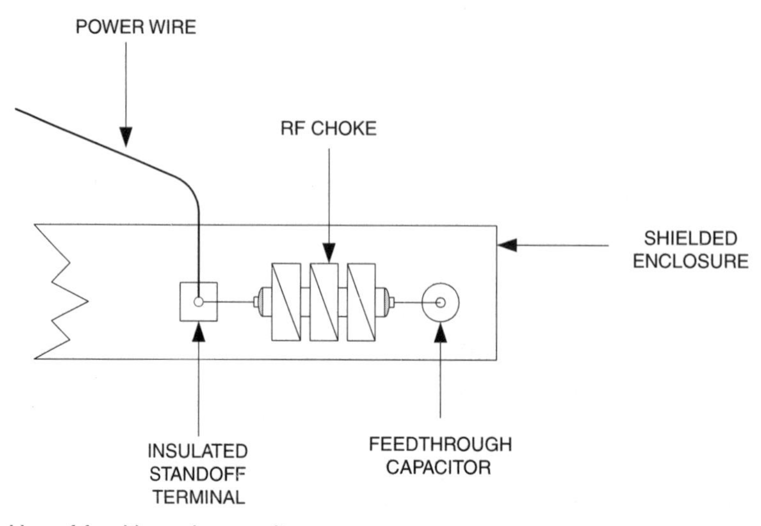

Figure 16.5 Use of feedthrough capacitor.

over straight feedthrough capacitors. Also shown in Fig. 16.4 are several different forms of circuit symbols used for feedthrough capacitors in circuit diagrams.

There are several different ways to use a feedthrough capacitor. One method is to simply pass it through the shielded compartment wall, and attach the wires to each side. In other cases, additional resistors or inductors are used to form a low-pass filter. Figure 16.5 shows one approach in which a radio frequency choke (RFC) is mounted external to the shielded compartment. This method is often used for tv and cable-box tuners. One end of the RFC is connected to the feedthrough capacitor, and the other to some other point in the external circuit (e.g. a stand-off insulator is shown here, as is common on tv/cable tuners). It is very important to keep the lead wire from the RFC to the feedthrough capacitor as short as possible to limit additional pick-up beyond the filtering.

Another approach is shown in Fig. 16.6. This method uses a separate shielded compartment inside the main shielded enclosure. Feedthrough capacitors (C1 and C2) are used to carry DC (or low frequency signals) into and out of the filter compartment. An inductor or RF choke, L1, is part of the filtering, so the combination L1–C1–C2 forms a low-pass pi-configuration filter.

One caution is in order: if L–C filters are used on both input and output signal lines of a circuit, then make sure they resonate on different frequencies. The reason is that they

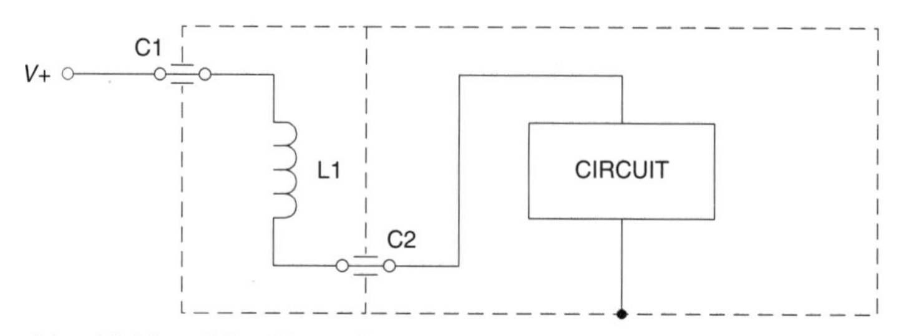

Figure 16.6 Double shield and feedthrough capacitor.

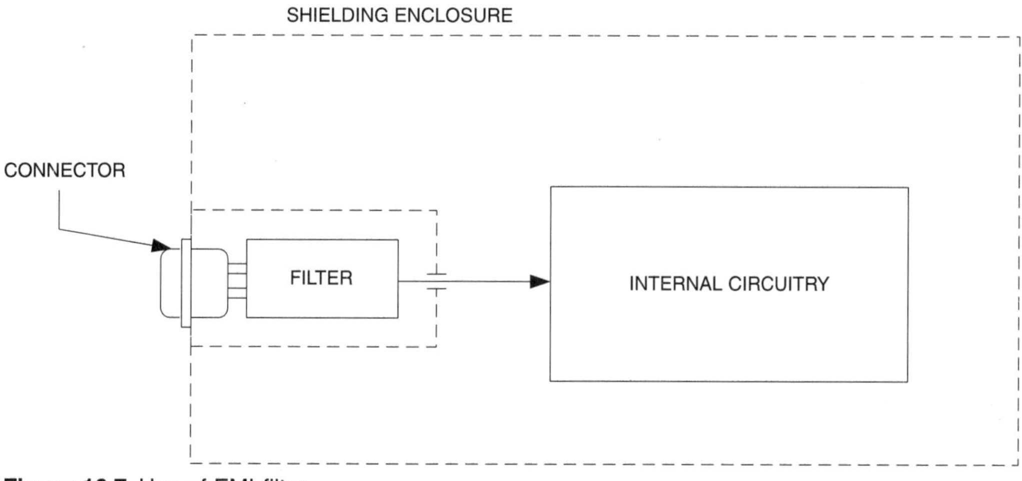

Figure 16.7 Use of EMI filter.

could form a tuned-input/tuned-output oscillator if the circuit being protected has sufficient gain at the filter's resonant frequency.

In some cases, connectors are bought that have filtering built in (Fig. 16.7). These products are usually described as EMI filtering connectors. Although most of them are designed to work with 120/220 volt AC power lines, others are available that work at higher frequencies.

One more approach is shown in Fig. 16.8. The filtering is performed by using a set of one or more ferrite beads slipped over the wire from the connector pin to the circuit board. Ferrite beads surrounding a wire act like a small value RF choke, so will filter (typically) VHF/UHF frequencies. It is common to see these beads on RF equipment, but they are also found on digital devices as well.

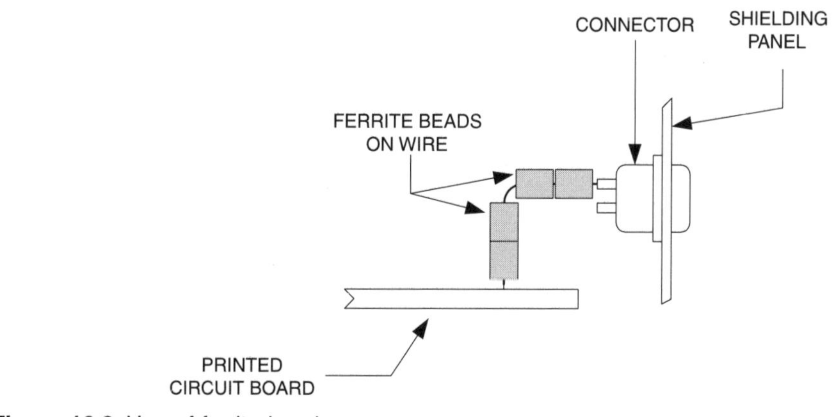

Figure 16.8 Use of ferrite beads on connector.

General guidelines

Thus far we've looked at a number of different filtering approaches to protecting equipment. Now let's consider the general guidelines:

1 Always shield the circuit. A circuit that is not shielded cannot be adequately protected by filtering. There is simply too much chance of direct pick-up of the EMI/RFI source by the components and wires of the circuit.
2 Apply filtering to the DC power lines entering or leaving the circuit's shielded enclosure.
3 Use the minimum filtering necessary for accomplishing the level of protection required. It does no good to add one more section of filtering when the job is done properly, but does add cost, complexity and opportunities for failed components (which will keep the service technician busy, if nothing else).
4 If it is necessary to filter signal input and output lines, use the minimum values of capacitance or inductance consistent with the degree of protection needed. Keep the cut-off frequency well away from the frequencies the circuit normally uses.

17 Noise cancellation bridges

Whether you operate a radio receiver, or some piece of scientific or medical instrumentation, noise interferes with acquiring desired signals. Noise is bad. After all, radio reception and other forms of signals acquisition are essentially a game of *signal-to-noise ratio* (SNR). The actual values of the desired signal and noise signal are not nearly as important as their ratio. If the signal is not significantly stronger than the noise, then it will not be properly detected.

Getting rid of noise battering a signal is a major chore. Although there are a number of different techniques for overcoming noise, the method described herein can be called the 'invert and obliterate' approach.

Figure 17.1 shows the basic problem and its solution. The signal from the main antenna is a mixture of the desired signal, and a locally generated noise signal. This noise signal is usually generated by the 60 Hz alternating current (AC) power lines, or machinery and appliances operating from the 60 Hz AC lines. The noise signal is not confined to 60 Hz, but will extend into the VHF region because of harmonic content. The noise spikes will appear every 60 Hz from the fundamental frequency up to about 200 MHz or so, although the harmonics become weaker and weaker at progressively higher frequencies. But in the VLF bands (where they are often overwhelming), AM broadcast band (AM BCB), and shortwave bands the noise signal can be tremendous. It will therefore cause a huge amount of interference.

The solution is to invert the noise signal, and combine it with the signal from the main antenna. When the phase inverted noise signal combines with the noise signal riding on the main signal, the result is cancellation of the noise signal, leaving the resultant main signal. What is needed is a noise sense antenna, a means for inverting the noise signal, and a summing circuit.

The phase inversion and summation functions of Fig. 17.1 are performed in a special *noise cancellation bridge* circuit. The main antenna is the antenna that is normally used with the receiver. It might be a dipole, vertical, beam, or just a random length of wire strung between two trees.

The noise sense antenna is optimized for pick-up of the noise source signal. One VLF radioscience observer told me via e-mail that he uses a 36 inch whip antenna mounted on his roof as the noise sense antenna. In some shortwave situations the sense antenna is a ten to 30 foot length of antenna wire running parallel to the power lines that are creating the noise. *Caution: Under no circumstances should you allow the sense antenna to touch the AC power lines, even if it breaks and whips around in the wind*. In the case of VHF noise reception

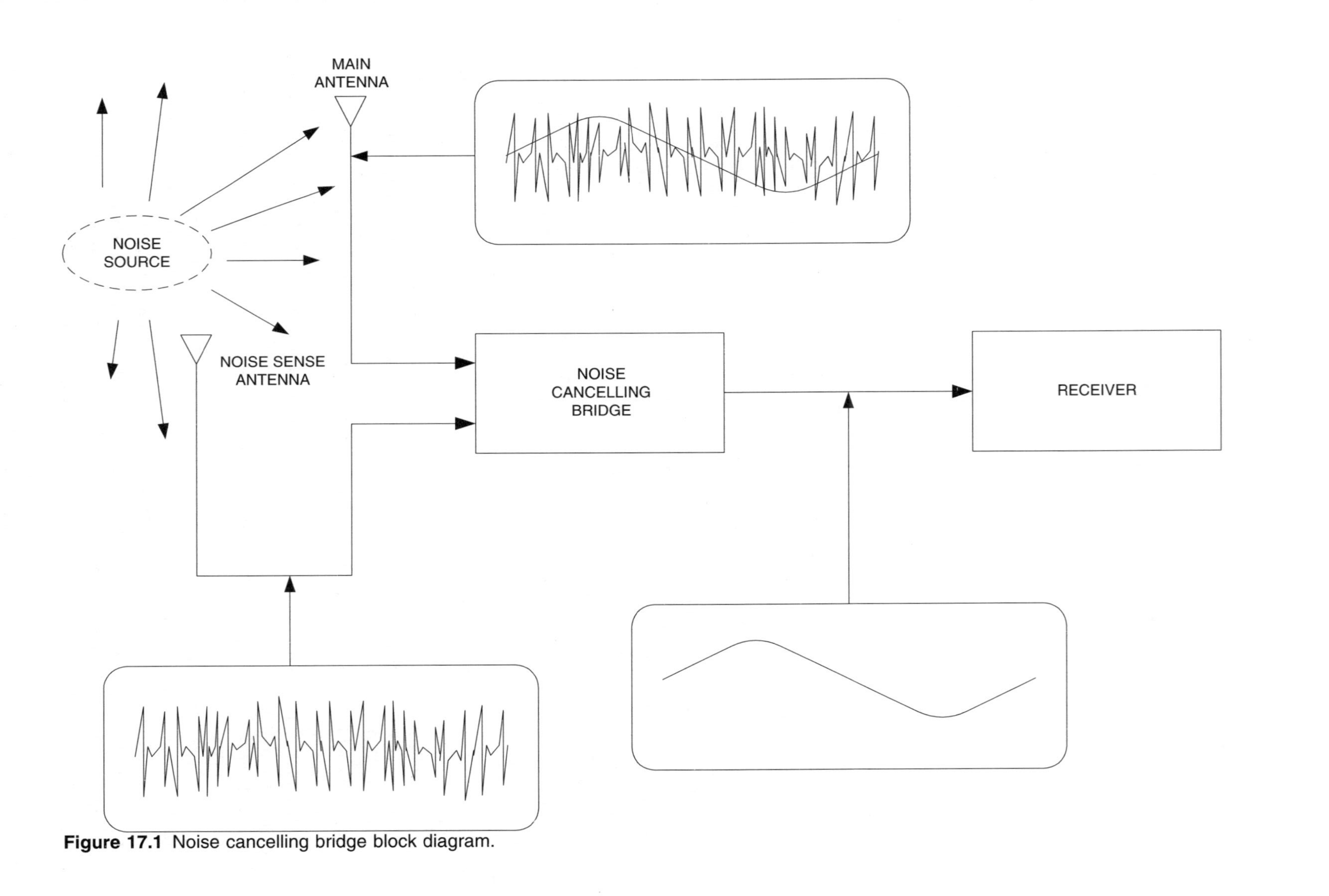

Figure 17.1 Noise cancelling bridge block diagram.

the sense antenna might be a two or three element beam (Yagi or Quad) aimed at the noise source. Other combinations are also possible, I presume.

One goal of the sense antenna is to make it highly sensitive to the local noise field, while being a lot less sensitive to the desired signal than the main antenna. Although both noise and desired signals appear in both antennas, the idea is to maximize the noise signal and minimize the 'desired' signal in the sense antenna, and do the opposite in the main antenna.

The design problems that must be overcome in producing the NCB are easy to see. First, it must either invert or provide other means for producing a 180 degree phase shift of the noise signal. It must also account for amplitude differences so that the inverted noise signal exactly cancels the noise component of the main signal. If the amplitudes are not matched, then either some of the original noise component will remain, or the excess amplitude of the inverted noise signal will transfer to the signal and become interference in its own right. The noise signal inversion can be accomplished by transformers, bridge circuits, RLC phase shift networks or delay lines.

A simple bridge circuit

Figure 17.2 shows a simple bridge circuit. I've used it at VLF on radioscience observing receivers, and others have used it on VHF receivers. The bridge consists of two transformers (T1 and T2). Transformer T1 is trifilar wound, i.e. it has three identical windings interwound with each other in the manner of Fig. 17.3. The black 'phasing dots' or 'sense dots' indicate one end of the windings, and will be used for wiring T1 into the circuit of Fig. 17.2.

Winding the toroid exactly as shown in Fig. 17.3 is a difficult task, so you might want to consider an alternative method. Select three lengths of enamelled wire (#18 AWG through #26 AWG can be used, but all wires should be the same size). In order to keep them straight in my mind as I work them, I select three different insulation colours from my wire rack.

Tie all of them together at one end, and insert that end into the chuck of a hand drill. I usually fasten the other ends into a bench vice, and back off until the wires between the vice and drill are about straight (more or less). Turn on the drill on a slow speed (slightly squeeze the trigger on variable speed drills), and let the drill twist them together. Keep this process going, being careful to not kink the wire (which happens easily!), until there are eight to 16 twists per inch (not critical).

Caution: Wear protective goggles or safety glasses when doing this job. I once let the drill speed get too high, the wire broke and I received a nasty lashing to the face . . . which could have damaged my eye except for the glasses.

Once the three-wire composite wire is formed, it can be wound onto the toroid form as if it were one wire. Before winding, however, separate the ends a bit, scrape off enough insulation to attach an ohmmeter probe, and measure both the continuity of each wire, and whether or not any two are shorted together. If the wires are wound too tight, then it's possible to break one wire, or breech the integrity of the insulation.

Note in Fig. 17.2 the way transformer T1 is wired. The main antenna signal from J1 is connected to the dotted end of one winding, while the sense antenna signal (J2) is applied to the non-dotted end of another of the three windings. These signals are transferred to the

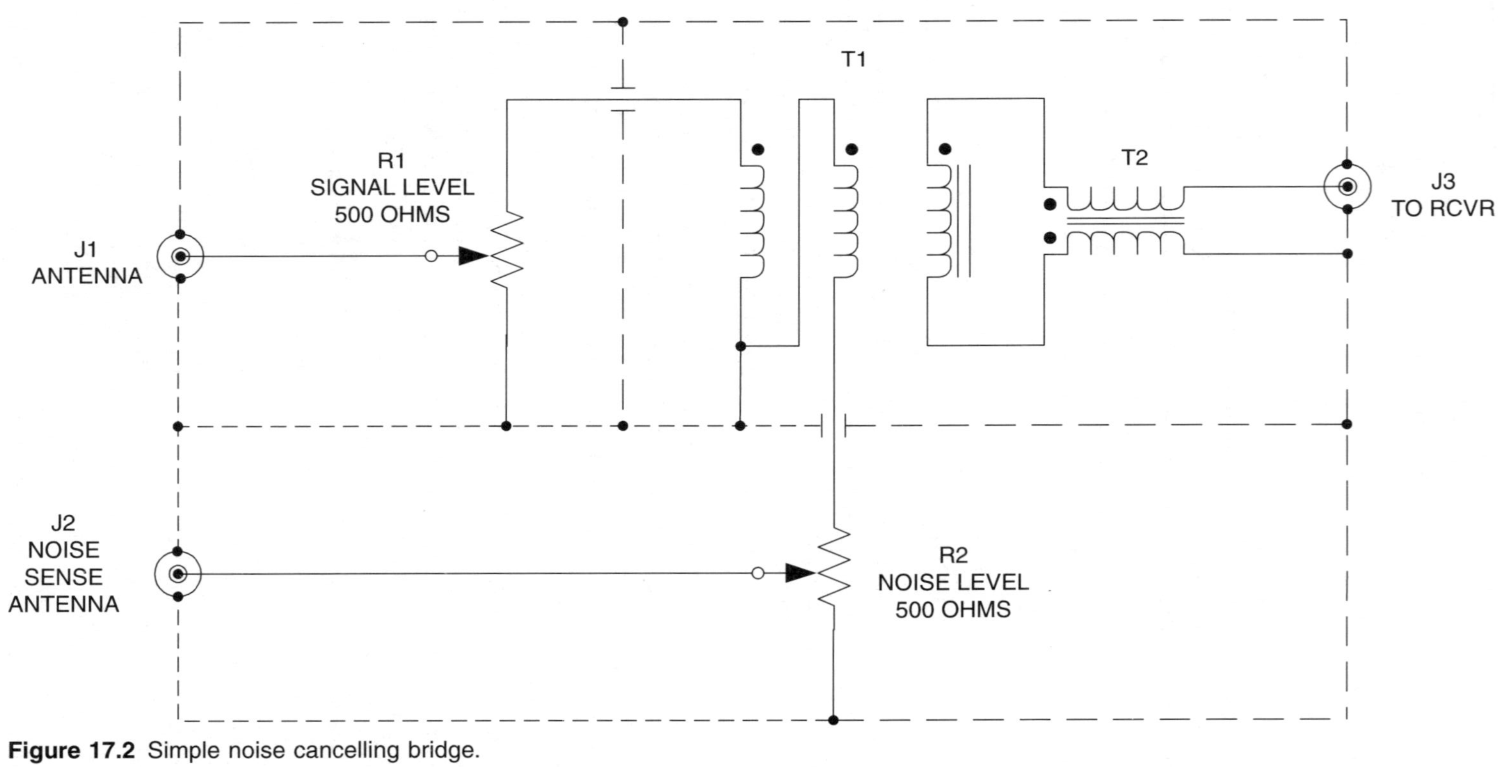

Figure 17.2 Simple noise cancelling bridge.

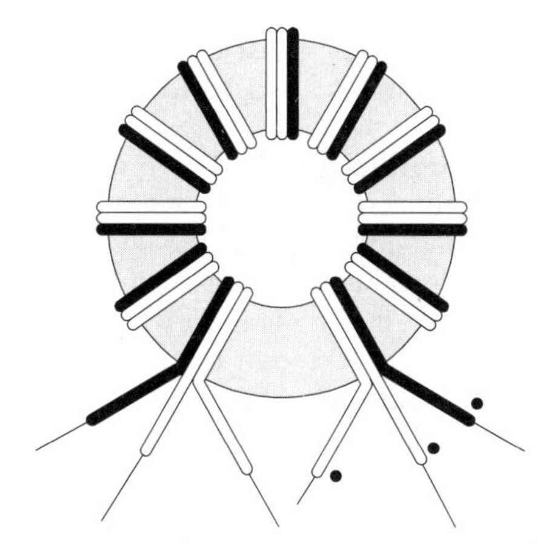

Figure 17.3 Transformer winding detail.

third winding, but because of their relative phasing they will be 180 degrees out of phase, thus cancelling the unwanted noise.

The composite output of T1 is applied to transformer T2. This transformer is inserted into the line as a *common mode choke*, so basically acts as a balun. Transformer T2 is built exactly like T1, but is bifilar (two windings) instead of trifilar.

Signal amplitudes from the two different antennas are controlled by a pair of 500 ohm potentiometers. The pots selected for R1 and R2 should be non-inductive (i.e. carbon or metal film, but *not* wirewound). In other words, rather ordinary potentiometers will work nicely. The wipers of both potentiometers are connected to their respective antenna jacks. The two ends are connected to T1 and ground, respectively.

Note that this circuit is not just built into a shielded box, but also in separate shielded compartments. Figure 17.4 shows a suitable form of building the circuits. A compart-

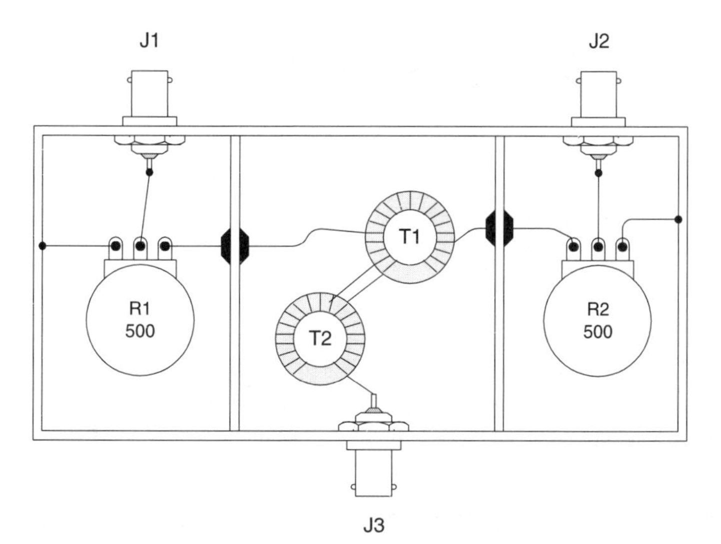

Figure 17.4 Physical layout of parts.

mented box such as made by SESCOM is used to hold the bridge. Small grommets mounted on the internal shield partitions are used to pass wires from one compartment to the other.

For VLF through shortwave, transformer T1 is wound with 16 turns of enamelled wired, and T2 is wound with 18 turns. Both can be wound on $\frac{1}{2}$ inch cores (T-50-xx or FT-50-xx), but it will be easier to use slightly larger forms such as FT-68-xx, T-68-xx, FT-82-xx and T-82-xx. Ferrite cores (FT-nn-xx) should be used in the AM BCB and below, while powered iron (T-nn-xx) can be used in the medium wave and shortwave bands.

Recommended ferrite types for VLF through the AM BCB include FT-82-75 and FT-82-77, medium wave units can be made using FT-82-61, and VHF units can be made using FT-67 or FT-68 (or their -50 and -68 equivalents). If powdered iron cores (T-nn-xx) are used, then select T-80-26 (YEL/WHT) for VLF, T-80-15 (RED/WHT) for AM BCB and low medium wave, either T-80-2 (RED) or T-80-6 (YEL) for medium wave to shortwave, and T-80-12 (GRN/WHT) for VHF. Again, their -50 and -68 equivalents are also usable. Some experimentation might be needed in specific cases depending on the local noise problem.

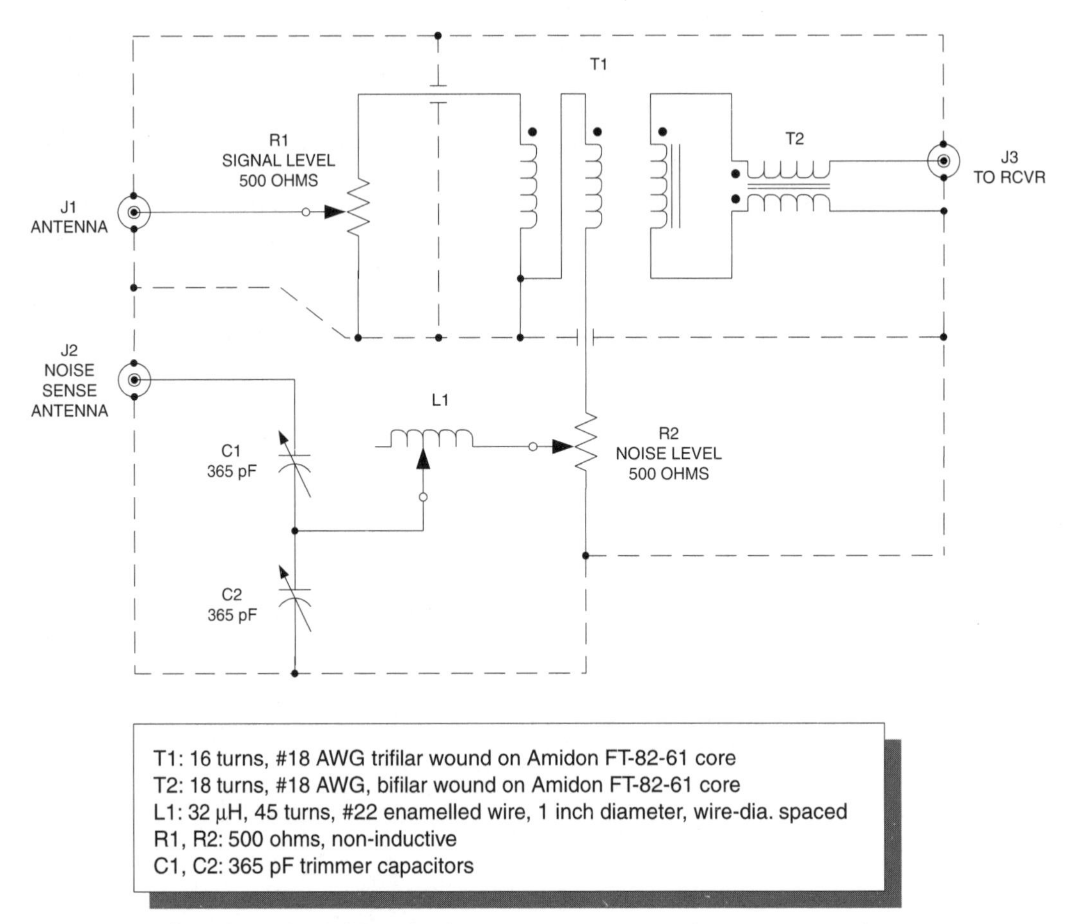

Figure 17.5 Complex noise bridge circuit.

Figure 17.5 shows a version of the noise cancellation bridge circuit made popular by William Orr (W6SAI) and William R. Nelson (WA6FQG) for amateur radio use (*Interference Handbook*, RAC Publications, PO Box 2013, Lakewood, NJ, 08701). It is built on the same principles as Fig. 17.2, but includes an L–C phase shift network consisting of L1, C1 and C2. The values are:

T1: 16 turns, #18 AWG trifilar wound on Amidon FT-82-61 core

T2: 18 turns, #18 AWG, bifilar wound on Amidon FT-82-61 core

L1: 32 μH, 45 turns, #22 enamelled wire, 1 inch diameter, wire-diameter spaced

R1, R2: 500 ohms, non-inductive linear taper potentiomer

C1, C2: 365 pF capacitors

The coil L1 should be wound with either enamelled wire or non-insulated solid wire so that it can be tapped.

To adjust this bridge, C1, C2 and the tap on L1 should be adjusted iteratively until the lowest possible noise signal is achieved. To do this trick, it is usually necessary to set R1 and R2 to a low setting, but not so low that both the noise and the signal disappear.

Parts can be a little difficult to obtain for RF projects, especially the capacitors. Ocean State Electronics (6 Industrial Drive, PO Box 1458, Westerly, RI, 02891, 1–401–596–3080 or fax 1–401–596–3590) stocks both new and used variable capacitors, as well as various inductors, toroid cores and other items of interest.

Radiated noise can be one of the most intractable electromagnetic interference (EMI) problems. These bridges are not a 'silver bullet' by any means, but they will perform sufficient noise reduction to make a significant difference in the signal-to-noise ratio . . . and that's what actually counts.

Bibliography

ARRL Handbook for Radio Amateurs CD-ROM Version 1.0 (1996). Newington, CT: ARRL.

ARRL Handbook for Radio Amateurs 1998 (CD-ROM Version). Chapter 16.

Bottom, V.E. (1982). *Introduction to Quartz Crystal Unit Design*. New York: Van Nostrand Reinhold Co.

Carr, Joseph J. (1996). *Secrets of RF Circuit Design 2nd Edition*. New York: McGraw-Hill.

Carr, Joseph J. (1997). *Microwave and Wireless Communications Technology.* Boston: Newnes.

Carr, Joseph J. (1998). *Practical Antenna Handbook 3rd Edition*. New York: McGraw-Hill.

Carver, Bill (1993). 'High Performance Crystal Filter Design', *Communications Quarterly,* Winter 1993.

Cohn, S. (1957). 'Direct Coupled Resonator Filters', *Proceedings of IRE*, February 1957.

Cohn, S. (1959). 'Dissipation Loss in Multiple Coupled Resonators', *Proceedings of IRE,* August 1959.

Colin (1958). 'Narrow Bandpass Filters Using Identical Crystals Design by the Image Parameter Method' (in French), *Cables and Transmission*, Vol. 21, April 1967.

DeMaw, D. and G. Collins (1981). 'Modern Receiver Mixers for High Dynamic Range', *QST,* May 1981; Newington, CT, USA: ARRL.

Dishal, (1965). 'Modern Network Theory Design of Single Sideband Crystal Ladder Filters', *Proceedings of IEEE,* Vol. 53, September 1965.

Gilbert, B. (1994). 'Demystifying the Mixer', self-published monograph.

Gottfried, Hugh L. (1958). 'An Inexpensive Crystal-Filter I.F. Amplifier', *QST,* February 1958.

Hagen, Jon B. (1996). *Radio-Frequency Electronics: Circuits and Applications*. Cambridge (UK): Cambridge Univ. Press.

Hamish (1962). 'An Introduction to Crystal Filters', *RSGB Bulletin*, January and February 1962.

Hardcastle, J.A. (1978). 'Some Experiments with High-Frequency Ladder Crystal Filters', *QST,* December 1978, and *Radio Communications* (RSGB) January, February, and September 1977.

Hardcastle, J.A. (1980). 'Ladder Crystal Filter Design', *QST,* November 1980, pp. 22–23.

Hardy, James (1979). *High Frequency Circuit Design*. Reston, VA: Reston Publishing Co. (Division of Prentice-Hall).

Hawker, P. (ed.) (1993) 'Super-Linear HF Receiver Front Ends', Technical Topics, *Radio Communication*, Sept. pp. 54–56.

Hayward, W. (1982). 'A Unified Approach to the Design of Crystal Ladder Filters', *QST*, May 1982, pp. 21–27.

Hayward, W. (1987). 'Designing and Building Simple Crystal Filters', *QST*, July 1987, pp. 24–29.

Hayward, W. (1994). *Introduction to Radio Frequency Design*. Newington, CT: ARRL.

Horowitz, P. and W. Hill (1989). *The Art of Electronics 2nd Edition*. New York: Cambridge University Press.

Joshi, S. (1993). 'Taking the Mystery Out of Double-Balanced Mixers', *QST*, Dec. 1993, pp. 32–36.

Kinley, R. Harold (1985). *Standard Radio Communications Manual: With Instrumentation and Testing Techniques*. Englewood Cliffs, NJ: Prentice-Hall.

Laverghetta, Thomas S. (1984). *Practical Microwaves*. Indianapolis, IN: Howard W. Sams.

Liao, Samuel Y. (1990). *Microwave Devices & Circuits*. Englewood Cliffs, NJ: Prentice-Hall.

Makhinson, J. (1993). 'High Dynamic Range MF/HF Receiver Front End', *QST*, Feb. 1993, pp. 23–28. Also see Feedback, *QST*, June 1993, p. 73.

Makhinson, J. (1995). 'Designing and Building High-Performance Crystal Ladder Filters', *QEX*, No. 155, January 1995.

Pochet, (1976/1977) 'Crystal Ladder Filters', Technical Topics, *Radio Communications*, September 1976, and *Wireless World*, July 1977.

'Refinements in Crystal Ladder Filter Design', *QEX*, No. 160, June 1995 (CD-ROM version).

Rohde, U. (1994a) 'Key Components of Modern Receiver Design', Part 1, *QST*, May 1994. Newington, CT, USA: ARRL, pp. 29–32.

Rohde, U. (1994b) 'Key Components of Modern Receiver Design', Part 2, *QST*, June 1994. Newington, CT, USA: ARRL, pp. 27–31.

Rohde, U. (1994c) 'Key Components of Modern Receiver Design', Part 3, *QST*, July 1994. Newington, CT, USA: ARRL, pp. 42–45.

Rohde, U. (1994d). 'Testing and Calculating Intermodulation Distortion in Receivers', *QEX*, July 1994. Newington, CT, USA: ARRL, pp. 3–4.

Rohde, U. and T. Bucher (1988). *Communications Receivers: Principles and Design*. New York: McGraw-Hill.

Rohde, U., J. Whitaker and T. Bucher (1988). *Communications Receivers 2nd edition*. New York: McGraw-Hill.

Sabin, William E. (1970). 'The Solid-State Receiver', *QST*, May 1970. Newington, CT, USA: ARRL, pp. 35–43.

Sabin, William E. (1996). 'The Mechanical Filter in HF Receiver Design', *QEX*, March 1996.

Sabin, William E. and Edgar O. Schoenike, (eds) (1998). *HF Radio Systems & Circuits 2nd Edition*. Atlanta: Noble Publishing.

Shrader, Robert L. (1975). *Electronic Communication 3rd Edition*. New York: McGraw-Hill.

Sykes, 'A New Approach to the Design of High-Frequency Crystal Filters', *Bell System Monograph 3180*, n.d.

Van Roberts and Walter B. (1953). 'Magnetostriction Devices and Mechanical Filters for Radio Frequencies Part I: Magnetostriction Resonators', *QST*, June 1953.

Van Roberts and Walter B. (1953). 'Magnetostriction Devices and Mechanical Filters for Radio Frequencies Part II: Filter Applications', *QST*, July 1953.

Van Roberts and Walter B. (1953). 'Magnetostriction Devices and Mechanical Filters for Radio Frequencies Part III: Mechanical Filters', *QST*, August 1953.
Vester, Benjamin H. (1959). 'Surplus-Crystal High-Frequency Filters: Selectivity at Low Cost', *QST*, January 1959, pp. 24–27.
Vizmuller, Peter (1995). *RF Design Guide*. Boston/London: Artech House.
Zverev, A.L. (1967). *Handbook of Filter Synthesis*. New York: John Wiley & Sons.

Index